CARDIOLOGY RESEARCH AND CLINICAL DEVELOPMENTS

CARDIOVASCULAR HEALTH AND CHRONOMICS

CARDIOLOGY RESEARCH AND CLINICAL DEVELOPMENTS

Additional books in this series can be found on Nova's website under the Series tab.

Additional e-books in this series can be found on Nova's website under the e-book tab.

CARDIOLOGY RESEARCH AND CLINICAL DEVELOPMENTS

CARDIOVASCULAR HEALTH AND CHRONOMICS

RAM B. SINGH
EDITOR

NOVA BIOMEDICAL

New York

For permission to use material from this book please contact us:
Telephone 631-231-7269; Fax 631-231-8175
Web Site: http://www.novapublishers.com

Library of Congress Cataloging-in-Publication Data

ISBN: 978-1-62948-976-6

Library of Congress Control Number: 2013958162

Published by Nova Science Publishers, Inc. † New York

CONTENTS

PREFACE

This book presents important research from around the world in the field of cardiovascular science. Particular emphasis is placed on chronomics, immunogenicity, diet and health, diabetes, coronary risk, and cardiovascular disease.

Chapter 1 – Elderly subjects have unique nutrient needs with revised recommended dietary reference intakes based on age, sex, ethnicity and special vulnerability to compromised nutrient status.

The food intake tends to decrease with advancing age due to decreased physical activity and metabolic rate. However vitamin; A,E,C, B vitamins and beta-carotine and minerals magnesium, calcium, selenium, chromium, zinc and copper as well as w-3 fatty acids needs either remain constant or increase. Omega-3 fatty acid, calcium, vitamin D and magnesium and antioxidant intakes are inversely associated with ageing and other complications like cardiovascular diseases, diabetes and dementia as well as cancers. It is pertinent to advise to eat 400g/day of fruits, vegetables and nuts and another 400g/day of whole grains in conjunction with 25-40g/day of canola oil or mustered oil, depending upon energy requirement, for prevention of cardiovascular disease and type 2 diabetes mellitus.

Chapter 2 – Exploring any non-photic as well as photic components in the spectrum of the incidence pattern of suicides, the authors find gender differences in Minnesota, USA, and Australia, notably in the para-annual range of the spectrum. A trans-tridecadal BEL cycle (also more liberally referred to as para-tridecadal) with a period of about 35 years is found in data pooled from both genders and separately for both men and women in Finland (1969-2007). When fitted separately as a single component, the estimated period and its 95% confidence interval (CI) (in years) is 35.5 [26.7, 44.3] overall, 36.4 [25.3, 47.5] for males, and 38.9 [28.5, 49.3] for females, with respective amplitudes (N/100,000) of 170, 145, and 29. Differences between a Schwabe cycle and an about 7-year component (possibly a harmonic of the Hale cycle) characterizing suicide patterns of Finnish men and women, respectively, remain to be examined in longer series. In Bulgaria (1929-1945), a far-transyear with a period of about 1.3 years is noteworthy as is a quinmensal component with an about 5-month period. In France (1979-2008), in males and females, suicides are characterized by about 10.0-year and 10.1-year cycles, respectively. In Japan (1987-1999), an about 12.7-year component is detected, albeit with a broad uncertainty. A better understanding of the origin of these patterns, notably in terms of the relative contribution of non-photic and photic (daily and yearly) cycles, may lead to better prediction of incidence dynamics and to novel interventions for the prevention of major depression and suicide.

Chapter 3 – The signature of an environmental decadal or longer cosmic cycle is found in each of 4 men's circulation. The assortment of pairs of environmental and physiological periods, τs, can differ among some variables and even for some of the two circadian rhythm characteristics of the same variable in the circulation of the same subject. As gauged by overlapping CIs (95% confidence intervals) of the τs in us and around us, the environmental pairing (e.g., of two subjects' diastolic blood pressure) can thus differ for the MESOR, M, a midline-estimating statistic of rhythm versus the circadian double amplitude, 2A, a measure of the extent of within-day change. A change in period length with the lengthening of a time series and/or advancing age is also recorded in the para-annual, para-decadal and para-tridecadal range of the spectrum. To whatever extent these findings can be extended to the population at large, the imaginary baselines reported in investigations involving interventions in aging will be amenable to replacement by cycles here demonstrated longitudinally, yet to be mapped on a population basis.

Chapter 4 – The authors complement an "Extended Consensus," focused on the individual's health, by documenting associations between the human heart and circulation and the environment near and far at the level of local and global populations in both space and time. Life in the atmosphere of the sun mimics its inanimate surroundings; its rhythmic components nearly match in length the cycles of the cosmos. Longitudinal records of the human blood circulation reveal heretofore unrecognized aspects of its variability. The view from a very few long and relatively dense series, covering mostly around-the-clock (up to) decades on individuals, reveals variability that cannot readily or necessarily be found in hundreds of thousands of individuals' around-the-clock 24-hour spotchecks and certainly not in single samples, unless the individuals' cycles are synchronized and assessed on a population basis.

Periods, τ, longer than 28 hours, that is infradian variations, like the circadians (with τ tentatively between 20 and 28 hours), have entered the genome and their characteristics await mapping. For those cycles that are already mapped, the authors consider further the extent to which the parameters and nonparametric endpoints of healthy infradian and circadian variability and their anomalies can become everyday measures of Sir William Osler's wear and tear, i.e., of Hans Selye's and Paul Rosch's stress and strain to the point of becoming gauges of distress (again complementing or replacing single samples or the mean values of dense sampling during single days). Criteria for recognizing and quantifying the alteration indicating when a load changes from benetensive to maletensive have yet to be identified, perhaps by monitoring blood pressure and heart rate in everyday life, replacing spotchecks in the clinic.

The probable role of cosmos-induced loads in pathogenesis, assessed, among others, as an effect of a magnetic storm in laboratory animals and in humans, awaits study by using superposed epochs and a subtraction and addition approach.

Chapter 5 – Over the past several years there is considerable interest among various experts and the lay public to find out the causes of sudden deaths in the trained athletes as well as in the clinical significance of cardiac symptoms as syncope and arrythmias. Catastrophic events in athletes are always unexpected, and although relatively uncommon, often achieve high visibility and convey a particularly devastating impact on the community. The most common causes of death in the young athletes are hypertrofic cardiomyopathy (need to be differentiated from “athlete’s heart”), congenital coronary arteries anomalies and commotio cordis. Older athletes can also harbor occult cardiac disease, but the most common cause of death is atherosclerotic coronary artery disease (CAD). In view of these findings, the guidelines for prevention, managment and rehabilitation were established, because the overall athlete population is in low risk dying for sudden cardiac death, but the risk exists.

Chapter 6 – Recently, many of the complexities associated with cardiovascular diseases (CVD) have been unlocked. However, despite these breakthroughs, CVD and its related complications are the leading contributors of morbidity and mortality worldwide, which indicates the shortcomings of current treatment regimens and the need for continued research. Published data within the field clearly indicates that CVD are built on inflammation and autoimmune platforms, though a strong, fundamental understanding of the mechanisms remains elusive. Areas such as the mechanisms underlying increased immunogenicity of self-proteins in the cardiovascular system, the roles of immunogenic auto-antigens in eliciting inflammatory autoimmune responses, and the immunosuppressive mechanisms involved in controlling inflammatory and autoimmune cardiovascular diseases remain to be well-understood. The authors will delve into these topics and the advancements made within the field in this review. Specifically, they will concentrate on the innate and adaptive immune responses mediating immunogenicity; the mechanisms of inflammation and autoimmunity in atherogenesis; the mechanisms of inflammation and autoimmunity in diabetic atherosclerosis; immunogenicity and stem cell therapy; as well as immunogenicity and immunosuppression. In depth examination and comprehension of these topics will provide insight into the recent progress of the field and bring to the forefront potentially novel therapeutic avenues.

Chapter 7 – Structures in time series can be viewed as consisting of chaos, trends and cycles (chronomes). Since Fourier's time, cycles have been harmonically resolved; since Gauss', the fitting of sinusoids can be carried out on unequidistant data. But the resolution of dynamics must also consider methods that do not depend on sinusoids. To the extent that biological processes are dissipative, they are subject to various changing nonlinear influences; hence they are not necessarily stable.

In principle the dynamics cannot always be sinusoidal, although cosine fitting is a useful approximation of the recurrent aspects of the given series.

The goal of this work was to elaborate a method allowing us: (1). To reveal the hidden periodicities having any form of their waves, (2). To restore their form not considering them as a sum of trigonometric functions, and (3). To evaluate the statistical significance of the profile's characteristics (parameters).

Chapter 8 – *Background.* Many experts have long accepted the brain-gut axis and its influence on obesity, insulin resistance and cardiovascular function. *Methods.* Internet search via Medline and Pub-Med and discussion with colleagues. *Results.* In experimental studies, it has been shown, how moderate intestinal inflammation affects complex behavior and how much treatment of such inflammation can influence factors from mind to matter, or brain to gut and the heart. It is likely that presence of long chain polyunsaturated fatty acids (PUFAs) in the duodenum, stimulates the release of cholecystokinin which in turn acts on cholecystokinin receptors, present on the abdominal vagus. After the food is ingested, it stimulates the secretion of incretins from the gut which increase the insulin secretion and initiate a gut-brain-liver axis in response to small amounts of triglycerides in the duodenum. Long chain PUFAs, particularly w-3 fatty acids, are cleaved from the dietary triglycerides by the gastrointestinal enzymes. Further studies indicated that a long chain fatty acid metabolite known as long chain fatty acid coenzyme A (LCFA CoA) is sensed by the gut. However, increased intake of trans fat, saturated fat and possibly w-6 fat and refined carbohydrates, particularly in presence of physical inactivity, may suppress gut-brain and liver axis by decreasing incretins, as well as cholecystokinin and leptin, resulting in insulin resistance and metabolic syndrome which is known to predispose cardiovascular diseases (CVDs). There is a rapid stimulation of leptin secretion from the gastric fundus on food ingestion, an effect which could be reproduced by cholecystokinin administration. Leptin gene expression and immune-reactivity have been reported in the gastric fundus. Leptin is known to enhance satiety–inducing effect of cholecystokinin indicating that cholecystokinin and leptin may function in concert with each other to induce satiety and regulate food consumption. Mediterranean diet and w-3 fatty acids have been reported to be protective against metabolic syndrome and CVDs but the exact mechanism of benefits is not known. It is possible that a part of the benefit may be due their protective effects on cell membrane phospholipids of the neuron, gut, liver and beta cells of pancreas apart from their direct effects on cardiovascular function.

Chapter 9 – These authors have discussed, for the last two decades, those environmental factors in general and nutrition in particular could be important in the pathobiology of epigenetic inheritance. This knowledge was unknown to Julian Huxley who published his landmark Monograph; 'Evolution: The Modern Synthesis in 1942.' This monograph brought Darwin's ideas into the 20th century and incorporated a knowledge of genes that was emerging in this century in the light of Gregor Mendel's experiments on inheritance (Monograph: *Experiments with Plant Hybrids)* at Hynčice (Vražné) in the now Czech Republic. In the mid-century, Barbara McClintock discovered transposable elements where parts of the genome can jump around and cause mutations or alter the gene expression, skewing Mendelian ratios and inheritance patterns. This report disrupted the predictable Mendelian system that went into building the Modern Synthesis.

In The Origin of Species, Darwin wrote: "Natural selection can act only by taking advantage of slight successive variations; it can never take a leap, but must advance by the shortest and slowest steps." It is possible, he believed that organisms gradually adapt to their environments *via* minute physical adjustments. Biologists have since found evidence that dramatic adaptations can also occur, for instance in the form of major morphological changes. These adaptations appear to be the cause of changes in structure from apes to man and various human races. It is not clear whether such adaptations may be responsible for the development of diseases or health or for development of man to humans.

The Modern Synthesis emphasizes that "populations containing some level of genetic variations evolve *via* changes in gene frequency, induced mostly by natural selection and phenotypic changes are gradual with speciation and diversification into higher taxonomic levels come about over long periods of change. It seems that these ideas remained largely unchallenged for more than 50 years from now, although phenotypic changes have occurred rapidly in the last 100 years. Furthermore, evolutionary biology, biochemistry, genomics, developmental biology, systems biology and the impact of the environment on genes concerning mechanism of evolution have grown significantly beyond the Modern Synthesis. It is possible that some lineages have greater 'evolvability' than others, independent of how much baseline genetic variation is present. In this connection, heritable phenotype variations may depend on the biology and biochemistry of the genes. Since some populations have more genetic variation than others and are expected to generate phenotypic variation more rapidly, the Modern Synthesis addresses 'evolvability' in a population which may not be a distinct trait, independent of genetic variation. The Modern synthesis also does not place emphasis on the influence of photosynthesis, flight, multicellularity which stands out against a backdrop of slow and steady evolution. It is therefore important that Modern Synthesis should be modified by the experts because Life Sciences are dynamic, and which continue to change.

Chapter 10 – An optimization approach for essential circadian timing of treatment of high blood pressure, diagnosed as one of the described vascular variability disorders, VVDs, is presented that specifically lowers the MESOR (M=Midline Estimating Statistic Of Rhythm) without inducing any other VVDs, such as CHAT (Circadian Hyper-Amplitude [A] Tension). This has been achieved using Cumulative sums and parameter tests applied to half-hourly automatic around-the-clock measurements for weeks at a time or longer. Hypotensive treatment (Rx) with a popular drug, at the wrong time was associated with CHAT, while the same dose when given at another time to the same patient was beneficial by lowering the M while leaving the 2A unaffected. In conclusion it is best to guard against an iatrogenic, excessive blood pressure overswing, systolic and/or diastolic, and against inducing other VVDs. "First do no harm" by not flying blind, i.e., by adequate surveillance of ambulatory blood pressure and heart rate measurements and concomitant sequential process control combined with statistical parameter testing.

Chapter 11 – *Background and Aim:* The aim of the present study is to assess the association of coronary risk factors with glucose levels among patients with type-2 diabetes mellitus.

Subjects and Methods: This study included 270 subjects with type 2 diabetes mellitus. The authors obtained primary risk factors, disease history, anthropometry, blood pressure, blood glucose and lipids by questionnaires following WHO guidelines and laboratory data by standard methods.

Results: Of 270 patients with type 2 diabetes, 43 percent of males and 50 percent of females are having family history of diabetes mellitus. None of the subjects fall in malnutrition category whereas, 30 percent were normal weight, 50 percent were overweight and 18 to 30 percent were obese. The risk factor prevalence rates are: systolic hypertension; 40%, diastolic hypertension; 60%, hypertriglyceridemia; 55-60% and hypercholesterolemia; 32-37%. Both fasting blood glucose (FBS) and post prandial blood glucose (PPBS) levels increased from<39 yrs age group to 40-54 yrs age group and then decreased in >55 yrs age group ($p<0.05$). Blood glucose levels were unaltered within the categories of BMI, while WHR elevated the FBS levels ($p<0.05$). Both FBS and PPBS elevated from latent diabetics to diabetics with 5-9 yrs duration, there after a decrease was observed in FBS alone. Sixty percent of the variance in FBS levels and 58 percent of the variance in PPBS was explained by independent variables. The ODD's of FBS are: 40-54 yrs age group (1.023, 95% CI: 1.011-1.034), >55 yrs (1.020, 95% CI: 1.009-1.032), duration of diabetes for 5-9 yrs (1.024, 95% CI:1.010-1.039), family history (1.084, 95% CI: 1.063-1.106), WHR>1 (1.028, 95% CI: 1.006-1.050), systolic hypertension (1.023, 95% CI:1.016-1.030), diastolic hypertension (1.022, 95% CI: 1.015-1.030). Similar ODD's were obtained for PPBS levels also.

Conclusion: In conclusion the results indicate that family history, duration of the disease and blood pressure are significant risk factors in type-2 diabetes mellitus. Hence preventive strategies warranted towards the management of hyperglycemia in type 2 diabetes to prevent vascular disease.

Chapter 12 – Research evidence suggests that high plasma fibrinogen represents a major risk factor for cardiovascular diseases. The increased occurrence of CVD necessitates the need to analyze fibrinogen levels in ischemic heart disease (IHD) patients and study its relationship with physiological and behavioral risk factors. The present study analyses plasma fibrinogen levels in relation to known cardiac risk factors, in 72 patients (63 men and 9 women) afflicted with myocardial infarction. Parameters studied were anthropometry, plasma fibrinogen and lipid profile. Anthropometric data indicated a lower muscle mass status in majority of the subjects. Although the body mass index (BMI) was normal in majority of subjects, waist-to-hip ratio (WHR) was higher than normal range. Clinical data revealed atherogenic lipid profile in 58% of the subjects. Mean plasma fibrinogen values were 3.8g/L, high normal (>3.0g/L) in 28% and above normal (>4.0g/L) in 30% of the patients. Fibrinogen levels were higher ($p<0.01$) in patients with diabetes, while they were lower in alcohol users ($p<0.05$). Also, mean fibrinogen, cholesterol, LDL, triglycerides, WHR and BMI were higher in smokers than in non-smokers. The results show that fibrinogen level may be the risk factor of IHD in presence of known CVD risk factors. Since most of the risk factors are amenable to changes in lifestyle, it is worthwhile to modulate the levels of these parameters to achieve reductions in plasma fibrinogen.

Chapter 13 – *Purpose:* The combination effect of *Gongronema latifolium* (GL) and *Nauclea latifolia* (NL), promising vegetables used in traditional management of metabolic disorders on serum lipid components and some liver marker enzymes was investigated in diabetic rat models. *Method:* The study design consisted of thirty six (36) rats divided into 6 groups of 6 rats each. Whereas groups 1 and 2, non diabetic and diabetic controls received placebo treatment, groups 3-5 respectively received 200mg/kg b.w. of GL, NL, and 100mg/kg b.w. each of GL and NL per day, for 21 days. The 6th group received subcutaneous insulin, 5IU/kg b.w. per day. Thereafter, the animals were sacrificed, and serum prepared from blood collected was used to assay lipid components and liver function enzymes, using standard analytical kits. *Result:* Measured blood glucose in diabetic animals decreased significantly from initial by 66.34%, 18.12% and 67.73% respectively upon treatment with GL, NI and GL plus NI. Whereas diabetes induction caused significant increases ($p<0.05$) in total cholesterol (TC), 54.42% and low density lipoprotein (LDL), 55.0% compared to the normal control (NC), treatment with extracts of GL or NI significantly decreased ($p<0.05$) these by (58.70% and 24.79%, TC) and (71.70% and 33.38%, LDL) respectively. However, the combined extracts produced more steady reduction (59.70% and 70.26, TC and LDL), similar to the NC. Also the amino transferases (ALT and AST) activities which increased by 66.83% and 72.87% in the DC rats indicating hepatotoxicity, secondary to hyperglycemia, became reduced upon treatment with GL, NI and GL plus NI. Again, the extent of reduction was better with combined extracts (85.58% and 53.77%), and compared favorably with NC. *Conclusion:* Lower dosages of the two extracts when combined provide higher efficacy in protection against atherosclerosis and hepatotoxicity in diabetes. Hence the need to adopt this strategy in bio-prospecting for antidiabetic natural products.

Chapter 14 – *Background:* Concerns about the health of night shift workers has long been debated. Shift work is associated with the disruption of circadian rhythms, where a person's internal body clock is in conflict with the rotating night shift schedule which causes internal desynchronization, sleep disturbances resulting in to increased risk of cardiovascular diseases and diabetes and other physiological disorders.

Objectives: The authors investigated the circadian pattern of blood pressure and heart rate in night shift workers and day shift controls, and whether the changes in circadian pattern produced by rotating night shift are reversible in due course of time.

Method: 14 healthy nursing professionals, aged 20-40 year, performing day and night shifts and 14 subjects as controls, performing day duty were recruited in the study. Circadian patterns of blood pressure and heart rate were evaluated in night shift workers during the work shift (Night & Day shift) and in controls.

Result: Night shift workers showed a very interesting altered circadian pattern of double amplitude when subject went back to the day shift. An extremely significant different pattern of SBP and DBP double amplitude was found between day shift and controls ($p<0.01$).

Conclusion: Effect of rotating night shift develops later due to rotating shift and desynchronization. Acrophase pattern was clinically significant when studied in individual subjects in different shifts with controls. Alterations in Acrophase were persistent during night as well as day shift due to incomplete recovery and ecphasia was very common among night shift workers.

Chapter 15 – *Objective:* Obesity is a well known risk factor for cardiovascular diseases (CVD). The present study is aimed to assess the relationship of obesity and abdominal obesity on blood pressure in an adult Indian population.

Subjects and Methods: Cross-sectional survey was conducted among students and staff members of the university. Six hundred and sixty adult subjects (303 males + 357 females) untreated for hypertension in the age range of 21-60 years. Physical assessment included height, weight, circumferences of waist and hip, systolic blood pressure (BP) and diastolic BP besides the information on demographic variables. Body mass index (BMI) was calculated as weight in kg/height in meter^2 (kgm^{-2}), waist-to-hip ratio (WHR) as waist circumference/hip circumference. Logistic regression analysis was carried out apart from one way anova and correlations. Categorical variables are shown in percentages and continuous variables as mean±S.D.

Results: Men are found to have higher abdominal obesity (0.90±0.07) than women (0.84±0.08) ($p<0.05$) while no difference in BMI. The prevalence of hypertension increased with age and BMI quartiles. The indicators of adiposity (BMI, WHR) were positively associated with blood pressure in males, while in females only BMI shown a positive association. Men with higher WHR are 2.988 times, and women with higher WHR are 1.177 times at risk to develop hypertension. The odds of hypertension were more than six fold among the elderly in male sex (OR=6.213: 95%CI 1.815, 21.273), but in females the odds of hypertension in elderly is only two fold (OR= 2.423: 95%CI 0.801, 7.334). The odds of hypertension rose steadily with increase in BMI reaching 7.579 (95CI; 1.510, 38.046) in males and 15.56 (95%CI; 1.883, 128.526) in females with BMI >25 kgm^{-2}. Adjustment for age decreased the odds of hypertension in males and increased in females in the BMI category of >25 kgm^{-2}, while no change in the remaining quartiles.

Conclusion: These findings suggest a linear relation of adiposity with Blood pressure.

Chapter 16 – The Studies made at the Halberg Chronobiology Center, Department of Laboratory Medicine and Pathology, University of Minnesota, have demonstrated that chronobiology is an essential tool for quantifying and understanding variability in diastolic (D) and systolic (S) blood pressure (BP), heart rate (HR) and pulse pressure and thus for identifying subjects at risk of myocardial infarction or stroke, and for choosing the appropriate treatment and timing thereof (chronotherapeutics).

Chapter 17 – According to the late Reinhart Koselleck's "Futures Past", history falls into the categories of "The Space of Experience" and "The Horizon of Expectation." The authors here add "and into the category of an inferential statistical estimation of the uncertainties involved in any claim of experience and/or expectation whenever possible." Koselleck concludes that lasting structures are concealed by conventions and must be discovered and investigated if historical experience is to be transformed into historical science. The view adapted herein is that for history to be scientific, first it has to examine claims, whenever the data permit and when time series are involved, meta-analyses are an opportunity to re-examine the past. This can be done by testing the hypotheses postulated and estimating the parameters involved in the light of all the information available at a given time in history and re-examining further data accumulated in the interim. The task of validating or invalidating claims is particularly pertinent in the case of cycles, in the light of accumulating added evidence from lengthening time series, notably those in space weather in an era of satellites and computers. From this practical viewpoint, the authors find that, as yet, a consideration of any history of effects of space weather on human affairs is included in neither the space of experience nor the horizon of expectation of most contemporary historians, the late William L. Langer being an exception. Second in interpreting history completely requires scholars to start learning the simple fact (with wide-reaching consequences) that, as Yohsuke Kamide puts it, we live in the sun's atmosphere and there is much that we cannot see in it, and (we add) it importantly involves, as a yet unknown factor, many human affairs, including whether or not we die suddenly by an electrical accident of the heart (which brings the topic immediately within the scope of this journal), by our own hand or by the hand of others. Here lies a challenge for those concerned with the heart and circulation, reflecting in time series the cycles of our cosmos, in which we evolved.

Chapter 18 – It seems that there are no better words to remember Salvador than those written by the Lord of Time - Dr Franz Halberg: For members of the BIOCOS worldwide, including Minnesota friends, Salvador remains an integral part of what we do. After Mexico City, Minnesota was his second home, whether he enthusiastically described what he did to Earl E. Bakken, the developer of the implantable cardiac pacemaker or analyzed data from an Italian woman isolated from society for over 100 days, to find what Earl had described as "free-running." His real home, of course, was and remains worldwide and beyond, the sphere of the human mind, what Edouard Le Roy, Pierre Teilhard de Chardin and above all Vladimir Ivanovich Vernadsky called the noösphere, the sphere of human creativity, complementing Eduard Suess's (the Viennese geophysist) biosphere. Salvador helped take us a step beyond the noösphere by examining both the basic rules and their applications. These were steps toward introducing time (chronos) into the noösphere. By seeking new transdisciplinary cycles to quantitatively assess the otherwise qualitative noosphere, we can measure, apparently for the first time in statistical inferential terms, the domain of the human mind by its mood and its creativity, manifested in the occurrence at about 500-year intervals of prominent historians, poets, and physicians. In classical Greek, "noos" and "nous" both describe the human mind; the Attic Greek "nous" serves to portmanteau it into "chromos," time, and "sphairos," sphere, into a chronousphere. On the basic side, Sal's enthusiasm and his stamina supporting his inquiry into the unknown laid the basis of a time-qualified feedsideward (rather than time-unqualified feedback) in the pineal-hypothalamic-pituitary-adrenal-cellular network first and in the cosmos next. His last contributions were data for examining the role of the cosmos in modulating human blood pressure.

Chapter 19 – In 2003, a special volume of papers dealing with non-invasive cardiology was dedicated to Bohumil Fiser, MD, the authors' dear friend, and, his relatively young age notwithstanding, the authors' mentor and supporter. In that volume, GC and JS laid the basis for a chrononeonatology. Another of us (OS) emphasized the need for long-term, preferably lifetime monitoring of blood pressure and heart rate as a concern for everybody and hence for government and ethics committees. Bohumil Fiser was on many decision-making bodies, always there whenever and wherever he was needed, whether the task on hand was small or of nationwide or worldwide scope.

In the honorable tradition of his fellow-countryman Jan Evangelista Purkinje, he served as a subject and as a principal investigator.

In: Cardiovascular Health and Chronomics
Editor: Ram B. Singh
ISBN: 978-1-62948-976-6

Chapter 1

DIET AND LIFESTYLE GUIDELINES AND DESIRABLE LEVELS OF RISK FACTORS FOR PREVENTION OF CARDIOVASCULAR DISEASE AND DIABETES AMONG ELDERLY SUBJECTS. A REVISED SCIENTIFIC STATEMENT OF THE INTERNATIONAL COLLEGE OF CARDIOLOGY AND INTERNATIONAL COLLEGE OF NUTRITION-2011

R. B. Singh[1,*], Adarsh Kumar[2], N. S. Neki[3], Daniel Pella[4], S. S. Rastogi[5], T. K. Basu[6], S. N. Jan Fedacko[4], Lekh Juneja[8], Takahashi Toru[9], K. Otsuka[10], Fabien De Meester[11], D. W. Wilson[12] and Teodora Handjieva-Darlenska[13]

[1]Formerly President, International College of Cardiology,
Halberg Hospital and Research Institute, Moradabad, India
[2]President, International College of Cardiology, Head, Department of Cardiology,
Government Medical College, Amritsar, India
[3]Department of Medicine, Government Medical College, Amritsar, India
[4]Department of Sports Medicine, Safaric University, Kosice, Slovakia
[5]President, International College of Nutrition,
Diabetes and Endocrine Research Center, Delhi, India
[6]Department of Food and Nutrition, University of Alberta, Edmonton, Canada
[7]Agriculture and Agri-Food Canada, Lethbridge, Canada
[8]Taiyo Kagaku, Yokkaichi, Japan
[9]Graduate School of Human Environment Science, Fukuoka Women's University, Japan
[10]Department of Medicine, Tokyo Womens Medical University, Tokyo, Japan
[11]The Tism Tsoum Institute, Belgium

[*]Address for correspondence: Prof Dr RB Singh, MD, FICN, FICC. Halberg Hospital and Research Institute, Civil Lines, Moradabad-10(UP) 244001, India; Email: rbs@tsimtsoum.net, drkk@dataone.in

[12]School of Medicine and Health, Durham, UK
[13]Department of Pharmacology, Medical University in Sofia, Bulgaria

ABSTRACT

Elderly subjects have unique nutrient needs with revised recommended dietary reference intakes based on age, sex, ethnicity and special vulnerability to compromised nutrient status.

The food intake tends to decrease with advancing age due to decreased physical activity and metabolic rate. However vitamin; A,E,C, B vitamins and beta-carotine and minerals magnesium, calcium, selenium, chromium, zinc and copper as well as w-3 fatty acids needs either remain constant or increase. Omega-3 fatty acid, calcium, vitamin D and magnesium and antioxidant intakes are inversely associated with ageing and other complications like cardiovascular diseases, diabetes and dementia as well as cancers. It is pertinent to advise to eat 400g/day of fruits, vegetables and nuts and another 400g/day of whole grains in conjunction with 25-40g/day of canola oil or mustered oil, depending upon energy requirement, for prevention of cardiovascular disease and type 2 diabetes mellitus.

Keywords: Ageing, older population, foods, nutrition, chronic diseases

INTRODUCTION

Non-communicable diseases such as cardiovascular diseases (CVD), diabetes mellitus, metabolic syndrome and cancer have become a major health problem in the Western world and these are rapidly increasing in the developing world with ageing of the population [1-3]. Elderly populations have unique nutrient needs with revised recommended dietary reference intakes based on age, and special vulnerability to compromised nutrient status. The food intake tends to decrease with advancing age to compensate for reduced energy requirements, however vitamin; A,E,C, B vitamins and beta carotine and minerals magnesium, calcium, selenium, chromium, zinc and copper as well as w-3 fatty acids needs either remain constant or increase. Omega-3 fatty acid and minerals; magnesium intakes are inversely associated with ageing [1-3].

The overconsumption of energy and nutrients must be avoided. Specific nutritional needs of elderly people were addressed in the Food Guide Pyramid for Older Adults, a modification of the 1990 Food Guide Pyramid for Americans. In 2005, the US Department of Agriculture (USDA) released MyPyramid (Figure 1), which includes an Internet-based program allowing individuals to calculate food-based dietary guidance based on their specific clinical variables. However, older adults tend to be less comfortable with Internet use, so a graphic representation for a Modified MyPyramid for Older Adults is intended for use in this age group as an adjunct to the current Web-based MyPyramid. Such approaches are no good for the developing world where most elderly do not use internet. Singh et al. have published several diet and lifestyle guidelines for prevention of cardiovascular diseases and diabetes among adults [4-6].

These guidelines may be modified in context of their lower food intake and increased requirement of nutrients in the elderly population. It seems that elderly need more fruits, vegetables, nuts and whole grains as well as more w-3 fatty acid rich oils for prevention of chronic diseases in general and brain degeneration in particular [1-3].

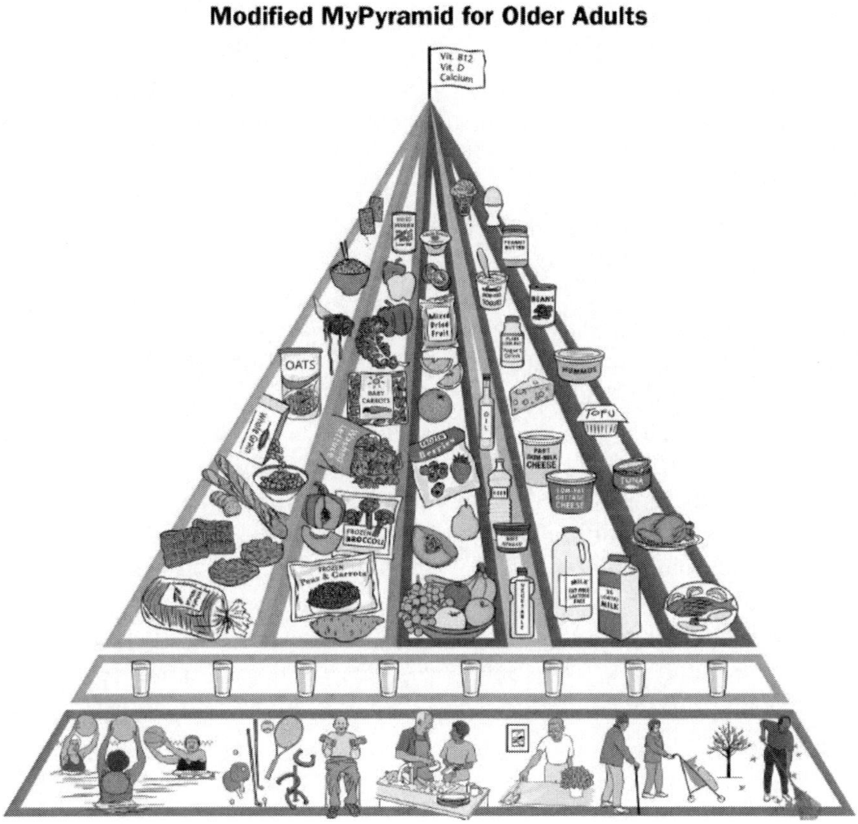

Figure 1. Modified MyPyramid for Older Adults.

How Elderly Differ from Adults

Non-communicable diseases (NCD) are much more common among elderly than among adults. Prevention of NCDs has been shown to be related to specific nutrients, so the key element in planning diets for the elderly should be nutrient based dietary guidelines [1-3]. In order to reduce the burden of chronic diseases by nutritional means, nutrient based guidelines, become, essential to develop. Elderly people have distinctly different metabolic processes that do not allow for easy extrapolation of nutrient needs from results of experiments performed on younger adults. Because adults over 70 years consume less food than younger people, primarily due to decreases in energy expenditure, which is important to emphasize for old people. Special considerations regarding nutritional needs of the elderly population are that food intake tends to decrease with advancing age, because of reduced energy needs associated with lower energy expenditure in physical activity and basal metabolic rate. However, requirements for vitamins and minerals either remain stable or increase.

UNITED STATES FOOD PYRAMID

A Modified Food Guide Pyramid was developed for adults aged 70 years and older to address their unique dietary challenges in 1999. Compared with the original USDA Food Guide Pyramid, the modifications included narrowing the base to reflect lower energy needs of older adults because of changes in body composition and metabolic rate. Selected food icons with nutrient-dense examples were replaced to help reconcile decreased food intake with unchanged or increased recommended dietary allowances. Addition of a fiber icon was done in appropriate food categories to facilitate achieving adequate intakes to promote optimal bowel function. A row of glasses at the base of the pyramid were included to remind older adults to maintain adequate fluid intakes; and placement of a flag at the top to alert some older adults that their healthcare provider should consider recommending vitamins B_{12} or D or calcium supplements. Looking upon nutrients as modulators of chronic disease risk represents an important paradigm shift. These nutrients are known to decrease homocysteine (B vitamins and folic acid) for reducing risk of atherosclerotic cardiovascular disease and prevent the bone demineralization and osteoporosis (calcium and vitamin D). The newly evolving Recommended Dietary Intakes in the world would define nutrient intake levels that should decrease an individual's risk of developing chronic diseases and degenerative conditions that are related to specific nutrients and associated with negative functional outcomes [7-9]· There is increased requirement of antioxidant nutrients vitamins C and E and selenium for the prevention of atherosclerotic CVDs and/or certain site-specific cancers and of the carotenoids, zeaxanthin and lutein, and the prevention of age-related macular degeneration as well as cancers [1-10]

THE NEW AMERICAN GUIDELINES

The 2010 Dietary Guidelines is available at www.dietaryguidelines.gov. Further details can be seen by visiting; www.health.gov/dietary guidelines and www.healthfinder.gov /prevention. The new guidelines advise that certain basic factors be taken into account when making daily food choices, such as age, ethnicity, and risk factor for diet-related chronic illnesses like diabetes and heart disease. The Dietary Guidelines also call for significant, specific reductions in food components like salt, added sugars, processed flour, solid fats, and saturated fats according to varying factors. The most notable of these specified restrictions is for sodium (salt) intake for persons who are age 51 or older; African American; or have hypertension, diabetes, or chronic kidney disease from the regular 2,300mg to 1,500mg.The Guidelines further change course by not only listing foods that should be encouraged, but also listing foods that should be reduced, such as grain-based deserts, regular cheese, dairy deserts, and even chicken dishes. The USDA supplies multiple pie-charts that show where the average American is going wrong with their diets. Continuous listings of statistics involving diet-related chronic illnesses appears to be the most attention-grabbing element of the Dietary

Guidelines for Americans 2010. According to the document, 72 percent of men and 64 percent of women in the United States are overweight or obese, 37 percent of the population has cardiovascular disease, and 35 percent of the adult population suffers from pre-diabetes. In fact, the prevalence is much greater among elderly populations which indicates that the need for nutritional modulators of chronic diseases is greater among elderly subjects compared to adults. There is no advice on adverse effects of w-6 fatty acids and beneficial effects of w-3 fatty acids in the American guidelines.

How to Provide Functional Foods to Elderly?

Recent studies indicate that Mediterranean diet, Indo-Mediterranean diet and Japanese diet are rich sources of protective nutrients as well as are known to decrease morbidity and mortality among various population groups [11-28]. Apart from these diets, consumption of nuts, chocolates, Mediterranean soups, fish as well as fish oil have been found to reduce the risk of cardiovascular diseases and diabetes [10-20]. Therefore these foods should be preferred among elderly populations in both developing and developed countries for prevention of NCDs.

To evaluate the association of chocolate consumption with the risk of developing cardiometabolic disorders, systematic review and meta-analysis of randomised controlled trials, cohort, case-control, and cross sectional studies carried out in human adults, in which the association between chocolate consumption and the risk of outcomes related to cardiometabolic disorders were reported [10]. Data were extracted by two independent investigators, and a consensus was reached with the involvement of a third. The primary outcome was cardiometabolic disorders, including coronary heart disease and stroke, diabetes, and metabolic syndrome. A meta-analysis assessed the risk of developing cardiometabolic disorders by comparing the highest and lowest level of chocolate consumption.

From 4576 references seven studies met the inclusion criteria (including 114,009 participants). None of the studies was a randomised trial, six were cohort studies, and one a cross sectional study. Large variation was observed between these seven studies for measurement of chocolate consumption, methods, and outcomes evaluated. Five of the seven studies reported a beneficial association between higher levels of chocolate consumption and the risk of cardiometabolic disorders. The highest levels of chocolate consumption were associated with a 37% reduction in cardiovascular disease [relative risk 0.63 (95% confidence interval 0.44 to 0.90)] and a 29% reduction in stroke compared with the lowest levels. Based on observational evidence, levels of chocolate consumption seem to be associated with a substantial reduction in the risk of cardiometabolic disorders. Further experimental studies are required to confirm a potentially beneficial effect of chocolate consumption.

Pathogenesis of Diet and Risk of Diseases

There is evidence that Western diet may be important in the pathogenesis of non-communicable diseases and all cause mortality [21-28]. The beneficial or adverse effects of certain foods may be due to their effects on inflammation, free fatty acids, hyperglycemia, hyperinsulinemia, oxidative stress, hypertriglyceridemia, hypercholesterolemia and high oxidized low density lipoprotein cholesterol (LDL) as well as on oxidized high density lipoprotein cholesterol (HDL), which are considered to be highly atherogenic [21-24]. These risk factors and other biological factors cause an adverse impact on vascular biology resulting in to endothelial dysfunction and inflammation which may be independently athero-thrombotic [29-33]. Hormone replacement therapy (HRT), oral contraceptive intake in women, anti-inflammatory effects of statins [29] and postprandial endothelial dysfunction and inflammation are indicator of vascular index responsible for atherosclerosis. In women, oestrogen therapy has been used to lower their cardiovascular risk in post-menopausal women, prior to HRT trials, showing no beneficial effects. Oestrogens are known to decrease LDL and increase HDL but they also increase proinflammatory C-reactive protein, coagulation and lipotrotein (a), which are known for their adverse effects on atherothrombosis [29-33]. It seems that vascular biological states appear to be most important determinant of atherothrombosis compared to circulating LDL, although most expert give greater importance to circulating LDL. Infact C-reactive protein has been found to be as predictive of subsequent coronary events as LDL on treatment with statins [29]. The concentration of oxidized LDL particularly in the vascular cells, depends on the internal antioxidant environment which is determined by dietary patterns. Dietary patterns consisting of high intakes of vegetables, fruits, legumes, fish, poultry, and whole grains, have been found to be protective against risk from cardiovascular disease (CVD) and malignant diseases [21-26]. Dietary patterns closer to a Western diet have adverse effects on these biomarkers and risk of CVD, diabetes and cancer. Other primary risk factors such as sedentary behaviour, tobacco intake, alcoholism and stress are also known to increase the risk of chronic diseases of affluence.

THE CONCEPT OF PALEOLITHIC DIET AND THE YING YANG TSIM TSOUM CONCEPT

About 40,000 years ago, our genes appear to be similar to the genes of our ancestors during the Paleolithic period, the time when our genetic profile was established. Man appears to live in a nutritional environment which completely differs from that for which our genetic constitution was selected. However, it was only during the last 100-160 years that dietary intakes have changed significantly, causing increased intake of saturated fatty acids (SFA) and linoleic acid, and decrease in w-3 fatty acids, from grain fed cattle, tamed at farm houses, rather than meat from running animals. The food and nutrient intake among hunter-gatherers and during Paleolithic period are given in Tables 1, 2. There is marked reduction in consumption of w-3 fatty acids, vitamins and minerals and proteins and significant increase in the intakes of carbohydrates (mainly refined), fat (saturated, trans fat, linoleic acid) and salt compared to Paleolithic period.

The Columbus concept of diet means that humans evolved on a diet that was low in saturated fat and the amount of w-3 and w-6 fatty acids was quite equal [23,40]. Nature recommends to ingest fatty acids in a balanced ratio (polyunsaturated: saturated=w-6:w-3=1:1)as part of dietary lipid pattern in monounsaturated fatty acids (P:M:S=1:6:1). These ratios represent the overall distribution of fats in a natural untamed environment.

(www.columbus-concept.com). The Columbus foods include egg, milk, meat, oil , and bread, all rich in w-3 fatty acids, similar to wild foods, consumed about 160 years ago from now. Blood lipid composition does reflect one's health status: (a) circulating serum lipoproteins and their ratio provide information on their atherogenicity to blood vessels and (b) circulating plasma fatty acids, such as w-6/w-3 fatty acid ratio, give indication on pro-inflammatory status of blood vessels.; (a) and (b) are phenotype-related and depend on genetic, environmental and developmental factors. As such, they appear as universal markers for holistic health. Blood cholesterol is central to this approach. Its 3D-representation shows how circulating lipoproteins affect blood vessels integrity upon their circulating throughout the body. Of major importance appear the essential dietary nutrients (essential amino acids, fatty acids, antioxidant vitamins and minerals) and the functional component of the regimen (diet, sport, spiritualism, etc).The Ying Yang Tsim Tsoum (YYTT) Concept is an extension of the Columbus Concept (YYCC). While YYCC is a 1-D approach that establishes the basis for the "essential" components of the diet, YYTT is a 2-D approach that focuses on the "functional" components of the lifestyle; stress, sedentary behaviour, meditation. YYCC and YYTT are complementary and do not compete against each other. Chronobiology is an important aspect of YYTT because it is based on mind- brain- body interactions (http://www.tsimtsoum.net/introduction_07.php). It seems that the YYTT approach favors the analysis of memes and genes as well as of (hu)man's influence on nature and also possibly the influence of nurture and environment on human health and behaviour. In several studies, breathing exercises, meditation and yoga have been found to influence brain function, cardiovascular function as well as biomarkers [32-34]. Experimental studies [35-37] indicate that ω3 PUFAs play an important role in neuronal structure and function. The brain is quite rich in ω3 PUFAs and several studies suggest a role for ω3 PUFAs in neurotransmitter synthesis, degradation, release, reuptake and binding [35-37].

Table 1. Food and nutrient intake among hunter-gatherer and western population

Food and nutrient	Hunter-gatherer	Western population
Energy density	Low	High
Protein	IIigh	Low–moderate
Animal	High	Low–moderate
Vegetable	Very low	Low–moderate
Carbohydrate	Low–moderate(slowly absorbed)	Moderate–rapidly absorbed
Fiber	High	Low
Fat	Low	High
Animal	Low	High
Vegetable	Very low	High
Total w-3	High (2.3g/day)	Low (0.2g/day)
Ratio w-6:w-3	Low 2.4	High 15-20
Vitamins and minerals	high	low

Modified from ref [19,20].

Fatty acids belong to the phospholipid group and, consequently, are part of all biological membranes. The membrane's fluidity, which is of crucial importance for its functioning, depends on its lipid components. Phospholipids built up of chains of polyunsatured fatty acids increase the membrane fluidity because, by binding some chains, double bonds prevent them from compacting themselves perfectly.

In addition, membrane fluidity is determined by the phospholipids/free cholesterol ratio, as cholesterol increases membrane viscosity.

Table 2. Estimated fatty acid consumption in the late Paleolithic period

Sources	Fatty acids(g/day) en 35.65/day
Plants	
Linoleic acid	4.28
Alpha-linoleic acid	11.40
Animal	
Linoleic acids	4.56
Alpha-linolenic acid	1.21
Total	
Linoleic acid	8.84
Alpha linolenic acid	12.60
Animal	
Arachidonic acid(w-6) (AA)	1.81
Eicosapentaenoic acid(w-3)(EPA)	0.39
Docosatetraenoic acid(w-6) (DTA)	0.12
Docosapentaenoic acid(w-3)(DPA)	0.42
Docosahexaenoic acid(w-3)(DHA)	0.27
Ratios of w-6/w-3	
Linoleic acid/alpha linolenic acid	0.70
AA+DTA/EPA+DPA+DHA	1.79
Total w-6/w-3	0.79

Modified from Eaton et al. in ref [19,20].

DHA deficit is associated with dysfunctions of neuronal membrane stability and transmission of serotonin, norepinephrine and dopamine, which might be related to the aetiology of the mood and cognitive dysfunction of depression [38, 39]. On the other hand, EPA is essential to balance the immune function and physical health by reducing the proportion of arachidonic acid (AA, C20:4ω6) in cell membrane and prostaglandin E2 (PGE2) synthesis. A diet based on a high proportion of essential polyunsatured fatty acids allows a higher incorporation of cholesterol in the membranes to balance their fluidity, which, in turn, would contribute to lower blood cholesterol levels and inflammation and lower risk of CVD and psychological disorders [37-39]. These interactions of diet, neurotransmitters, neuronal function and other body functions and dysfunctions provide a basis for YYTT concept and the mind-body connection.

New Biomarkers for CVD and Dietary Patterns

Several epidemiological studies, have reported an inverse relationship between consumption of fruits and vegetables and the incidence of coronary artery disease (CAD) [21-

24]. Only a few randomized, controlled trials, using fruits, vegetables, legumes, fatty acids, and nuts are available, showing significant reduction in mortality and cardiovascular events [25-27]. There is further evidence that inflammation could be an independent mechanism for pathogenesis of atherothrombosis, resulting into heart attack or acute coronary syndrome (ACS) [28-30]. Therefore, statins are being used for primary prevention of cardiovascular events in patients with increased high sensitive C-reactive proteins (hsCRP), without much consideration of serum cholesterol levels (29). Moreover, LDL cholesterol becomes atherogenic only, when it is modified into oxidized LDL cholesterol, which is most attractive to macrophages in the pathogenesis of atherosclerosis. It seems that statins not only reduce LDL but also decrease oxidized LDL which may be more important for prevention of CVD. The results of the JUPITER (Justification for the Use of statins in Primary prevention: an Intervention Trial Evaluating Rosuvastatin) trial add up to a formidable amount of information about the beneficial effects of statins in healthy subjects with high hsCRP [29]. High levels of CRP as well as other proinflammatory cytokines; interleukin-6 (IL-6), IL-1, IL-18 and tumor necrosis factor-alpha are also risk factors of CVD, diabetes, brain degeneration, psychological disorders and cancer [30-39]. The JUPITER study screened almost 90, 000 healthy volunteers aged 50 years or more for participation in this trial among healthy subjects. One in five [17, 802] of those screened were included in the trial. Treatment with statin is not indicated in such individuals because they had no history of CAD, and their LDL-cholesterol levels were below the usual threshold for treatment of 130 mg/dL (3.33 mmol/L). Those who achieved a dual treatment target of LDL cholesterol lower than 1.8 mmol/l (70 mg/dl) and hsCRP lower than 2 mg/l had a 65% reduction in CVD events, compared with a 33% reduction in patients who achieved one or neither target. The authors reported that participants who achieved an LDL cholesterol level of below 1.8 mmol/l (70 mg/dl) had a 55% reduction in CVD events and those who achieved a hsCRP reduction of below 2 mg/l a 62% reduction, which is slightly greater benefit compared to LDL reduction. However, patients who achieved the more stringent target of a LDL cholesterol level of below 1.8 m mol/l (70 mg/dl) and a hsCRP reduction of below 1 mg/l had a 79% reduction in CVD events. This shows that decrease in inflammation in conjunction with LDL reduction has greater benefit at the same level of LDL reduction, which appears to be due to a further decrease in oxidized LDL. There is a need to measure oxidized LDL as well as total LDL to find out the mechanism of this interaction.

A meta-analysis, quantitatively assessed the relation between fruit and vegetable intake and incidence of CAD in all cohort studies which reported relative risks (RRs) and corresponding 95% confidence interval (CI) of CAD with respect to frequency of fruit and vegetable intake [22]. These studies, included a total of 278 459 individuals (9143 CAD events) with a median follow-up of 11 years. The individuals who had less than 3 servings/day of fruit and vegetables, the pooled RR of CAD was 0.93 (95% CI: 0.86–1.00, P¼0.06) while those with more than 5 servings/ day , the RR was 0.83 (0.77–0.89, P<0.0001).This meta-analysis of prospective cohort studies demonstrated that increased consumption of fruit and vegetables, less than 3 to more than 5 servings/day is related to a 17% reduction, whereas increased intake to 3–5 servings/day is associated with a smaller and borderline significant reduction in CAD risk. These results provide evidence supporting that 5 or more servings per day of fruit and vegetables, are needed to protect from CVD. In a large, prospective, observational study (24) involving 72,113 female nurses who were free of CAD, stroke, diabetes, and cancer, factor analysis identified 2 dietary patterns from data collected on serial food frequency questionnaires. One pattern, called prudent, was characterized by a high consumption of vegetables, fruit, legumes, fish, poultry, and whole grains. The other pattern, called Western, corresponded to a high consumption of red meat, processed meat, refined grains, french fries, sweets, and desserts. Individuals were classified by their level of adherence to both the prudent diet and the Western diet. After baseline data collection in 1984, follow-up lasted 18 years, during which time 6011 deaths occurred; 3319 [52%] as a result of cancer; 1154 [19%] resulting from CVD; and 1718 (29%) resulting from other causes. In multivariable analyses, there was a 17% lower risk of total mortality among those who were most adherent to the prudent diet (highest versus lowest quintile of adherence), a 28% lower risk of CVD mortality, and 30% lower mortality from non-CVD, non-cancer causes. Interestingly, there was no significant relationship between the prudent dietary pattern and cancer. A comparison of the highest and lowest quintiles of adherence showed that consumption of the Western diet was associated with increased total mortality (21%), CVD mortality (22%), cancer mortality (16%), and mortality from non-CVD, non-cancer causes (31%). Hence, except for cancer, risk relationships for the prudent and Western dietary patterns appear to be the inverse of each other: Mortality thus was increased as adherence to the prudent diet decreased and adherence to the Western diet increased.

The relationship between dietary patterns and risk of acute coronary syndrome (ACS) was examined in a standardized case-control study INTERHEART, involving participants from 52 countries. [41] Using the principle-component analysis technique, the authors identified 3 major dietary patterns; Oriental (high intake of tofu and soy and other sauces), Western (high in fried foods, salty snacks, eggs, and meat), and prudent (high in fruit and vegetables). Consistent with previous studies in single within-population cohort studies, the authors found an inverse association between the prudent pattern score and risk of ACS and a significant positive association between the Western pattern score and increased risk of ACS. The Oriental pattern was not significantly associated with risk. The investigators constructed a dietary risk score based on 7 food items on the food-frequency questionnaire (meat, salty snacks, fried foods, fruits, green leafy vegetables, cooked vegetables, and other raw vegetables) and found that a higher score (indicating a poor diet) was strongly associated with ACS risk: Those in the highest quartile of the score had nearly 2-fold increased risk, even after adjustment for established coronary risk factors. In sensitivity analyses, the investigators found a consistent association for the composite diet score between men and women and across different regions of the world (North America, western Europe, Australia, central Europe, Middle East, Africa, south Asia, southeast Asia, China, and South America). On the basis of an arbitrary cut point of the score (top 3 quartiles versus the bottom quartile), the investigators estimated that 30% of MI could be explained by unhealthy diets worldwide.

Although Western-style changes in food patterns are widely believed to adversely influence risk of chronic diseases, few epidemiological studies have directly linked dietary patterns and mortality from coronary heart disease. The INTERHEART study is the first large study to quantify eating patterns in all geographic regions of the world. It provides evidence that despite different food habits in various populations, reproducible patterns can be found in diverse regions of the world. These findings are important because there has been a concern that dietary patterns derived through a data-driven approach such as principle-component analysis may be highly unstable and non-reproducible because of very different eating habits in different populations.

In recent years, dietary-patterning analysis has been increasingly used as an alternative method to traditional single-nutrient analysis because it can assess cumulative effects of the overall nutrients in the diet. Habitual intake patterns are typically quantified by statistical methods such as factor or cluster analysis or diet-quality indexes based on prevailing dietary recommendations or healthful traditional diets; e.g., the Mediterranean diet, and Indo-Mediterranean diet. Principle-component analysis is commonly used to define dietary patterns using food consumption information to identify common underlying dimensions (factors or patterns) of food intake. The method aggregates specific food items based on the degree to which these food items are correlated with each other. A summary score for each pattern is then derived and can be used to examine relationships between various eating patterns and outcomes of interest such as coronary heart disease and other chronic diseases. Previous validation studies found that 2 major patterns (the prudent and Western patterns) identified through principle-component analysis of food consumption data assessed by food frequency questionnaires were reproducible over time and correlated reasonably well with the patterns identified from diet records.

In a more recent study [42], 74,886 women 38 to 63 years of age in the Nurses' Health Study, a cohort study of female nurses, without a history of cardiovascular disease and diabetes were followed up from 1984 to 2004. The authors computed an Alternate Mediterranean Diet Score from self-reported dietary data collected through validated food frequency questionnaires administered 6 times between 1984 and 2002. Relative risks for incident CAD, stroke, and combined fatal cardiovascular disease were estimated with Cox proportional-hazards models adjusted for cardiovascular risk factors. During 20 years of follow-up, 2391 incident cases of CAD, 1763 incident cases of stroke, and 1077 cardiovascular disease deaths (fatal CAD and strokes combined) were ascertained. Women in the top Alternate Mediterranean Diet Score quintile were at lower risk for both CAD and stroke compared with those in the bottom quintile (relative risk [RR], 0.71; 95% CI, 0.62 to 0.82; *P* for trend<0.0001 for CHD; RR, 0.87;95% CI, 0.73 to 1.02; *P* for trend=0.03 for stroke). Cardiovascular disease mortality was significantly lower among women in the top quintile of the Alternate Mediterranean Diet Score (RR, 0.61; 95% CI, 0.49 to 0.76; *P* for trend <0.0001) indicating that Mediterranean diet may be protective against CVD. The consistent association observed between the Western or unhealthy dietary pattern (high in animal products, salty snacks, refined starches and sugar and fried foods and low in fruits and vegetables) and ACS risk in different regions of the world from the INTERHEART study and other studies [21, 24, 41, 42] provides reliable evidence of the adverse effects of globalization on human nutrition and chronic disease risk, but the evidence is indirect because these studies did not specifically assess the impact of global trade and marketing on food consumption patterns, and their effects on inflammation, endothelial dysfunction, across different countries. [21, 24, 41, 42] Despite this weakness, most recent studies suggest that the current trend of dietary convergence toward a typical Western diet is likely to play a role in the globalization of obesity, CVD, diabetes and cancer [38-44]. These studies further emphasize the role of anti-inflammatory, hypocholesterolemic and hypoglycemic agents in the pathogenesis and prevention of heart attack [21-29].

The endothelium plays a major role in regulating vascular tone, mainly by secreting the potent vasodilator nitric oxide (NO), which is antiatherogenic. NO is synthesized from its precursor, L-arginine (Arg), by endothelial NO Synthase (NOS) [24].It is possible that L-arginine levels are under influence of dietary amino acid arginine which is rich in nuts. NOS is competitively inhibited by asymmetric dimethylarginine (ADMA), an endogenous compound that is elevated in renal failure, cardiovascular disease (CVD), and diabetes mellitus. A low ratio of Arg to ADMA (Arg/ADMA ratio) is also a marker of endothelial dysfunction. Prospective investigations of ADMA have highlighted its role as a predictor of CVD events or death in patients with established coronary artery disease, advanced renal failure, or other high-risk conditions. There is evidence that NO levels are under influence of dietary w-3 fatty acids, physical activity and meditation and may be mediated by inflammation [43-46].

Chronobiology and the YYTT Concept

Study of biological functions according to time structure that are usually under influence of environmental factors, are included in chronobiology [47-50]. Chronobiology establishes a strong relation of mind with brain and their effects on body functions and dysfunctions. The YYTT concept is based on such interactions of mind, brain and various organs of our body. We detect blood pressure overswinging and other vascular variability disorders (VVDs), including a high blood pressure, that is an elevation of the MESOR, for midline estimating statistic of rhythm, i.e., MESOR-hypertension, that is also a VVD. [47] Moreover this is a principal point, other VVDs can complicate a high blood pressure or may be induced by treating a high blood pressure. Most of the conventionally unrecognized VVDs change a risk of stroke or of a myocardial infarction within the next 6 years from ~5% to near 100% and can be eliminated, sometimes by changing only the time of taking the medication. A recent international consensus, guided by leaders in cardiovascular physiology, recognizes that 7-day/24-hour blood pressure and heart rate monitoring interpreted chronobiologically can detect new risks associated with a very high likelihood of stroke [47]. It also recognizes that over one-third of 72 million Americans said to be treated for hypertension can have a great, conventionally unrecognized risk increase that can often be removed by adjusting the timing of antihypertensive drugs. If and only if analyzed chronobiologically, computer-implemented self-surveillance of blood pressure and heart rate by 7-day/24-h monitoring detects conditions to which conventional treatment is now blind [47-50].

Availability of Foods

Availability of prudent wild foods at reasonable cost, time taken in preparation, taste and aggressive publicity of western foods by the food industry are major problems that need attention. Food industry in general is not very interested in preparing health foods that are known to decrease mortality and morbidity, because their primary goal is to sell the product at a competitive price, to gain profit and prevent losses. Wild foods as the amount of wild fish, animal foods is rapidly getting impossible to get, because the oceans are overfished. Therefore the goal is to have sufficient cultivated wild type of foods with characteristics of wild type ones, as there is a limited and decreasing supply of real wild foods. We are now in a situation where the public health policies are clear for recommending prudent dietary patterns and providing health foods to the community. However, the food policies of the government and of the farming and food industries are geared to completely different goals. The growing demands on the farmers and the food industry for a healthier diet are now being recognized, as marketing opportunity for developing healthy foodstuffs, with balanced amount of fatty acids, low glycemic foods and other nutrients in the diet.

World Health Organization and Food and Agriculture Organization should join hands with international scientific societies like the World Heart Federation, International College of Cardiology, American Heart Association, European Society of Cardiology, International College Of Nutrition and International Union of Nutritional Sciences, and collaborate with World Mind-Body Council, and YYTT Institute http://www.tsimtsoum.net/introduction_07.php, to formulate guidelines for manufacturing of health foods. This non-profit organization has expertise, how to develop wild type of foods for prevention of diseases and enhance high quality of productive life. These foods are rich in micronutrients, antioxidants, vitamins and minerals, while providing high w-3 fatty acids, monounsaturated fatty acids, proteins and slowly absorbed carbohydrates. Long term, randomized, controlled clinical trials are necessary to provide a scientific proof regarding the safety and efficacy of these foods in the prevention of diseases. These trials should be supported by the government agencies and international organizations such as European Union grants, because private companies are not capable in providing large amount funds, necessary for such trials.

In one such trial, [23] it has been demonstrated that whole grain- enriched hypocaloric diet is superior to refined grain group in decreasing risk of CVD in patients with metabolic syndrome. However, no study has referred to w-6/w-3 fatty acid ratio of the two diets, which could have been a more important factor for decreasing inflammation due to increased content of w-3 fat and low w-6 fat [24]. Wild - whole grains rich in w-3 fatty acids, antioxidants and magnesium may provide even better results compared to whole grain grown by modern farming.

The western diet is rich in w-6 fatty acids and low in w-3 fatty acids, fiber, phyto-estrogens, minerals and antioxidants. CVD, diabetes mellitus, cancer, autoimmune diseases, rheumatoid arthritis, asthma and depression are associated with increased production of thromboxane A2, leucotrienes, interleukins-1 and 6, tumor necrosis factor-alpha and C-reactive proteins. Increased dietary intake of w-6 fatty acids is known to enhance all these risk factors as well as atherogenicity of cholesterol which have adverse proinflammatry effects resulting into thrombosis and CAD [23-28].

The second quarter of the day, particularly 8.00-11.00 AM may be associated with activation of neurohormones and proinflammatory cytokines; interleukin- 6, and 18, resulting into neuroendocrine dysfunction, which worsens the internal environment of our body predisposing greater prevalence of cardiovascular events due to mind-brain and body interactions [24-28]. There may be oxidative stress, hyperglycemia, hyperinsulinemia, hypertriglyceridemia, increased levels of catecholamines, free fatty acids and cortisol which may cause endothelial dysfunction and rupture of plaque resulting in to recurrent cardiovascular events and VVDs. [47-50]

Recent studies [24-28, 40-43] indicate that eating high fat, refined carbohydrate rich fast foods (western diet), can produce a similar proinflammatory state in our body, with increase in oxidative stress, free fatty acids and triglycerides; resulting into endothelial dysfunction. It is therefore, logical to avoid western diet in patients with CVD, and to administer Indo-Mediterranean foods which may be beneficial to vascular endothelium and myocardium [22-28]. There is no precise and proven guideline for dietary advice in patients with CAD, which may be protective against recurrent cardiac events. A Mediterranean soup (tomatoes, grapes/raisins, vegetables; carrot, spinach, walnuts, almonds+lin/chia seeds and olive oil) or yogurt containing, walnuts, almonds, raisins, could be prepared for ready use, for nonpharmacological intervention, among patients of CVD. These foods appear to be protective against metabolic syndrome [23]. Such recipes have been commonly used in the Indo-Mediterranean diet heart study and Indian experiment of infarct survival [26-28].

Therefore for prevention of metabolic syndrome, or CAD, or cardiovascular and all cause mortality, eating 400g/day of fruits and vegetables along with 400g/day of legumes and other whole grains, and 50g/day of almonds and walnuts, in conjunction with 25-50g/day of Columbus oil (olive oil 88 % + flax seed oil 12.0%) P:S= w-6:w-3:=1:1 Patent Application WO2005020698 may be protective against all cause mortality and morbidity. Since egg intake up to one daily has no adverse effects on coronary risk factors and proinflammatory factors, including egg in the breakfast along with Mediterranean soup, in place of refined starches and meat, may be a prudent choice in the primary prevention of CVD. Designer eggs with modified fatty acid composition and increased content of w-3 fatty acids appear to be good for prevention of metabolic syndrome. [24] Designer bread and chicken rich in w-3 fatty acids may be another prudent substitute for whole grains and meat, due to their most beneficial chemical composition causing slow absorption without glycemia.

Although, most workers working on dietary patterns do not mention the nutrient content of their prudent diet [21, 22, 33-35, 40] but one single difference is the w-3 fatty acids, apart from other micronutrients, which is rich in fruits, leafy vegetables, nuts and whole grains. It would be very interesting to know the role of refined starches and sugar, large meals, decreased intake of fruits, vegetables, whole grains and nuts and serum nitrite levels as risk predictors of CVD in these cohorts [21, 22, 35].

Fruits, vegetables, nuts, whole grains, animal foods rich in w-3 fatty acids are slowly absorbed and may prevent the increase in free fatty acids, and inflammation, which is a characteristic of Columbus foods (www.columbus-concept.com) and therefore such foods may decrease the risk in CVD as well as metabolic syndrome [23-28, 33-35, 39-44]. Omega-3 fatty acids can regulate leptin gene expression and the concentrations of anandamides in the brain, which in turn binds to endogenous cannabinoid receptors and regulate food intake and satiety and weight gain [24]. It seems that suboptimal nutrition is an important factor in the global burden of CAD. Cohort studies [21, 22, 38-40] add to this evidence by identifying the association between an unhealthy dietary pattern and risk of ACS in 52 countries around the world and among women in the United States [21, 22, 38]. By defining a common feature of nutrition transitions with their accompanying risks to public health, the authors underscore the importance of developing an effective and comprehensive set of health policies that address globalization and its impact on obesity and diet-related chronic diseases. Indeed, the findings from recent studies indicate [23-28, 38-40] that a common set of dietary recommendations based on availability of wild type of foods can be made to prevent chronic diseases on a worldwide basis. These foods are known to have anti-inflammatory effects and can modulate about 125 genes, most of them proinflammatory, which express due to increased consumption of western meals [43, 44]. Such a public health strategy should emphasize replacing saturated, *trans* fats and w-6 fats with unsaturated fats from natural vegetable oils rich in w-3 fatty acids (canola oil, grapeseed oil), and replacing refined grain products, starches and sugar with whole grain products, legumes, fruits, and vegetables, fish and nuts. [23-28] International College of Cardiology and International College of Nutrition in association with Columbus

Paradigm Institute have emphasized to develop guidelines [40] to include wild type of foods in the dietary patterns as well as moderate physical activity and yogasan, which may be highly protective in the prevention of morbidity and mortality due to CVD, diabetes and Cancer (Table 3).

Table 3. Dietary guidelines and desirable level of risk factors for populations

Factors	Desirable Values
Energy (k calories/day)	1900-2300
Total Carbohydrate (k calories/day)	65.0
Complex Carbohydrate ((k calories/day)	55.0
Total Fat (k calories/day)	21.0
Saturated Fatty Acids (k calories/day)	7.0
Polyunsaturated Fatty Acids (k calories/day)	7.0
Polyunsaturated/Saturated Fat Ratio	1.0
n 6/n-3 Fatty Acid Ratio	1:1
Dietary Cholesterol (mg/day)	100
Whole Grains (wheat, rice, corn, legumes) (g/day)	400-500
Fruit, vegetables and nuts (g/day)	400-500
Salt (g/day)	<6.0
Brisk Walking (km/day)	>9.0
Meditation/pranayam (minutes/day)	30.0
Body Mass Index (kg/m^2)	
Range	19.0-23.0
Average	21.0
Waist-Hip Girth Ratio	
Male	<0.88
Female	<0.85
Serum Total Cholesterol (mg/dl) (4.42 mmol/L)	<170
Mild Hypercholesterolemia (mg/dl) (4.42-5.20 mmol/L)	170-200
Hypercholesterolemia (mg/dl) (>5.20 mmol/L)	>200
Low Density Lipoprotein Cholesterol (mg/dl) (2.32 mmol/L)	<90
Borderline High (mg/dl) (2.32-2.84 mmol/L)	90-110
High (mg/dl) (2.84 mmol/L)	>110
Triglycerides (mg/dl) (1.7 mmol/L)	<150
High Density Lipoprotein Cholesterol (mg/dl) (0.9 mmol/L)	>40 men, >50women
Blood Pressure (mmHg)	<125/80
Drug therapy in view of high risk of diabetes and CAD.	Amblodipine, ACE-I , receptor blockers. Fish oil, aspirin, statins
Moderate physical activity and	30 min daily
Meditation and yogasan, pranayam.	30 min daily

Modified from Indian Consensus Group, J Nutr Environ Med, 1996, Singh et al. 2009(Ref [4-6]).

KEY POINTS, USFD AND INTERNATIONAL COLLEGE OF CARDIOLOGY

- Older adults have no or less access and familiarity with computers vs younger adults and are less adept at obtaining Web-based information.
- The Modified Food Guide Pyramid for adults aged 70 years and older has a narrower base to reflect lower energy needs of older adults because of changes in body composition and metabolic rate, hence functional food advice should be available.
- Selected food icons are replaced with functional food examples so that decreased food intake can accommodate unchanged or increased recommended dietary allowances for vitamins, minerals, antioxidants, amino acids and other crucial nutrients.
- Fiber icons in appropriate food categories emphasize adequate fiber intake to preferably from natural resources, to promote optimal bowel function.
- Older adults may be prone to overconsumption of energy and nutrients, which is addressed by a new category of recommendations: tolerable upper intake levels, or the highest average daily nutrient intake that is likely to pose no risk for adverse health effects to almost all individuals in the general population.
- Folate and sodium tend to be overconsumed in the older population.
- The proposed Modified MyPyramid for Older Adults is in a format consistent with the MyPyramid graphic, but it is not intended to substitute for MyPyramid (a multifunctional, Web-based program providing individualized, food-based dietary guidance and supplemental information regarding food choices and preparation, based on sex, body weight, height, and level of physical activity).
- The new graphic version of MyPyramid specifically targets relatively healthy individuals aged 70 years and older who are reasonably active and live independently. Consistent with current guidelines, it portrays a diet rich in fruits, vegetables, whole grains, low-fat and nonfat dairy products, legumes, fish, and lean meats.
- The Modified MyPyramid for Older Adults specifically emphasizes functional foods whole grains; variety within the grains group; variety and nutrient density; vegetables and fruits that are easy to prepare (e.g., frozen foods); low-fat and nonfat dairy products including reduced lactose alternatives in the milk group; low saturated fat and trans fat choices in the oils group; and low saturated fat and vegetable options in the meat and beans group. There is a need to avoid w-6 rich oils and substitute with w-3 rich foods.
- Because of risks associated with alcohol consumption in elderly people, alcohol is not included as an integral component of the Modified MyPyramid for Older Adults but if and only if desired, red wine 100-300 ml on alternate day may be advised.
- The focus is on functional nutrient-rich and fiber-rich foods within each group, recommending food sources of nutrients rather than supplements, and use of fluid and physical activity icons.
- The base of the pyramid has a row of glasses to remind older adults that it is crucial to remain well hydrated.

- In the second row, pictures of various physical activities emphasize the need for regular exercise but take care that outdoor games played by adults can break the bones.
- A flag on top of the pyramid highlights the possible need for supplemental forms of calcium and vitamins D and B_{12} because of increased requirements for these nutrients associated with aging.
- The Modified MyPyramid for Older Adults is a graphic intended to improve understanding and use of MyPyramid by the elderly population, who may be less comfortable with obtaining Web-based information.

In brief, eat 400g/day of fruits, vegetables and nuts and another 200-400 g/day of whole grains in conjunction with 25-40g/day of w-3 fatty acid and flavanoid rich oils (blend of mustard and olive oil) for prevention of NCDs [4-15]. Reduce daily sodium intake to less than 2,300 milligrams (mg) and further reduce intake to 1,500 mg among persons who are 51 and older and those of any age who are African American, Asians or have hypertension, diabetes, or chronic kidney disease. The 1,500 mg recommendation applies to about half of the World population, including children, and the majority of adults. Consume less than 7 percent of calories from saturated fatty acids by replacing them with monounsaturated and polyunsaturated fatty acids. Consume less than 300 mg per day of dietary cholesterol. Keep trans fatty acid and w-6 fatty acid consumption as low as possible, especially by limiting foods that contain synthetic sources of trans fats, such as partially hydrogenated oils, and by limiting other solid fats and w-6 rich oils. Reduce the intake of calories from solid fats and added sugars. Avoid the consumption of foods that contain refined grains, especially refined grain foods that contain solid fats, added sugars, and sodium. A Mediterranean soup containing carrot (100g), tomatoes (150g), walnuts (50g), almonds (50g), raisins (50g) and coriander leaves as well as olive oil and salt for taste and flavor, appears to be the best functional food dish, for health and prevention of diseases among elderly [51].

If alcohol is consumed, it should be consumed in moderation—up to one drink alternate day for women and two drinks alternate day for men—and only by adults of legal drinking age. One drink is defined as 12 fluid ounces of regular beer (5% alcohol), 5 fluid ounces of wine (12% alcohol), or 1.5 fluid ounces of 80 proof (40% alcohol) distilled spirits. One drink contains 0.6 fluid ounces of alcohol. Strong evidence from observational studies has shown that moderate alcohol consumption is associated with a lower risk of cardiovascular disease. Moderate alcohol consumption also is associated with reduced risk of all-cause mortality among middle-aged and older adults and may help to keep cognitive function intact with age. However, it is not recommended that anyone begin drinking or drink more frequently on the basis of potential health benefits because moderate alcohol intake also is associated with increased risk of breast cancer, violence, suicides, drowning, and injuries from falls and motor vehicle crashes.

Older adults should follow the adult guidelines [4,5]. When older adults cannot meet the adult guide-lines, they should be as physically active as their abilities and conditions will allow. Older adults should do exercises that maintain or improve balance if they are at risk of falling. Older adults should determine their level of effort for physical activity relative to their level of fitness. Older adults with chronic conditions should understand whether and how their conditions affect their ability to do regular physical activity safely. Moderate-intensity physical activity: Aerobic activity that increases a person's heart rate and breathing to some extent. On a scale relative to a person's capacity, moderate-intensity activity is usually a 5 or 6 on a 0 to 10 scale. Brisk walking, dancing, swimming, or bicycling on a level terrain are examples. Vigorous-intensity physical activity: Aerobic activity that greatly increases a person's heart rate and breathing. On a scale relative to a person's capacity, vigorous-intensity activity is usually a 7 or 8 on a 0 to 10 scale. Jogging, singles tennis, swimming continuous laps, or bicycling uphill are examples. Muscle-strengthening activity: Physical activity, including exercise that increases skeletal muscle strength, power, endurance, and mass. It includes strength training, resistance training, and muscular strength and endurance exercises. Bone-strengthening activity: Physical activity that produces an impact or tension force on bones, which promotes bone growth and strength. Running, jumping rope, and lifting weights are examples. Regular physical activity, yogasan, meditation and pranayam [16] each for 30 minutes, may be protective against NCDs and may provide better spiritual, mental, social and physical wellbeing of the elderly population.

ACKNOWLEDGMENTS

To International College of Cardiology and International College of Nutrition for logistic support.

REFERENCES

[1] Singh RB, Rastogi V, Singh R, et al. Magnesium and antioxidant vitamins status and risk of complications of ageing in an elderly urban population. *Magnesium. Res.*, 1996; 9: 299-306.

[2] Singh RB, Rao RS, Thakur AS, Srivastav S, Niaz MA, Shinde SN. Prevalence and risk factors of cognitive deficits and dementia in relation to socioeconomic class in an elderly population of India. *J. Anti Ageing Med.*, 1999; 2: 141-8.

[3] Singh RB, Rastogi SS, Rastogi V, et al. Blood pressure trends, plasma insulin levels, and risk factors in rural and urban elderly populations of north India. *Coron. Artery Dis.*, 1997; 8: 463-8.

[4] Singh RB, Rastogi SS, Rao PV, et al. Diet and lifestyle guidelines and desirable levels of risk factors for the prevention of diabetes and its vascular complications in Indians: a scientific statement of the International College of Nutrition. *J. Cardiovasc. Risk*, 1997; 4: 201-8.

[5] Singh RB, Rastogi SS, Rao PV, et al. Diet and lifestyle guidelines and desirable levels of risk factors for the prevention of diabetes and its vascular complications in Indians: a scientific statement of the International College of Nutrition. *J. Cardiovasc. Risk,* 1997; 4: 201-8.

[6] Singh RB and Indian Consensus Group. Indian consensus for prevention of hypertension and coronary artery disease. A scientific statement of the Indian Society of Hypertension and International College of Nutrition. *J. Nutr. Environ. Med.,* 1996; 6: 309-18.

[7] Siri PW, Vewrhoef P, Kok FJ. Vitamin B6, B12, and folate: association with plasma total homocysteine and risk of coronary atherosclerosis. *J. Am. Coll Nutr.*, 1998; 17(5);435-441.

[8] Chapuy MC, Arlot ME, Duboeuf F, Brun J, Crouzet B, Arnaud S, Delmas PD, Meunier PJ. Vitamin D3 and calcium to prevent hip fractures in elderly women. *N Engl. J. Med.*, 1992;327:1637-1642.

[9] Food and Nutrition Board, Institute of Medicine. How Should the Recommended Dietary Allowances Be Revised? Washington, DC. National Academy Press,1994.

[10] Buitrago-Lopez A, Sanderson J, Johnson L, Warnakula S, Wood A, Di Angelantonio E, Franco OH. Chocolate consumption and cardiometabolic disorders: systematic review and meta-analysis.BMJ. 2011 Aug 26;343:d4488. doi: 10.1136/bmj.d4488.

[11] Singh RB, Pella D, Kartikey K, DeMeester F, et al., and the Five City Study Group. Prevalence of obesity, physical inactivity and undernutrition, a triple burden of diseases, during transition in a middle income country. *Acta Cardiol.,* 2007, 62:119-127.

[12] Singh RB, Fedacko J, Pella D, Macejova Z et al. Prevalence and risk factors of prehypertension and hypertension in five Indian citie. *Acta Cardiol.,* 2011; 66:29-37.

[13] Singh RB, De Meester F, Wilkzynska A. The Tsim Tsoum Approaches for Prevention of Cardiovascular Disease. Cardiology Research and Practice 2010; ID 824938, 18 pages; doi:10.4061/2010/824938 s.

[14] Singh RB, Fadecko, J, Pellad, D, De Meester F, Moshiri M, Aroussy WE. Superfood dietary approaches for acute myocardial infarction. *World Heart J.,* 2010;2: 13-23.

[15] Viola Vargova, Viola Mechirova, Jan Fedacko, Rafai Ryber, Daniel Pella, Agnieska Wilczynska, Fabien De Meester, Ram B Singh. Can nut consumption modulate cardiovascular disease?report of a case and review of literature.*The Open Nutra J.*, 2011;

[16] Singh RB, Singh Sushav, Fedacko J, Pella D, De Meester F. Pranayam Yoga Breathing Practices Modulate the Mind-Body Connection. *Int. J.disability Human Dev.,* 2009;8:141-153.

[17] Pella D, Otsuka K, Singh RB. Metabolic syndrome: A disease of the brain. *The Open Nutr J.,* 2011;4;107-118.

[18] De Meester F. Progress in lipid nutrition. In Simopoulos AP, De Meester F (eds), A Balanced Omega-6/Omega-3 Fatty acid Ratio.Cholesterol and Coronary Heart Disease. *World Rev. Nutr. Diet,* Basel, Karger, 2009; 100:110-121.

[19] Simopoulos AP.Evolutionary aspects of the dietary omega-6/omega-3 fatty acid ratio: medical implications. In Simopoulos AP, De Meester F (eds), A Balanced Omega-6/Omega-3 Fatty acid Ratio.Cholesterol and Coronary Heart Disease. *World Rev. Nutr. Diet,* Basel, Karger 2009; 100:1-21.

[20] Eaton SB, Konner M, Shostak M. Stone agers in the fast lane: chronic degenerative diseases in evolutionary perspective. *Am J Med.*, 1988; 84: 739-749.

[21] He FJ, Nowson CA, Lucas M and MacGregor GA. Increased consumption of fruit and vegetables is related to a reduced risk of coronary heart disease: meta-analysis of cohort studies. *J. Human Hyperten,* 2007; 21, 717–728.

[22] Katcher HI, Legro RS, Kunselman AR, Gillies PJ, Demers LM, Bagshaw DM, Kris-etherton PM. The effects of whole grain- enriched hypocaloric diet on cardiovascular

disease risk factors in men and women with metabolic syndrome. *Am. J. Clin. Nutr.*, 2008, 87:79-90.

[23] Singh RB, DeMeester F, Mechirova V, Pella D, Otsuka K. Fatty acids in the causation and therapy of metabolic syndrome. In *Wild Type Foods in Health Promotion and Disease Prevention*, editors Fabien De Meester and RR Watson, Humana Press, NJ 2008,p 263-284.

[24] Heidemann C, Schulze MB, Franco OH, van Dam RM, Mantzoros CS, Hu. FB. Dietary Patterns and Risk of Mortality From Cardiovascular Disease, Cancer, and All Causes in a Prospective Cohort of Women. *Circulation,* 2008;118:230-237.

[25] De Logeril M, Salen P, Martin JL, Monjaud I, Delaye J, Mamelle N. Mediterranean diet, traditional risk factors and the rate of cardiovascular complications after myocardial infarction. Final report of the Lyon Diet Heart Study. *Circulation,* 1999;99:779–785.

[26] Singh RB, Dubnov G, Niaz MA, Ghosh S, Singh R, Rastogi SS, Manor O, Pella D, Berry EM. Effect of an Indo-Mediterranean diet on progression of coronary disease in high risk patients: a randomized single blind trial. *Lancet,* 2002,360:1455-1461.

[27] Singh RB, Rastogi SS, Verma R, Bolaki L, and Singh R. An Indian experiment with nutritional modulation in acute myocardial infarction. *Am. J. Cardiol.,* 1992, 69: 879-885.

[28] Singh RB,Pella D,DeMeester F..What to eat and chew in acute myocardial infarction. *Eur. Heart J.,* 2006, 27:1628-29.

[29] Ridker PM, Danieison E, Fonseka FAH, Genest J, Gotto AM, Kastelein JJP, Koenig W, Libby P, Lor AJ, MacFadyen JG, Nordestgaard BG, Shepherd J, Willerson JT. Reduction in C-reactive protein and LDL cholesterol and cardiovascular event rates after initiation of rosuvastatin: a prospective study of the JUPITER trial. *Lancet,* March 2009doi:10.1016/S0140-6736(09)60447-5.

[30] Russo, G.L.Dietary n - 6 and n - 3 polyunsaturated fatty acids: From biochemistry to clinical implications in cardiovascular prevention. *Biochemical Pharmacology,* 2009, 77: 937 – 946.

[31] Rival, Y., Puech, L., Taillandier, T., Benacteau, N., Rouquette, A., Lestienne, F., Dupont-Passelaigue, E., Le Roy, I., Patoiseau, JF, Junqu Ã. PPAR activators and COX inhibitors selectively block cytokine-induced COX-2 expression and activity in human aortic smooth muscle cells. *Eur. Jour Pharmacol.*, 2009, 606: 121 – 129.

[32] Paul-Labrador M, Polk D, Dwyer JH, Velasquez I, Nidich S, et al. Effects of a randomized controlled trial of transcendental meditation on components of the metabolic syndrome in subjects with coronary heart disease. *Arch. Intern. Med.*, 2006; 166:1218-24.

[33] Bernardi I, Spadacini G, Bellwon J, Hajric R, Roskamm H, et al. Effect of breathing rate on oxygen saturation and exercise performance in chronic heart failure. *Lancet,* 1998;351:1308-11.

[34] Bernardi I, Gabutti A, Porta C, Spicuzza I. Slow breathing reduces chemoreflex response to hypoxia and hyercapnia and increases baroreflex sensitivity. *J. Hypertens,* 2001;19:2221-9.

[35] Bernardi I, Wdowczyc-Szulc J, Valenti C, Castoldi S, Passino C, et al. Effects of controlled breathing, mental activity and mental stress with or without verbalization on heart rate variability. *J. Am. Coll Cardiol.*, 2000;35:1462-9.

[36] Delion S, Cahlon S, Guilloteau G, et al. Chronic dietary alpha-linolenic acid deficiency alters dopaminergic and serotinergic neurotransmission in rats. *J. Nutr.*, 1994; 124:266-76.

[37] Delion S, Cahlon S, Guillotea.UG, et al. Alpha-linolenic acid deficiency alters age-related changes of dopaminergic and serotoninergic neurotransmission in the rat frontal cortex. *J. Neurochem.,* 1996; 66:1582-91.

[38] Hibbeln JR, Salem N. Dietary polyunsaturated fatty acids and depression: when cholesterol does not satisfy. *Am. J. Clin. Nutr.*, 1995; 62:1-9.

[39] Garkan, A., Emingil, G., Saygan, B.H., Atilla, G., KÃse, T., Baylas, H.,Berdeli, A. Angiotensin-converting enzyme (ACE), angiotensinogen (AGT), and angiotensin II type 1 receptor (AT1R) gene polymorphisms in generalized aggressive periodontitis. *Archives of Oral Biology*, 2009;54: 337 – 344.

[40] Devaraj S, Singh U, and Ishwarlal Jialal I. Is CRP entering into Olympus of the Gods? *Clin. Chem.,* 2009; 55:229-38.

[41] Singh RB, De Meester F, Pella D, Basu TK, Watson R. Globalization of dietary wild foods protect against cardiovascular disease and all cause mortalities? A Scientific statement from the International College of Cardiology, Columbus Paradigm Institute and the International College of Nutrition. *The Open Nutraceuticals Journal*, 2009, 2, 42-45.

[42] Iqbal R, Anand S, Ounpuu S, Islam S, Zhang X, Rangarajan S, Chifamba J, Al-Hinai A, Keltai M, Yusuf S, on behalf of the INTERHEART Study Investigators. Dietary patterns and the risk of acute myocardial infarction in 52 countries: results of the INTERHEART study. *Circulation,* 2008; 118: 1929–1937.

[43] Fung, TT, Rexrode, KM, Mantzoros, CS, JoAnn E. Manson, JE, Willett, WC and Hu FB. Mediterranean diet and incidence of and mortality from coronary heart disease and stroke in women. *Circulation,* 2009; 119: 1093 - 1100.

[44] Vogel RA. 21. Eating vascular biology and atherosclerosis: a lot to chew on. *Eur. Heart J.*, 2006, 27:13-14.

[45] Esposito K, Giugiliano D. Diet and inflammation: a link to metabolic and cardiovascular diseases. *Eur. Heart J.,* 2006,27: 15-20.

[46] Kelishadi, R., Mirghaffari, N., Poursafa, P., Gidding, S.S. Lifestyle and environmental factors associated with inflammation, oxidative stress and insulin resistance in children. *Atherosclerosis,* 2009, 203: 311 – 319.

[47] Boger RH, Sullivan LM, Schewedhelrn E, Wang T et al.Plasma asymmetric dimethylarginine and incidence of cardiovascular disease and Death in the Community. *Circulation*, 2009;119:1592-1600.

[48] Halberg F, Cornelissen G, Halberg J, Schwartzkopff O. Prehypertension and other variabilities also await treatment. *Am. J. Med.,* 2007; 120:e19-e20.

[49] Singh RB, Cornelissen G, Weydahl A, Schwartzkopff O et al. Circadian heart rate and blood pressure variability considered for research and patient care. *Int. J. Cardiol.*, 2003; 87: 9-28.

[50] Singh RB, Gupta A, Pella D, Kumar A.Circadian Cardiology. *World Heart J.,* 2008;1:101-126.

[51] Dubnov Gal, Pella D, Singh RB. The effect of an alpha-linolenic acid rich diet on the circadian rhythm of cardiac events. *World Heart J.,* 2008,1:49-56.

[52] Singh RB, Chaudhuri J, De Meester F, D W Wilson. Development of Mediterranean soup for enteral nutrition and for prevention of cardiovascular diseases. *The Open Nutra Jour.,* 2012;5:90-98.

In: Cardiovascular Health and Chronomics
Editor: Ram B. Singh
ISBN: 978-1-62948-976-6

Chapter 2

SPACE AND EARTH WEATHER MIRRORED IN PATTERNS OF SUICIDE INCIDENCE

Germaine Cornélissen[1,*]**, *Borislav D. Dimitrov***[2]**,**
Franca Carandente[3] ***and Franz Halberg***[1]

[1]Halberg ChronobiologyCenter, University of Minnesota, Minneapolis, MN, US
[2]Department General Practice, RoyalCollege of Surgeons in Ireland, Dublin, Ireland
[3]University of Milan, Milan, Italy

Support: U.S. National Institutes of Health (GM-13981) (FH);
University of Minnesota Supercomputing Institute (GC, FH)

ABSTRACT

Exploring any non-photic as well as photic components in the spectrum of the incidence pattern of suicides, we find gender differences in Minnesota, USA, and Australia, notably in the para-annual range of the spectrum. A trans-tridecadal BEL cycle (also more liberally referred to as para-tridecadal) with a period of about 35 years is found in data pooled from both genders and separately for both men and women in Finland (1969-2007). When fitted separately as a single component, the estimated period and its 95% confidence interval (CI) (in years) is 35.5 [26.7, 44.3] overall, 36.4 [25.3, 47.5] for males, and 38.9 [28.5, 49.3] for females, with respective amplitudes (N/100,000) of 170 [119, 222], 145 [98, 192], and 29 [18, 40]. Differences between a Schwabe cycle and an about 7-year component (possibly a harmonic of the Hale cycle) characterizing suicide patterns of Finnish men and women, respectively, remain to be examined in longer series. In Bulgaria (1929-1945), a far-transyear with a period of about 1.3 years is noteworthy as is a quinmensal component with an about 5-month period. In France (1979-2008), in males and females, suicides are characterized by about 10.0-year and 10.1-year cycles, respectively. In Japan (1987-1999), an about 12.7-year component is detected, albeit with a broad uncertainty. A better understanding of the origin of these patterns, notably in terms of the relative contribution of non-photic and photic (daily and yearly) cycles, may lead to better prediction of incidence dynamics and to novel interventions for the prevention of major depression and suicide.

---

* Germaine Cornelissen <corne001@tc.umn.edu>

Introduction

Among conditions shown to be affected by space weather, insofar as they exhibit transyears (cycles with a period of about 1.3 years) characteristic of solar wind speed [1], are sudden cardiac death [2-4], terrorism [5] and suicide. It is hence of interest to examine the incidence patterns of all three conditions in the search for a possibly common underlying factor. A depressive mood has already been documented to independently relate to stroke and cardiovascular events [6], a further reason to study the incidence patterns of the extreme depression of suicide, the topic of this paper. It is written in the steps of Düll and Düll who in 1938 found by superposed epochs on 45,190 cases that the incidence of human suicide increases in association with the occurrence of solar calcium plages [7]. Evidence based on correlation analysis is much weaker and potentially misleading in view of the various cycles characterizing both incidence patterns of suicide and natural environmental variables.

Background

While positive affect (PA) gauged by the PANAS scale [8] is characterized by a large about 24-hour (circadian) component, negative affect (NA) is prominently 7-day (circaseptan) periodic [9]. An extreme negative affect, the severe depression leading to suicide, in turn is characterized by a non-random distribution along the scales of a day, a week and a year [10-16], illustrated in Figures 1A-C for an Italian cohort [11].

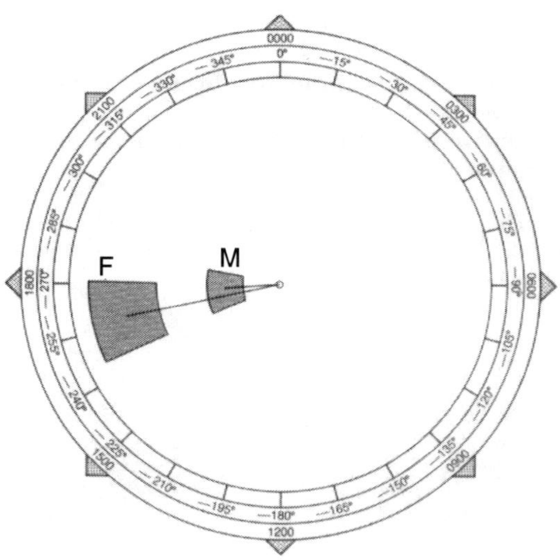

key	p	PR	no.	mesor ±SE		amplitude ±SE		acrophase ±SE	
M	0.01	34	111	5	±0.5	2.3	±0.69	−265°	±17°
F	0.001	50	340	14	±1	6.5	±1.4	−258°	±13°

(a)

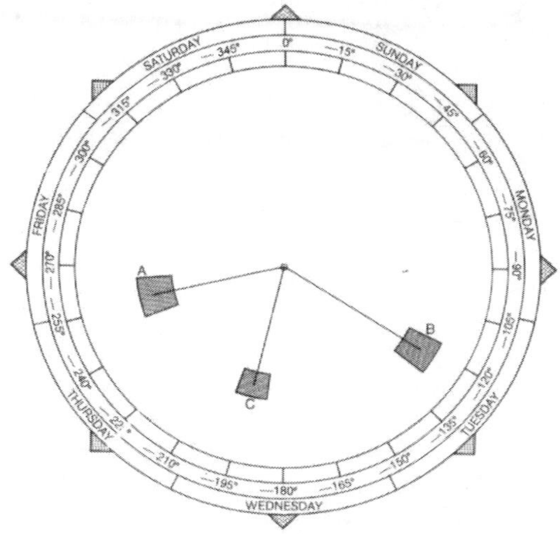

key	p	PR	no.	mesor	± SE	amplitude ± SE		acrophase ± SE	
A age: 14–22 years	0.005	93	114	16	± 0.28	2.95	± 0.40	− 257°	± 8°
B age: 23–40 years	0.002	95	140	19	± 0.28	3.53	± 0.39	− 126°	± 6°
C age: > 40 years	0.002	95	86	12	± 0.21	2.73	± 0.30	− 194°	± 6°

(b)

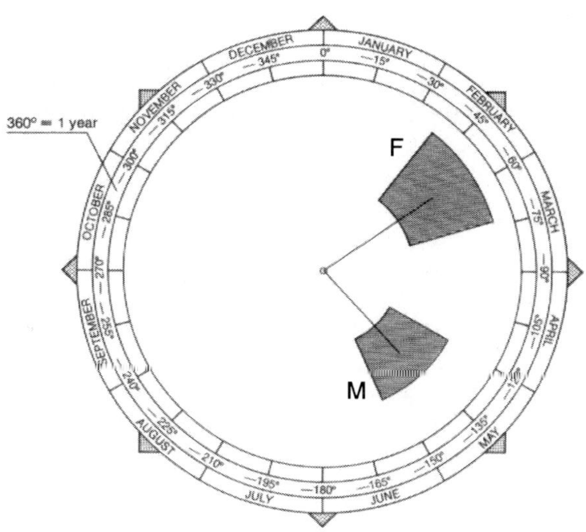

key	p	PR	no.	mesor ± SE		amplitude ± SE		acrophase ± SE	
M	0.039	51	111	4.4	± 0.21	0.95	± 0.31	− 137°	± 19°
F	0.034	53	340	11.9	± 0.27	1.17	± 0.39	− 56°	± 18°

(c)

Figure 1. (A) Circadian, (B) circaseptan and (C) circannual rhythm of attempted suicides (N: 431, 14-75 years of age, 19780627 to 19801111 in Milan, Italy[11]. © Halberg.

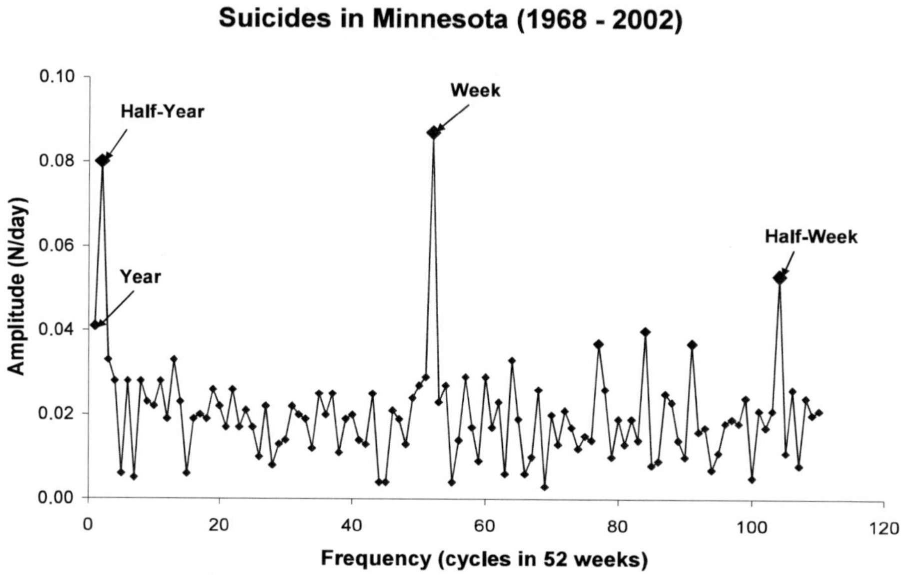

Figure 2. Population-mean cosinor spectrum of the daily incidence of suicides in Minnesota (1968-2002). Data analyzed over consecutive years and summarized across years at each frequency, considering results from consecutive years as replications. Spectral peaks with larger diamonds are detected with statistical significance (P<0.05). © Halberg.

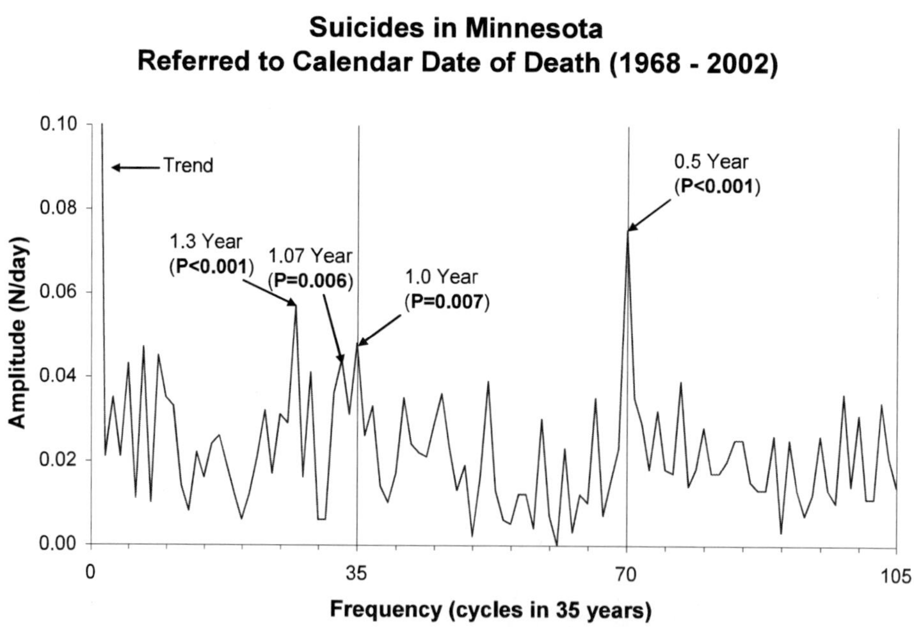

Figure 3. Infradian least squares spectrum of suicides in Minnesota (1968-2002). © Halberg.

The importance of a 1-year synchronized human circannual system in the incidence pattern of suicide by calendar date of death was documented in 1973 as a consistent phenomenon in Minnesota, USA, during 1941-1967 [10].

The yearly rhythms in the incidence of suicides, accidents and heart disease were consistent in Minnesota during the span investigated in terms of both their reproducibility as statistically significant phenomena in the consecutive decades investigated and in terms of differences in phase among these annual incidence patterns [10].

In a follow-up study of suicide (by calendar date of death) in Minnesota from 1968-2002, [14-16], the largest spectral peaks corresponded to periods of half a year and a week, when assessed by population-mean cosinor, using consecutive yearly records as replicates (Figure 2). Focus on the low-frequency region of the least squares spectrum finds the half-year to be very prominent (Figure 3), in keeping with a geomagnetic association [17, 18]. The latter approach revealed an about 1.3-year component, a far-transyear (defined as 1.2 years $\leq [\tau - CI] < [\tau + CI] < 1.9$ years, where τ is the period and CI is the conservative 95% confidence interval of τ; [cf. 11-17]), with an amplitude larger than the 1-year component, but smaller than the half-yearly cycle. An about 1.07-year near-transyear (defined as 1.00 year $< [\tau - CI] < [\tau + CI] < 1.20$ years) was also detected with statistical significance in Minnesota (Figure 3).

The wide ranges of frequency used for the definitions of transyears stem from the wide drifts in their frequency assessed concomitantly in different geographic locations and in the frequency of their environmental counterparts, such as those observed in solar wind speed [2-4] and the antipodal geomagnetic index *aa*. Similar frequency changes of transyear components are seen for various physiological variables, some recorded over decades [19]. The definitions also account for geographic differences in the average periods of major spectral peaks characterizing conditions such as sudden cardiac death [3, 19], among others [20-23].

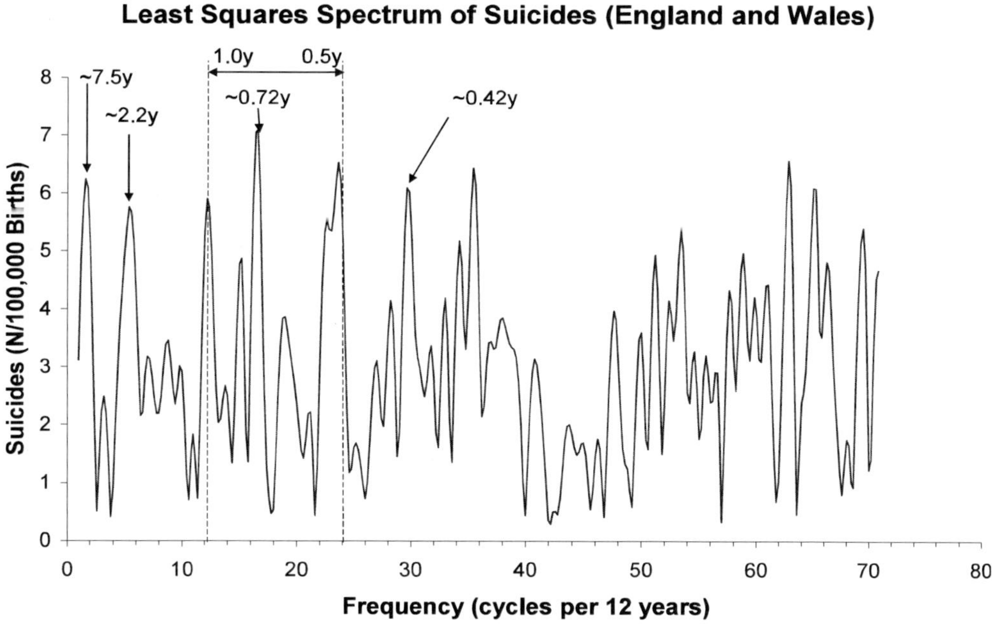

Figure 4. Suicides by time of birth in England and Wales, irrespective of gender. About 0.727-year component validated nonlinearly (CI: 0.703, 0.751). © Halberg.

Analyses of data recorded separately for males and females have also shown interesting gender differences in the patterns of incidence of suicide. In Australia, suicides are characterized by a prominent year and a lesser half-year in men, whereas these components are not detected with statistical significance in women [24]. By contrast, suicides in women are characterized by about 0.563-year (trans-half-year; CI: 0.560-0.567) and about 1.214-year (transyear; CI: 1.193-1.236) components. The latter are likely of solar origin, in keeping with the far- and near-transyears found in Minnesota [14-16]. In Australia, an about 9.2-year cycle is shared by men (9.16 years; CI: 8.50-9.82) and women (9.19 years; CI: 8.37-10.01) [24]. Suicides by the time of birth (rather than by time of death) in England and Wales also show an about 0.42-year (quinmensal) cycle similar to that found in solar flares, along with a prominent cis-year of 0.73 year, a geomagnetic half-year, and a calendar year (Figure 4) [16].

METHODS

Against this background, it was of interest to re-examine data collected and provided by one of us (BDD) on monthly suicides in Bulgaria from 1929 to 1945 [25], yearly suicides in Finland from 1969 to 2007 [26], Japan from 1987 to 1999, and France from 1979 to 2008. The extended cosinor [27, 28; cf. 13] including Marquardt's nonlinear algorithm [29] was used for analysis.

RESULTS

In Bulgaria, the amplitude of a possibly global solar cycle [30] of 15.46 years (CI: 12.00-18.91) was 78% that of the fixed calendar yearly component, while the fixed half-year had only an amplitude of 28% that of the 15.46-year cycle (Figure 5).

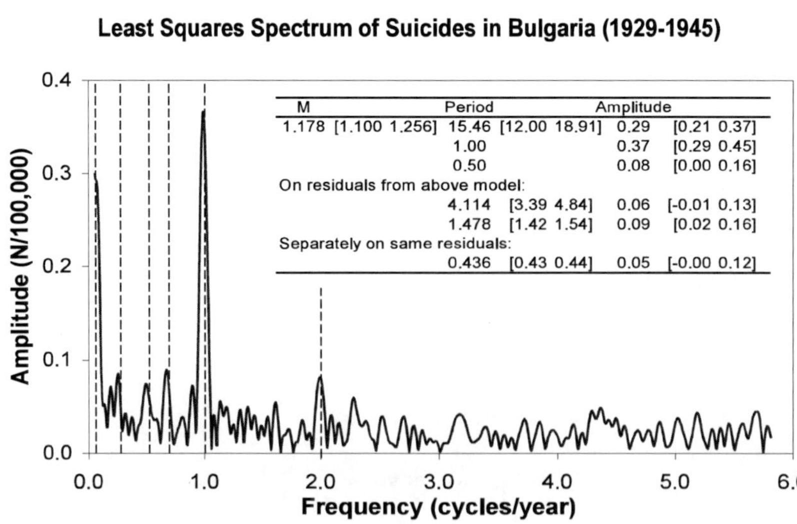

Figure 5. Extended cosinor analyses revealing nonphotic components in suicides in Bulgaria. © Halberg.

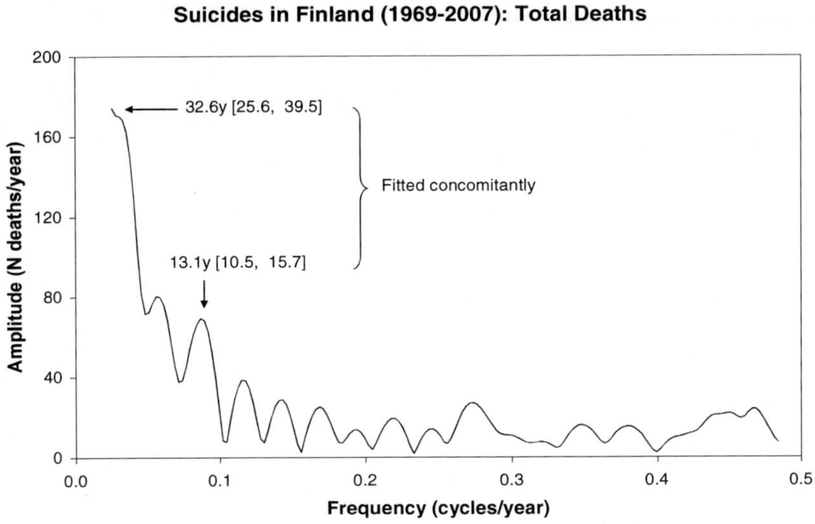

Figure 6. Extended cosinor analyses suggest nonphotic components in suicides in Finland (total deaths). © Halberg.

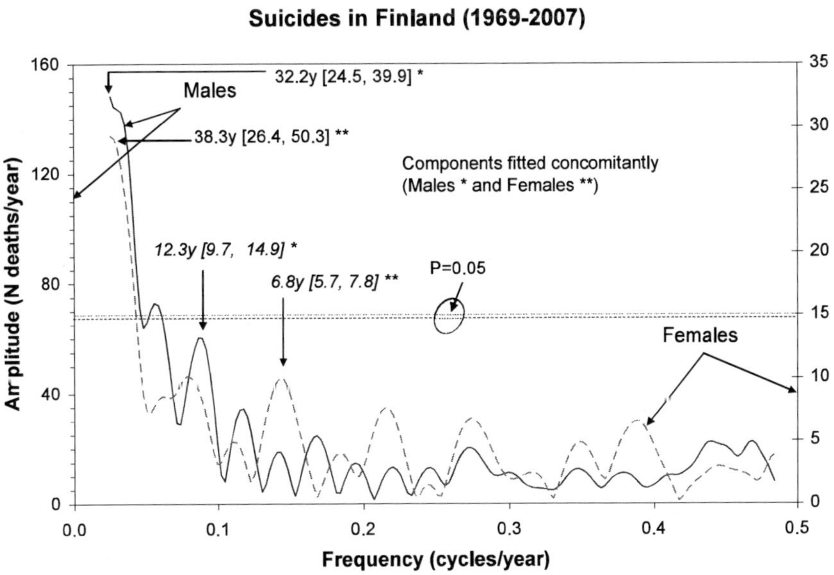

Figure 7. Extended cosinor analyses suggest nonphotic components in suicides in Finland (men and women considered separately). © Halberg.

A nonlinear analysis of the residuals from this model revealed the presence of a far-transyear of 1.48 years (CI: 1.42-1.54) and of an about 4.11-year cycle (CI: 3.39-4.84) (Figure 5). A separate nonlinear analysis of the same residuals also validated the presence of a cis-half-year (quinmensal) of 0.436 year (CI: 0.43-0.44) (Figure 5). Transyears [1] and cis-half-years [31-69] have been documented in physics and in a broad context [70] in the biosphere [71-73].

Table 1. Nonlinear cosinor analyses of yearly suicide in Finland: 39 years (y) (1969-2007)*

	Period (y) [95% CI]	Amplitude (N/100,000 persons) [95% CI]
Trial τ = 35.0 y		
Total (T)	35.50 [26.75, 44.25] y	170 [119, 222]
Males (M)	36.42 [25.32, 47.51] y	145 [98, 192]
Females (F)	38.88 [28.48, 49.29] y	29 [18, 40]
Trial □ = 12.5 y		
T	11.45 [8.59, 14.31] y	69 [-33, 171] NS; [5, 133] 1P
M	11.39 [8.64, 14.13] y	60 [-27, 148] NS; [5, 115] 1P
F	12.49 [8.35, 16.63] y	NS
Trial y = 7 y		
F	6.96 [5.72, 8.20]	NS
Composite model: 35.0 y + 12.5 y (T/M) or 7 y (F)		
T	32.56 [25.62, 39.49]	175 [119, 232]
	13.14 [10.53, 15.74]	59 [1, 118]
M	32.17 [24.47, 39.87]	143 [90, 196]
	12.33 [9.73, 14.92]	48 [-8, 104] NS; [20, 76] 1P
F	38.34 [26.35, 50.34]	29 [16, 42]
	6.77 [5.72, 7.81]	8 [-5, 21] NS; [1, 15] 1P

*CI: (conservative) confidence interval; 1P: 1-parameter limits.

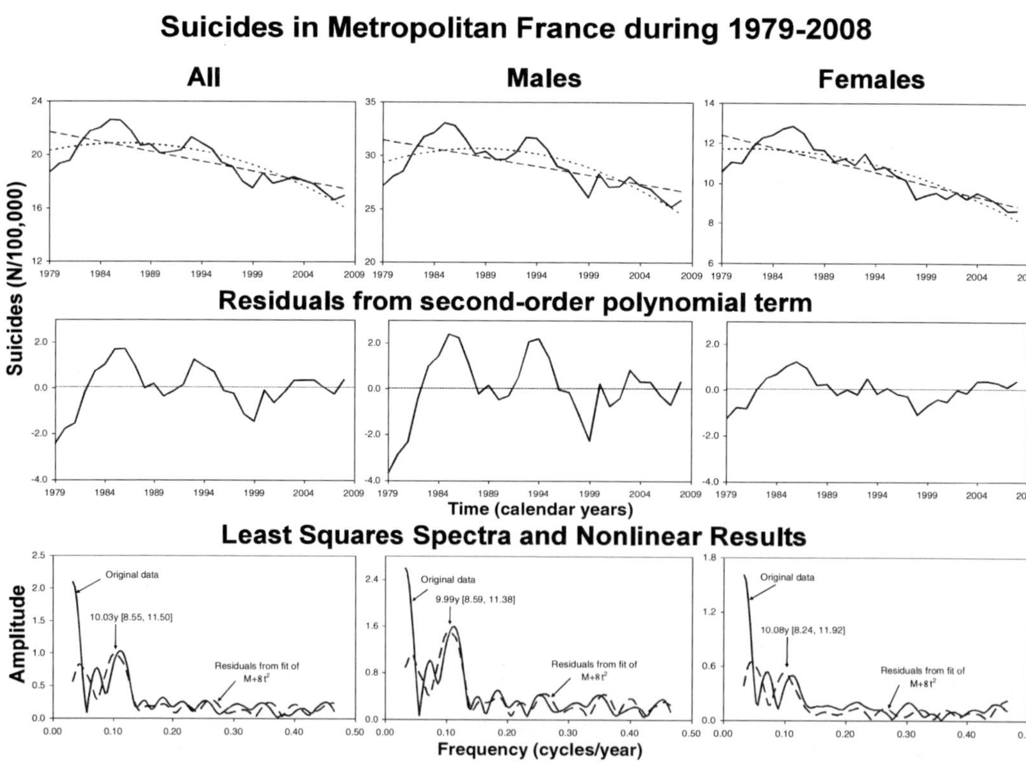

Figure 8. Extended cosinor analyses suggest nonphotic components in suicides in metropolitan France. © Halberg.

In 39 years (1969-2002) of yearly data from Finland, a trans-tridecadal cycle was prominent (Table 1, Figures 6 and 7), but it is very close to the length of the series. With this qualification, a composite model of about 32.2-year and 12.3-year components reached statistical and borderline statistical significance, respectively, for men (Figure 7), whereas for women, the composite model consisted of a statistically significant about 38.3-year trans-tridecadal and an about 6.8-year cycle of borderline statistical significance (Figure 7).

In France (1979-2008), suicides show an about 10.03-year cycle (CI: 8.55-11.50) overall (Figure 8). It is similar in males and females with estimated periods of 9.99 years (CI: 8.59, 11.38) and 10.08 years (CI: 8.24, 11.92), respectively. In a brief series (1987-1999) of yearly data from Japan, an about 12.7-year cycle (CI: 7.3-18.2) was found.

DISCUSSION

Clearly, longer and denser data are needed for these particular and other geographic comparisons. The analyses in hand suffice to show that both genders are influenced in the decadal range by the earth and the cosmos. As in Minnesotan and in Australian women, non-photic para-annual cycles are present in addition to the seasons, the reproducibility of the latter's association with suicide notwithstanding.

Near- and far-transyears can coexist with a prominent one-year synchronized component, and can even replace it for a while in some cases, as in the case of blood pressure in one of us [2]. The congruent non-photic components in physiology may be damped but not lost when the corresponding environmental cycles can no longer be demonstrated [2]. While the biospheric cycle can resonate with the environmental cycle [2, cf. 74] it is most likely built into the population, just as the circadian is documented in the genome of individuals [70, 75, 76]. Quinmensals that have been predicted [31] and found [32] and extensively investigated in physics [33-69] and in the biosphere [71-73] are to be examined, among others, in longer series with an addition and subtraction (remove and replace) approach. The latter has proved helpful in dealing with the built-in nature of transyears [2, 19] and circaseptans [74].

The possibility of a trans-tridecadal cycle of suicide in Finland deserves further scrutiny in view of the country's high latitude and the reported dependence on latitude of suicides [77-79]. A similar latitude-dependence has been reported for melatonin [80]. Retinal melanopsin receptors have been reported to be directly linked to the suprachiasmatic nuclei [81]. The latter may act via the serotonergic system on seasonal mood changes, as noted by Swaab [82]. Examination of suicide rates in near-arctic regions may be particularly pertinent to assess non-photic influences since a BEL cycle was found to characterize the planetary geomagnetic index Kp and several features of the interplanetary magnetic field such as proton temperature and plasma speed [83], polar regions being more strongly affected by geomagnetic disturbances. Data from rising latitudes in Greenland [84] would be particularly valuable.

REFERENCES

[1] Richardson JD, Paularena KI, Belcher JW, Lazarus AJ. Solar wind oscillations with a 1.3-year period. *Geophys. Res. Lett.,* 1994; 21: 1559-1560.

[2] Halberg F, Cornélissen G, Katinas G, Tvildiani L, Gigolashvili M, Janashia K, Toba T, Revilla M, Regal P, Sothern RB, Wendt HW, Wang ZR, Zeman M, Jozsa R, Singh RB, Mitsutake G, Chibisov SM, Lee J, Holley D, Holte JE, Sonkowsky RP, Schwartzkopff O, Delmore P, Otsuka K, Bakken EE, Czaplicki J, International BIOCOS Group. Chronobiology's progress: season's appreciations 2004-2005. Time-, frequency-, phase-, variable-, individual-, age- and site-specific chronomics. *J. Appl. Biomed.*, 2006; 4: 1-38. http://www.zsf.jcu.cz/vyzkum/jab/4_1/halberg.pdf.

[3] Halberg F, Cornélissen G, Otsuka K, Fiser B, Mitsutake G, Wendt HW, Johnson P, Gigolashvili M, Breus T, Sonkowsky R, Chibisov SM, Katinas G, Siegelova J, Dusek J, Singh RB, Berri BL, Schwartzkopff O. Incidence of sudden cardiac death, myocardial infarction and far- and near-transyears. *Biomed & Pharmacother,* 2005; 59 (Suppl 1): S239-S261.

[4] Cornélissen G, Halberg F, Rostagno C, Otsuka K. A chronomic approach to cardiac arrhythmia and sudden cardiac death. *The Autonomic Nervous System* ,2007; 44: 251-254.

[5] Halberg F, Cornélissen G, Sothern RB, Katinas GS, Schwartzkopff O, Otsuka K. Cycles tipping the scale between death and survival (= "life"). *Progress of Theoretical Physics,* 2008; Suppl. 173: 153-181.

[6] Yamanaka G, Otsuka K, Hotta N, Murakami S, Kubo Y, Matsuoka O, Takasugi E, Yamanaka T, Shinagawa M, Nunoda S, Nishimura Y, Shibata K, Saitoh H, Nishinaga M, Ishine M, Wada T, Okumiya K, Matsubayashi K, Yano S, Ishizuka S, Ichihara K, Cornélissen G, Halberg F. Depressive mood is independently related to stroke and cardiovascular events in a community. *Biomed & Pharmacother,* 2005; 59 (Suppl 1): S31-S39.

[7] Düll B, Düll T. Erd- und Sonnenphysikalische Vorgaenge in ihrer Bedeutung fuer Krankheits- und Todauslosung. *Nosokomeion*, 1938; 9: 103-120.

[8] Watson D, Clark LA, Tellegen A. Development and validation of brief measures of positive and negative affect: the PANAS scales. *J. Person Soc. Psychol.,* 1988; 54: 1063-1070.

[9] Cornélissen G, Watson D, Mitsutake G, Fiser B, Siegelova J, Dusek J, Vohlidalova L, Svacinova H, Halberg F. Mapping of circaseptan and circadian changes in mood. *Scripta medica,* 2005; 78: 89-98.

[10] Halberg F. Laboratory techniques and rhythmometry. In: Mills JN. (Ed.) *Biological Aspects of Circadian Rhythms.* London/New York: Plenum Press; 1973. p. 1-26.

[11] De Maio D, Carandente F, Riva C. Evaluation of circadian, circaseptan and circannual periodicity of attempted suicides. *Chronobiologia,* 1982; 9: 185-193.

[12] Houwelingen CAJ van, Beersma DGM. Seasonal changes in 24-h patterns of suicide rates: a study on train suicides in The Netherlands. *J. Affect Disord.,* 2001; 66: 215-223.

[13] Refinetti R, Cornélissen G, Halberg F. Procedures for numerical analysis of circadian rhythms. Biological Rhythm Research 2007; 38 (4): 275-325. http://dx.doi.org/10.1080/09291010600903692.

[14] Cornélissen G, Palermo J, Carandente F, Siegelova J, Fiser B, Dusek J, Halberg F. Chronomics: chronome of suicides in Minnesota (1968-2002). *Abstract, Noninvasive methods in cardiology,* 2.11.2004, Congress MEFA, Brno, Czech Republic, 2004, p. 8.

[15] Halberg F, Cornélissen G, Panksepp J, Otsuka K, Johnson D. Chronomics of autism and suicide. *Biomed. & Pharmacother,* 2005; 59 (Suppl 1): S100-S108.

[16] Cornélissen G, Halberg F. Chronomics of suicides and the solar wind. *Br. J. Psychiatry,* 2006; 189: 567-568.

[17] Chapman S, Bartels J. *Geomagnetism.* 3rd ed. Oxford: Clarendon Press; 1962. 1049 pp.

[18] Cornélissen G, Halberg F, Pöllmann L, Pöllman B, Katinas GS, Minne H, Breus T, Sothern RB, Watanabe Y, Tarquini R, Perfetto F, Maggioni C, Wilson D, Gubin D, Otsuka K, Bakken EE. Circasemiannual chronomics: half-yearly biospheric changes in their own right and as a circannual waveform. *Biomed & Pharmacother,* 2003; 57 (Suppl 1): 45s-54s.

[19] Cornélissen G, Masalov A, Halberg F, Richardson JD, Katinas GS, Sothern RB, Watanabe Y, Syutkina EV, Wendt HW, Bakken EE, Romanov Y. Multiple resonances among time structures, chronomes, around and in us. Is an about 1.3-year periodicity in solar wind built into the human cardiovascular chronome? *Human Physiology*, 2004; 30 (2): 86-92.

[20] Kovac M, Mikulecky M. Secular rhythms and Halberg's paraseasonality in the time occurrence of cerebral stroke. *Bratisl Lek Listy,* 2005; 106 (2): 423-427.

[21] Mikulecky M, Florida PL. Daily birth numbers in Davao, Philippines, 1993-2003: Halberg's transyear stronger than year. Abstract, 26th Seminar, Man in HisTerrestrial and Cosmic Environment, Upice, Czech Republic, May 17-19, 2005.

[22] Kovac M, Mikulecky M. Time sequence of epileptic attacks from the point of view of possible lunisolar connections. International Conference on the Frontiers of Biomedical Science:Chronobiology, Chengdu, China, September 24-26, 2006, p. 175-179.

[23] Mikulecky M. Reanalysis of natality in South Brazil. Halberg's paraseasonality dominating again. International Conference on the Frontiers of Biomedical Science: Chronobiology, Chengdu, China, September 24-26, 2006, p. 187-188.

[24] Halberg F, Cornélissen G, Berk M, Dodd S, Henry M, Wetterberg L, Nolley E, Beaty L, Siegelova J, Fiser B, Wolff C, BIOCOS project. Solar signatures in Australian suicide incidence: gender differences in prominence of photic vs. nonphotic spectral components. In: Halberg F, Kenner T, Fiser B, Siegelova J (Eds.) Proceedings, Noninvasive Methods in Cardiology, Brno, Czech Republic, October 4-7, 2008. p. 44-62. http://web.fnusa.cz/files/kfdr2008/sbornik_2008.pdf.

[25] Dimitrov BD. Cyclicity of variations in the frequency of suicides in Bulgaria. *Medicina Legalis Baltica,* 1999; 8 (11): 26-31.

[26] Dimitrov BD, Atanassova PA, Rachkova MI. Cyclicity of suicides may be modulated by internal or external ~11-year cycles: an example of suicide rates in Finland. *Sun and Geosphere,* 2009; 4 (2): 50-54.

[27] Halberg F. Chronobiology: methodological problems. *Acta med. rom.*, 1980; 18: 399-440.

[28] Cornélissen G, Halberg F. Chronomedicine. In: Armitage P, Colton T (Eds.) Encyclopedia of Biostatistics, 2nd ed. Chichester, UK: John Wiley and Sons Ltd; 2005. p. 796-812.

[29] Marquardt DW. An algorithm for least-squares estimation of nonlinear parameters. *J. Soc. Indust. Appl.Math.,* 1963; 11: 431-441.

[30] Makarov VI, Sivaraman KR. New results concerning the global solar cycle. *Solar Physics,* 1989; 123: 367-380.

[31] Wolff CL. The rotational spectrum of g-modes in the sun. *Astrophys. J.,* 1983; 264: 667-676.

[32] Rieger A, Share GH, Forrest DJ, Kanbach G, Reppin C, Chupp EL. A 154-day periodicity in the occurrence of hard solar flares? *Nature,* 1984; 312: 623-625.

[33] Kiplinger AL, Dennis BR, Orwig LE. Detection of a 158-day periodicity in the solar hard X-ray flare rate. *Bull. Am. Astronom. Soc.*, 1984; 16: 891.

[34] Bogart RS, Bai T. Confirmation of a 152-day periodicity in the occurrence of solar flares inferred from microwave data. *Astrophys J.,* 1985; 299: L51-L55.

[35] Dennis BR. Solar hard X-ray bursts. *Solar Physics,* 1985; 100: 465-490.

[36] Bai T, Sturrock PA. The 152-day periodicity of the solar flare occurrence rate. *Nature,* 1987; 327: 601-604.

[37] Lean JL, Brueckner GE. Intermediate-term solar periodicities -- 100-500 days. *Astrophys. J.,* 1989; 337: 568-578.

[38] Özgüç A, Ataç T. Periodic behavior of solar flare index in solar cycles 20 and 21. *Solar Physics,* 1989; 123: 357-365.

[39] Bai T, Cliver EW. A 154 day periodicity in the occurrence rate of proton flares. *Astrophys. J.*, 1990; 363: 299-309.

[40] Carbonell M, Ballester JL. A short-term periodicity near 155 day in sunspot areas. *Astron Astrophys,* 1990; 238: 377-381.

[41] Dröge W, Gibbs K, Grunsfeld JM, Meyer P, Newport BJ, Evenson P, Moses D. A 153 day periodicity in the occurrence of solar flares producing energetic interplanetary electrons. *Astrophys J. Suppl. Ser.,* 1990; 73: 279-283.

[42] Lean J. Evolution of the 155-day periodicity in sunspot areas during solar cycles 12 to 21. *Astrophys. J.*, 1990; 363: 718-727.

[43] Silverman SM. The 155-day solar period in the sixteenth century and later. *Nature* 1990; 347: 365-367.

[44] Bai T, Sturrock PA. The 154-day and related periodicities of solar activity as subharmonics of a fundamental period. *Nature,* 1991; 350: 141-143.

[45] Kile JN, Cliver EW. A search for the 154 day periodicity in the occurrence rate of solar flares using Ottawa 2.8 GHz burst data, 1955-1990. *Astrophys J.,* 1991; 370: 442-448.

[46] Verma VK, Joshi JC, Uddin W, Paliwal DC. Search for a 152-158 days periodicity in the occurrence rate of solar flares inferred from spectral data of radio bursts. *Astron Astrophys. Suppl Ser.*, 1991; 90: 83-87.

[47] Bai T. The 77 day periodicity in the flare rate of cycle 22. *Astrophys J.*, 1992; 388: L69-L72.

[48] Carbonell M, Ballester JL. The periodic behaviour of solar activity: the near 155-day periodicity in sunspot areas. *Astron Astrophys,* 1992; 255: 350-362.

[49] Verma VK, Joshi JC, Paliwal DC. Study of periodicities of solar nuclear gamma ray flares and sunspots. *Solar Physics,* 1992; 138: 205-208.

[50] Bai T, Sturrock PA. Evidence for a fundamental period of the sun and its relation to the 154 day complex of periodicities. *Astrophys. J.,* 1993; 409: 476-486.

[51] Cane HV, Richardson IG, von Rosenvinge TT. Interplanetary magnetic field periodicity of ~153 days. *Geophys Res. Lett.,* 1998; 25: 4437-4440.

[52] Oliver R, Ballester JL, Baudin F. Emergence of magnetic flux on the Sun as the cause of a 158-day periodicity in sunspot areas. *Nature,* 1998; 394: 552-553 doi:10.1038/29012.

[53] Ballester JL, Oliver R, Baudin F. Discovery of the near 158 day periodicity in group sunspot numbers during the eighteenth century. *Astrophys. J.,* 1999; 522: L153-L156.

[54] Lou YQ. Rossby-type wave-induced periodicities in flare activities and sunspot areas or groups during solar maxima. *Astrophys. J.,* 2000; 540: 1102-1108.

[55] Hady AA. Analytical studies of solar cycle 23 and its periodicities. *Planetary and Space Science,* 2002; 50: 89-92.

[56] Krivova NA, Solanki SK. The 1.3-year and 156-day periodicities in sunspot data: wavelet analysis suggests a common origin. *Astron Astrophys,* 2002; 394: 701-706.

[57] Ballester JL, Oliver R, Carbonell M. The near 160 day periodicity in the photospheric magnetic flux. *Astrophys J.,* 2002; 566: 505-511.

[58] Han Yanben, Han Yonggang. Time variation of the near 5-month period of sunspot numbers. *Chinese Sci. Bull.,* 2002; 47 (23): 1967-1973.

[59] Prabhakaran Nayar SR, Radhika VN, Revathy K, Ramadas V. Wavelet analysis of solar, solar wind and geomagnetic parameters. *Solar Phys.,* 2002; 208: 359-373.

[60] Bai T. Periodicities in solar flare occurrence analysis of cycles 19-23. *Astrophys. J.*, 2003; 591: 406-415.

[61] Ballester JL, Oliver R, Carbonell M. Return of the near 160 day periodicity in the photospheric magnetic flux during solar cycle 23. *Astrophys. J.,* 2004; 615: L173-L176.

[62] Knaack R, Stenflo JO, Berdyugina SV. Evolution and rotation of large-scale photospheric magnetic fields of the Sun during cycles 21-23: Periodicities, north-south asymmetries and r-mode signatures. *Astron Astrophys,* 2005; 438: 1067-1082.

[63] Chowdhury P, Ray PC. Periodicities of solar electron flare occurrence: analysis of cycles 21–23. *Mon NotRoy Astronom. Soc.*, 2006; 373: 1577-1589.

[64] Chowdhury P, Ray PC, Ray S. Periodicity of 155 days in solar electron fluence. *Ind. J. Phys.,* 2008; 82: 95-104.

[65] Dimitropoulou M, Moussas X, Strintzi D. Enhanced Rieger-type periodicities' detection in X-ray solar flares and statistical validation of Rossby waves' existence. *Mon. Not. R Astron. Soc.,* 2008; 386: 2278-2284.

[66] Javaraiah J, Ulrich RK, Bertello L, Boyden JE. Search for short-term periodicities in the sun's surface rotation: a revisit. *Solar Phys.,* 2009; 257: 61-69.

[67] Vaquero JM, Trigo RM, Vazquez M, Gallego MC. 155-day periodicity in solar cycles 3 and 4. New Astronomy 2009. doi: 10.1016/j.newast.2009.11.004.

[68] Zaqarashvili TV, Carbonell M, Oliver R, Ballester JL. Magnetic Rossby waves in the solar tachocline and Rieger-type periodicities. arXiv:0911.4591v1 [astro-ph.SR] 24 Nov 2009.

[69] Fischbach E, Jenkins JH, Buncher JB, Gruenwald JT, Sturrock PA, Javorsek D II. Evidence for solar influences on nuclear decay rates. arXiv:1007.3318v1 [hep.ph] 20 Jul 2010.

[70] Halberg Franz, Cornélissen G, Katinas G, Syutkina EV, Sothern RB, Zaslavskaya R, Halberg Francine, Watanabe Y, Schwartzkopff O, Otsuka K, Tarquini R, Perfetto P, Siegelova J. Transdisciplinary unifying implications of circadian findings in the 1950s. *J. Circadian. Rhythms,* 2003; 1: 2. 61 pp. www.JCircadianRhythms.com/content /pdf/1740-3391-2-3.pdf.

[71] Cornélissen G, Tarquini R, Perfetto F, Otsuka K, Gigolashvili M, Halberg F. About 5-month cycle in human circulating melatonin: signature of weather in extraterrestrial space? Poster presentation, Fourth UN/ESA/NASA/JAXA Workshop on the International Heliophysical Year 2007 and Basic Space Science: "First Results from the International Heliophysical Year 2007", Sozopol, Bulgaria, June 2-6, 2008.

[72] Halberg F, Cornélissen G, Schwartzkopff O. Quo vadis chronomics 2008: Measuring variability in us, among us and around us. In: Halberg F, Kenner T, Fiser B, Siegelova J (Eds.) Proceedings, Noninvasive Methods in Cardiology, Brno, Czech Republic, October 4-7, 2008. p. 16-25. http://web.fnusa.cz/files/kfdr2008/sbornik_2008.pdf.

[73] Cornélissen G, Halberg F, Sothern RB, Hillman DC, Siegelova J. Blood pressure, heart rate and melatonin cycles synchronization with the season, earth magnetism and solar flares. *Scripta med.*, 2010; 83: 16-32.

[74] Cornélissen G, Halberg F, Wendt HW, Bingham C, Sothern RB, Haus E, Kleitman E, Kleitman N, Revilla MA, Revilla M Jr, Breus TK, Pimenov K, Grigoriev AE, Mitish MD, Yatsyk GV, Syutkina EV. Resonance of about-weekly human heart rate rhythm with solar activity change. *Biologia,* (Bratislava) 1996; 51: 749-756.

[75] Refinetti R. Circadian Physiology. 2nd ed. Boca Raton, Florida: CRC Press; 2006. 700 pp.

[76] Koukkari WL, Sothern RB. Introducing Biological Rhythms: A primer on the temporal organization of life,with implications for health, society, reproduction and the natural environment. New York: Springer; 2006. 655 pp.

[77] Durkheim E. Le Suicide: Etude de Sociologie. Paris, Presses Universitaires de France, 1897.

[78] Terao T, Soeda S, Yoshimura R, Nakamura J, Iwata N. Effect of latitude on suicide rates in Japan. *The Lancet*, 2002; 360: 1892.

[79] Lawrynowicz AEB, Baker TD. Suicide and latitude in Argentina: Durkheim upside-down. *Am. J. Psychiatry,* 2005; 162: 1022.

[80] Wetterberg L, Bratlid T, Knorring Lv, Eberhard G, Yuwiler A. A multinational study of the relationships between nighttime urinary melatonin production, age, gender, body size, and latitude. *Eur. Arch. Psychiatr. Neurosci.,* 1999; 249: 256-262.

[81] Ruby NF, Brennan TJ, Xinmin X. Role of melanopsin in circadian response to light. *Science,* 2002; 298: 2211-2213.

[82] Swaab DF. The Human Hypothalamus: basic and clinical aspects. Part II: Neuropathology and the human hypothalamus and adjacent structures. In: Handbook of clinical neurology. MJ Aminoff, F Boller, DF Swaab (Series Eds.). Elsevier, 2004.

[83] Cornélissen G, Prabhakaran Nayar SR, Czaplicki J, Siegelova J, Mendoza B, Halberg F. Brueckner-Egeson-Lockyer (BEL) cycle in heliogeomagnetics. In: Halberg F, Kenner T, Fiser B, Siegelova J (Eds.) *Proc., Noninvasive Methods in Cardiology,* Brno, Czech Republic, October 4-7, 2008. p. 106-115. http://web.fnusa.cz/files/kfdr2008 /sbornik_2008.pdf.

[84] Bjorksten KS, Kripke DF, Bjerregaard P. Accentuation of suicides but not homicides with rising latitudes of Greenland in the sunny months. *BMC Psychiatry,* 2009; 9: 20.doi:10.1186/1471-244X-9-20.

In: Cardiovascular Health and Chronomics
Editor: Ram B. Singh

ISBN: 978-1-62948-976-6

Chapter 3

SIGNATURES OF SPACE WEATHER IN THE AGING HUMAN BLOOD CIRCULATION?

Franz Halberg*[*,1], Robert B. Sothern[1], Yoshihiko Watanabe[2], Dewayne Hillman[1], William R. Best[3], Othild Schwartzkopff[1] and Germaine Cornélissen[1]

[1]Halberg Chronobiology Center, University of Minnesota, Minneapolis, MN, US
[2]Tokyo Women's Medical University, Medical Center East, Tokyo, Japan
[3]Hines VA Hospital, Hines, IL, US

ABSTRACT

The signature of an environmental decadal or longer cosmic cycle is found in each of 4 men's circulation. The assortment of pairs of environmental and physiological periods, τs, can differ among some variables and even for some of the two circadian rhythm characteristics of the same variable in the circulation of the same subject. As gauged by overlapping CIs (95% confidence intervals) of the τs in us and around us, the environmental pairing (e.g., of two subjects' diastolic blood pressure) can thus differ for the MESOR, M, a midline-estimating statistic of rhythm versus the circadian double amplitude, 2A, a measure of the extent of within-day change. A change in period length with the lengthening of a time series and/or advancing age is also recorded in the para-annual, para-decadal and para-tridecadal range of the spectrum. To whatever extent these findings can be extended to the population at large, the imaginary baselines reported in investigations involving interventions in aging will be amenable to replacement by cycles here demonstrated longitudinally, yet to be mapped on a population basis.

Keywords: Blood pressure, cycles, cosmos, decadal, heart rate, paratridecadal

[*] Franz Halberg <halbe001@tc.umn.edu>

BACKGROUND ON CYCLES AROUND US

Charles Egeson [1] (aka John J. George) [2] found a 33-year cycle in rainfall and winds in Australia by 1889, presenting actual numerical data for Lord Francis Bacon's anecdotal statement [3] that "it is observed in the Low Countries [...] that every five and thirty years the same kind and suit of years and weathers comes about again; as great frosts, great wet, great droughts, warm winters, summers with little heat, and the like; and they call it the Prime". Egeson's findings were followed within months by the publication of climatic and related data, on different variables such as environmental temperature, ice-free rivers, cold winters and the wine harvest as well as rainfall in Europe, reported as 33- or 35-year cycle lengths by Eduard Brückner [4]. William J.S. Lockyer in 1901 [5] published a 35-year cycle in the changes of the length of the about 11-year (undecennian) periodicity of sunspots, described as such by Samuel Heinrich Schwabe [6], recorded earlier by Christian Horrebow [7] and François Arago, as reported by Manuel J. Johnson [8] (who awarded the gold medal of the Royal Astronomical Society to Schwabe). It was promptly confirmed by Liznar [9] and cited more recently by Hoyt and Schatten [10] and constituted the subject of focused scholarly contributions by Stehr and von Storch [11, 12]. Nonetheless, about 33- or 35-year cycles are mostly forgotten and not readily known to the public at large. By contrast, the about 11-year Horrebow-Schwabe cycle in sunspot numbers and the Hale cycle in sunspot bipolarity [13] have been extensively investigated; a global cycle [14] has also been reported in physics and all await the detection of further biospheric counterparts [15-21].

A recent set of meta-analyses [2] of Egeson's and Brückner's original data taken off original summary graphs allows the demonstrations, among others, in the series analyzed as a whole, of a τ that may fall outside the 30-40-year range but with CIs (95% confidence intervals) of the τs overlapping this range [22-24]. Hence, these transtridecadals were defined as τs with CIs overlapping the 30-40-year range, whether or not the τ itself falls into it. The initials of **B**rückner, **E**geson and **L**ockyer, the discoverers of this cycle, led to the acronym BEL; *transtridecadal* is derived from *trans* (= over, to the other side of; i.e., longer than) and *tridecadal* (= of 30 years). When the CI of τ also falls outside the 30-40-year range, but is very near to it, differing by perhaps no more than 1 year from the 30-40-year range, we propose the term "paratridecadal", with *para-* denoting "beside, alongside of". Other non-photic infradian cycles have also been described [15, 16, 18-20].

Subjects and Methods

RBS, a clinically healthy man, starting at ~20 years of age self-measured a number of psychophysiological variables for over 40 years, about 5-6 times a day. WRB, taking hypotensive medication, was nearly 52 years of age when he began taking single morning measurements of body weight and systolic and diastolic blood pressure [25, 26] to which he subsequently added measurements of heart rate. He prolonged a series of these morning measurements from 31 to 36 years. YW, who was unmedicated for his blood pressure throughout, produced a half-hourly ~23-year series starting at ~34 years of age. FH, a man on hypotensive medications, was ~68 years of age at the start of ~23 years of **c**hronobiologically-interpreted systematic **a**utomatic half-hourly **b**lood **p**ressure and heart rate **m**onitoring

(C-ABPM) with interruptions. Four of the series of blood pressure (from RBS, FH, YW and WRB) and three of heart rate allowed a test of the hypothesis of at least one selective assortment [18, 20] of cycles of similar lengths -- of reciprocals only in this sense -- in us and around us in the individuals investigated. Another subject, GSK, a man treated with hypotensive medication, ~72 years of age at the start, took half-hourly automatic measurements with instruments from Colin Medical Instruments (Komaki, Japan) and from A&D (Tokyo, Japan), both monitors also used by FH and YW. The extended cosinor [27-29], including Marquardt's algorithm [30], served for analyses by least-squares, Figure 1A, to obtain the endpoints shown in Figure 1B, shown without indispensable uncertainties for clarity's sake.

Time Structure Change with Lengthening of Series

An elderly man (GSK) [20], self-monitoring his systolic blood pressure at half-hourly intervals for a first ~4.8-year span, had a CI of an about-yearly rhythm that overlapped the calendar year length. This overlap did not occur when his measurements covered a span of ~6.5 and ~7 years (10b-c in Figure 2); he then demonstrated a near-transyear, Figure 2, defined as 1.00 year $< [\tau - \text{CI}] < [\tau + \text{CI}] <$ 1.20 years.

A clinically healthy man, RBS, aging from ~20 to over 60 years, carried out self-studies around the clock, with on the average about 5-6 measurements per day [24; cf.15, 16, 20]. RBS' series of systolic blood pressure, analyzed in four successive ~9-year spans of the para-annual spectral region, underwent changes in period length, Figure 3.

In two of the sections, but not in two others, the CI of the τ overlapped the calendar year. The average of these about-yearly τs covered, with its broad CI, the precise 1-year length, whereas this was not the case, as also shown in Figure 3, when the series is analyzed as a whole. At that time, in the decadal-multidecadal spectral region, RBS's systolic and diastolic blood pressure showed a periodicity of about two decades, a didecadal, whereas the heart rate revealed a cycle longer than 3 decades, the transtridecadal BEL, Figure 4. Decadals and didecadals have been extensively mapped in the biosphere, notably in epidemiologic data, Figure 5 [17, 18, 20].

The finding of a reciprocal period in physiology, not only for the para-tridecadal BEL [17, 22-24] but also for the ~decadal Horrebow-Schwabe [6, 7] and the ~didecadal Hale [13] cycle of sunspot numbers and bipolarity, respectively, led to inquiries about the number of τs of decadal and multi-decadal length in and around us, as a premise to the investigation of the patterns of any associations between these internal and any external cycles [17]. A selective assortment of τs and of their means (MESORs, from curve-fitting, Figure 1, usually more accurate and more precise than the arithmetic means) and double amplitudes (2As, measures of the extent of within-day change computed by a linear-nonlinear cosine fit) in variables in us and around us, that are congruent in the sense of overlapping CIs of their τs, irrespective of the point estimate of τ was thus found in a man, RBS, and awaited scrutiny in others herein, as do, with the addition of 5 years in Table 1, the results of three 31-year series of morning self-measurements by WRB [25, 26].

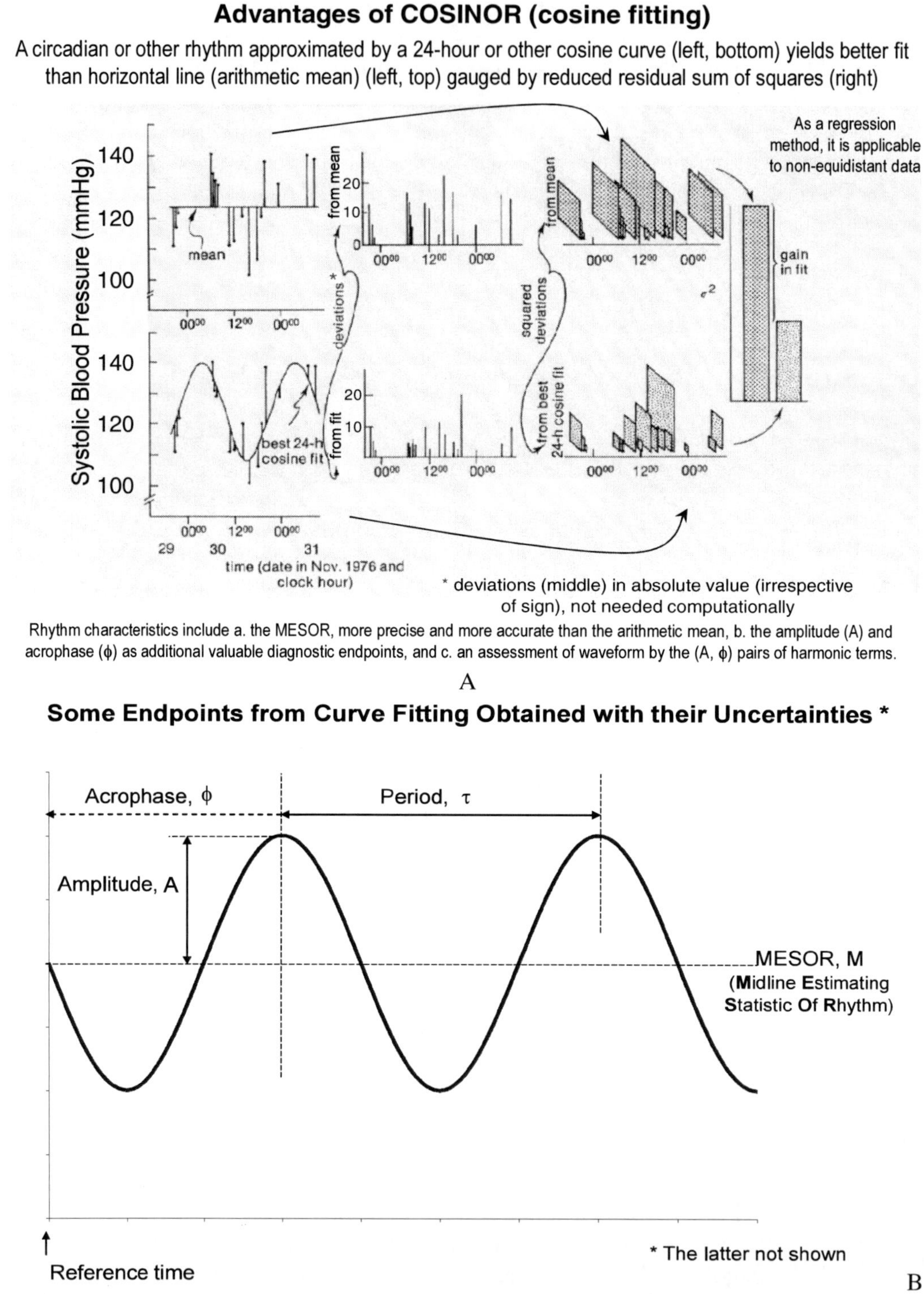

Figure 1. Gauss's method of least-squares (A) when applied to data with a circadian rhythm yields not only a better estimate of the overall operating value [26], but also new endpoints of the dynamics of change (B). © Halberg.

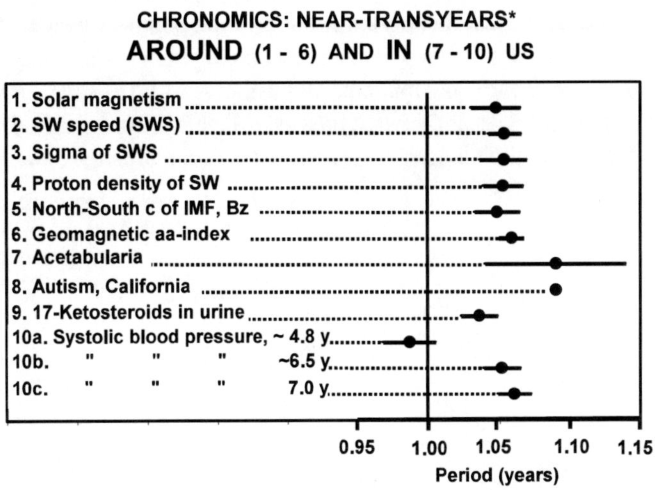

***as separate components and probably not as sidelobes of circannual variation, stemming from amplitude- and/or phase-modulation by components with a lower frequency.**

1 - data from National Oceanic and Atmospheric Administration (www.noaa.gov), May 10, 1975 - April 30, 2002. 2- 4 = SW = Solar wind; 2- 5 data from OMNI 2 series (1963-2003); 3 - Standard deviation of SW; 5 = Bz = North-South component of interplanetary magnetic field (IMF); 6 = 131 years (1868-1998); 7 = oxygen production (1980-1994); 8 = Quarterly incidence, 10.5 y (1994-2004); 9 = daily excretion for 15 y of urinary steroidal metabolites (17-ketosteroids, CH, October 25 1948 - October 22 1963); 10a = half-hourly record from March 31, 1998; from a man (GSK), 72 years old at start of measurements; 10b - same data series, extended to December 10, 2004; 10c - extended to April 1, 2005. An initially lacking near-transyear appears with lengthening of series and/or age.

Circle = period length (point estimation); horizontal line crossing the circle = 95% confidence interval (CI) of spectral component; dashed line: CI of a spectral component not validated nonlinearly.

Figure 2. It is indispensable to specify the calendar span and thereby also the length of a time series upon which the resolution of a (set of) spectral component(s) with period(s) (τ) rests. Thus, based on a series of ~4.5 years (y), one finds a component of systolic blood pressure (10a) with a τ of 1y since its CI (95% confidence interval) overlaps the exact 1y τ. In row 10b below, based on an ~6.5-y record, the resolution of a near-transyear (with a τ between 1.0 and 1.2y and a CI overlapping neither of these lengths) becomes possible, perhaps as a function of age. In another case (RBS), examined over decades as a function of age (not shown), transyears become more prominent with age in some variables. A still longer record will have to confirm that during an even longer span at a still later age, as in FH, there may be only transyears and no component with a CI of τ overlapping the calendar year length. As to mechanisms, it seems pertinent that the only endocrine record covering 15y of observation of an adult's 17-ketosteroid excretion reveals globally no yearly component, only a near- and two far-transyears, the latter not shown. The finding of mutually reinforcing physical and biological near-transyears, defined by $1.00y < \tau < 1.20y$ (see above), is the message of this graph that illustrates the uncertainties of aeolians by the horizontal bars representing CIs of τs. These near-transyears in us, acquired by evolution, reinforce the reality of those around us from which they likely derive. We all stacked data in the past over 1y and other τs such as a day; yet whenever possible, we should get an estimate of the period involved before stacking! This was a sine qua non well over 50y ago for circadians and has become one for circannuals and in particular for aeolians, such as the near-transyear, here depicted as calendar-time-location- and series-length-dependent near-transyears. Lengthening of a series (and/or age?) may change a circannual rhythm into a near-transyear, the former but not the latter covering the precise 365.75y length with the CI of the τ. But the differential diagnosis between a desynchronized circannual rhythm and a transyear will have to await womb-to-tomb monitoring. © Halberg.

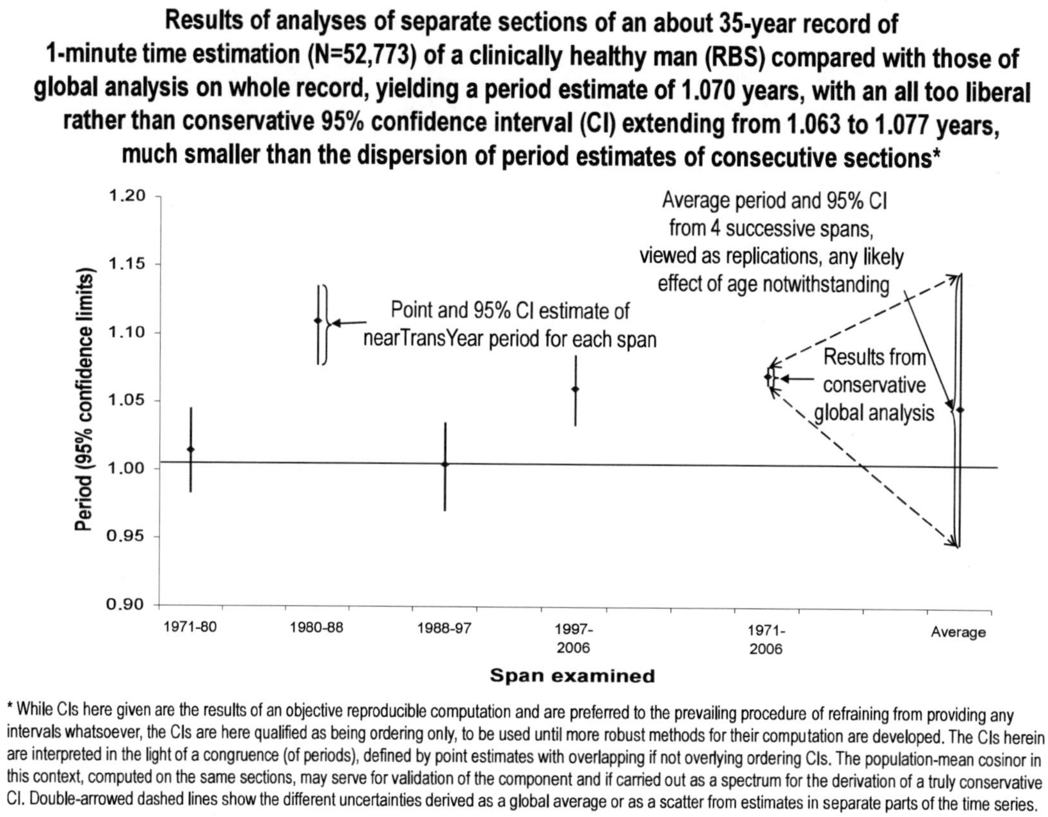

Figure 3. Complementary global and local analyses of the same time series of systolic blood pressure yield different estimates of uncertainty. © Halberg.

These changed drastically while, by contrast, for RBS, a remarkable reproducibility was found, notably of about 10-year and longer cycles, as seen for 11 variables in Figure 6. A correlation coefficient of over 0.8 (P<0.001) is the same whether or not results on a noisy variable are included.

Table 2, underlying Figure 6, compares the length of certain decadal periods published in 2000 [31] based on ~30 years of data with the results of the mostly 43.5-year series. Note that for 8 of 10 or 11 variables the CIs of the τs estimated over a decade apart overlap each other, as they also happen to do for sunspots in the two spans compared (where the 43.5-year record includes the data of the first 30 years, just as in the case of WRB in Table 1). Table 3 summarizes a selective assortment of circadian characteristics with known environmental periods in two cases with the selective assortment involving specific rhythm characteristics, M vs. 2A.

Table 1. Time series of WRB, a man ~52 years of age at start of self-measurements, first for 31 years, with the addition of 5 more years thereafter

Time series length	29 Apr 1974 [HR: 13 Sep 1993]-31 Dec 2004 (~ 31 y; HR: ~11 y)		→ 31 Dec 2009 (~36 y/16 y for HR)			
Trial τ:	35y (or 32.17y with linear trend, LT)		35y		35y + linear trend	
Variable	τ [95% CI]	A [95% CI]	τ [95% CI]	A [95% CI]	τ [95% CI]	A [95% CI]
SBP	30.56 [29.24, 31.87]	9.03 [8.46, 9.60]	31.92 [30.38, 33.47]	8.06 [7.55, 8.57]		
DBP	31.68 [23.94, 39.42] (+LT)	4.62 [2.63, 6.61]			22.06 [21.23, 22.90]	2.85 [2.51, 3.18]
HR			22.36 [17.96, 26.75]	2.48 [2.12, 2.84]		
BW	41.23 [31.25, 51.20] (+LT)	3.31 [1.80, 4.81]	24.23 [23.62, 24.84]	1.77 [1.64, 1.89]	24.40 [23.71, 25.09]	1.76 [1.63, 1.90]
Trial τ:	15y		15y		15y + linear trend	
	τ (y) [95% CI]	A [95% CI]	τ (y) [95% CI]	A [95% CI]	τ (y) [95% CI]	A [95% CI]
SBP	12.55 [11.73, 13.37]	2.66 [1.93, 3.39]	13.32 [12.35, 14.29]	1.83 [1.17, 2.49]	13.87 [12.67, 15.07]	1.76 [1.06, 2.46]
DBP	16.29 [15.79, 16.79]	4.29 [3.89, 4.68]	18.40 [17.88, 18.92]	4.27 [3.86, 4.87]	22.05 [21.23, 22.90]	2.85 [2.51, 3.18]
HR HR + LT	12.68 [10.68, 14.68] 5.90 [5.34, 6.45]	2.25 [1.76, 2.75] 1.70 [1.13, 2.27]	22.36 [17.96, 26.75]	2.48 [2.12, 2.84]		
BW BW + LT	12.52 [12.18, 12.87] 12.58 [12.20, 12.96]	1.23 [1.08, 1.38] 1.23 [1.07, 1.39]	13.25 [12.91, 13.59]	1.14 [1.01, 1.27]	13.24 [12.84, 13.64]	1.14 [1.00, 1.29]
Trial τ:	7y		7y		7y + linear trend	
	τ (y) [95% CI]	A [95% CI]	τ (y) [95% CI]	A [95% CI]	τ (y) [95% CI]	A [95% CI]
SBP	6.14 [6.00, 6.28]	3.39 [2.66, 4.11]	7.92 [7.58, 8.85]	1.99 [1.35, 2.63]	7.60 [7.28, 7.91]	2.11 [1.44, 2.78]
DBP DBP + LT	6.21 [6.05, 6.37] 6.26 [6.08, 6.44]	2.01 [1.55, 2.47] 1.55 [1.17, 1.94]	7.44 [7.18, 7.69]	1.62 [1.17, 2.06]	6.34 [6.18, 6.50]	1.31 [0.97, 1.66]
HR HR + LT	12.68 [10.68, 14.68] 5.90 [5.34, 6.45]	2.25 [1.76, 2.75] 1.70 [1.13, 2.27]	7.62 [7.00, 8.24]	1.26 [0.83, 1.70]	5.69 [5.32, 6.05]	1.30 [0.88, 1.71]
BW BW + LT	6.06 [5.96, 6.16] 6.09 [5.98, 6.19]	1.01 [0.86, 1.17] 1.05 [0.89, 1.22]	6.65 [6.55, 6.76]	0.92 [0.79, 1.06]	6.62 [6.51, 6.73]	0.92 [0.77, 1.97]
Trial τ:			6y		6y + linear trend	
			τ (y) [95% CI]	A [95% CI]	τ (y) [95% CI]	A [95% CI]
SBP			6.08 [5.94, 6.20]	2.84 [2.19, 3.48]	6.02 [5.89, 6.15]	2.94 [2.26, 3.62]
DBP			6.04 [5.85, 6.21]	1.46 [1.00, 1.91]	6.34 [6.18, 6.50]	1.31 [0.97, 1.66]
HR			7.62 [7.00, 8.24]	1.26 [0.83, 1.70]	5.69 [5.32, 6.05]	1.29 [0.88, 1.71]
BW			6.65 [6.55, 6.76]	0.92 [0.79, 1.06]	6.62 [6.51, 6.73]	0.92 [0.77, 1.07]

Table 2. Long periods (τ) in ~30- vs. ~43.5-year (y) records of RBS* and in sunspots during span documented for RBS

			2010: ~43.5y		
Variable	N at ~43y	MESOR (SE)	τ (CI)	2A (CI)	Acrophase (CI)
Sunspots (Wolf N) (span analyzed)	30693		10.48 (10.43, 10.52)**	88 (86, 89)	-224° (-224, -225)
Mood	64704	4.16 ± 0.01	11.60 (11.46, 11.74)	0.38 (0.35, 0.42) ^{c}	-282° (-278, -188)
Vigor (arbitrary units)	64761	2.03 ± 0.01	^{n} 9.44 (9.36, 9.51) ^{c}	0.23 (0.21, 0.24) ^{c}	-343° (-339, -346)
Temperature (°F)	65290	97.90 ± 0.01	^{c} 10.40 (10.32, 10.49) $_{n}$	0.12 (0.10, 0.13) ^{c}	-9° (-6, -13)
Peak expiratory flow (L/min)	63627	688.59 ± 0.14	^{n} 10.69 (10.64, 10.74) $_{c}$	21.34 (20.58, 22.10) ^{n}↑	-13° (-11, -15)
1-minute time estimation (seconds)	63517	63.98 ± 0.04	^{n} 11.16 (11.08, 11.23) $_{c}$	3.59 (3.43, 3.76) ^{c}	-48° (-45, -51)
Respiratory rate (breaths/min)	65340	11.70 ± 0.01	^{n} 11.18 (11.13, 11.22) $_{c}$	1.62 (1.56, 1.66) ^{n}↑	-83° (-81, -85)
Diastolic blood pressure (BP) (mmHg)	64478	83.40 ± 0.02	^{n} 11.23 (11.13, 11.32) $_{c}$	1.92 (1.80, 2.05) ^{n}↓	-156° (-152, -159)
Mean arterial BP (mmHg)	64478	97.50 ± 0.02	^{n} 11.45 (11.37, 11.53) $_{d}$	2.08 (1.97, 2.19) ^{c}↓	-107° (-104, -110)
Systolic BP (mmHg)	64478	125.28 ± 0.03	^{n} 11.99 (11.91, 12.06) $_{n**}$	2.59 (2.47, 2.71) ^{c}↓	-340° (-338, -343)
Heart rate (HR) (beats/min)	65347	70.43 ± 0.04	^{n} 14.57 (14.52, 14.62) $_{c}$	9.85 (9.65, 10.0) ^{n}↑	-325° (-324, -326)
Eye-hand coordination (seconds)	51084	21.98 ± 0.02	^{n} 20.27 (20.11, 20.42) $_{c}$	2.50 (2.44, 2.57) ^{n}↑	-167° (-165, -268)
			Reference 1997: ~30y		
			τ (CI)	2A (CI)	
Sunspots (Wolf N) (span analyzed)			10.27 (10.06, 10.50)	131.94 (118.82, 145.06)	
Mood			11.50 (10.11, 13.41)	0.38 (0.14, 0.64)	
Vigor (arbitrary units)			8.48 (7.33, 9.64)	0.14 (0.04, 0.24)	
Temperature (°F)			8.36 (7.66, 9.06)	0.14 (0.06, 0.22)	
Peak expiratory flow (L/min)			11.74 (10.36, 13.11)	7.68 (3.56, 11.80)	
1-minute time estimation (seconds)			10.29 (9.38, 11.37)	2.16 (0.88, 3.44)	
Respiratory rate (breaths/min)			13.73 (10.13, 17.32)	0.38 (0.06, 0.70)	
Diastolic blood pressure (BP) (mmHg)			10.98 (10.09, 11.87)	3.68 (2.28; 5.08)	
Mean arterial BP (mmHg)			10.48 (9.50, 11.46)	3.10 (1.66, 4.54)	
Systolic BP (mmHg)			10.21 (9.25, 11.36)	2.48 (1.08, 3.86)	
Heart rate (HR) (beats/min)			16.53 (14.12, 18.94)	5.22 (3.18, 7.24)	
Eye-hand coordination (seconds)			17.20 (7.65, 26.75)	0.56 (; 1.34)	

*RBS, man, ~20.5 years of age at start of ~5-6-daily self-measurements in clinical health; data are edited by removing measurements during illness or transmeridian trips. Analyses by extended cosinor; decimals given as computational results only. CI: 95% confidence interval given in ( ). c: congruent by overlapping CIs; n: not congruent. Note that congruence with the concurrent sunspot cycle, shown on the left of the τ column in 2010, was found for only one variable out of 11. Congruence of the τ in 2010 with the τ in 1997 (the latter shown in the bottom half of the grid) was found in 8 cases out of 11, as indicated on the right of the τ column in 2010. The amplitudes in 2010 vs. 1997 were congruent in 6 cases, not congruent in 5, in view of changes with age. ↑ increase; ↓ decrease.

**During the years from 196705 to 201011 (~43 years), the spectrum of Wolf numbers also exhibited an 18.70 (18.21, 19.30)-year component. The amplitude of this component, however, was smaller than that of the 10.54-yearly component: 7.79 (5.88, 9.70) vs. 57.14 (55.11, 59.18). A third component in sunspots was 5.51 (5.48, 5.55) years with an amplitude of 12.66 (10.76, 14.56).

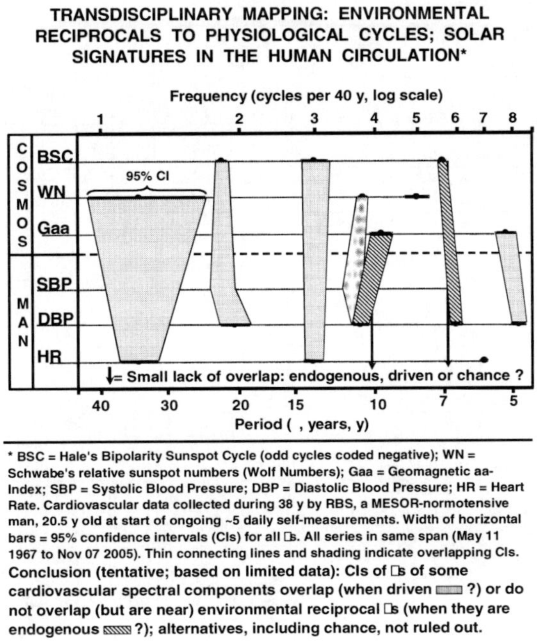

Figure 4. Selective physiological-environmental pairing of the variables of a clinically healthy man (RBS). © Halberg.

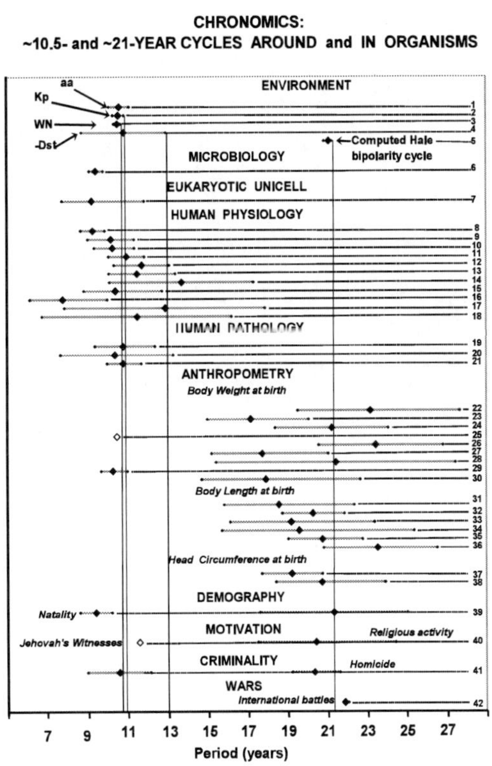

(A)

Figure 5. (Continued)

Chronomics: ~10.5- and ~21-year cycles in and around us

	Line				Period (years) Lower limit*	Best Fit	Upper limit*	Series duration Dates	Years	Number of data	Geographic site
Environment	1	aa = Antipodal Geomagnetic Index			10.12	**10.63**	11.13	1890-1999	110	1 / year	
	2	Kp = Planetary Geomagnetic disturbance			10.32	**10.58**	10.85	1932-1999	68	1 / month	
	3	WN = Wolf relative sunspot number			10.37	**10.54**	10.70	1890-1999	110	1 / year	
	4	-Dst = Equatorial geomagnetic disturbance			8.75	**10.85**	12.96	1973-1999	27	"	
					10.48	**10.51**	10.55	1700-1999	300	"	
	5	Bipolarity "Hale Cycle" **			20.86	**21.10**	21.26	1890-1999	110	"	
					21.42	**21.428**	21.43	1700-1999	300	"	
Biology	6	Prokaryotes: Air Bacterial Sectoring			9.12	**9.45**	9.81	1970-1982	13	3,744	Italy
	7	Eukaryotes: Unicellular Algal O_2 Production			7.79	**9.24**	11.87	1980-1991	11	324	Germany
Physiology***	8	Mood (RBS)			10.11	**11.50**	13.41	1966-1998	33	~5 / day	USA
	9	Time (1-Minute) Estimation (RBS)			9.38	**10.29**	11.37	1966-1998	33	"	"
	10	Urinary 17-ketosteroid excretion (CH)			8.70	**9.30**	9.90	1948-1963	15	1 / day	Denmark
	11	Peak Expiratory Flow (RBS)			10.36	**11.74**	13.11	1966-1998	33	~5 / day	USA
	12	Respiratory Rate (RBS)			10.13	**12.50**	17.32	1966-1998	33	"	"
	13	Systolic Blood Pressure - SBP (RBS)			9.05	**10.21**	11.36	1966-1998	33	"	"
	14	Standard Deviation of SBP (YW)			8.85	**10.43**	12.76	1987-1998	11	~48 / day	Japan
	15	Diastolic Blood Pressure - DBP (RBS)			10.09	**10.98**	11.87	1966-1998	33	~5 / day	USA
	16	Standard Deviation of DBP (YW)			6.18	**7.82**	10.02	1987-1998	11	~48 / day	Japan
	17	Heart Rate - HR (YW)			9.54	**12.93**	17.91	1987-1998	11	"	"
	18	Standard Deviation of HR (YW)			8.27	**11.52**	16.22	1987-1998	11	"	"
Pathology	19	Myocardial Infarction			10.00	**10.80**	11.70	1960-1996	37	129,205	USA
	20	Leptospirosis			9.40	**10.80**	12.40	1949-1995	47	2,907	Slovakia
	21	Diabetes			7.70	**10.40**	13.30	1985-1995	11	1,369	"
Anthropo-		***Body Weight***	***Boys***								
metry	22		Minnesota		19.53	**23.19**	27.67	1963-1998	36	2,136,745	USA
at birth	23		Alma-Ata Russians		14.99	**17.17**	20.07	1946-1998	53	9,056	Kazakhstan
	24		"	Kazakhs	18.39	**21.24**	24.05	1946-1998	53	3,459	"
	25		Moscow			**10.49**		1874-1985	112	5,987	Russia
			Girls								
	26		Minnesota		20.58	**23.46**	26.83	1963-1998	36	1,039,464	USA
	27		Alma-Ata Russians		15.21	**17.75**	21.06	1946-1998	53	9,105	Kazakhstan
	28		"	Kazakhs	15.44	**21.45**	27.45	1946-1998	53	3,448	"
	29		Moscow		9.70	**10.29**	11.01	1874-1985	112	5,840	Russia
			Both genders								
	30		Denmark		14.71	**17.94**	22.68	1973-1994	22	1,166,206	Denmark
		Body Length	***Boys***								
	31		Alma-Ata Russians		15.82	**18.58**	22.38	1946-1998	53	9,026	Kazakhstan
	32		Moscow		18.76	**20.28**	21.86	1874-1985	112	5,976	Russia
			Girls								
	33		Alma-Ata Russians		16.13	**19.20**	23.39	1946-1998	53	9,105	Kazakhstan
	34		"	Kazakhs	15.72	**19.60**	25.40	1946-1998	53	3,485	"
	35		Moscow		19.05	**20.76**	22.78	1874-1985	112	5,976	Russia
			Both genders								
	36		Denmark		20.81	**23.55**	26.55	1973-1994	22	1,166,206	Denmark
		Head Circumference									
	37		***Boys***	Moscow	17.71	**19.23**	20.75	1874-1985	112	5,976	Russia
	38		***Girls***	"	18.42	**20.73**	23.95	1874-1985	112	5,820	"
Demography	39	Birth rate			8.63	**9.43**	10.23	1940-1996	57	57	USA
					17.61	**21.33**	25.05	1940-1996			
Motivation	41	Religious activity of Jehovah's Witnesses			17.52	**20.44**	24.45	1950-1999	50	328,572#	Worldwide
										5,653,987##	
Criminality	40	Homicide			8.99	**10.58**	12.16	1900-1998	99	99	USA
					19.23	**20.35**	21.62				
Wars	42	International battles			21.87	**21.96**	22.06	599BC-1957	2556	2556	Worldwide

* 95% confidence limit; not shown if cycle is not statistically significant.

** Computed by changing the sign of WN at each WN minimum.

***RBS - Dr. Robert B. Sothern, CH - Dr. Christian Hamburger, YW - Dr. Yoshihiko Watanabe.

in 1950, ## in 1999, pool of 103 plus other unspecified number of sites.

(B)

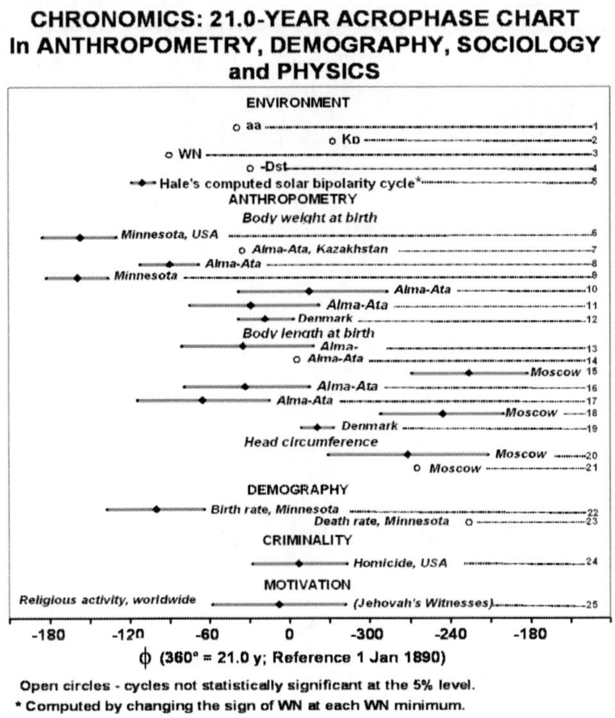

(C)

Figure 5. Decadal and didecadal map of periods (A) and its sources (B) with corresponding phases (C). © Halberg.

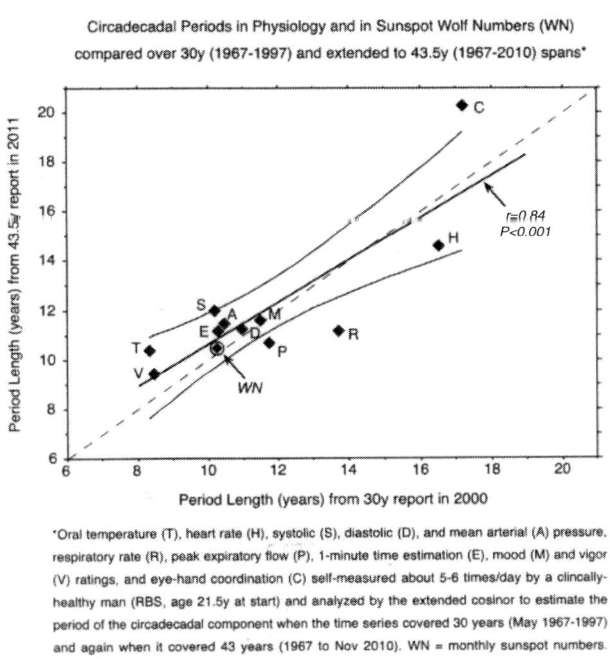

Figure 6. Limited reproducibility of some decadal or longer cycles in RBS. © Halberg.

Table 3. Long infradian psychophysiological periods, τ, differ among variables and even between circadian characteristics (MESOR, M, vs. double amplitude, 2A) in the circulation of two men*

Subject/variable (unit)	Data analyzed	τ (years) (CI)	2A (CI)
FH			
SBP (mmHg)	Original	11.9 (11.7, 12.1)	8.90 (8.44, 9.38)
	M (24-h average)	11.8 (11.4, 12.1)	9.46 (8.42, 10.50)
	2A (24-h average)	11.0 (10.5, 11.5)	3.98 (3.20, 4.76)
DBP (mmHg)	Original	14.4 (14.0, 14.7) 7.5 (7.1, 7.8)	5.81 (5.24, 6.39) 1.64 (1.13, 2.15)
	M (24-h average)	14.5 (13.8, 15.1)	5.06 (4.28, 5.84)
	2A (24-h average)	10.4 (10.1, 10.8)	3.26 (2.72, 3.80)
HR (beats/min)	Original	29.9 (29.0, 30.8) 7.6 (7.5, 7.7)	19.70 (19.28, 20.10) 5.54 (5.14, 5.94)
	M (24-h average)	29.9 (27.4, 32.3) 7.6 (7.4, 7.9)	19.62 (19.04, 20.18) 5.20 (4.12, 6.26)
	2A (24-h average)	26.4 (20.4, 32.4) 5.2 (5.0, 5.4)	1.80 (1.26, 2.34) 1.64 (1.10, 2.18)
YW			
SBP (mmHg)	Original	9.98 (9.82, 10.14)	5.94 (5.58, 6.30)
	M (21-day average)	9.85 (8.89, 10.80)	4.88 (3.06, 6.70)
	2A (21-day average)	19.36 (13.96, 24.77) 7.61 (7.14, 8.08)	2.32 (0.60, 4.02) 4.46 (2.80, 6.12)
DBP (mmHg)	Original	33.1 (31.5, 34.6) 10.5 (10.0, 11.0)	11.74 (11.22, 12.26) 1.84 (1.52, 2.18)
	M (21-day average)	*34.8 (25.1, 44.5)* 10.6 (8.2, 13.0)	11.32 (8.26, 14.36) 1.98 (0.62, 3.36)
	2A (21-day average)	*21.3 (15.1, 27.4)*	1.28 (0.36, 2.20)
HR (beats/min)	Original	32.7 (27.1, 38.2) 11.3 (10.6, 12.0)	7.40 (7.06, 7.76) 3.58 (3.10, 4.06)
	M (21-day average)	29.4 (26.2, 49.1) 11.8 (7.3, 16.4)	7.94 (6.36, 9.52) 3.36 (0.32, 6.40)
	2A (21-day average)	27.0 (20.8, 33.0) 7.04 (6.22, 7.86)	6.08 (4.90, 6.44) 1.44 (0.56, 2.54)

FH, ~68 years, and YW, ~34 years of age at start of ~23 years of half-hourly around-the-clock measurements of systolic (S) and diastolic (D) blood pressure (BP) and heart rate (HR) data sections, as analyzed by the extended nonlinear cosinor in unedited original values for both men, for FH also in 24-hour averages, and for YW in edited 21-day MESORs (short for midline-estimating statistic of rhythm) and as 2A. CI: 95% confidence interval. Some τs found are longer than the span analyzed and hence are particularly tentative; all τs await validation in additional data.

Some Selective Assortment by MESOR vs. Double Amplitude

FH's systolic blood pressure shows a circadecadal cycle in both his rhythm-adjusted mean, the MESOR, M, and in his 2A, assessed by the fit of a 24-hour cosine curve, Figure 7A. This length corresponds roughly to the average length of the Horrebow-Schwabe cycle of sunspot numbers [6, 7; cf.32]. An about 11-year cycle also characterizes FH's diastolic 2A, Figure 7B. His diastolic M in turn shows a longer than 11-year cycle, as apparent from the non-overlapping CIs of the τ extending from 13.8 to 16.1 years vs. the τ for the 2A of 10.4

(10.1, 10.8) years. The τ of the series consisting of averages (M) tends toward the length of a global cycle of solar activity [14]. Figure 7C is in keeping with a para-tridecadal cycle for the M and the 2A of heart rate, an interpretation supported by CIs of these τs extending into the 30- to 40-year range and further by the fact that the τ was found in the output (not input) of the nonlinear cosinor program. Nonetheless, the detection of the trans-tridecadal is qualified by the fact that the series analyzed is much shorter than the length of the estimated τ and will have to be checked by follow-up data, FH's lifespan and motivation permitting.

Table 3 summarizes the analyses of the original data and of the circadian characteristics M and 2A, of FH and of YW, another (untreated) man, also with a series of ~23 years, around-the-clock with few and short gaps. A trans-tridecadal seems most prominent in diastolic blood pressure of YW, although the series is again too short for an unqualified estimate, Figure 8. When a more complex model is fitted to the diastolic blood pressures of YW, a circadecadal is also statistically significant, although its double amplitude is only one fifth of the trans-tridecadal 2As. In FH with automatic measurements every 30 minutes with interruptions, and in RBS with an average of ~5-6 unequidistant self-measurements/day, the variable exhibiting the most prominent para-tridecadal period is heart rate, whereas in YW's original data it is in diastolic blood pressure. When YW's data are separately analyzed for M and 2A, the M of diastolic blood pressure as well as both the M and 2A of heart rate are fitted with a para-tridecadal τ. The situation differs further in another (fourth) man (WRB) with ~36 years of data only on awakening (rather than around-the-clock), Table 1. In WRB, with a ~36-year series, a BEL is no longer prominent in the trans-tridecadal range, in two of three variables investigated, diastolic blood pressure and body weight, as it was with only ~31 years of data [25, 26]. In ~36 years of data from WRB as well, there is selective assortment of environmental-physiological cycle pairing with systolic blood pressure revealing a consistent para-tridecadal cycle, Table 1.

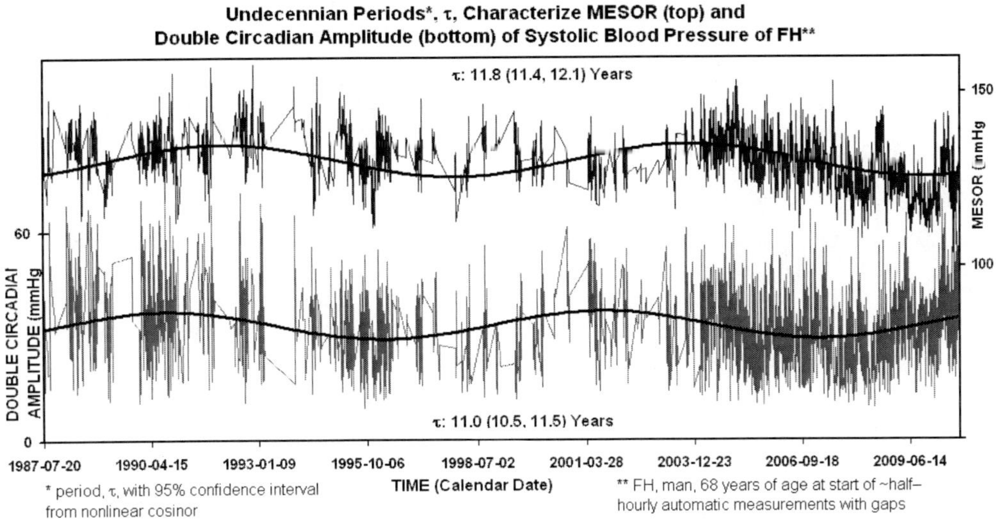

(A)

Figure 7. (Continued)

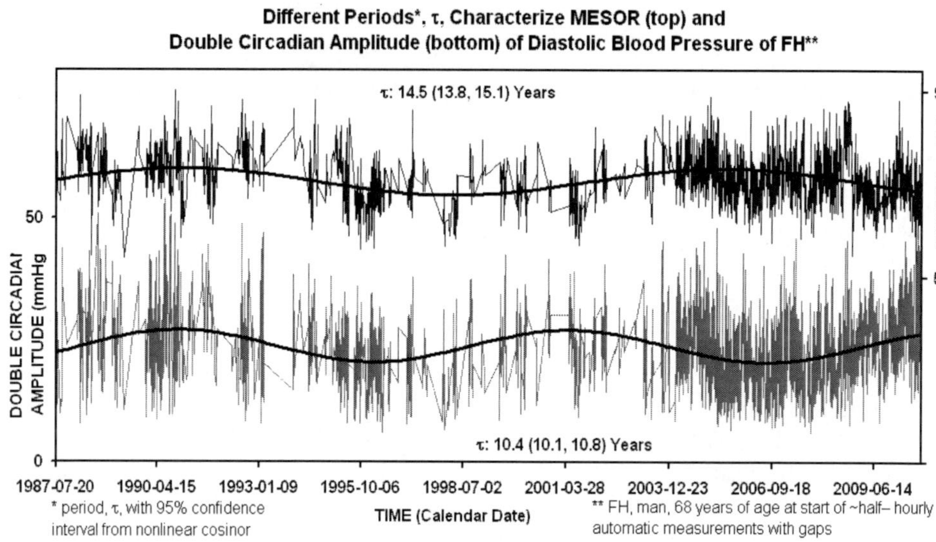

(B)

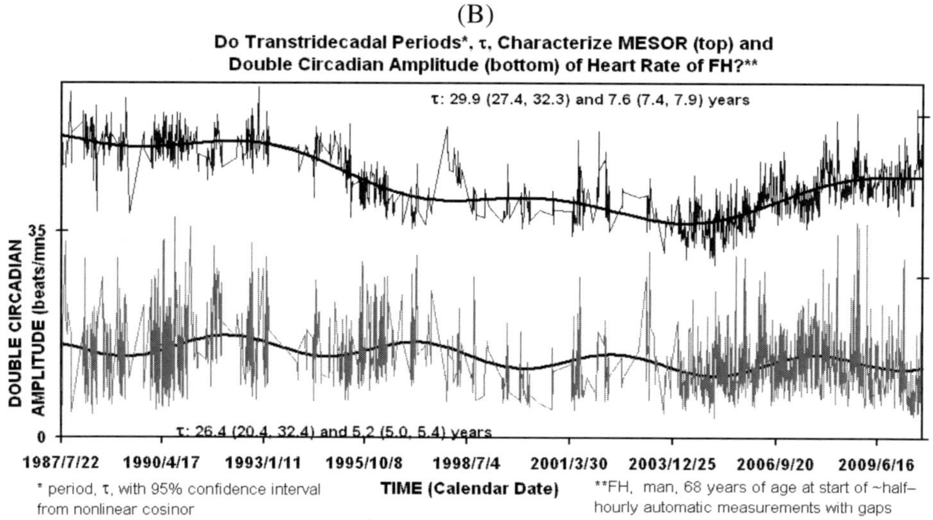

(C)

Figure 7. Systolic (A) and diastolic (B) blood pressure and heart rate (C) of FH, an elderly man on hypertensive medication. © Halberg.

Discussion

The congruences of periods (i.e., the overlapping CIs of their τs), of course, are no indication of causal relations. The latter have been explored for about 7-day [33] and for about 1.3-year [20] rhythms by a subtraction-addition approach. Thus, the loss in solar activity of a circaseptan component was associated with a damping (not a loss) of this component in the heart rate of 5 men [33]. Likewise, the loss of a broad, 1.3 year centered far-transyear band in the speed of the solar wind was associated (after a lag of about one far-transyear) with a narrowing and damping (again no complete loss) of the far-transyear in the systolic blood pressure of FH [20]. Such an approach (remove and/or replace, as possible, or at least by addition and/or subtraction) is much more demanding of data for a prospective study in the case of multi-decadal vs. shorter physiologic cycles and, of course, also depends on the cooperation of the cosmos.

It will have to be implemented, for the time being, retrospectively, on the four available long physiological and on the long epidemiologic time series, such as natality, mortality and morbidity, until lifelong vascular monitoring on additional subjects becomes possible on the one hand, prospectively, and on the other hand the environment happens to cooperate by transiently "removing" the detectability of a trans-tridecadal cycle, among others.

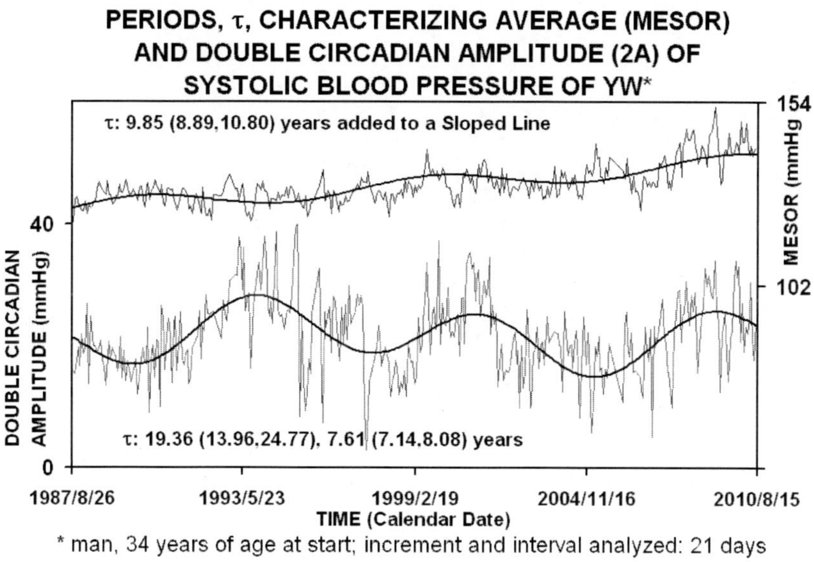

(A)

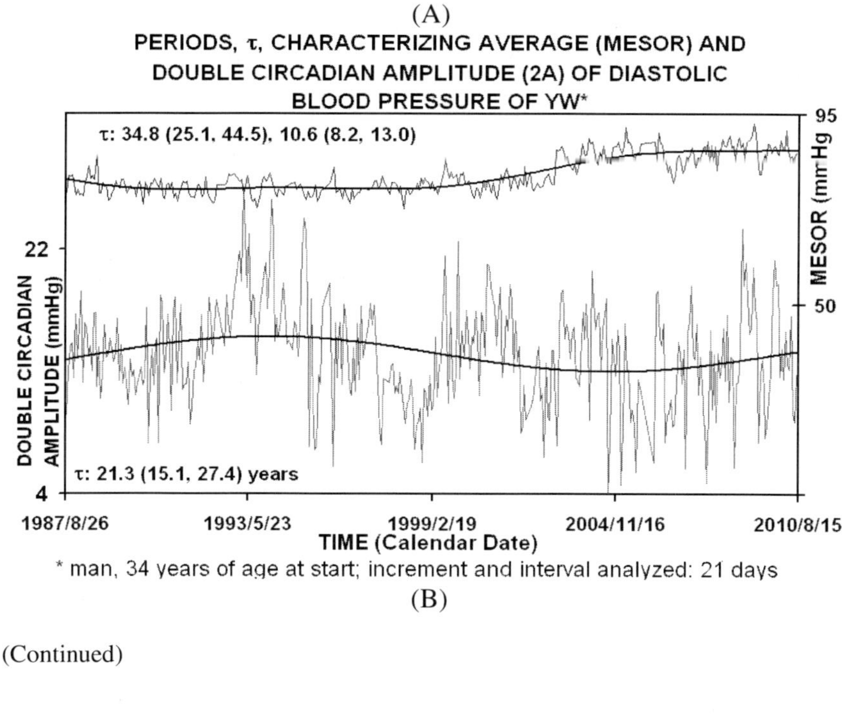

(B)

Figure 8. (Continued)

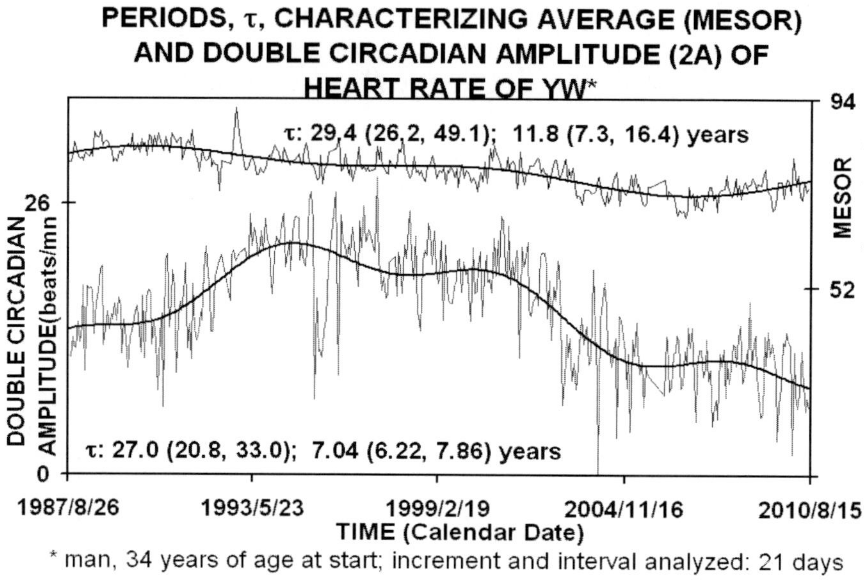

(C)

Figure 8. Systolic (A) and diastolic (B) blood pressure and heart rate (C) of YW, a male cardiologist, depicting cycles during the development of true white-coat hypertension for a while occurring only on Monday to Saturday when he sees patients (and dons a white coat), but not on Sundays at home, and eventually occurring on Sundays as well. © Halberg.

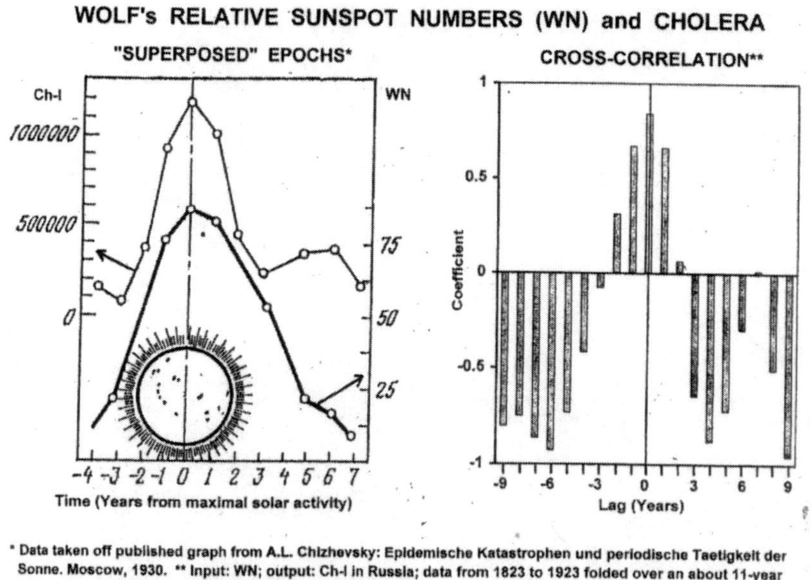

Figure 9. Data from A.L. Chizhevsky on cases of cholera recorded in Moscow between 1823 and 1923, originally folded over an about 11-year solar activity cycle, showing a peak incidence coinciding with maximal solar activity (left). Data taken off the published graph, analyzed by cross-correlation, corroborate that the largest correlation corresponds to lag of zero (right). Should the original 100-year series have been available for analysis, a measure of uncertainty could have complemented the average curves (left). © Halberg.

IN 1965 (left): APPARENTLY DESYNCHRONIZED BIOLOGIC YEAR¥
IN 2004: SEVERAL TRANS-YEARS# (1-YEAR ASYNCHRONIZED COMPONENTS)
CHARACTERIZE HUMAN 17-KETOSTEROID (17-KS) EXCRETION*

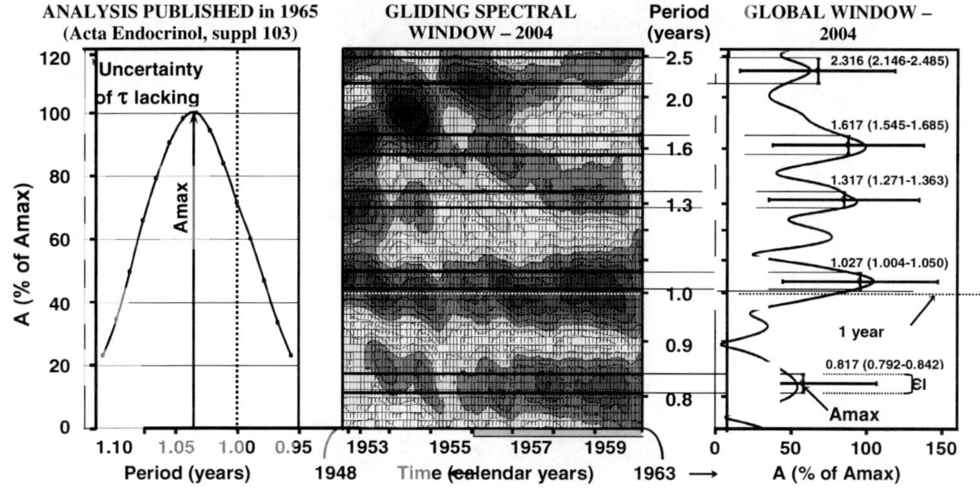

Transyears are oscillations with 95% confidence intervals (CIs) of their period (τ) between 1 and 2 years, overlapping neither of these τs nor one another. Cisyears are oscillations with periods between 1 and 0.5 year with CIs overlapping neither the 1-year nor the 0.5-year τ, as seen for a τ of ~ (but not) = 0.8 y. * ~ 15-year record, daily data (with gaps) from a clinically healthy man (CH, 43 – 58 y), N = 3719. Gliding interval = 8 years (y), increment = 1 month, longest trial τ = 2.265 y, shortest τ = 0.75 y, harmonic increment = 0.05. Darker shading corresponds to larger amplitude (A). Horizontal shaded bar starting in 1956 indicates span of intermittent androgen Rx; data during Rx omitted from analysis.

In 2010, we add a quinmensal, an about 5-month cycle.

(A)

THE SPECTRUM OF STEROIDAL METABOLITES BROADENS AS BIOLOGICAL RESONANCES WITH NEW ENVIRONMENTAL CYCLICITIES ARE RECOGNIZED AND OLD BIOLOGICAL FINDINGS ARE BETTER RESOLVED*

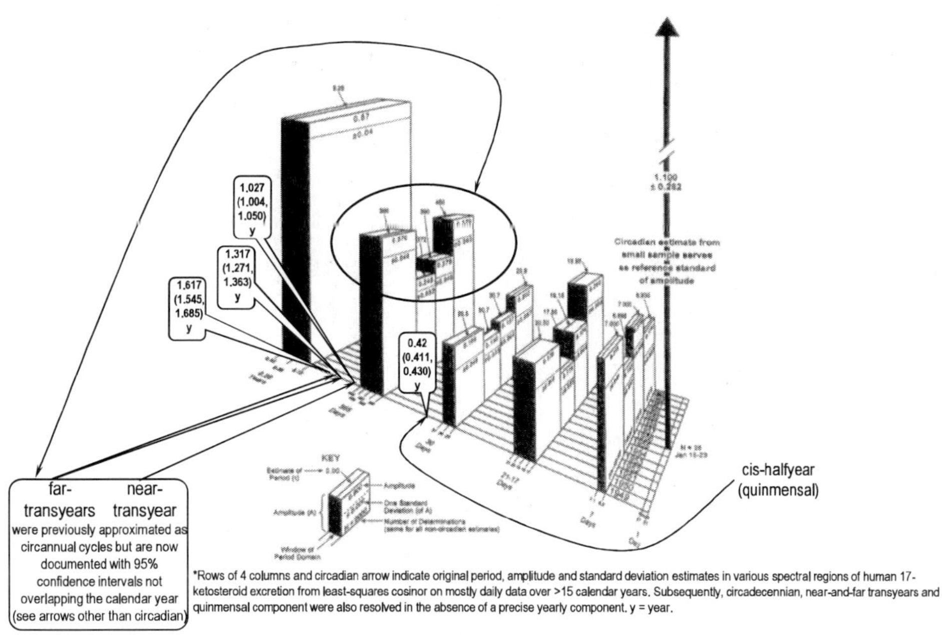

*Rows of 4 columns and circadian arrow indicate original period, amplitude and standard deviation estimates in various spectral regions of human 17-ketosteroid excretion from least-squares cosinor on mostly daily data over >15 calendar years. Subsequently, circadecennian, near-and-far transyears and a quinmensal component were also resolved in the absence of a precise yearly component. y = year.

(B)

Figure 10. (Continued)

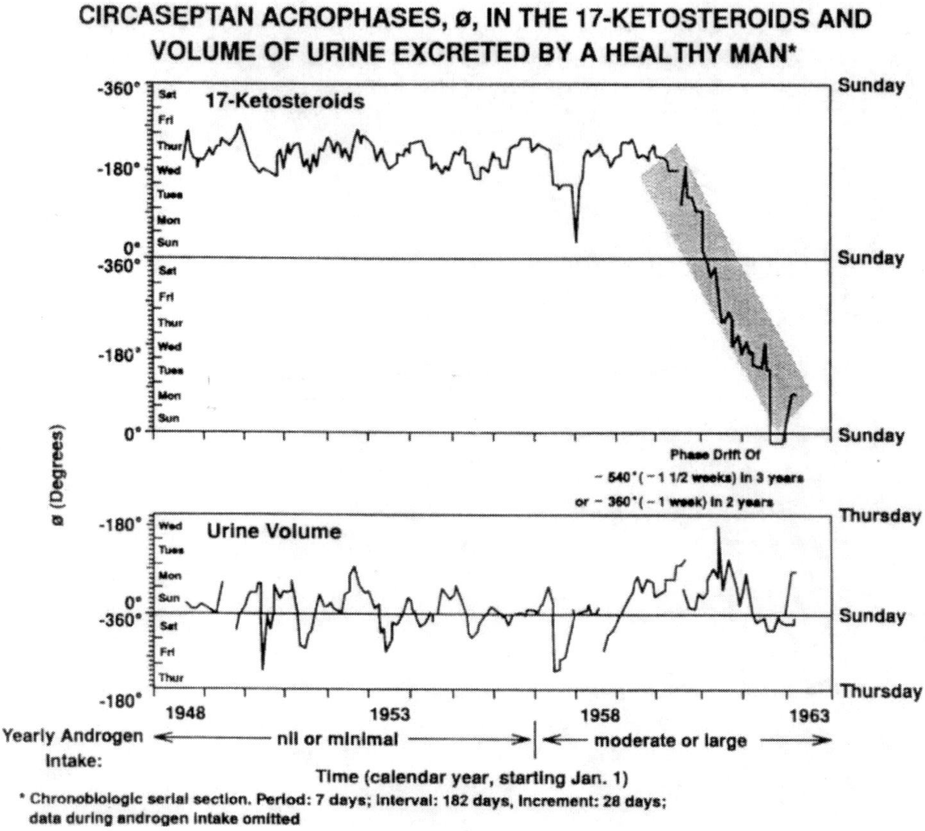

(C)

Figure 10. The spectrum of human 17-ketosteroid excretion was originally (mis)interpreted as having a desynchronized circannual rhythm (A, left), which was reinterpreted as representing a set of transyears [42] (A middle and right), inserted into a broader spectrum (B) replacing the circannual and adding a para-decadal component (B, left). Note that during the last 3 years, the about 7-day rhythm was desynchronized from the societal week (C). © Halberg.

The BEL, also found in environmental temperatures [17], is certainly pertinent to discussions of global warming [17, 22]. A pertinent study thus far has considered solar cycle length without its cyclicity [34]. Ignoring the variability of different solar cycles, such as that of the decadal sunspot cycle, can yield controversial results [36] while their assessment opens new perspectives, notably in economic and military-political affairs [17; cf. 31].

The findings in Figure 4 mimic, for different variables of one individual, RBS, a different assortment with prominent environmental cycles. Moreover, Figures 7 and 8 and Table 3 suggest, apparently for the first time, the possibility of a different assortment with reciprocal environmental periods of the M and the 2A of the same variable in the same person, FH or YW.

Any extrapolation to periods longer than that covered by the data series is tentative, but may be sought further in individuals' longitudinal blood pressure and heart rate monitoring by repeated passes over the accumulating data. Definitive conclusions await extension of the individual's time series to replicated cycles and to groups of individuals studied with lifetime outcomes, yet RBS's now ~43-year record with, on the average, 5-6 measurements/day is unique in terms of combined density and length. Unquestionably, the assumption of baselines for the interpretation of data on aging is unwarranted [36], notably when the use of this term serves to justify a single sample taken before the start of an intervention as the reference standard in aging. Table 2 compares the double amplitude of a trans-tridecadal cycle in RBS, in his current ~43.5-year series, with that of the circadian component. This double amplitude is seen to be almost as large (~90%) as the circadian in the same data, the latter clipped by the limitation to, on the average, 5-6 values/day only during waking. The presence of many near-decadal periods, published when the series covered ~31 years [31], and again in the current series of 43.5 years, Table 2, should prompt determining circadecennian cycle lengths in each variable investigated as a reference standard for investigations of aging.

It must also be realized that these reference standards evolve with age. That some of the variables show a period length corresponding to a global near-17-year solar activity cycle deserves further scrutiny on RBS and others. The change in τ, perhaps with age, was first realized when the addition of a few years' data to a time series transformed a calendar year component to a near-transyear, Figure 2.

As cited by Sigel [37], Chizhevsky presented a circadecadal sunspot period by summarizing the incidence patterns of cholera over a 100-year span, a finding supported by a cross-correlation [38], Figure 9. The data of Figure 5 are, with those of Chizhevsky, the background for those presented herein which extend decadals and multi-decadals to four men with a related environmental-physiological cycle pairing for at least one variable in each. Figures 7 and 8 carry these selective pairings to different circadian characteristics of at least one variable of the same two men. Santorio Santorio knew the value of long-term records, discovering insensible perspiration by their use [39, 40] but had no computers to analyze the uncertainties of periods that he may have noted in his also decades-long series of measurements (that we could thus far not track down, intensive inquiries notwithstanding). That para-decadal τs also apply to human metabolism is suggested by WRB's data on body weight, Table 1, and by a para-decadal spectral component in the excretion of human steroidal metabolites by a clinically healthy man [35; cf. 41], Figure 10. Editors of journals should notice that there are no "baselines", only reference cycles to be taken into account by scholars of aging, of history and of all other human affairs.

REFERENCES

[1] Egeson C. *Egeson's weather system of sun-spot causality: being original researches in solar and terrestrial meteorology.* Sydney: Turner and Henderson; 1889. 63 pp.

[2] Halberg F, Cornélissen G, Bernhardt K-H, Sampson M, Schwartzkopff O, Sonntag D. Egeson's (George's) transtridecadal weather cycling and sunspots. *Hist. Geo Space Sci.* 2010; 1: 49-61.

[3] Bacon F. LVIII. Of Vicissitude of Things. In: Essays, Civil and Moral (1597), included in Eliot CW. (Ed.) *The Harvard Classics.* New York: P.F. Collier and Son; 1909. p. 144.

[4] Brückner E. *Klimaschwankungen seit 1700 nebst Beobachtungen über die Klimaschwankungen der Diluvialzeit. Wien und Olmütz: E. Hölzel*; 1890. 324 pp. (Penck A, Hrsg. Geographische Abhandlungen, Band IV.)

[5] Lockyer WJS. The solar activity 1833-1900. *Proc. Roy Soc Lond* 1901; 68: 285-300.

[6] Schwabe H. Sonnen-Beobachtungen im Jahre 1843. *Astronomische Nachrichten* 1844; 21: 254-256 (no. 495).

[7] Thiele ThN. De Macularum Solis antiquioribus quibusdam observationibus Hafniae institutis. *Astronomische Nachrichten* 1859; 50: 259-261.

[8] Johnson MJ. [Address delivered by the President, M.J. Johnson, Esq., on presenting the Medal of the Society to M. Schwabe.] *Monthly Notices*, Royal Astronomical Society 1857; 17: 126-132.

[9] Liznar J. Über die 33-jährige Periode der Sonnenflecken, *Meteorologische Zeitschrift* 1902; 19: 237-238.

[10] Hoyt DV, Schatten KH. *The Role of the Sun in Climate Change.* New York/Oxford: Oxford University Press; 1997. 279 pp.

[11] Stehr N, von Storch H. (Eds.) (Stehr B, Gamlin G, trans*). Eduard Brückner: the sources and consequences of climate change and climate variability in historical times.* Dordrecht/Boston: Kluwer Academic Publishers; 2000. 338 p.

[12] Stehr N, von Storch H. *Eduard Brückner's ideas -- relevant in his time and today.* Geesthacht: GKSS-Forschungszentrum Geesthacht GmbH; 2006. XXXV and 393 pp.

[13] Hale GE. Sun-spots as magnets and the periodic reversal of their polarity. *Nature* 1924; 113: 105-112.

[14] Makarov VI, Sivaraman KR. New results concerning the global solar cycle. *Solar Physics* 1989; 123: 367-380.

[15] Halberg F, Cornélissen G, Schack B, et al. Blood pressure self-surveillance for health also reflects 1.3-year Richardson solar wind variation: spin-off from chronomics. *Biomed & Pharmacother* 2003; 57 (Suppl 1): 58s-76s.

[16] Cornélissen G, Masalov A, Halberg F, et al. Multiple resonances among time structures, chronomes, around and in us. Is an about 1.3-year periodicity in solar wind built into the human cardiovascular chronome? *Human Physiology* 2004; 30 (2): 86-92.

[17] Halberg F, Cornélissen G, Sothern RB, Czaplicki J, Schwartzkopff O. Thirty-five-year climatic cycle in heliogeophysics, psychophysiology, military politics, and economics. Izvestiya, Atmospheric and Oceanic Physics 2010; 46 (7): 844-864. (Backtranslation from *Geophysical Processes and Biosphere* 2009; 8 [2]: 13-42.)

[18] Halberg F, Cornélissen G, Wilson D, et al. Chronobiology and chronomics: detecting and applying the cycles of nature. *Biologist* 2009; 56 (4): 209-214.

[19] Cornélissen G, Halberg F, Breus T, et al. Non-photic solar associations of heart rate variability and myocardial infarction. *J. Atmos Solar-Terr Phys.* 2002; 64: 707-720.

[20] Halberg F, Cornélissen G, Katinas G, et al. Chronobiology's progress: season's appreciations 2004-2005. Time-, frequency-, phase-, variable-, individual-, age- and site-specific chronomics. *J Appl. Biomed.* 2006; 4: 1-38. *http://www.zsf.jcu.cz /vyzkum/jab/4_1/halberg.pdf*

[21] Cornélissen G, Grambsch P, Sothern RB, Katinas G, Otsuka K, Halberg F. Congruent biospheric and solar-terrestrial cycles. *J. Appl. Biomed.* 2011; 9: 1-40. DOI 10.2478/v10136-0 09-0023-7.

[22] Halberg F, Cornélissen G, Czaplicki J, Prabhakaran Nayar SR, Siegelova J. Brückner-Egeson-Lockyer (BEL) climate cycle in original Brückner's, Lockyer's and follow-up data. In: Halberg F, Kenner T, Fiser B, Siegelova J. (Eds.) *Proceedings, Noninvasive*

Methods in Cardiology, Brno, Czech Republic, October 4-7, 2008. p. 74-89. *http://web.fnusa.cz/files/kfdr2008/sbornik_2008.pdf*

[23] Cornélissen G, Prabhakaran Nayar SR, Czaplicki J, et al. Brückner-Egeson-Lockyer (BEL) cycle in heliogeomagnetics. In: Halberg F, Kenner T, Fiser B, Siegelova J. (Eds.) *Proceedings, Noninvasive Methods in Cardiology*, Brno, Czech Republic, October 4-7, 2008. p. 106-115. *http://web.fnusa.cz/files/kfdr2008/sbornik_2008.pdf*

[24] Sothern RB, Katinas GS, Fiser B, Siegelova J, Cornélissen G, Halberg F. A transtridecadal cycle in human heart rate: Selective infradian, notably multidecadal solar-physiologic BEL congruences. In: Halberg F, Kenner T, Fiser B, Siegelova J. (Eds.) *Proceedings, Noninvasive Methods in Cardiology,* Brno, Czech Republic, October 4-7, 2008. p. 204-213. http://web.fnusa.cz/files/kfdr2008/sbornik_2008.pdf

[25] Cornélissen G, Best WR, Kanabrocki EL, Halberg F. Transyears complement or replace a calendar-yearly spectral component in human physiology. *Proceedings, International Conference on the Frontiers of Biomedical Science: Chronobiology,* Chengdu, China, September 24-26, 2006, p. 130-131.

[26] Halberg F, Cornélissen G, Best WR. A transtridecadal BEL cycle in human blood pressure and body weight. In: Halberg F, Kenner T, Fiser B, Siegelova J. (Eds.) *Proceedings, Noninvasive Methods in Cardiology*, Brno, Czech Republic, July 7-10, 2009. p. 257-270. *http://web.fnusa.cz/files /kfdr2009/sbornik_2009.pdf*

[27] Halberg F. Chronobiology: methodological problems. *Acta med rom* 1980; 18: 399-440.

[28] Cornélissen G, Halberg F. Chronomedicine. In: Armitage P, Colton T. (Eds.) *Encyclopedia of Biostatistics,* 2nd ed. Chichester, UK: John Wiley and Sons Ltd; 2005. p. 796-812.

[29] Refinetti R, Cornélissen G, Halberg F. Procedures for numerical analysis of circadian rhythms. *Biological Rhythm Research* 2007; 38 (4): 275-325. *http://dx.doi.org/10.1080/09291010600903692*

[30] Marquardt DW. An algorithm for least-squares estimation of nonlinear parameters. *J. Soc. Indust. Appl. Math* 1963; 11: 431-441.

[31] Halberg F, Cornélissen G, Otsuka K, et al. Cross-spectrally coherent ~10.5- and 21-year biological and physical cycles, magnetic storms and myocardial infarctions. *Neuroendocrinol. Lett.* 2000; 21: 233-258.

[32] Cornélissen G, Halberg F, Sonkowsky R, et al. Meta-analysis of Horrebow's and Schwabe's scholarship with a view of sampling requirements. In: *Proceedings, Noninvasive Methods in Cardiology*, Brno, Czech Republic, July 7-10, 2009. p. 141-158. *http://web.fnusa.cz/files/kfdr2009/sbornik_2009.pdf*

[33] Cornélissen G, Halberg F, Wendt HW, et al. Resonance of about-weekly human heart rate rhythm with solar activity change. *Biologia* (Bratislava) 1996; 51: 749-756.

[34] Friis-Christensen E, Lassen K. Length of the solar cycle: an indicator of solar activity closely associated with climate. *Science* 1991; 254: 698-700.

[35] Halberg F, Cornélissen G, Watanabe Y, et al. Near 10-year and longer periods modulate circadians: intersecting anti-aging and chronoastrobiological research. *J. Gerontol. A Biol. Sci. Med. Sci.* 2001; 56: M304-M324.

[36] Halberg F, Sothern RB, Cornélissen G, Czaplicki J. Chronomics, human time estimation, and aging. *Clinical Interventions in Aging* 2008; 3 (4): 749-760. *http://www.dovepress.com/articles.php?article_id=2608*

[37] Sigel F (Dreier W, Lerche D, Übers.; Göring H, Wissenschaftl. Red. der deutschsprachigen). *Schuld ist die Sonne. Thun/Frankfurt am Main: Harri Deutsch*; 1979. 215 pp.

[38] Halberg F, Cornélissen G, Otsuka K, Katinas G, Schwartzkopff O. Essays on chronomics spawned by transdisciplinary chronobiology: Witness in time: Earl Elmer Bakken. *Neuroendocrinol. Lett.* 2001; 22: 359-384.

[39] Santorio S. De *Statica Medicina*. Hagae-Comitis: ex typographia A. Vlaco; 1657.

[40] Fidanza F. The "De Medicina Statica" by Santorio Santorio. *Nutrition* 1989; 5 (4): 226-228.

[41] Halberg F, Engeli M, Hamburger C, Hillman D. Spectral resolution of low-frequency, small-amplitude rhythms in excreted 17-ketosteroid; probable androgen induced circaseptan desychronization. *Acta endocrinol.* (Kbh) 1965; 50 (Suppl 103): 5-54.

[42] Halberg F, Cornélissen G, Schwartzkopff O. Quo vadis chronomics 2008: Measuring variability in us, among us and around us. In: Halberg F, Kenner T, Fiser B, Siegelova J. (Eds.) *Proceedings, Noninvasive Methods in Cardiology*, Brno, Czech Republic, October 4-7, 2008. p. 16-25. *http://web.fnusa.cz/files/kfdr2008/sbornik_2008.pdf*

In: Cardiovascular Health and Chronomics
Editor: Ram B. Singh

ISBN: 978-1-62948-976-6

Chapter 4

TIME STRUCTURES (CHRONOMES) OF THE BLOOD CIRCULATION, POPULATIONS' HEALTH, HUMAN AFFAIRS AND SPACE WEATHER*

Franz Halberg[1]*, Germaine Cornélissen[1], Rollin McCraty[2], Jerzy Czaplicki[3,4] and Abdullah A. Al-Abdulgader[5]

[1]Halberg Chronobiology Center, University of Minnesota, Minneapolis, MN, US
[2]Institute of HeartMath, Boulder Creek, CA, US
[3]CNRS, Institut de Pharmacologie et de Biologie Structurale, Toulouse, France
[4]University of Toulouse, Toulouse, France
[5]Prince Sultan Cardiac Center, Al Ahasa, Saudi Arabia

ABSTRACT

We complement an "Extended Consensus" [1], focused on the individual's health, by documenting associations between the human heart and circulation and the environment near and far at the level of local and global populations in both space and time [2, 3]. Life in the atmosphere of the sun [4] mimics its inanimate surroundings; its rhythmic components nearly match in length the cycles of the cosmos [5-15]. Longitudinal records of the human blood circulation reveal heretofore unrecognized aspects of its variability. The view from a very few long and relatively dense series, covering mostly around-the-clock (up to) decades on individuals, reveals variability that cannot readily or necessarily be found in hundreds of thousands of individuals' around-the-clock 24-hour spotchecks and certainly not in single samples, unless the individuals' cycles are synchronized and assessed on a population basis.

Periods, τ, longer than 28 hours, that is infradian variations, like the circadians (with τ tentatively between 20 and 28 hours), have entered the genome and their characteristics await mapping. For those cycles that are already mapped, we consider further the extent to which the parameters and nonparametric endpoints of healthy infradian and circadian variability and their anomalies can become everyday measures of Sir William Osler's

* Update on chronomics, the first in a series of presentations at the 3rd International Conference on Advanced Cardiac Sciences "King of Organs 2010" (September 27-30, 2010, Hofuf, Saudi Arabia) from the Halberg Chronobiology Center at the University of Minnesota, Minneapolis, MN, USA

wear and tear, i.e., of Hans Selye's and Paul Rosch's stress and strain to the point of becoming gauges of distress (again complementing or replacing single samples or the mean values of dense sampling during single days). Criteria for recognizing and quantifying the alteration indicating when a load changes from benetensive to maletensive have yet to be identified, perhaps by monitoring blood pressure and heart rate in everyday life, replacing spotchecks in the clinic.

The probable role of cosmos-induced loads in pathogenesis, assessed, among others, as an effect of a magnetic storm in laboratory animals [16-18] and in humans [19], awaits study by using superposed epochs and a subtraction and addition approach [2, 3].

1. Preamble

We set the stage for Paul Rosch, who has devoted a lifetime to problems of stress, as Hans Selye and many others did, without concern for cycles. Rosch also wrote a scholarly treatise on electromagnetism [20], although again a novel spectrum of cycles was not in his focus. Cycles, however, underlie both stress and electromagnetism, much more than is apparent in pertinent papers. Magnetic fields, like electric ones, constitute a usually ignored yet ever-present and greatly varying background to life; they can represent a time-varying stress which can accumulate to unacceptable strain that can eventually harm. By resolving the many aeolian (i.e., nonstationary) frequencies involved, chronobiology, the study of time structures (chronomes) in the biosphere and chronomics, analyzing interactions between internal and external chronomes, are each tools for examining the "subtle energies" that act frequency-dependently and that, from a circadian and lunar viewpoint, were of great interest to Frank A. Brown Jr. [21] and Franklin Barnwell [22] and are currently in the focus of Rollin McCraty, who with Rosch prompted our participation at the meeting; we are indebted to both.

We are at a site where a geomagnetic recording station, tuned to the resonances in the ionosphere and geomagnetic field line resonances, was just installed by Abdullah al-Abdulgader and Rollin McCraty, which may complement (for a unified science) the systematic monitoring that Gauss, Humboldt and Sabine started in the nineteenth century to measure geomagnetism worldwide, Figure 1A.

Stations yet to be built and aligned on a global scale for the study of congruences among periods in and around us, as one of several approaches used by us, constitute an indispensable step toward an inferential statistical cross-spectral (as well as humanistic) assessment of coherence. In this context, Saudi Arabia could provide a model for the world. A study designed to investigate correspondences in the activity of biosphere with those found in long-term recordings of heart rate variability and blood pressure is planned, which in addition to the geomagnetic and solar indicators intends to include a comparison with the locally collected Schumann Resonance data.

Arabic medicine during the Islamic Golden Age (c. 9th century-c. 16th century CE) has already generated Ibn al-Nafis'[1], Figure 1B, experimental approach to physiology [23, 24]. His endeavor "in determining the use of each organ" led him to discover the pulmonary circulation. His motto, "[W]e shall rely necessarily on verified examination, disregarding

[1] This paper is dedicated to the memory of Ibn al-Nafis (1213-1288), who discovered the movement of blood from the right to the left ventricle of the heart via the lungs. George Sarton (1884-1956), the father of the history of science, wrote: "... his importance will increase enormously, for he must be considered one of the main forerunners of William Harvey and the great physiologists of the Middle Ages" [24].

whether our opinions will agree or disagree with those of our predecessors", is timely, applying to those who wish to replace a spotcheck medicine that flies blind between office visits, to examine each variable, beginning with heart rate and blood pressure, longitudinally. In al-Nafis' spirit, in this forum we describe some of the reciprocal cycles others and we encountered in and around us, also alluding to ongoing studies and problems. While their amplitudes can be relatively small, these cycles constitute potential tipping factors leading to several hard events, including sudden cardiac death and suicide by individuals as well as crime and terrorism in populations.

2. METHOD

Table 1. Comparison of Methods Applied to Noisy Test Series with Two Components with Close Periods -- Only Two Approaches (7 and 8) Resolve Both Components† $Y_i = 100 + 10 \cos(2\pi t/24 - \pi) + 2 \cos(2\pi t/24.8 - \pi) + 5\,R$, where i = 1, …, 336 ($\Delta t$=1 hour; T=14 days) and R is uniformly distributed with zero mean and range =1 (-0.5 < value < +0.5)

Method		Period (h)#	MESOR	Amplitude	Acrophase (degrees)	Detection		N of components detected
N	Kind					Test	P	
1#	"Cosinor"-SC[1]	24.00	100.18	10.47	-98	F(2,333) = 2951.51	<0.001	1
2#	Cosine_fit	24.00	100.18	11.53	-98	GoodFit = 0.001	<0.001	1
3*	Fourier	23.93					<0.001	1
4*	Lomb-Scargle	24.00				Robustness = 96%	<0.001	1
5*	Enright	24.00				Robustness = 96%	<0.001	1
6	ARIMA.MLE[2] in SPLUS§	24.00	100.15 (99.8,100.5)	10.45 (10.11,10.80)	-96.5(-94.5,-101)		<0.001	>12
7a	Linear step 1 (L)[3]	24.00	100.18 (99.98,100.37)	10.47 (10.20,10.74)	-98(-97,-100)	PR=95%	<0.001	
7b	Nonlinear step 2 (NL)	23.97 (23.83,24.10)	100.13 (99.83,100.43)	9.75(7.67,11.84)	-91(-78,-105)		<0.05@	
		24.63 (24.04,25.21)		2.49(0.34,4.63)	-98(-47,-149)		<0.05@	2
8	Simulated Annealing (GOSA software)[4]	23.96 (23.89,24.03)	100.4 (100.1,100.7)	9.6(8.4,10.8)	-94(-83,-105)		<0.05	
		24.6(24.3,24.9)		2.5(1.3,3.7)	-94(-60,-128)		<0.05	2
9	Wavelets	24.00	Time evolution noted but not quantified by point-and-interval estimates.					

Acrophase expressed in (negative) degrees, with 360°≡period length, and 0° set to 00:00 at start of series. # Period is anticipated trial period. * Only one peak detected.

GoodFit: Goodness of Fit. @ From non-overlap of zero-amplitude by 95% Confidence Interval (CI).

† Items 1-8 from Refinetti R, Cornélissen G, Halberg F. Procedures for numerical analysis of circadian rhythms. Biological Rhythm Research 2007; 38 (4): 275-325. http://dx.doi.org/10.1080/09291010600903692; item 9 courtesy of Prof. S.R. Prabhakaran Nayar, Department of Physics, Kerala University, Kariavattom, Trivandrum, India.

1SC: South Carolina: cosinor restricted to 24-hour component, by contrast to Minnesota's cosinor, which includes least squares fit of multiple components consisting of cosine functions with anticipated components not necessarily equal to 24 hours, originally called 'single cosinor' to indicate analyses of individual time series (and NOT of a single component fit), soon thereafter extended to include the chronobiologic serial section and the combined linear-nonlinear rhythmometry, among others; all procedures involving the least squares fit of COSINe curves often displayed vectORially along polar coordinates and checked against chronograms and data stacked for the period(s) found, i.e., as plexograms, for each cosinor-isolated component (periods are estimated with their respective confidence intervals).

2 Structure of residuals found to be correlated and modeled with an ARMA (1,1) model, with AR coefficient = 0.78 (0.49, 0.89) and MA coefficient = 0.54 (0.23, 0.75).Results reported at fixed anticipated period of 24 hours.

3 Least squares spectrum detects peak at anticipated period of 24 hours and small sidelobes with periods of 26.7 and 21.8 hours resolved nonlinearly as presence of second component with period slightly longer than 24 hours.

4 Czaplicki J, Cornélissen G, Halberg F. GOSA, a simulated annealing-based program for global optimization of nonlinear problems, also reveals transyears. J Applied Biomedicine 2006; 4: 87-93. http://www.zsf.jcu.cz/vyzkum/jab/4_2/czaplicki.pdf

6, 7 and 8: 95% CIs in parentheses; 7a and 7b are part of the more extensive, as need be, MN-cosinor.

§ Chatfield C. The analysis of time series: an introduction. London: Chapman and Hall; 1984.

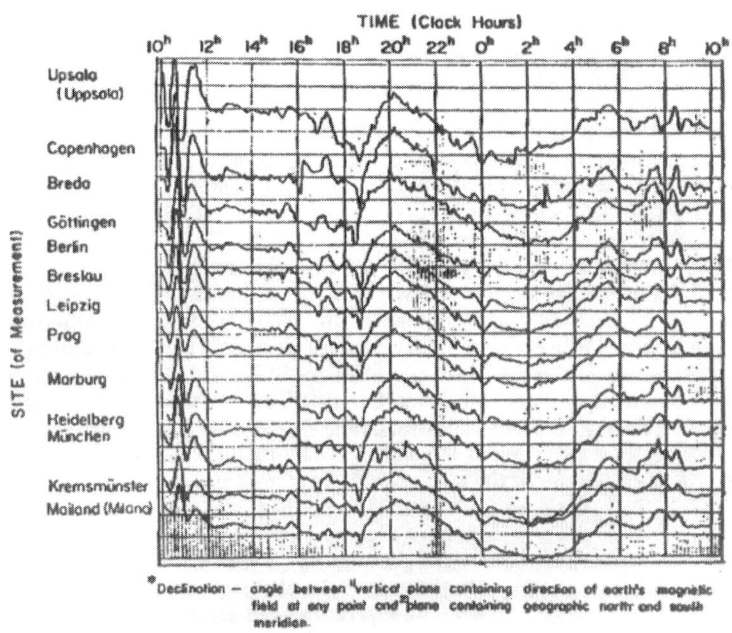

Figure 1A. The "harmony of curves" in itself can definitively be documented but not quantified in instances such as the record presented without any inferential statistics in the pre-computer era by Carl Friedrich Gauss, discoverer of the least-squares method, and Wilhelm Weber in Biermann K-R (Ed.) Briefwechsel zwischen Alexander von Humboldt und Carl Friedrich Gauß: Zum 200. Geburtstag von C.F. Gauß im Auftrage des Gauß-Komitees bei der Akademie der Wissenschaften der DDR (Berlin: Akademie-Verlag; 1977; 202 pp. + plates).

Figure 1C introduces least squares, used by the extended cosinor to get, in the presence of partly rhythmic time series, a usually more accurate or more precise mean in the case of unequidistant data in particular. Furthermore, in addition to the superiority of the MESOR, M, a **m**idline-**e**stimating **s**tatistic **o**f **r**hythm, as compared to an arithmetic mean, the methodology used also yields for each τ amplitudes, acrophases, Figure 2A, and the (A,ϕ) pairs of harmonics as new endpoints used for diagnosis and treatment [25-27]. With appropriate checks for non-stationarities that can yield spurious spectral peaks (Figures 2B-2D), one or several τs can also be resolved by the extended linear-nonlinear cosinor (Figure 2E) and by other methods, Table 1.

Figure 1B. Ibn al-Nafis (1213-1288), a dominant figure of the Islamic Golden Age (c. 9th century-c. 16th century CE) [23, 24].

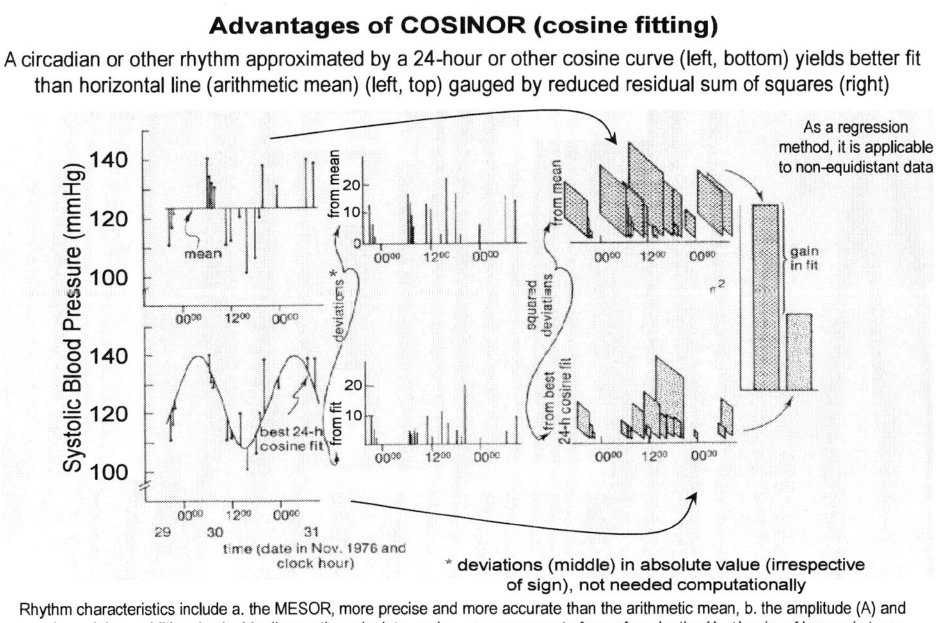

Figure 1C. Gauss' least-squares method in the inferentially extended cosinor [25, 26] provides a usually more accurate and more precise average than the arithmetic mean, a task easily accomplished in the era of computers. © Halberg.

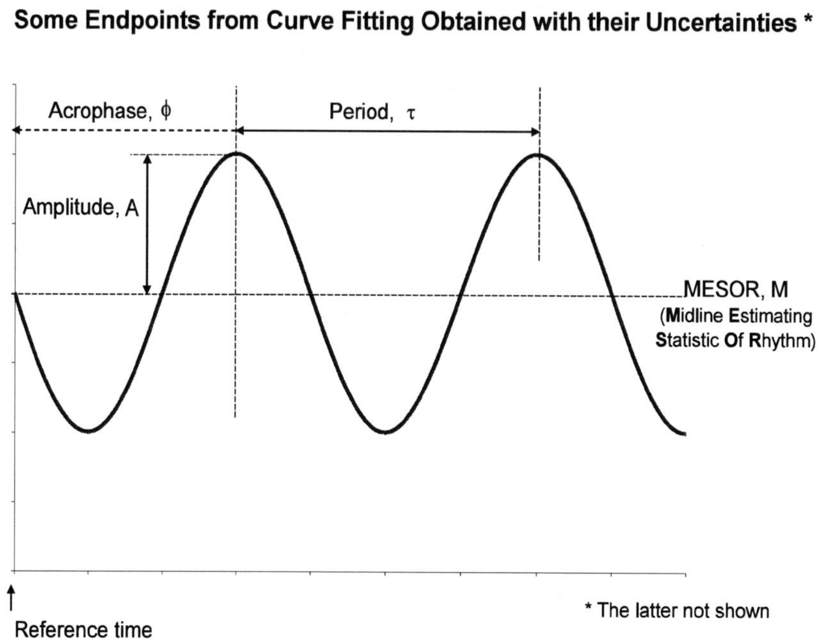

Figure 2A. Diagnostically useful endpoints from the cosinor, each underlying a *v*ascular *v*ariability *a*nomaly, VVA [1]. © Halberg.

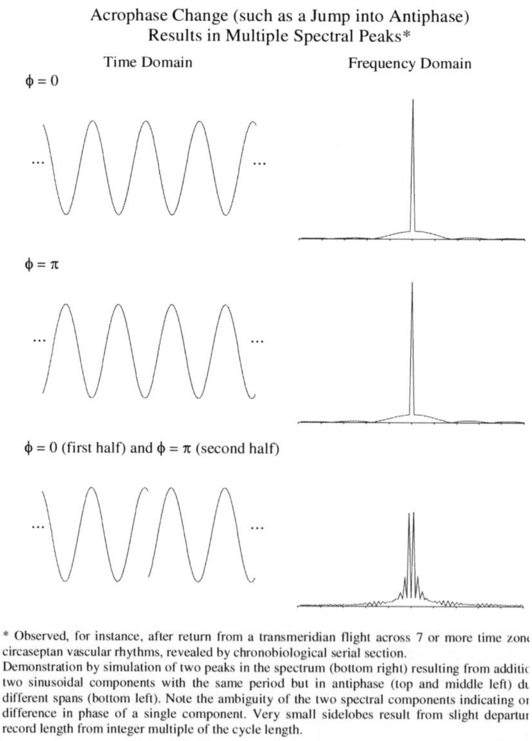

Figure 2B. The presence of a phase jump (e.g., by 180°) of a cosine curve with a fixed period is spuriously associated with two spectral peaks. © Halberg.

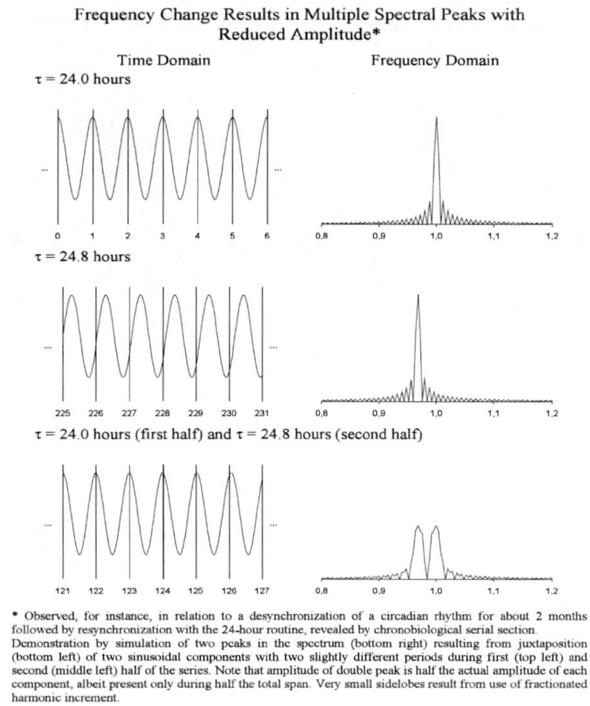

Figure 2C. The occurrence of a change in period length (e.g., from 24.0 to 24.8 hours in mid-course of a 124-day record) of a single cosine component is spuriously associated with two spectral peaks with amplitudes smaller than that of actual signal. © Halberg.

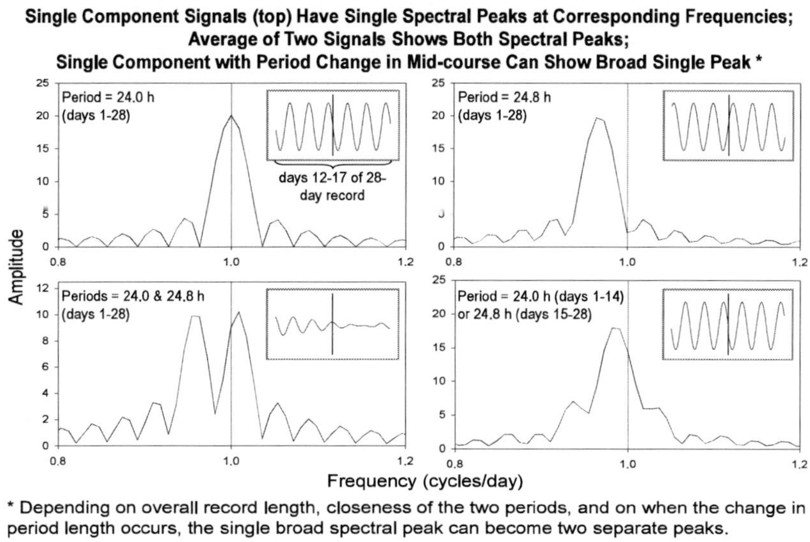

Figure 2D. The occurrence of a change in period length (e.g., from 24.0 to 24.8 hours in mid-course of a 28-day record) of a single cosine component is spuriously associated with a single broad spectral peak with a reduced amplitude and sidelobes. Whether a single broad peak or two spectral peaks are observed depends on the length of the record, the extent of the difference in period length and the time when the switch in period occurs in the record. © Halberg.

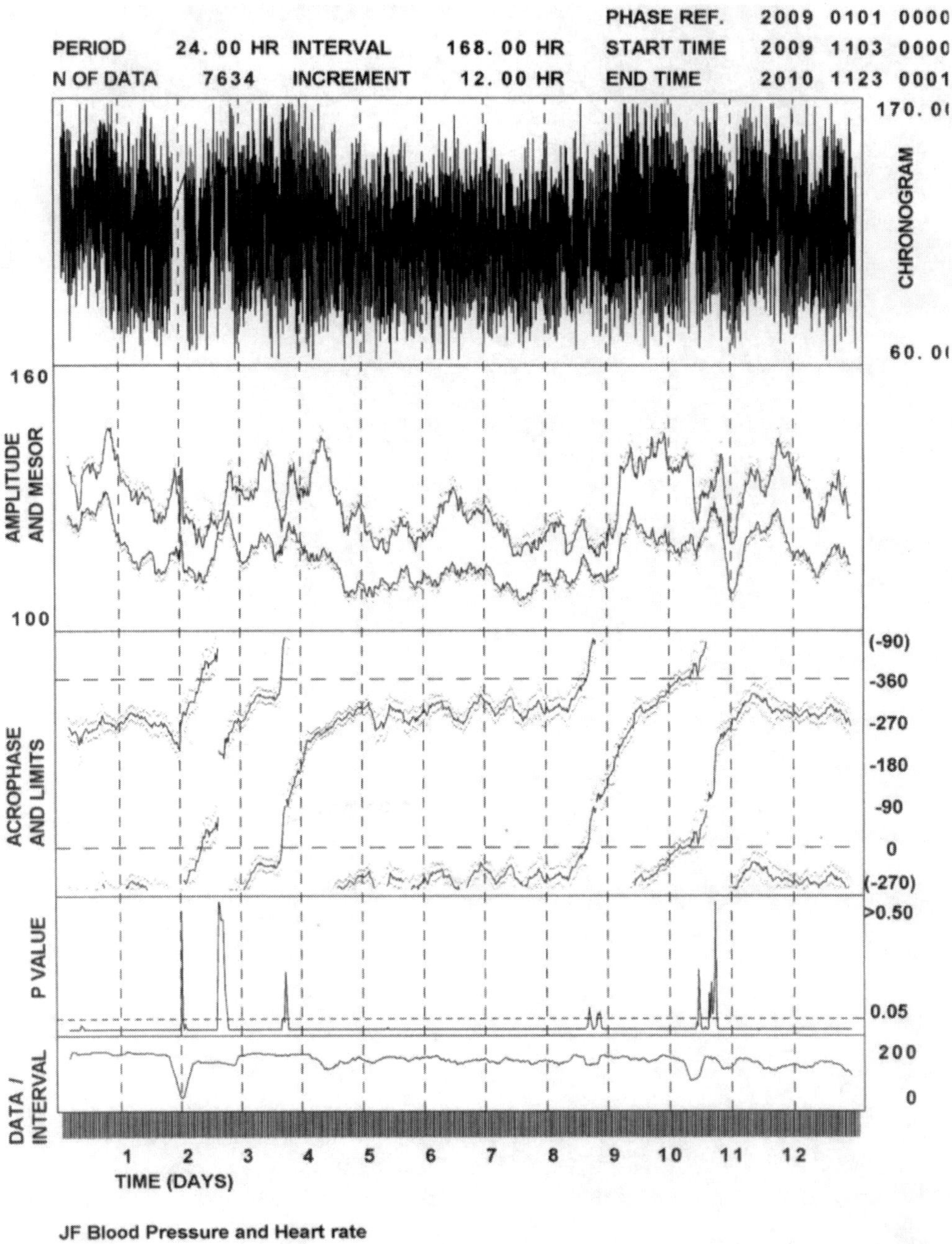

Figure 2E. The systolic blood pressure of JF, a woman 61-62 years of age, suggests alternating 24.0-hour synchronization (up to full moon 2) and circadian desynchronization of acrophases (in row 3) (up to full moon 5), followed by 24.0-hour resynchronization lasting several months, alternating with renewed circadian desynchronization (after full moon 8) and 24.0-hour resynchronization (after full moon 11). This chronobiological serial section shows the dominant behavior but does not reveal the coexistence of two circadian periods, demonstrated in Figure 2F. Vertical dashed numbered lines indicate consecutive full moons. © Halberg.

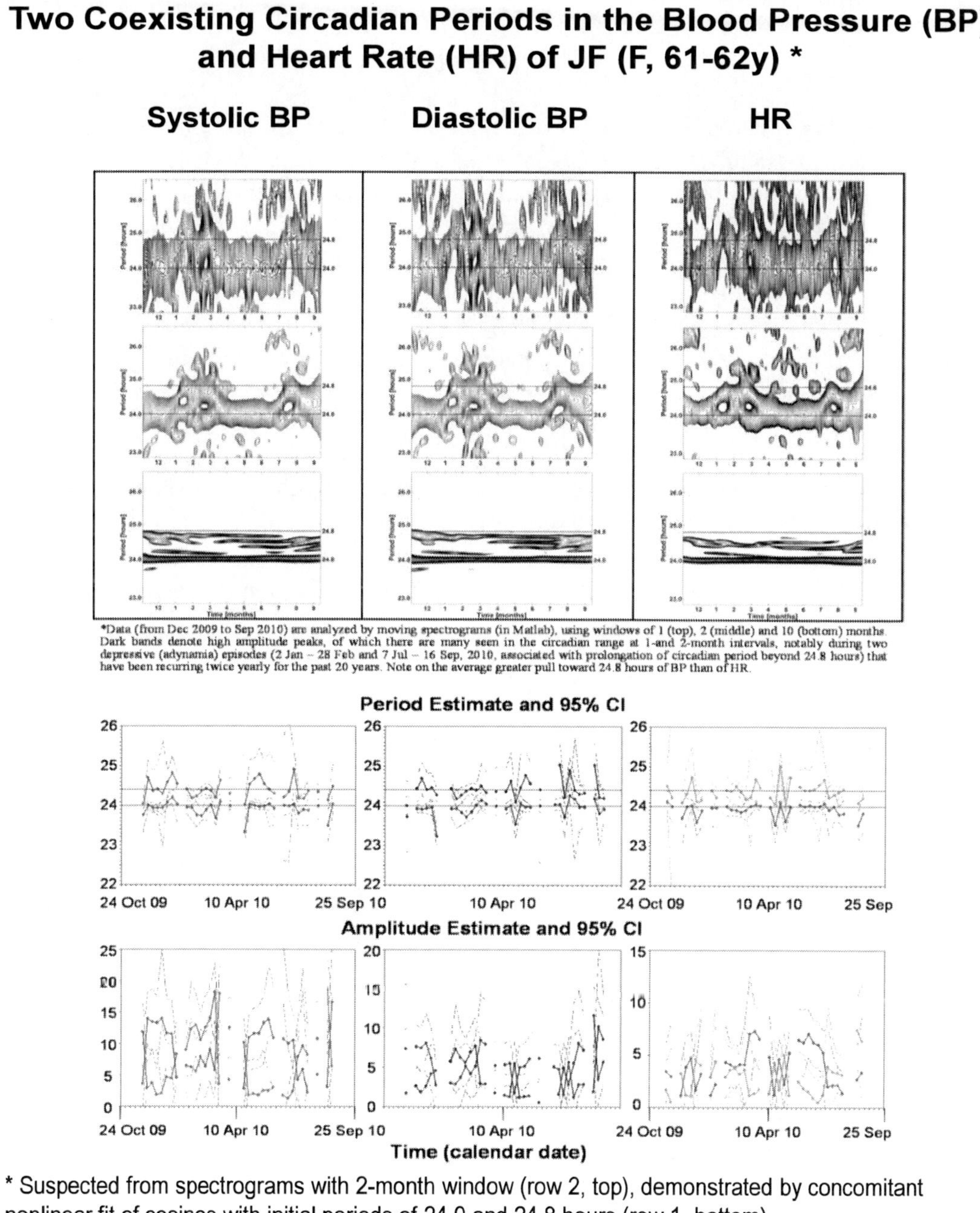

Figure 2F. A change in the relative prominence of 2 coexisting circadian components has been observed in a 61-62-year old woman (JF) who suffers from adynamia episodes lasting 2-3 months and recurring twice a year, using two complementary approaches, namely spectrograms (top section) and nonlinear fit of two components (bottom section). In the third row (top section), components with different frequencies coexist, demonstrating responses to both the solar and lunar day glocally in time, with parameters estimated with their uncertainties in the last two rows. Spectrograms from J Czaplicki. © Halberg.

3. Diverse Cycles in a Unified Science Reinforce Each Other

In 1600 the physician William Gilbert [28] gave physics its first scientific treatise [29]. In 1995, perhaps to return the favor, Juan Roederer [30] asked physicists for "unselfish cooperation" in helping biomedicine so that it can detect "physiological harbingers and their possible correlations with space weather". Elsewhere, Roederer [31] as well as others [2, 3, 32-34] noted cycles in and around us with periods, τ, that are reciprocal by their similar length (not arithmetically). In the interim, old paradecadal (Figure 3A) [35, 36] and novel biospheric spectral components (Figure 3B) [34] mirror environmental quasi-periodicities, among others, in the speed of the solar wind. Some of the τs are shorter than (cis = on this side of) the year or half-year or of other reference periodicities; others are longer than a year (trans = beyond), i.e., they are transyears [2, 3, 33, 34, 36-43]. In the spectrum of the incidence of sudden cardiac death in Minnesota, USA [37, 43], with hot summers and often-harsh winters, transyears can coexist with and even replace, at least for a while, the effect of the seasons, as they also do in Tokyo, Japan [42].

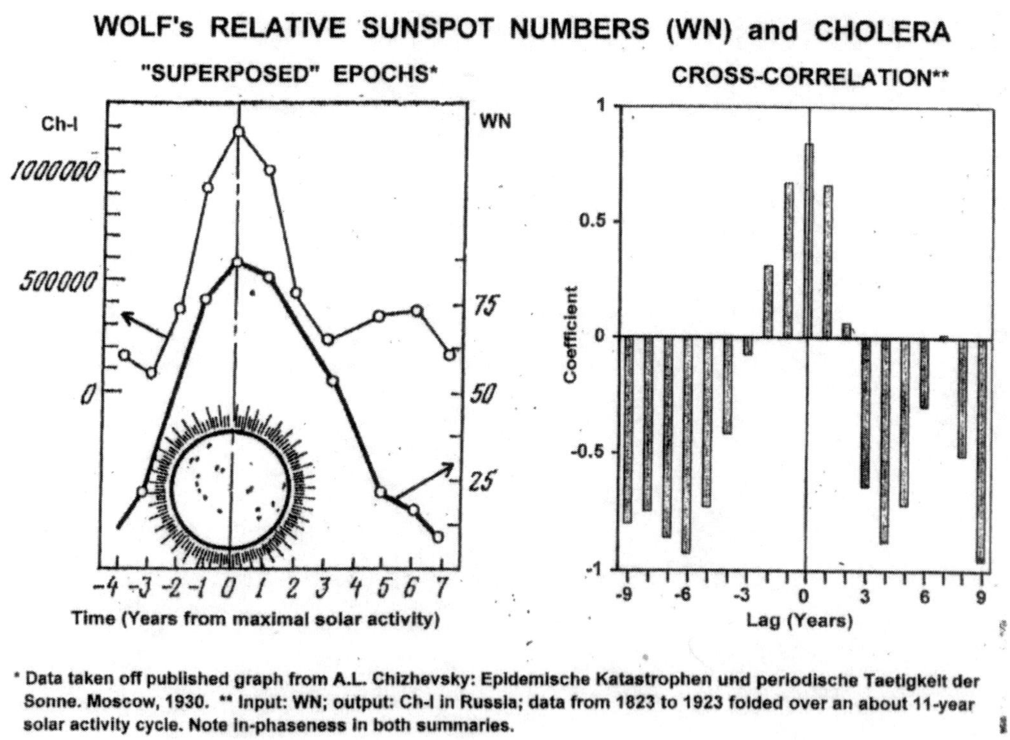

Figure 3A. Cholera in Moscow characterized by a decadal cycle revealed by superposed epochs summarized by Alexander Leonidovich Chizhevsky, according to Sigel [35]. Data from Chizhevsky on cases of cholera recorded in Moscow between 1823 and 1923, originally folded over an ~11-year solar activity cycle, showing a peak incidence coinciding with maximal solar activity (left). Data taken off the published graph, analyzed by cross-correlation, corroborate that the largest correlation corresponds to a lag of zero (right). Should the original 100-year series have been available for analysis, a measure of uncertainty could have complemented the average curves (left). © Halberg.

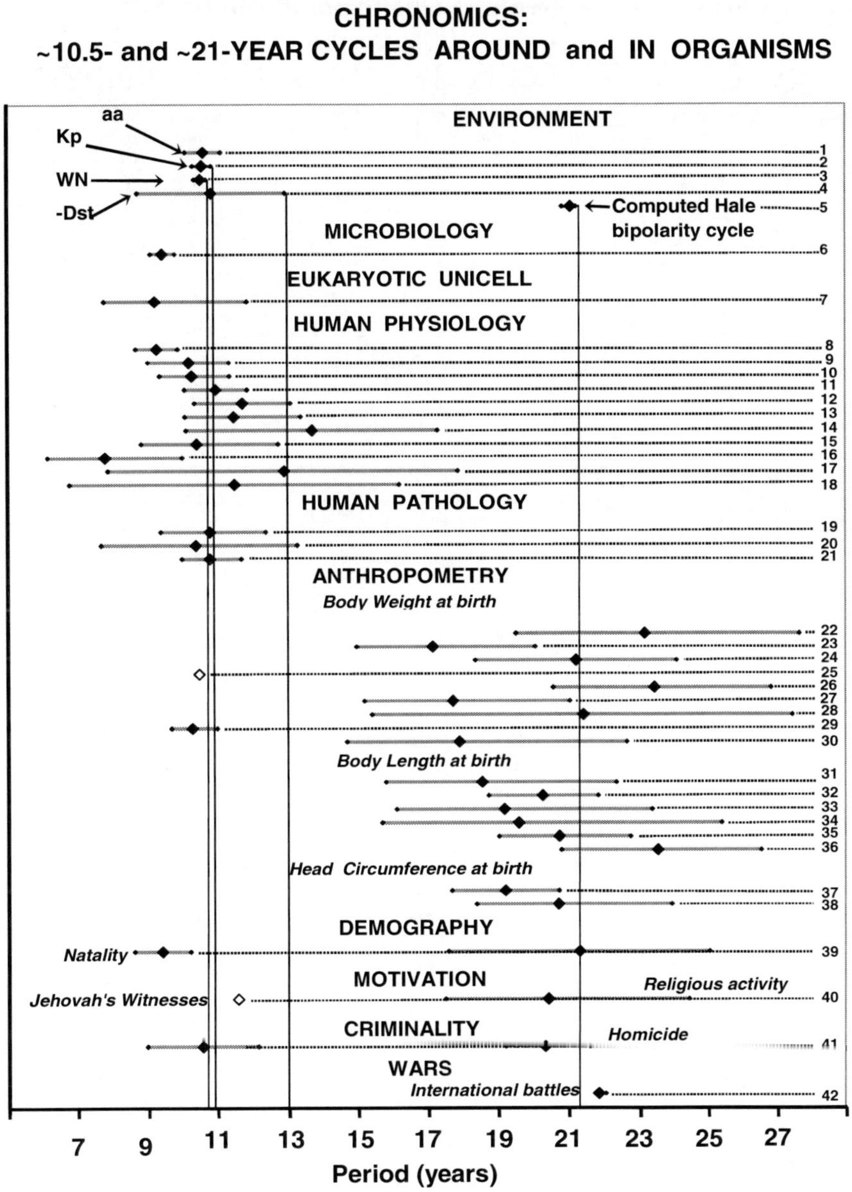

Figure 3B. Paradecadals and paradidecadals, ~10- and ~20-year cycles similar in length to the Schwabe and Hale solar activity cycles characterize a host of biological variables, from micro-organisms to human physiology, pathology, anthropometry and sociology, assessed from population statistics, yet critically dependent upon also-available individual "pilot studies". Such self-experimentation is indispensable for a scrutiny of mechanisms underlying population cycles (see HUMAN PHYSIOLOGY, above, and Figures 2E, 2F, 6-9, 17 and Tables 3 and 4). © Halberg. Desynchronization occurred from both the societal week, to which urine volume remained synchronized, and from a 6.75-day component of the earth's magnetism [33]. The biospheric cycles with novel or forgotten environmental nonphotic counterparts bear on physiology, as shown in Figure 5 for ~5-month cycles (quinmensals or cis-half-years) and on many other human affairs, as in the case of ~10-, ~21- and ~35-year cycles, Figures 3 and 4 [50].

Chronomics: ~10.5- and ~21-year cycles in and around us

	Line				Period (years) Lower limit*	Best Fit	Upper limit*	Series duration Dates	Years	Number of data	Geographic site
Environment	1	aa = Antipodal Geomagnetic Index			10.12	**10.63**	11.13	1890-1999	110	1 / year	
	2	Kp = Planetary Geomagnetic disturbance			10.32	**10.58**	10.85	1932-1999	68	1 / month	
	3	WN = Wolf relative sunspot number			10.37	**10.54**	10.70	1890-1999	110	1 / year	
	4	-Dst = Equatorial geomagnetic disturbance			8.75	**10.85**	12.96	1973-1999	27	"	
					10.48	**10.51**	10.55	1700-1999	300	"	
	5	Bipolarity "Hale Cycle" **			20.86	**21.10**	21.26	1890-1999	110	"	
					21.42	**21.428**	21.43	1700-1999	300	"	
Biology	6	Prokaryotes: Air Bacterial Sectoring			9.12	**9.45**	9.81	1970-1982	13	3,744	Italy
	7	Eukaryotes: Unicellular Algal O_2 Production			7.79	**9.24**	11.87	1980-1991	11	324	Germany
Physiology***	8	Mood (RBS)			10.11	**11.50**	13.41	1966-1998	33	~5 / day	USA
	9	Time (1-Minute) Estimation (RBS)			9.38	**10.29**	11.37	1966-1998	33	"	"
	10	Urinary 17-ketosteroid excretion (CH)			8.70	**9.30**	9.90	1948-1963	15	1 / day	Denmark
	11	Peak Expiratory Flow (RBS)			10.36	**11.74**	13.11	1966-1998	33	~5 / day	USA
	12	Respiratory Rate (RBS)			10.13	**12.50**	17.32	1966-1998	33	"	"
	13	Systolic Blood Pressure - SBP (RBS)			9.05	**10.21**	11.36	1966-1998	33	"	"
	14	Standard Deviation of SBP (YW)			8.85	**10.43**	12.76	1987-1998	11	~48 / day	Japan
	15	Diastolic Blood Pressure - DBP (RBS)			10.09	**10.98**	11.87	1966-1998	33	~5 / day	USA
	16	Standard Deviation of DBP (YW)			6.18	**7.82**	10.02	1987-1998	11	~48 / day	Japan
	17	Heart Rate - HR (YW)			9.54	**12.93**	17.91	1987-1998	11	"	"
	18	Standard Deviation of HR (YW)			8.27	**11.52**	16.22	1987-1998	11	"	"
Pathology	19	Myocardial Infarction			10.00	**10.80**	11.70	1960-1996	37	129,205	USA
	20	Leptospirosis			9.40	**10.80**	12.40	1949-1995	47	2,907	Slovakia
	21	Diabetes			7.70	**10.40**	13.30	1985-1995	11	1,369	"
Anthropo-		***Body Weight***	***Boys***								
metry	22		Minnesota		19.53	**23.19**	27.67	1963-1998	36	2,136,745	USA
at birth	23		Alma-Ata Russians		14.99	**17.17**	20.07	1946-1998	53	9,056	Kazakhstan
	24		"	Kazakhs	18.39	**21.24**	24.05	1946-1998	53	3,459	"
	25		Moscow			**10.49**		1874-1985	112	5,987	Russia
			Girls								
	26		Minnesota		20.58	**23.46**	26.83	1963-1998	36	1,039,464	USA
	27		Alma-Ata Russians		15.21	**17.75**	21.06	1946-1998	53	9,105	Kazakhstan
	28		"	Kazakhs	15.44	**21.45**	27.45	1946-1998	53	3,448	"
	29		Moscow		9.70	**10.29**	11.01	1874-1985	112	5,840	Russia
			Both genders								
	30		Denmark		14.71	**17.94**	22.68	1973-1994	22	1,166,206	Denmark
		Body Length	***Boys***								
	31		Alma-Ata Russians		15.82	**18.58**	22.38	1946-1998	53	9,026	Kazakhstan
	32		Moscow		18.76	**20.28**	21.86	1874-1985	112	5,976	Russia
			Girls								
	33		Alma-Ata Russians		16.13	**19.20**	23.39	1946-1998	53	9,105	Kazakhstan
	34		"	Kazakhs	15.72	**19.60**	25.40	1946-1998	53	3,485	"
	35		Moscow		19.05	**20.76**	22.78	1874-1985	112	5,976	Russia
			Both genders								
	36		Denmark		20.81	**23.55**	26.55	1973-1994	22	1,166,206	Denmark
		Head Circumference									
	37		***Boys***	Moscow	17.71	**19.23**	20.75	1874-1985	112	5,976	Russia
	38		***Girls***	"	18.42	**20.73**	23.95	1874-1985	112	5,820	"
Demography	39	Birth rate			8.63	**9.43**	10.23	1940-1996	57	57	USA
					17.61	**21.33**	25.05	1940-1996			
Motivation	41	Religious activity of Jehovah's Witnesses			17.52	**20.44**	24.45	1950-1999	50	328,572"	Worldwide
										5,653,987""	
Criminality	40	Homicide			8.99	**10.58**	12.16	1900-1998	99	99	USA
					19.23	**20.35**	21.62				
Wars	42	International battles			21.87	**21.96**	22.06	599BC-1957	2556	2556	Worldwide

* 95% confidence limit; not shown if cycle is not statistically significant.
** Computed by changing the sign of WN at each WN minimum.
***RBS - Dr. Robert B. Sothern, CH - Dr. Christian Hamburger, YW - Dr. Yoshihiko Watanabe.
" in 1950, "" in 1999, pool of 103 plus other unspecified number of sites.

Figure 3C. Key to Figure 3B. © Halberg.

These nonphotic cycles bear not only on whether one is unaffected by (versus dies from) an electrical accident of the heart (sudden cardiac death) but also whether death is brought about by one's own hand (suicide) [44-46] or by the action of others (crime, terrorism) [3].

The variable sun, interplanetary space, the atmosphere and earth all undergo a spectrum of non-photic and photic cycles, many of them mirrored in the biosphere. Paraphrasing Roederer [30], the physicists' cooperation may also serve their own discipline by transdisciplinarily supporting the reality of questioned or with notable exceptions [47, 48; cf. 49] forgotten periods of weather in space or on earth that are found in living things [50]. Among others [51], an about (~) 35-year or paratridecadal climate cycle, Figure 4 [50; cf. 52] is a case in point. In addition, the case of an ~5-month or quinmensal cycle, Figure 5 [53-55], was predicted [56] and found first by physicists [57]. Further, in transyears, Figure 6 [37-43; cf. 3], the aeolian behavior of both solar wind speed and systolic blood pressure, and the effect of the former on the latter, are seen.

The photic changes with day and night and the seasons are unmistakable in the entire biosphere: some degree of endogenicity has been documented by modern molecular biology for what the senior author has called circadians over half a century ago [51]. Partly built-in 7-day cycles are also documented by their desynchronization for 3 years -- after prior 7-day societal synchronization for the preceding 12 years, of the urinary 17-ketosteroids, breakdown products of steroidal hormones (essential for survival and reproduction) [58].

Multiple cycles of similar length are said to be congruent by overlying or even only overlapping uncertainties, CIs (95% confidence intervals) of their τs. By not insisting on identical point estimates of their τ, one allows for measurement and other error. Congruences in τ (not necessarily in phase) are also found among biospheric and environmental 7-day periods (circaseptans) and at multiples and submultiples of 7 days (multiseptans), at ~5 (and/or ~6) months, the quinmensals perhaps related to a beat of rotations at different solar latitudes [56; cf. 57] and the semiannuals to geomagnetics [29, 59]. Congruent near-transyears (1.0 year < [τ - CI] < [τ + CI] < 1.2 years) and far-transyears (1.2 years < [τ - CI] < [τ + CI] < 1.9 years) have wide ranges of periods. The ranges of τ depend on the extent of the τ's drifting in time on the one hand and on wide concurrent differences in period length in various geographic locations of the same biospheric reciprocal counterpart to a component of the solar wind's speed such as sudden cardiac death [37, 43] on the other hand. Many biospheric cycles of ~11 or ~22 years (decadals and didecadals), Figure 3, and longer ones approaching or exceeding 30 years (paratridecadals), Figure 4, have their approximate counterparts in sunspots, in variables of the interplanetary magnetic field, Table 2[2], and/or in geomagnetics.

[2] It is always desirable in time series analyses to cover more than several cycles of a given rhythm investigated. Extrapolating from a few cycles or just one, or even from part of a cycle, has indeed led to controversy. Inferences from very extensive data in the USA regarding the lack of an effect of geomagnetics indeed were drawn [86-87, cf. 88] when they did not detect any effect of magnetic storms upon the cardiovascular system based on data covering less than a circadecadal solar cycle. Classic statistical methods were used on very large samples, one totaling an estimated 1 million and another over 14 million cases. The controversy, however, can be readily resolved by the use of chronomics, combining inferential statistical time series analyses with a subtractive and an additive (remove-and-replace) approach on longer series from the USA and elsewhere, once we realize that basing a study on less than a single solar cycle is equivalent to taking the pulse for one second! Nonetheless, there are occasions when, with emphasis on necessary caution, extrapolation as-one-goes becomes desirable before the data cover a full cycle, as in the case of FH when, in ~23 years of data, the output of computer analyses yielded an ~30-year cycle in heart rate that was not part of the input. For another person, RBS, an ~50-year cycle was discovered in ~5-6 self-ratings per day of mood recorded during ~43.5 years, since he was ~20.5 years of age. Waiting for a maximum of 3 covered cycles would mean waiting until RBS reaches 170 years of age (!). This may be highly desirable since he is a motivated, useful scientist, but may not be assuredly possible.

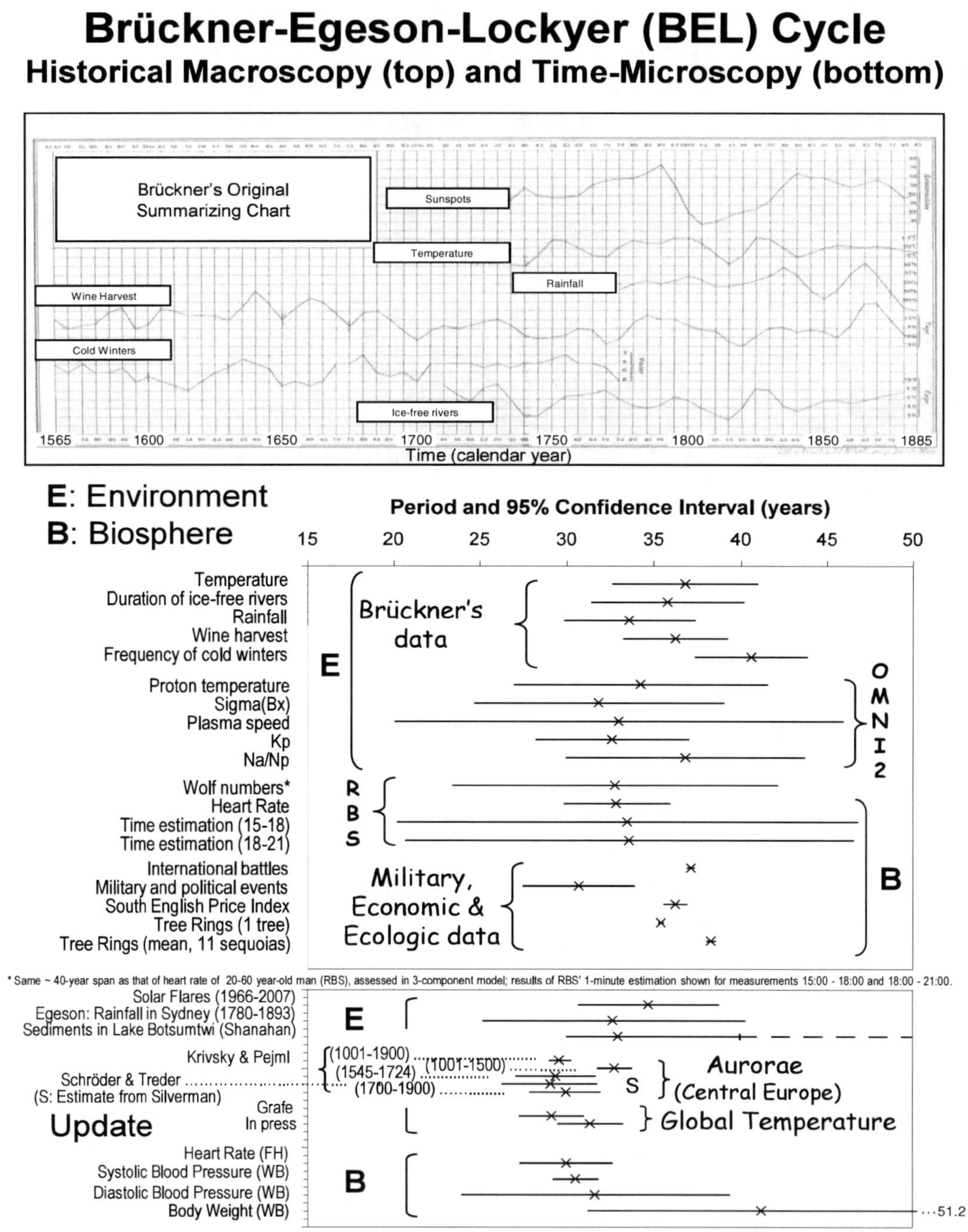

Figure 4. Paratridecadal BEL (**B**rückner-**E**geson-**L**ockyer) cycles, defined by a confidence interval of their period approaching or overlapping the 30-40-year range. © Halberg.

A selective assortment of biospheric-cosmic pairing is found in an individual's heart rate with a transtridecadal environmental cycle and for his blood pressure with an image of a didecadal Hale cycle of sunspot bipolarity, Figure 7.

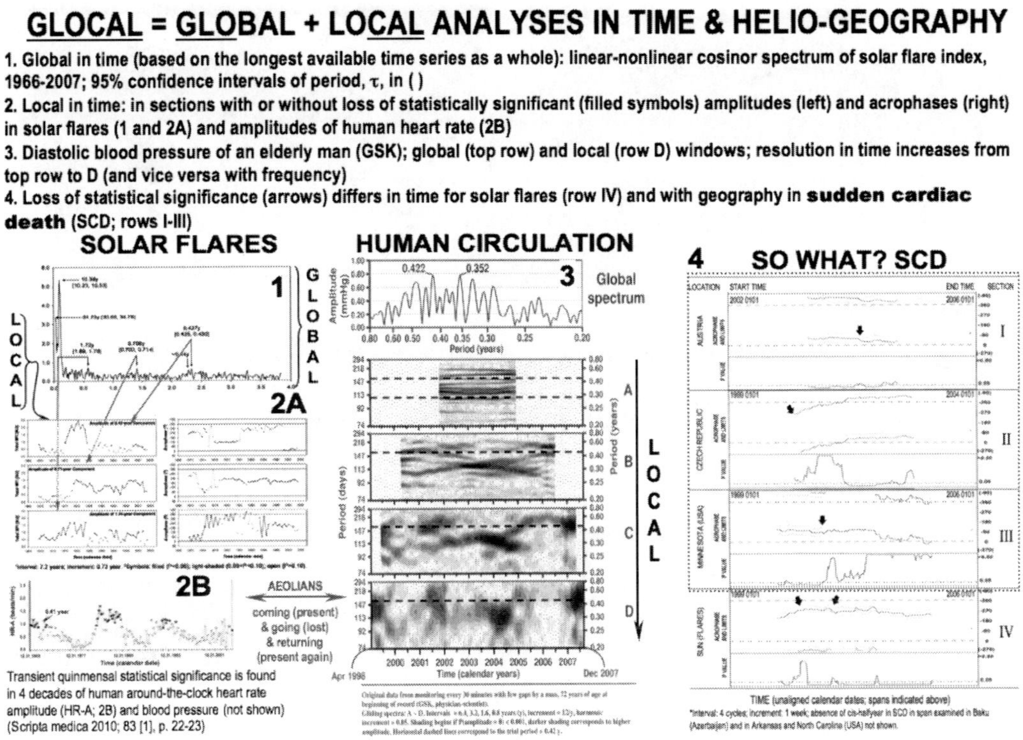

Figure 5. Quinmensals, ~5-month periods predicted by Charles Wolff [56] documented by Rieger et al. [57] in solar flares and found in the human circulation [53], in the excretion of 17-ketosteroids [54] and in the incidence pattern of sudden cardiac death (SCD) [37-43]. © Halberg.

A pull by the geomagnetic planetary index Kp of the heart rate's, but not of the blood pressure's ~7-day (circaseptan) phase [60], constitutes a selective assortment, as does, for the quinmensal, the parallelism of phases in 17-ketosteroid excretion and of Kp [59]. Some statistically significant environmental-biospheric congruences, Tables 3 and 4, are not likely due to chance, Figure 8 [61, 62]; their selective assortment is reminiscent of Mendel's laws of segregation and of the independent assortment of what became genes [63].

Subtractions, Figure 6, from the transdisciplinary spectrum of an environmental ~1.3-year or 7-day component, or their addition or both, Figure 9, may entail corresponding damping (but not loss) or amplification in the biosphere. Such observations validate biospherically the long-term occurrence of an environmental cycle to the point that it became genetically coded [63]. Damped persistence of the biospheric cycle upon removal of the environmental counterpart is hence a transdisciplinary tool in the service of physics, biology and beyond for an eventual unified science. Among others [64], Alexander Leonidovich Chizhevsky [65; cf. 35], Figure 10, Frank A. Brown Jr, Figure 11, Düll and Düll [89] and Stetson [90] foresaw all of this: Chizhevsky found the circadecadal sunspot cycle of ~11 years in military-political affairs [35, 65] and folded cases of cholera in Moscow from 1823 to 1923 over an 11-year cycle; a peak cholera incidence coincided with maximal solar activity, corresponding (in our metaanalysis) to maximal cross-correlation [36], Figure 3A. Breus and Chibisov have updated heliobiology [18, cf. 91].

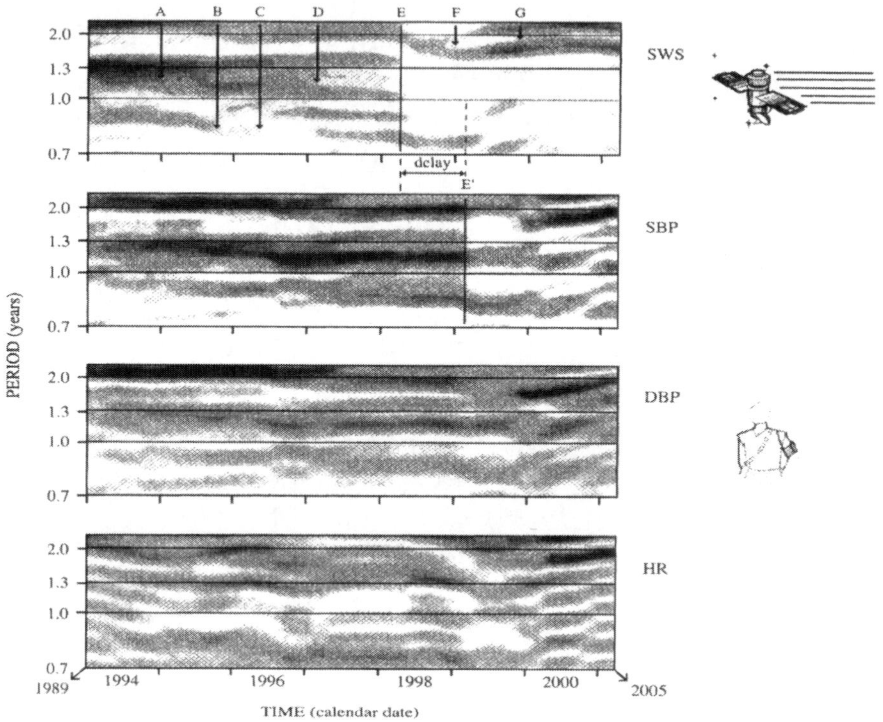

Figure 6. Time courses of the para-annual frequency structures in solar wind speed (SWS) (top) and of an elderly man's (FH) systolic and diastolic blood pressure and heart rate, SBP, DBP and HR (rows 2-4, respectively), examined by gliding spectral windows, constitute an example of the subtractive or "remove" approach. The aeolian behavior shared by SWS and SBP (drifting, bi- or trifurcation, disappearance and reappearance of a frequency and, when present, the waxing and waning of its amplitude) and the fading and narrowing, but persistence of an ~1.3-year component are in keeping with external driving of a built-in (since dampened but persisting) component in the individual. © Halberg.

Table 2. Brückner-Egeson-Lockyer (BEL) cycle in heliogeomagnetics*

Variable	Period (years) [95% CI]	Amplitude (%M)** [95% CI]
Interplanetary magnetic field		
ProtonTemperature	34.28 [26.99, 41.57]	13.62 [6.56, 20.68]
sigma(Bx)	31.87 [24.70, 39.04]	6.81 [3.41, 10.71]
Plasma Speed	33.04 [20.10, 45.97]	2.47 [0.10, 4.83]
Planetary geomagnetic index		
Kp	32.65 [28.27, 37.03]	12.74 [8.58, 16.86]

* Monthly means from 1963 to 2003 (41 years), analyzed by extended cosinor with nonlinear least squares, using 33.0 years as a trial period. By the criterion of a CI of the period overlapping the 30-40-year range and, if so, of a positive CI of the amplitude, five time series in the OMNI2 database qualify as compatible with a BEL cycle. Na/Np (alpha/proton ratio: http://nssdcftp.gsfc.nasa.gov /spacecraft_data/omni/omni2.text) also had a period of 36.84 years [29.97, 43.72]; 22 other variables converge to periods shorter than 30 years or longer than 40 years; three others do not reach statistical significance, and the remaining 10 do not converge, i.e., do not allow a period estimate with the trial period used. Zürich (Wolf) sunspot numbers for the span from 1745 to 2003 yield a period of 29.07 [27.92, 30.22] years.

** %M: percentage of MESOR.

Table 3. Some congruent* periods of helio-geomagnetics (columns 1 and 2), the estimation of 1-minute by a healthy man over 3.5 decades (column 3) and terrorism (bottom, bold)

	Period (years) (CI=95% confidence interval)			
Variable:	solar wind	aa	1-min estimation	
	15.6 (15.2, 16.0)			
		10.85 (10.78, 10.92)		
	9.54 (9.38, 9.70)			
			8.71 (8.52, 8.90)	
		5.31 (5.38, 5.35)		
	3.56 (3.52, 3.60)		4.11 (4.06, 4.16)	
			2.81 (2.76, 2.84)	
	2.17 (2.15, 2.19)			
		1.92 (1.89, 1.95)*	1.98 (1.94, 2.01)*	E
			1.85 (1.82, 1.88)*	
	*1.69 (1.67, 1.72)**	*1.71 (1.69, 1.74)**	*1.71 (1.68, 1.74)**	*SE*
	1.60 (1.58, 1.62)*			
	*1.52 (1.50, 1.54)**		*1.54 (1.52, 1.56)**	*S*
	*1.39 (1.37, 1.41)**	*1.39 (1.37, 1.41)**		
	*1.32 (1.31, 1.34)**			
	1.24 (1.23, 1.26)*		*1.26 (1.25, 1.27)**	S
			1.16 (1.15, 1.18)*	
	*1.06 (1.05, 1.07)**	*1.06 (1.04, 1.07)*	*1.06 (1.05, 1.07)**	*SE*
		0.99 (0.98, 1.01)*	*0.99 (0.98, 1.01)**	*SE*
	0.91 (0.90, 0.92)*			
	0.83 (0.82, 0.84)*		*0.82 (0.81, 0.83)**	**S**
	*0.72 (0.71, 0.73)**	*0.72 (0.71, 0.73)**	*0.713 (0.708,0.719)*	*SE*
		0.599 (0.598,0.600)		
	0.559 (0.557,0.561)		0.561 (0.558,0.564)	S
		0.548 (0.547,0.549)		
	0.524 (0.522,0.526)			
	0.500 (0.499,0.501)	0.500 (0.499,0.501)		
		0.437 (0.436,0.438)	0.440 (0.438,0.442)	E
	0.485 (0.483,0.487)			
	0.425 (0.423,0.427)			
	0.409 (0.407,0.411)			
	0.355 (0.354,0.356)			
		0.341 (0.340,0.342)	0.339 (0.338, 0.340)	

* Based on weekly, otherwise on daily measurements. Congruence defined by overlying or overlapping 95% confidence intervals, given in ( ). Congruence in the last column is designated as pertaining to the Sun (S) or Earth (E).

The transyear of 1.34 years, with CI [1.31, 1.37], of terrorism worldwide is congruent with environmental periods and time courses show similarities that are resolved with hypothesis testing and estimation of the uncertainty of the acrophase in Figure 3c. When more data on terrorism are analyzed, the transyear period (in the footnote to Table 2) is of 1.28 [1.26-1.29] year length overlapping with a period of 1.26 years in 1-minute estimation and solar wind speed.

Table 4. Spectral components in human violence (MIPT) and in (possibly involved) solar (SWS), terrestrial (aa) and psychophysiological variables*

Trial period (y)	MIPT**			SWS			aa			1MTE			HR		
25	27.81	25.54	30.08	NS			NS			25.92	23.68	28.16	NS		
11	9.67	9.17	10.17	9.48	9.24	9.73	9.40	9.09	9.70	NS			NS		
3	2.88	2.82	2.94	2.90	2.83	2.97	NS			2.82	2.76	2.87	2.75	2.68	2.83
1.7	1.69	1.67	1.72	1.69	1.66	1.72	1.71	1.69	1.74	1.71	1.68	1.74	NS		
1.3	1.34	1.32	1.37**	1.32	1.31	1.34	1.32	1.30	1.33	1.26	1.24	1.27	1.25	1.23	1.27
1	NS			0.998	0.985	1.012	0.989	0.979	0.999	0.994	0.982	1.007	0.994	0.987	1.001
0.5	0.496	0.493	0.498	NS			0.500	0.499	0.502	NS			0.502	0.499	0.505
Trial period (y)	SBP			DBP			Mood			Vigor			Oral Temp		
25	24.20	23.02	25.38	21.22	20.12	22.32	23.06	19.11	27.02	NS			NS		
11	11.01	10.65	11.37	10.54	10.15	10.93	11.27	10.81	11.73	8.91	8.60	9.22	11.00	10.50	11.50
3	3.02	2.94	3.10	3.22	3.12	3.33	2.71	2.63	2.78	2.97	2.86	3.08	2.85	2.77	2.94
1.7	1.68	1.65	1.71	1.69	1.66	1.73	1.74	1.71	1.77	1.71	1.68	1.74	NS		
1.3	1.31	1.29	1.33	NS			1.38	1.36	1.40	NS			NS		
1	1.002	0.998	1.006	1.007	1.002	1.012	1.018	1.007	1.029	1.032	1.019	1.044	1.015	1.007	1.023
0.5	0.500	0.498	0.502	0.500	0.497	0.504	0.502	0.499	0.504	0.502	0.498	0.504	0.504	0.500	0.508

*MIPT: Memorial Institute for the Prevention of Terrorism (the presence of counterparts of statistically significant components in MIPT is investigated in other time series); SWS: solar wind speed; aa: antipodal geomagnetic index; 1MTE: 1-minute time estimation; HR: heart rate; SBP: systolic blood pressure; DBP: diastolic blood pressure.

1MTE, HR, SBP, DBP, mood, vigor and oral temperature from RBS, a clinically healthy man, ~20.5 years old at the start of record of ~5-6 daily self-measurements on May 11, 1967, and continuing for over four decades to the present day.

When found, counterparts are listed here.

**When added daily data were accessed in MIPT and analyzed 3 months after our first use of the MIPT database (in January 2008 rather than October 2007), the about 1.3-year period changed to 1.28 [1.26-1.29] years length for MIPT, becoming congruent with 1MTE and HR. Second decimals are computational results to which no importance is necessarily implied in view of the drifting of the periods.

Analyses (not tabulated above) of much shorter time series of a) helioseismologic probing of the solar interior at the sun's equator yield a period of 1.36 [1.27, 1.45] years and b) of Gallup Poll's approval ratings of George W. Bush from February 2001-December 2007 (A: approval; D: disapproval; N: no opinion) show for A, a period of 1.46 [1.33, 1.58] years, for D one of 1.42 [1.29, 1.54] years and for N one of 1.75 [1.51, 1.99] years. These findings hint at the importance of the about 1.3-year component, while the about 1.7-year component is also seen in cosmic rays. The finding in polls (b) supports the pervasive influence of the cosmos, overriding the influence of the seasons, since a calendar-yearly component is absent.

The extension of the mapping of the resonances in the human heart and mind of helio-, interplanetary and geomagnetics is part of an international transdisciplinary effort on The BIOsphere and the COSmos (BIOCOS) [2, 3].

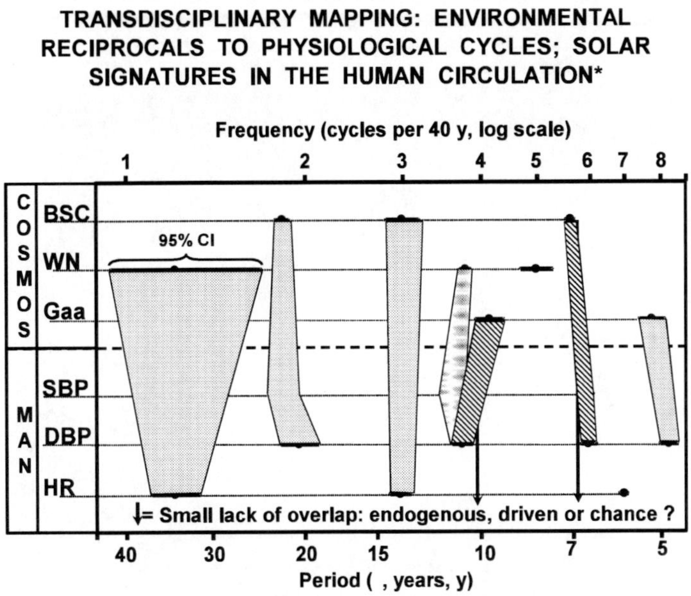

* BSC = Hale's Bipolarity Sunspot Cycle (odd cycles coded negative); WN = Schwabe's relative sunspot numbers (Wolf Numbers); Gaa = Geomagnetic aa-Index; SBP = Systolic Blood Pressure; DBP = Diastolic Blood Pressure; HR = Heart Rate. Cardiovascular data collected during 38 y by RBS, a MESOR-normotensive man, 20.5 y old at start of ongoing ~5 daily self-measurements. Width of horizontal bars = 95% confidence intervals (CIs) for all □s. All series in same span (May 11 1967 to Nov 07 2005). Thin connecting lines and shading indicate overlapping CIs.

Conclusion (tentative; based on limited data): CIs of □s of some cardiovascular spectral components overlap (when driven ▒ ?) or do not overlap (but are near) environmental reciprocal □s (when they are endogenous ▨ ?); alternatives, including chance, not ruled out.

A

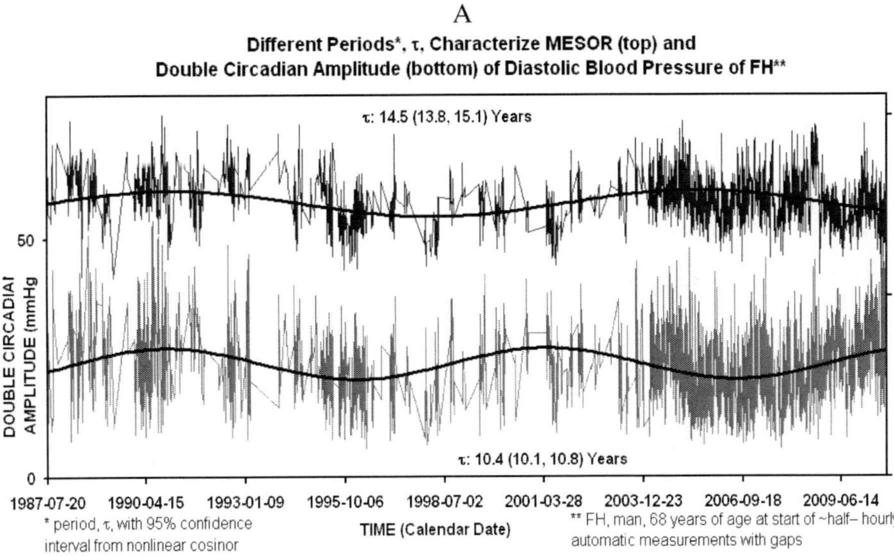

B

Figure 7. Selective environmental-psychophysiological pairing of variables (A) in a clinically healthy man, RBS, and of circadian characteristics, MESOR vs. circadian amplitude (B) in an elderly man on antihypertensive medication (FH). © Halberg.

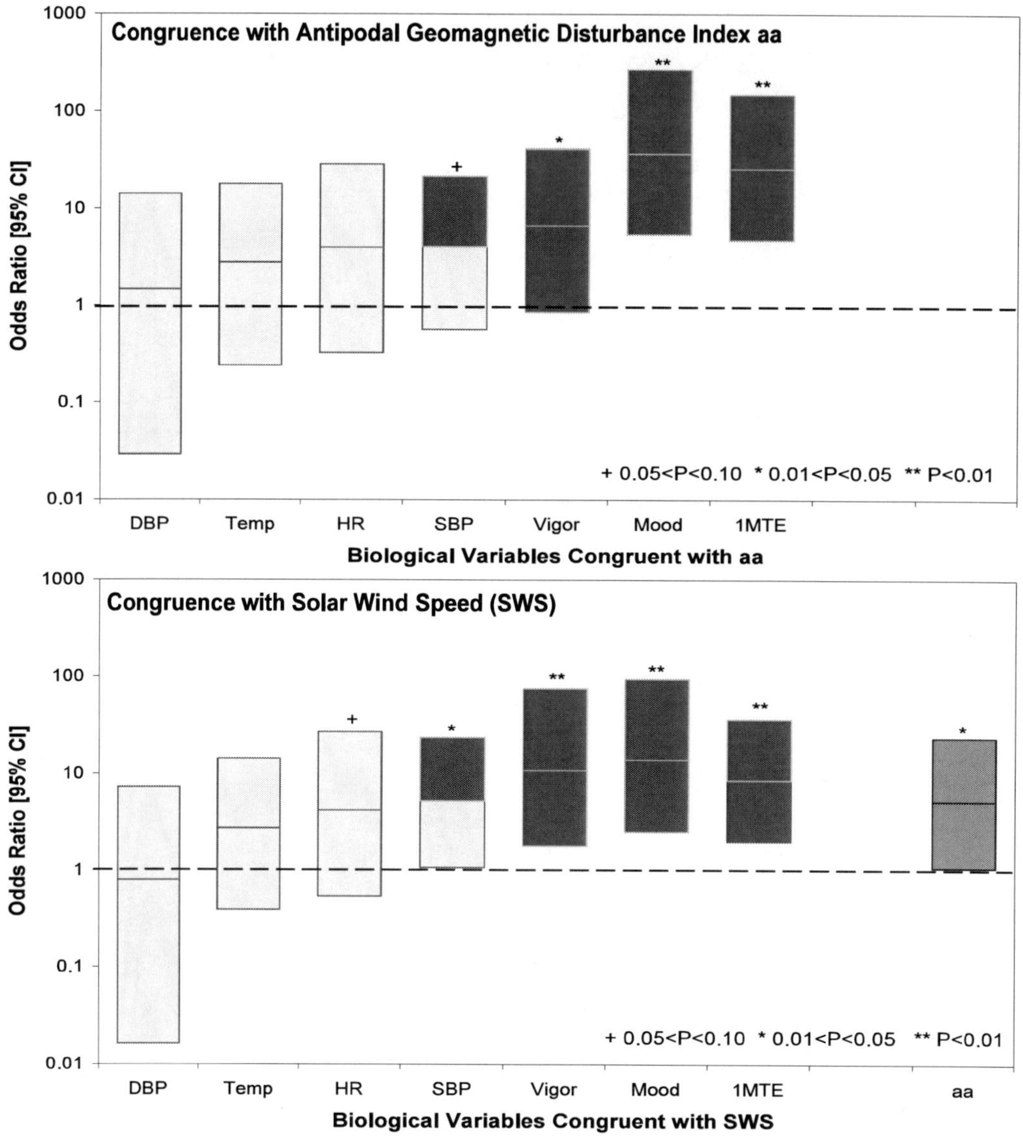

Figure 8. The association of RBS's psychophysiology with a true terrestrial or interplanetary magnetism more than matches the association of the latter two sharing common para-annual spectral components [61, 62] (see also Figure 17). © Halberg.

The question thus arises: Can environmental influences, whether direct ones or exerted via the weather in space acting upon terrestrial conditions, be monitored and thus investigated in human individuals, and if so, can a cost-effective self-surveillance be implemented for the goals of both the individual and society at large? In the case of the given proverbial person on the street or at home, the aim is to avoid massive brain attacks as well as transient ischemic attacks. In the case of the population, we aim at the prevention of terrorism and war as well as crime. The unseen factor "cosmos" can perhaps be assessed as a dividend from self-help by computer-implemented self-surveillance.

For an automatic analysis of the data of individuals, a website manned for special tasks arising in the course of the care or cure of individuals could provide the accumulating data for many aspects of research; its combination could be a first step toward a worldwide surveillance of how we are affected by the sun in whose atmosphere we live, starting at birth [92], Figure 12. Photic and non-photic spectral components determine both the deep history of the origins of living matter and the more recent past. During the current lifetimes of individuals, we should strive to optimize not only the habitat's temperature and light, but also more subtle factors that impinge upon us and can constitute loads that are best recognized so that we can develop countermeasures for associations that started with the changing resistance to cholera (Figure 3A) and more broadly to infection [66, 67; cf. 35] and with revolutions [65, 68] and have become apparent by the patterns of sudden cardiac deaths, Figure 13, suicides, Figure 14, and terrorism, Figures 15 and 16.

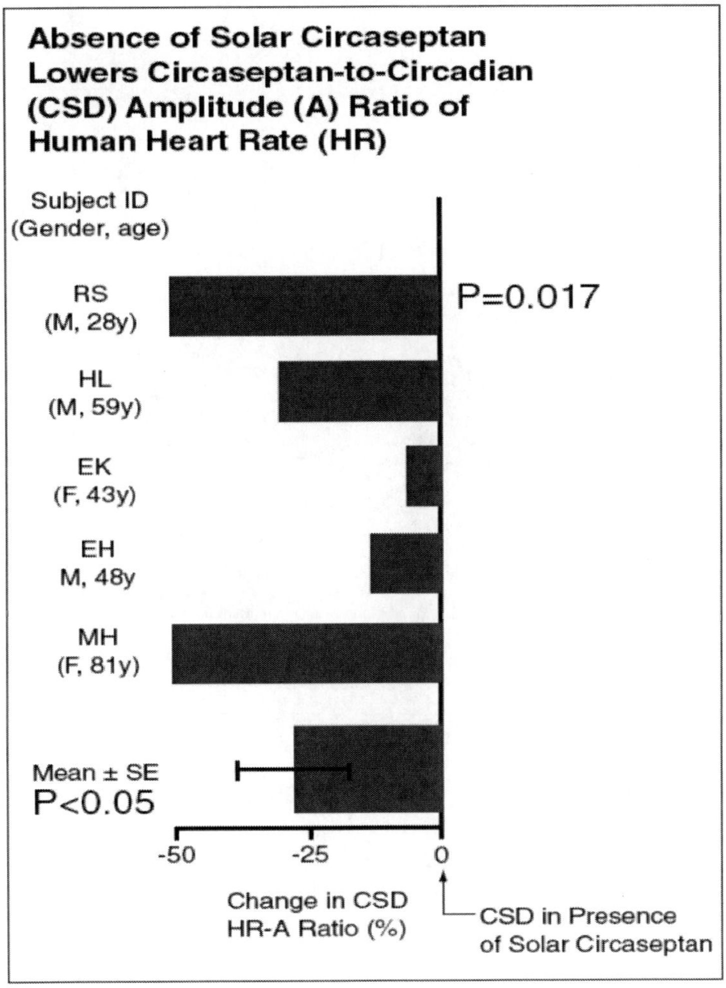

Figure 9. An ~7-day spectral component in the heart rate of five men is less prominent when the solar wind loses its counterpart of corresponding length. Implied, but not shown, is the persistence in the biosphere of an ~7-day component that can be amplified (driven) by a reciprocal component in solar activity [62, 63, 93]. © Halberg.

The criterion used herein to examine environmental-physiological associations based on congruence, defined by overlapping CIs, can be used to inquire about the extent to which the results may be due to chance, as compared to the reference standard of an association between interplanetary and terrestrial magnetism with long-sought results. The biological associations of the interplanetary (solar wind speed, SWS) or the terrestrial (antipodal index aa-gauged) magnetism more than match the relation of the chosen standard of interplanetary and terrestrial magnetism, Figure 17 [61, 62].

Figure 10. Alexander Leonidovich Chizhevsky (1887-1964).

Figure 11. Frank A. Brown Jr. (1908-1983)

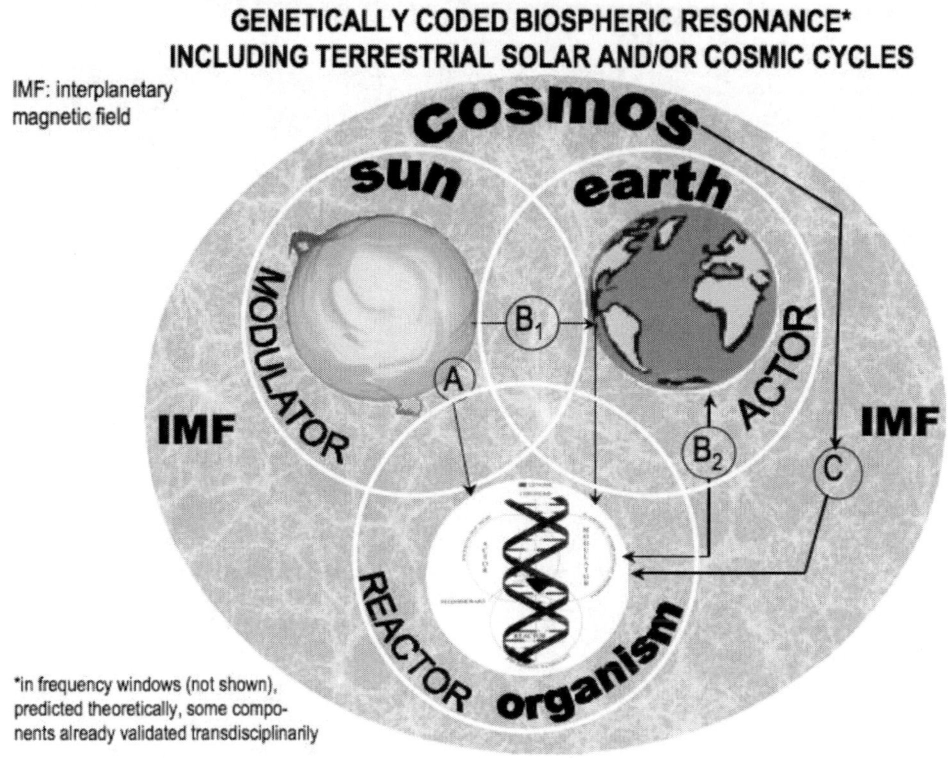

Figure 12. We need not ignore the many signatures of the cosmos with validated statistical significance of anticipated periods that characterize: a. dozens of decades-long time series of human blood pressure and heart rate; b. other physiology and psychology, including human mental functions; c. religious proselytism; d. suicide; e. sudden cardiac death; f. terrorist activity for the past 39 years; g. 2,556 years of international battles compiled by Raymond Holder Wheeler; h. military expenditures for training in non-medical science; i. non-medical (scientific) degrees earned; j. Gallup Polls; and k. political and military actions in nearly 200 years, meta-analyzed from the much broader treasure of data compiled by Alexander Leonidovich Chizhevsky. While chance can never be ruled out, it would be further greatly reduced by systematic lifetime monitoring of physiology in health, of pathology and disease, notably in archives to separate effects of sun and earth, many of which are beneficial. Other effects such as extreme cold and heat or extreme light can be met by countermeasures such as housing, heating and air conditioning. The task remains to develop countermeasures to those nonphotic effects that can be documented as harmful. While the earth is the immediate actor, and for that action the sun is the modulator, the biosphere reacts directly (Figure 17) to both terrestrial and solar factors; the roles of actor, reactor and modulator are continuously changing. In the greenhouse effect, organisms are the actors, as shown by a double-headed arrow (B_2). There may be other more subtle effects of synchronized human action upon the earth. Effects of the moon also are demonstrated [94]. Original drawing by Mary Sampson. © Halberg.

Everybody's task is to do preventive maintenance for one's own health, donating the data (deidentified when desired to protect privacy) to contribute to the maintenance of societal health. For these purposes, just as we learn geography from maps, imaging in time seems essential, the sooner and the earlier in life the better, all within the scope of a website for these purposes, Figure 18.

4. A CHRONOBIOLOGIC-CHRONOMIC WEBSITE

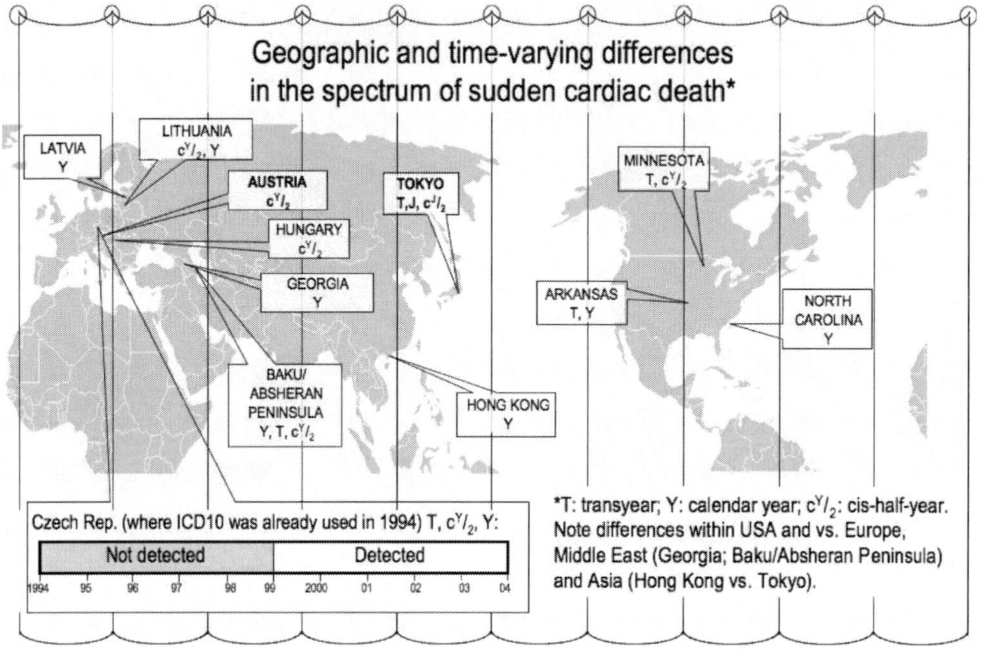

Figure 13. A curtain of uncertainty, because of limited available data, hides any time- and geographic (geomagnetic or dip-magnetic) site-specificity of various spectral aspects of sudden cardiac death. Thus, we find a transyear in Minnesota with a cis-half-year (cY/2) and both a calendar year and a transyear in Arkansas and the Czech Republic: at the latter site a cis-half-year, corresponding in length to an also-transient period of hard solar flares, is detected after but not before 1999. Whether other geographic differences in sudden cardiac death may also relate to any magnetic latitude deserves scrutiny. It is noteworthy in any event that cardiac arrhythmias can also transiently reveal a transyear or a cis-half-year, each in a different solar Schwabe cycle stage, as seen from the limited data in [37]. In 1998, Peter V. Minorsky (Annals of Botany 1998; 82: 133-140) re-evaluated "Latitudinal differences in coconut palms' foliar spiral direction" (FSD); he hypothesized that latitude-dependent biases in FSD, a non-Mendelian trait, will be associated with a temporally-varying component of the earth's magnetic field. Earth currents, which are measurable in trees, bias the difference of auxin (or auxin transport proteins) in young embryos, such that left-handed palm trees (LHP) are produced preferentially in the Northern Hemisphere and right-handed ones (RHP) in the Southern Hemisphere. He built his observations on a previously collected data base by Davis and Davis, revealing that the ratio of (LHP – RHP)/Total was better correlated with magnetic (dip) latitude than with geographic or geomagnetic (centered dipole) latitude. Most recently, Minorsky and Bronstein (Natural experiments indicate that geomagnetic variations cause spatial and temporal variations in coconut palm asymmetry. Plant Physiol 2006; 142: 40-44) referred to this classical case of morphological antisymmetry in which dextral and sinistral forms are not inherited and are equally common within a species, proposing that Pc1-induced earth currents may bias the diffusion of morphogens in coconut palm embryos, thereby giving rise to asymmetries of FSD. © Halberg.[3]

[3] Thus far analyses were on data from Argentina, Armenia, Austria, Azerbaijan, Belgium, Bulgaria, Canada, China, the Czech Republic, Denmark, the Republic of Georgia, Hong Kong, Hungary, India, Italy, Japan, Kazakhstan, Mexico, Poland, Peru, Romania, Russia, Saudi Arabia, Slovakia, Spain, Sweden, Ukraine and the USA.

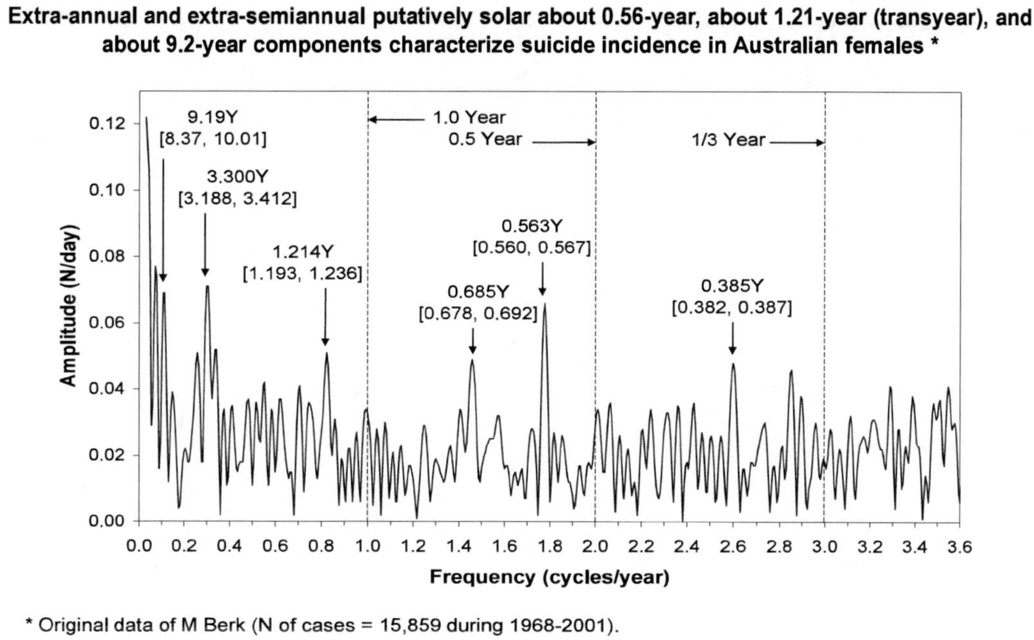

Figure 14A. Nonphotic para-annual prominence in the incidence spectrum of suicides by females in Australia. © Halberg.

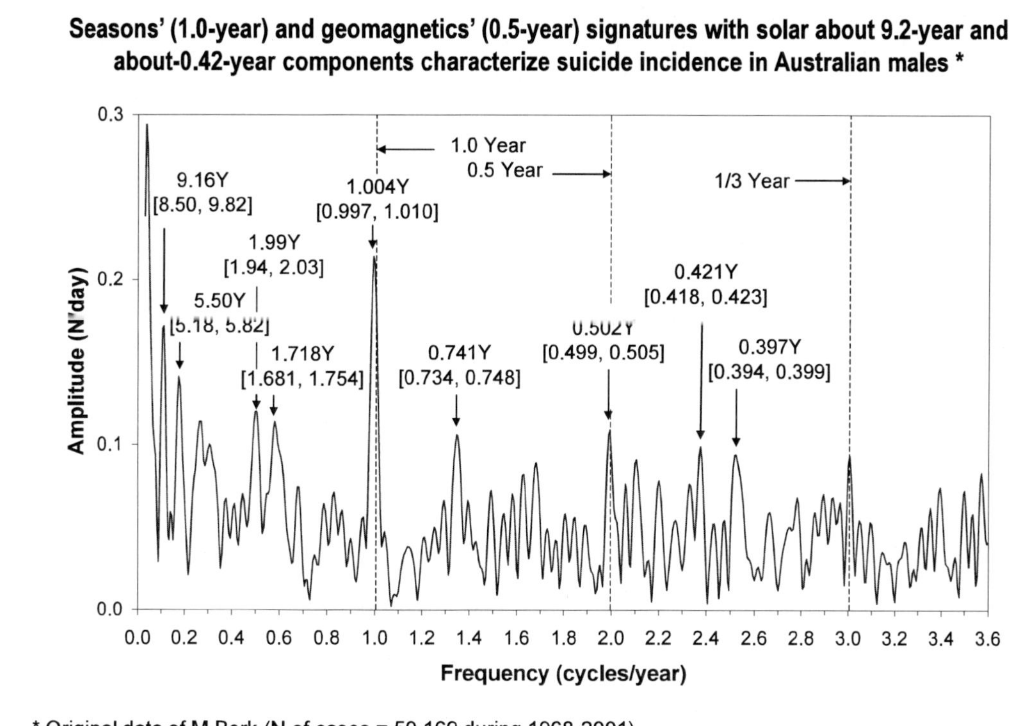

Figure 14B. Coexisting photic and nonphotic components in suicides by males in Australia. © Halberg.

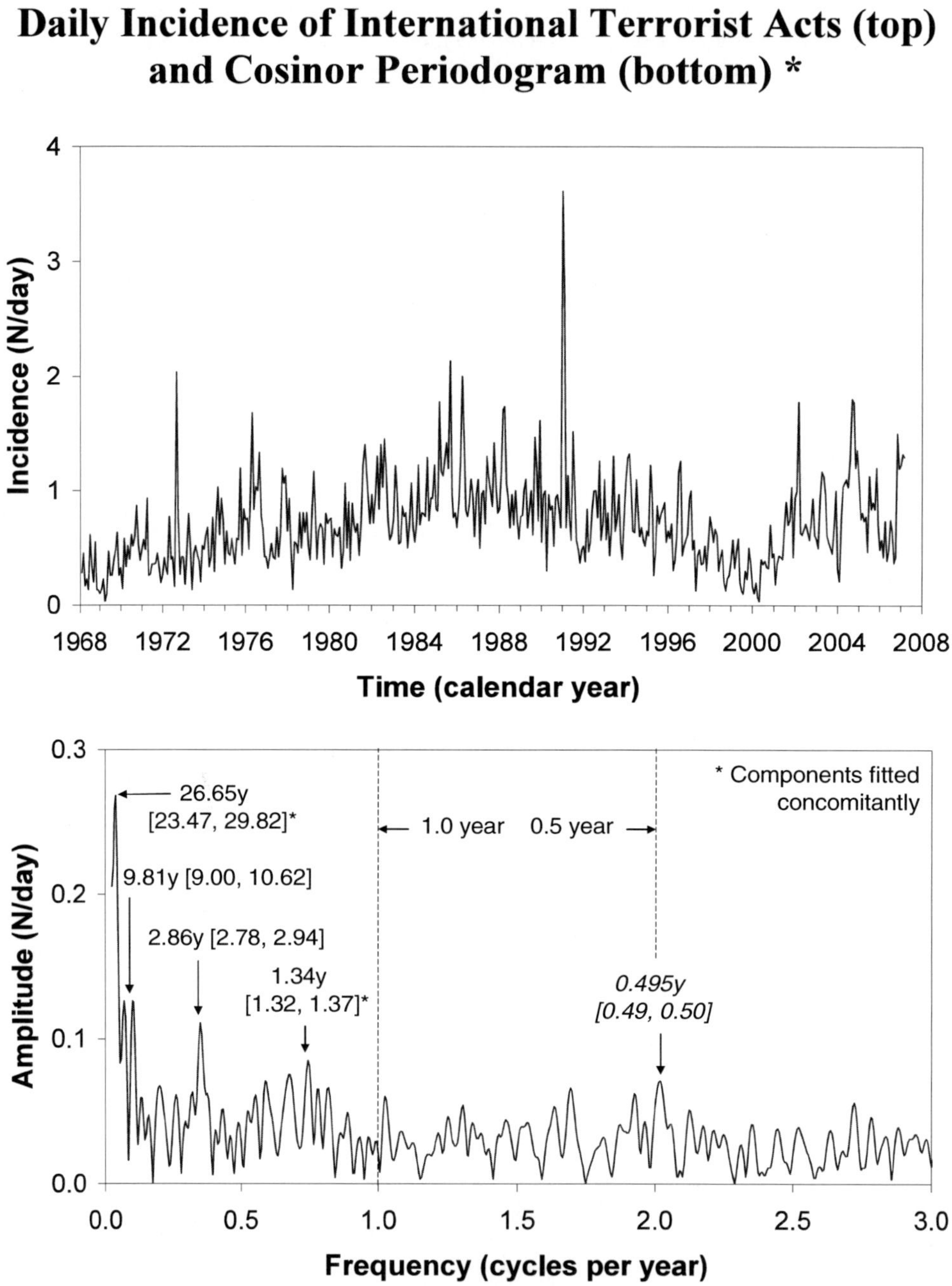

*From February 1968 to March 2007, daily incidence computed from monthly totals (adjusted for differences in the number of days per month). Data from MIPT Terrorism Knowledge Base (http://www.tkb.org/).

Figure 15. In data from the Memorial Institute for the Prevention of Terrorism's (MIPT) Terrorism Knowledge Base (1968-2008) (http://www.tkb.org/) (top), a transyear of ~1.3 years is detected in the absence of a calendar yearly component. © Halberg.

Chronobiology, the study of the time structures of living things (chronomes), including predictable rhythms, along with chaos and trends, and chronomics, the investigation of links between cycles in organisms and their environment, near and far, seen and unseen, can be taught early in life [69, 70]. Chronobiology and chronomics image in time, revealing alterations of variability that currently go unrecognized and concern many millions of people, including all those with high blood pressure.

There is a need to change from a current spotcheck-based health care to one based on time series for educated self-help by self-surveillance [95]. Cybercare [71] could use modern technology not only to enable what we do today with blood pressure and heart rate (spotchecks and around-the-clock 24-hour profiles) more readily and on a larger scale. Instead, it could also lead to great improvements (from a minimal 7-day to continuous surveillance around-the-clock and as-one-goes analyses) by new diagnoses that recognize and treat disease risks (revealed by altered variability) rather than established illness only.

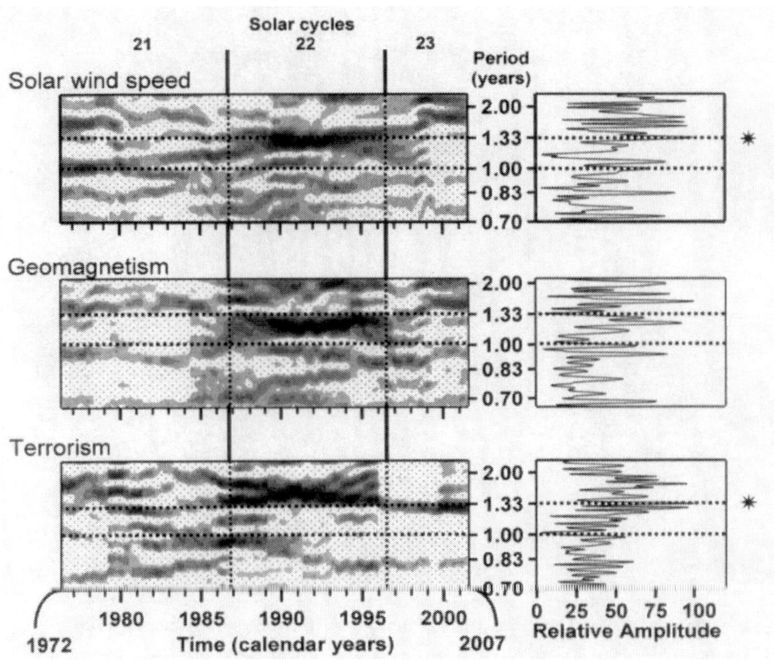

Figure 16. Just as removal and replacement of a gland led to endocrinology, so biological consequences of the loss of environmental spectral components are critical to chronomics, the study of chronomes (time structures) in and around us. Note neighboring frequencies waxing at or near a transyear component in solar wind speed (top row), geomagnetism (second row, aa), and in 39 years of terrorist activity (bottom row), an association supported by an independent method (by the statistical significance of the fit of a 1.3-year far-transyear to interplanetary (SWS) and terrestrial (aa) magnetism and to terrorism. © Halberg.

Ever-refined, miniaturized, affordable, instrumentation for data collection and transfer, to replace current functional tools and the development of a smart phone as a diary, psychological test device and data storer and transmitter, in combination or as a complement to an international multilingual chronobiologic-chronomic website, could be the first two steps toward this goal, for self-health service, education and multipurpose research.

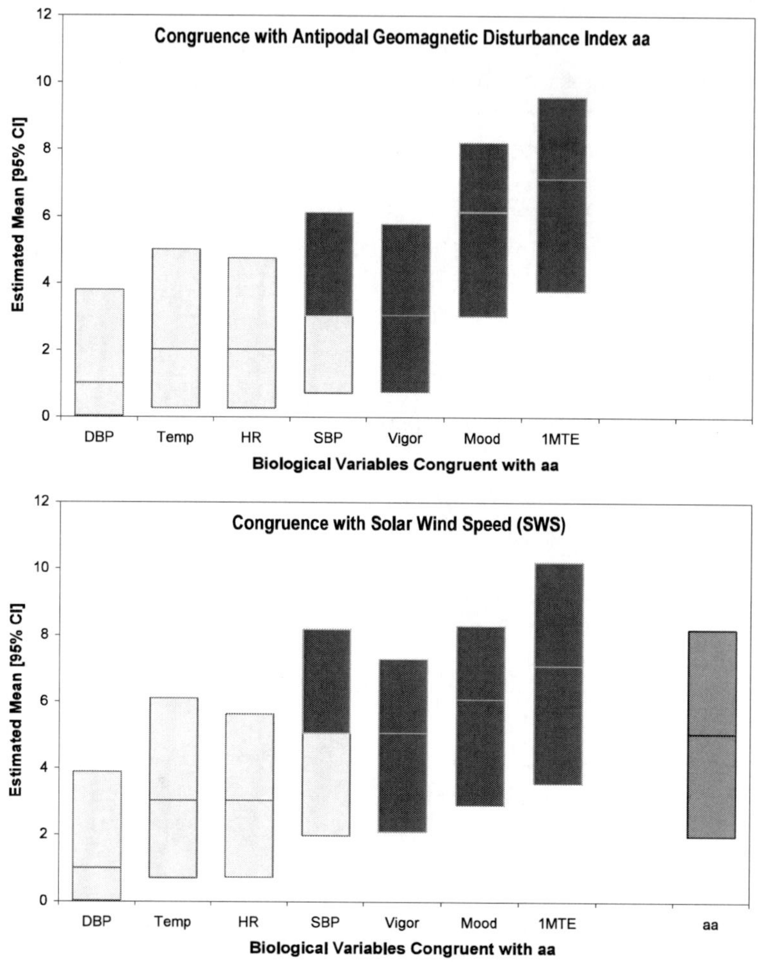

Figure 17. Some inferentially statistically validated congruences of human somatic and mental functions with the environment, gauged by the antipodal geomagnetic disturbance value (aa) (top) and solar wind speed (SWS) (bottom). While limited to a single clinically healthy case, RBS, a man aging during the study from ~20 to ~60 years of age, discussed in detail elsewhere (Tables 2 and 3 and Figure 8), the number of congruences found for the estimation of 1-minute and for mood in the spectral range investigated (from one cycle in 2.5 years to 3 cycles per year) more than equals that of the known association of helio- and geomagnetism, commented upon in 1852 by Gen. Sir Edward Sabine (on the basis of too few data) -- "... it is certainly a most striking coincidence, that the period, and the epochs of minima and maxima, which M. SCHWABE has assigned to the variation of the solar spots, are absolutely identical with those which have here been assigned to the magnetic variations" -- and extended to the solar wind by Marcia Neugebauer (Large-scale and solar-cycle variations of the solar wind. Space Science Reviews 1975; 17: 221-254) and qualified by Gordon Rostoker and Carl-Gunne Falthammar (Relationship between changes in the interplanetary magnetic field and variations in the magnetic field at the earth's surface. J Geophys Res 1967; 72, 5853-5863) and by Wolfgang Baumjohann and Gerhard Haerendel (Magnetospheric convection observed between 0600 and 2100 LT: solar wind and IMF dependence. J Geophys Res 1986; 90 [A7]: 6370-6378). When congruence is assessed by means of the odds ratio based on the non-central hypergeometric distribution (see Figure 8), mental functions show higher congruence than somatic functions; among the latter, systolic blood pressure is responsive, perhaps constituting a seemingly acceptable approximation for the mental functions. © Halberg.

4.1. Background

During the past decade in particular and in some cases for much longer, the Halberg Chronobiology Center at the University of Minnesota has served all comers from different continents[2] with cost-free analyses of time series, including sphygmochrons, summaries of blood pressure and heart rate for diagnostic purposes and more recently control charts, Figure 19 (cf. Figure 20), to assess changes in time series as a function, among others, of the timing of medication [1; cf. 72-74]. The Phoenix Study Group, composed of volunteering members of the Twin Cities chapter of the Institute of Electrical and Electronics Engineers (http://www.phoenix.tc-ieee.org) has developed scenarios [75] on how this current activity, on the international scale (on different continents) could be enlarged.

As hardware and software technologies advance, Larry A. Beaty of Phoenix [76, 77] expects to see improvements in the convenience, cost, sensitivity of sensors, and network connectivity of blood pressure monitors, to which he already contributed [78]. He continues to investigate a change from the current desideratum of 7-day/24-hour monitoring preferably to continued surveillance by an unobtrusive, affordable instrument, possibly on the wrist. He also started a website for use of the currently available analysis software by self-helpers in their homes, interested in self-surveillance-based, preventive health care. A manned international website would save the (if necessary de-identified) data for ongoing multipurpose research.

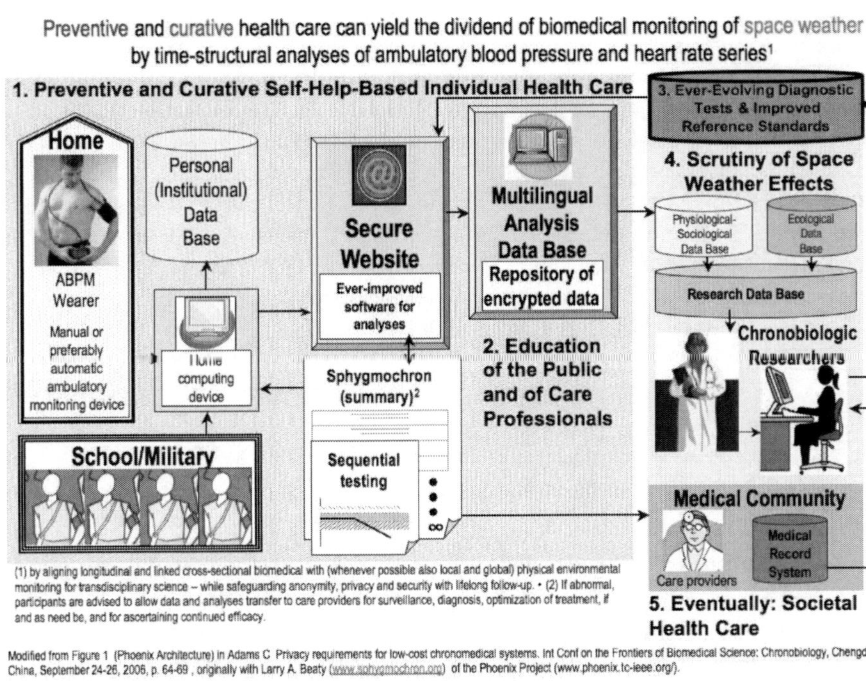

Figure 18. The Phoenix Group of volunteering electrical and electronic engineers from the Twin Cities chapter of the Institute of Electrical and Electronics Engineers (http://www.phoenix.tc-ieee.org) is planning on developing an inexpensive, cuffless automatic monitor of blood pressure and on implementing the concept of a website (www.sphygmochron.org) for collection and analysis of data collected with these instruments. © Halberg.

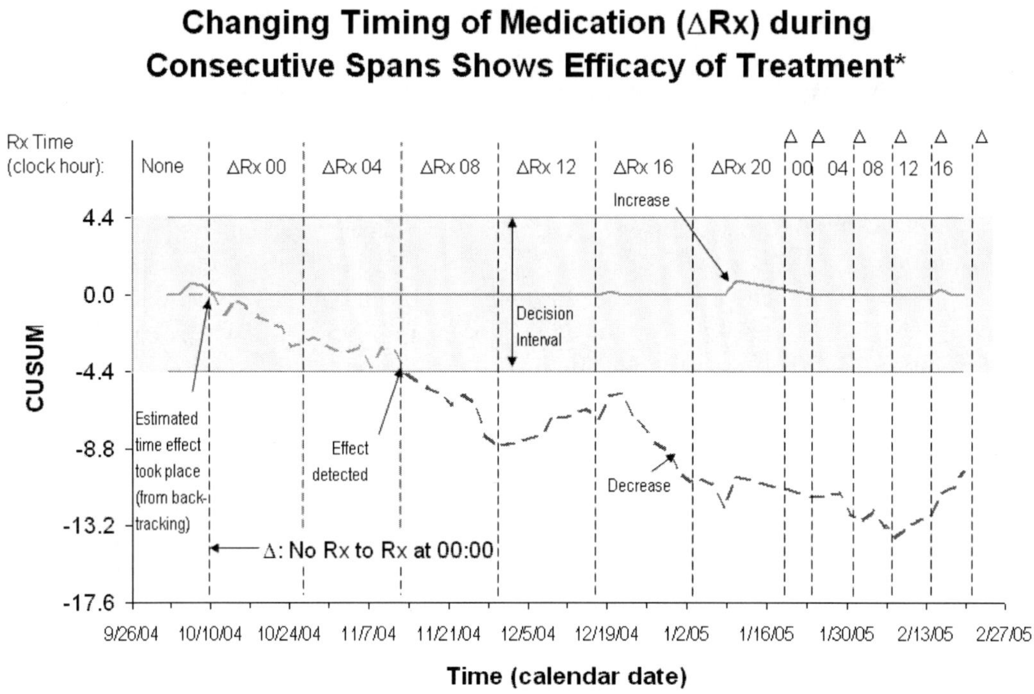

Figure 19A. Changing timing of medication (ΔRx) during consecutive spans shows whether and, if so, when treatment is most effective as an empirical approach to chronotherapy. Immediately after the diagnosis of MESOR-hypertension, one should ascertain that the treatment is effective. Optimization of treatment effects by timing can be achieved for the individual patient by systematically changing, e.g., advancing the time of treatment. Successful treatment of MESOR-hypertension can thus be assessed by a self-starting cumulative sum control chart to be offered by the website. To optimize his hypotensive treatment (Rx), a just-diagnosed 24-year-old individual (TT) switched his Rx first every 17 days by 4 hours and then mostly at shorter intervals. Note statistically significant decrease in MESOR, evidenced by the breakout outside the decision interval of the negative CUSUM line. With continued Rx, the blood pressure MESOR leaves the decision interval, indicating a statistically significant decrease but Rx-time-dependent in overall blood pressure. © Halberg.

4.2. What Is Needed

We propose beginning implementation and manning of an international website that could serve the multiple purposes of 1) automatic service to healthy or sick self-helpers as well as to the health care professionals, 2) personnel preparing materials and interesting subjects in life-long self-help in their self-surveilled health care, for 3) teaching how to interpret results of analyses that detect novel indicators of risk [1, 72] or disease, for 4) guiding any treatment, for 5) continuing research on improving reference standards and 6) on refining the procedures that now detect treatable conditions, some with risks greater than a high blood pressure in certain populations and, on the same data pool, as it already accumulates in some subjects for decades, 7) to monitor the Sun, in whose atmosphere we live and the broader cosmos to map on a population the many environmental cycles reflected in their biospheric counterparts, and 8) to investigate in continuously monitored individuals, any alterations with focus on any undesired consequences of environmental events and, as necessary, 9) for the development of countermeasures.

Specific components of the first website design addressed by Larry Beaty [76-78] are:

- making the sphygmochron analysis available to home users, and their health care providers,
- making blood pressure data from home users available to researchers,
- updating the analysis software in the light of results of new research as time progresses,
- providing a library of educational materials for home users and health care providers, and
- building a community of self-helping users consisting of those engaged in preventive health care as well as those being treated by health care providers.

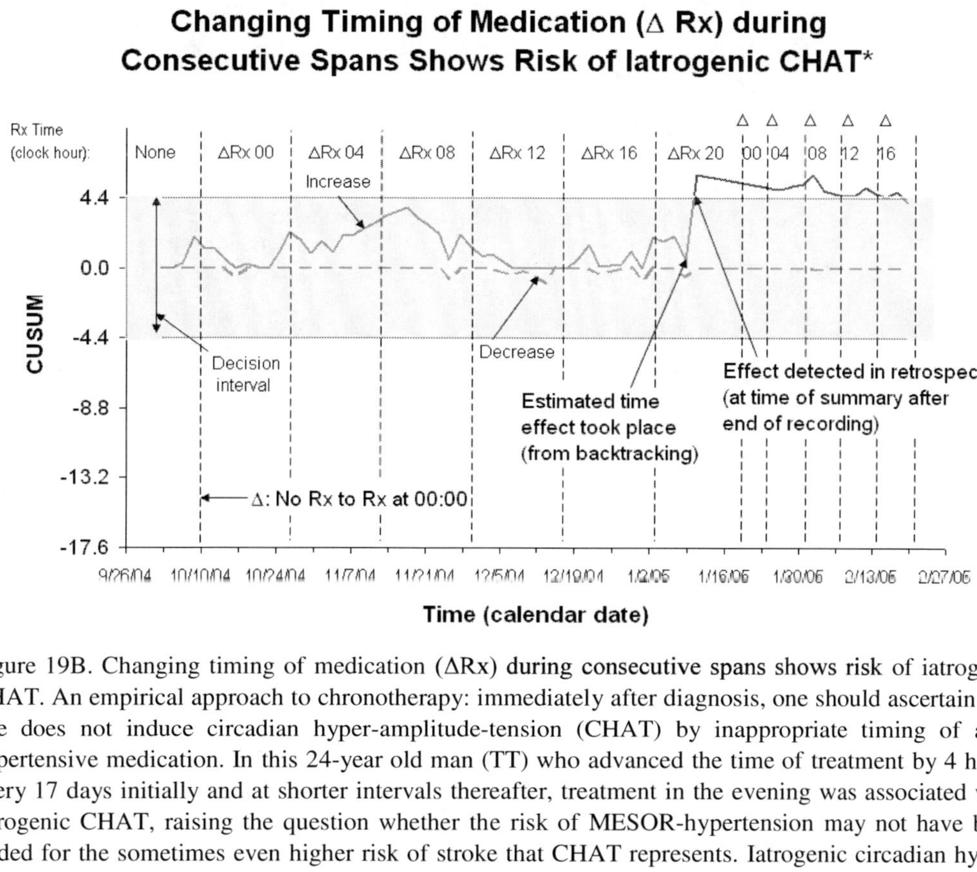

Figure 19B. Changing timing of medication (ΔRx) during consecutive spans shows risk of iatrogenic CHAT. An empirical approach to chronotherapy: immediately after diagnosis, one should ascertain that one does not induce circadian hyper-amplitude-tension (CHAT) by inappropriate timing of anti-hypertensive medication. In this 24-year old man (TT) who advanced the time of treatment by 4 hours every 17 days initially and at shorter intervals thereafter, treatment in the evening was associated with iatrogenic CHAT, raising the question whether the risk of MESOR-hypertension may not have been traded for the sometimes even higher risk of stroke that CHAT represents. Iatrogenic circadian hyper-amplitude-tension, CHAT, induced by treatment at 20:00 daily, was silent to office visits. TT may have traded benefit (lowering of the MESOR of blood pressure) for something worse (circadian overswinging of blood pressure). This danger applies to some hypertensives (who tend to have a large circadian amplitude of blood pressure) to whom treatment time is not specified by the care provider, as was the case for TT (or is specified for bedtime). A few others who took hypertensive medication at bedtime were also found to have CHAT. The figure also shows the assessability of otherwise undetected harm by as-one-goes sequential analysis. © Halberg.

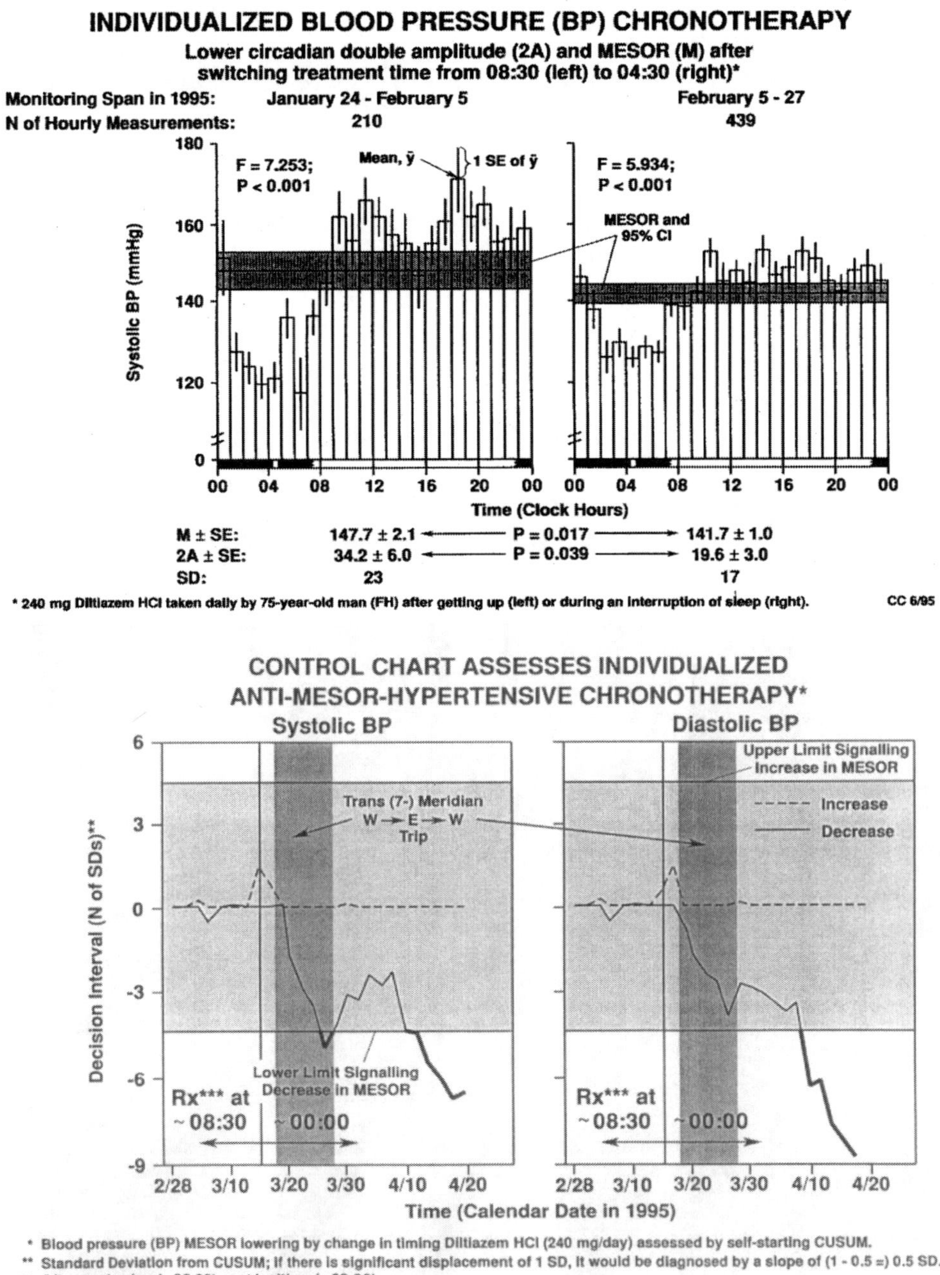

Figure 20. The usefulness of a chronobiologic approach is supported by the demonstration that treatment with antihypertensive drugs can be optimized by timing. The same dose of the same drug can have different effects on the MESOR and circadian amplitude of the same patient's blood pressure when it is administered daily at a different circadian stage as seen by the naked eye in the top graph. The practicality of a chronobiological approach lies in the availability of inferential statistical methods for the rigorous assessment of intervention effects applicable to the individual patient as shown in the bottom graph by using CUmulative SUM (CUSUM) control charts for personalized sequential testing. © Halberg.

Another larger set of website components yet to be developed aims at

- further temporally and spatially global and local (glocal) spectral and cross-spectral analyses of time-varying coherence among environmental, psychophysiological and epidemiological variables, including multiple component serial sections of these sets of multidisciplinary time series,
- at each of the known shared frequencies, by
- programs for automatically repeated multiple passes over each of the accumulating time series and of their coherence, if any, and
- subtraction and addition approaches.

4.3. Discussion

The website would contribute advanced transdisciplinary basic biological as well as medical research data bases, with the biological data base aiming at a clarification of biological associations of solar and terrestrial effects in human physiology. The medical data base would serve to gradually improve the service rendered by the website to self-care recipients by a refinement of the reference standards and of the harbingers of elevated risk of severe disease. Several updates to the existing sphygmochron software and additional analysis programs are planned for use by those who do continuous monitoring with repeated passes over the accumulating series and of those who do intermittent surveillance, repeating monitoring sessions (annually, on some other schedule and/or as warranted by analysis). To support users retrieving data directly from blood pressure monitors, the website will have to accept files as produced by many commercially-available devices and manufacturer-supplied analysis software, just one of several problems [76-78].

The user of the website must be computer-literate and able to perform analyses for multiple people, including those who need help with tasks involving computers. The first major scenario to be supported includes a computer-savvy friend or preferably family member accessing the website for his/her spouse, children, aged parents, people with disabilities, or others who fall within the individual's guardianship. The second scenario is for health care professionals or their staff members to be taught why and how to log in and submit data from patients. In the near term, the website supports running the sphygmochron analysis. In the long term, comparative analysis programs currently used at the Halberg Chronobiology Center at the University of Minnesota can also be run, improved and extended on the basis of ongoing experience.

The comparative analysis programs need two or more files of blood pressure data taken at different times in the person's life, or preferably a more or less continuous record, such as that taken by hypertensive opinion leaders from the time of diagnosis to their life's end. For instance, a former head of the then Hypertension/Endocrinology section of the U.S. National Institutes of Health who thereafter became director of that institution's Clinical Center believed that when hypertension is involved, one should not "fly blind" [79] (without continued surveillance by monitoring) [80].

Two major components already available as modules to be added to the website for comparative analyses of an individual accumulating continuous or intermittent record consist of parameter tests [81] and control charts [74, 82], the former triggered by the latter, Figure 18. Parameter tests can tell whether a change in (rhythm-adjusted) mean value or in the circadian dynamics has taken place with statistical significance. Diaries have yet to be developed to record times of getting up, going to bed, onset and end of exercise, meals and emotional events. They could perhaps be smart phones that also allow carrying out psychophysiological tests such as 1-minute estimation, eye-hand coordination, a positive and negative affect and vigor rating, along with transient storage and eventual transfer of data to the website. This desirable two-way tool may provide information to the participant, prompted by a statistically significant change in the record for which its clinical significance must be resolved as a function of whether it occurs spontaneously, in relation to a load, or in response to an intentional intervention such as the institution or modification of treatment. When data are collected continuously, cumulative sum (CUSUM) control charts can further help determine whether detected changes in circadian rhythm characteristics are temporally if not causally related to the given intervention, thereby enabling a rigorous assessment of treatment efficacy and safety.

Beaty writes that:

> "A small, well-focused library of materials related to sphygmochron analysis, heart and blood vessel physiology, cardiac and eventually other diseases, and treatments will be kept on the website. The library will contain sections for different types of users, including children, high school and college students, adult home users, and medical and health care students and professionals. The most important section might be the one for adult home users. It supports the desires of users with different backgrounds and levels of education to:
>
> - "Learn advanced interpretation of sphygmochron reports, and
> - "Research medical conditions, diseases, and treatments beyond the level of everyday 'household' words [76, 77].
>
> "To support building a community of self-helped home users interested in preventive and curative health care, a popular web mechanism currently employed by some internet users in self-help or self-directed educational situations is 'forum' software, which lets users communicate with each other via leaving publicly available messages directly on the website. The envisioned website will have such forums available early in its lifetime. The concept of community-building on the Internet still seems to be undergoing considerable change. We expect to watch for new trends in this area and adapt the website accordingly to accommodate desires of the users.
>
> "The website will make available information about obtaining a blood pressure monitor for ambulatory use (often called an Ambulatory Blood Pressure Monitor, ABPM, where 'ambulatory' means only that the instrument also functions during briefly interrupted walking), since many home users, especially those interested in the self-help style of medical care, might not know how to obtain ABPMs at reasonable cost, and will have questions about ABPM quality, testing and calibration, data off-loading software, and the logistics of wearing an ABPM around the clock for a week or much longer. A common question will be along the lines of 'This is different from current practice. How can this information help me, and how can my care provider use it?' We plan to encourage home users to monitor themselves for longer and longer time spans to get more and more basic and applied information from the analyses available on the website....
>
> "As additional models and categories of ABPMs become available over the next few years, the website will be updated with descriptions and other appropriate information, to

> help its users capitalize on those improvements in the convenience, cost, sensitivity of sensors, and connectivity of blood pressure monitors that will affect how 7-day/24-hour monitoring, or preferably much longer surveillance, analyzed as-one-goes, is done.
> "By 1880, Ignaz Zadek [83] had sufficient data to allow the demonstration in a meta-analysis of about-24-hour, about 84-hour and about-7-day changes in blood pressure [84]. In 1904, Janeway at Johns Hopkins Hospital insisted, before seeing a patient, on having enough data to evaluate periodic variations (note: variations is plural) [85]. He could do so on the basis of Zadek's record which allowed the demonstration of circadian and two infradian components of variability, Figure 21. In 2011, signatures of the about-10.5-year Schwabe cycle and 21-year Hale cycle, and even of the paratridecadal BEL cycle, are demonstrated in the human circulation, along with many other non-photic cycles, probably magnetoperiodisms, some of which also have signatures in sudden cardiac death [37], an association speaking for their importance."

Another important module of the website that needs development is an automated mechanism of several passes of data analyses aimed at assessing cycles with increasingly longer periods. On a short-term basis, ultradians and circadians can be determined. For ongoing analyses, rather than storing all original data, the salient rhythm characteristics are stored instead, serving for the determination of components with the next lower frequencies and for an assessment of changes with time in these components' characteristics. Such ongoing monitoring, analyzing and recycling could be the larger-scale supplement to the ongoing systematic mapping of rhythms of several frequencies in a few individuals. The examination of congruences with cycles in solar, interplanetary and other magnetics could yield a better understanding of environmental factors impacting human health. Studies by a subtraction-addition approach can be included opportunistically, triggered by concomitant environmental geomagnetic monitoring, preferably glocal (*global and local*) in geography and in time, by examining both the entire time series available at a given time (globally) and in its sections (locally in time).

Beaty concludes:

> Assessment of these different periodicities in long populations' and individual's records will be a dividend in basic science and space weather monitoring from self-help in health care that continuously assesses variability. This longitudinal surveillance could start for many hypertensives at the time of a chronomic diagnosis of abnormality to the time of life's end. The benefit for the individual involved would be a reduction in the number of episodes of variability disorders by their prompt detection and treatment (while otherwise an anti-hypertensive treatment might constitute the trade of a lesser risk such as hypertension for greater risk such as a circadian blood pressure overswing). In some cases, control of a silent disease (hypertension) cannot currently be achieved without bringing about an even greater also silent risk (overswing). The putative benefit of continuous surveillance is a critical remaining problem awaiting studies on populations. Every person with alterations in the variability of blood pressure and heart rate should continuously monitor and analyze, and thus serve his or her own health care as well as science. The website could, in combination with existing and preferably improved blood pressure monitors, mood and vigor assessment enable the public at large to determine the need for self-surveillance, and then implement continuous monitoring, individual-by-individual.

4.4. Problems

Larry Beaty, speaking for the Phoenix Study Group, composed of volunteering members of the Twin Cities chapter of the Institute of Electrical and Electronics Engineers (http://www.phoenix.tc-ieee.org), has well described the problem in advocating only 7-day/24-hour chronobiologic monitoring of blood pressure and heart rate in the USA. He notes that

> When we advocate that "healthcare should change with the help of modern technology", specifically talking about "a relatively inexpensive, largely self-implemented, continuous computer-aided surveillance" and associated website, and then suggest that people need to be routinely monitored for Vascular Variability Disorders (VVDs), notably those with a high blood pressure, a VVD in its own right, we are compelled to describe the current situation in the USA, since the implementation of these goals meets a major problem, namely that of fighting the current business model of the health care industry. This leads to the necessity of convincing "everybody" that it is all worthwhile.
>
> ... having inexpensive new technology is not enough to make a medical device successful in the market, no matter how novel or effective the device. The industry is rife with stories of technically successful devices that failed due to "poor marketing". What this phrase usually means is that there was not enough recognition that, at least in the USA, the marketing did not take the care providers' (and the companies that control them) business (read: profit) models into account well enough. Given a choice between two devices for diagnosis or treatment that are of equal merit in the USA, a care provider will often choose the device that generates more profit for him (or the company that controls his choices), regardless of which device is more costly [and, perhaps, some differences in quality will be ignored].
>
> In this atmosphere, merely putting out for display to the public new, inexpensive devices or systems that can detect vascular variability disorders, VVDs, or other conditions might not be enough to push a new device into use in everyday health care. Convincing government officials to make it a mandate might also not be enough, and the attempt to convince them might not even work as the competition for the attention of politicians is large.
>
> When one walks down the aisle of a local pharmacy and looks at the medical devices there, one sees eyeglasses, glucose meters and A1C diabetes tests, pregnancy tests, certain drugs, and of course blood pressure monitors. But while all these medical devices share the characteristic of now being directly available to the consumer after having once been largely exclusively in the domain of prescription-controlled healthcare, they also share another characteristic, which is that they are used in the home essentially the same way as they are used by clinicians.
>
> Here we are talking about using blood pressure monitors (those now available for home use, and those dependent on new technology still under development) in an importantly different way. The need to have blood pressure data for at least 7 days to detect VVDs and preferably for much longer for the validation of pre-hypertension of an altered circadian period or of chronobiologic pre-hypertension (rather than a single measurement of systolic blood pressure between 128-139 and/or of diastolic blood pressure of 80-89 mm Hg) is not currently recognized by the profession, notwithstanding the fact that it was advocated over a century ago by opinion leaders such as Theodore C. Janeway. The history of medical devices becoming widely used at home in a significantly different way from that in which they were used by clinicians may yet have to be written.
>
> Similarly, when one looks at the history of convincing politicians to enact healthcare mandates based on their own experiences, successful examples again seem sparse, and

attempts seem prone to failure. Thus, an enthusiastic mayor of Roseville, a suburb of Minneapolis and St. Paul, ordered a set of automatic blood pressure and heart rate monitors, convinced of both the need for their use in the community at large as such and for a 7-day record to start with. He had seen that he himself, on one of 7 days, had been very abnormal when he had to keep time in an ice hockey game and did not satisfy some husky players. By the time the instruments arrived, however, the mayor was no longer in office and his successor was not interested in the project. Fortunately for people in Japan, the project was picked up and has now been running in two cities there (Urausu, on the island of Hokkaido, and Tosa City, in Kochi on the island of Shikoku) for over 10 years.
... If the health care providers, politicians, educators, religious leaders, and others that we appeal to first do not see any obvious benefits in the short term (meaning, if they and their families are not affected by and cured of health problems such as chronobiologic pre-hypertension, a pre-metabolic syndrome, or another VVD or a combination thereof leading to a Vascular Variability Syndrome, VVS), then the idea of screening them for these conditions and for other diseases as a way of educating them and working towards a mandate for all to be screened in that way, will fail as they lose interest and go back to working on more short-term, tangible concerns in their lives.

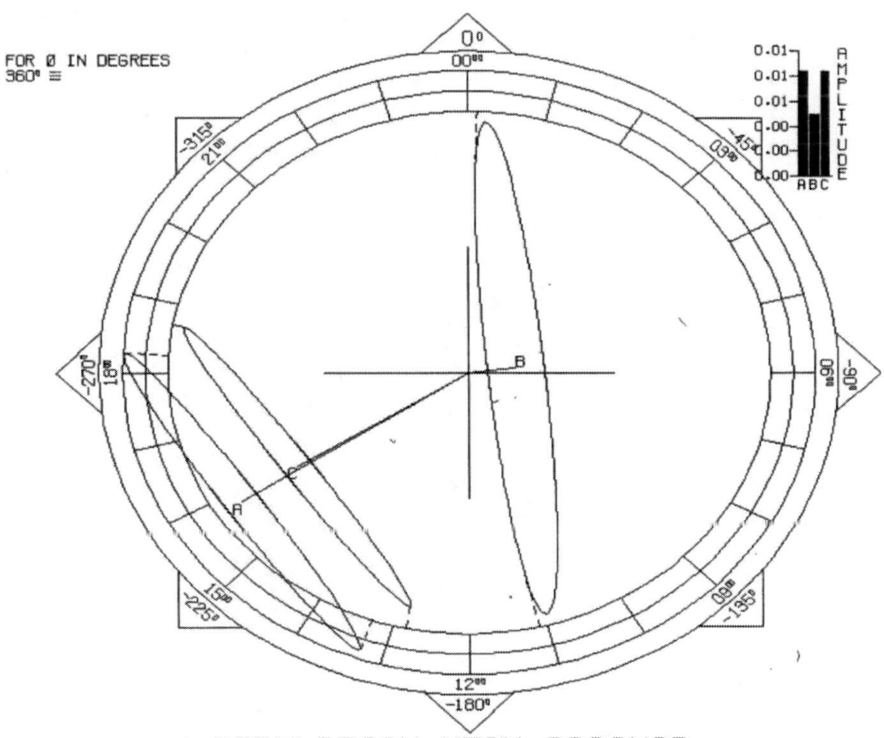

POPULATION MEAN COSINOR

KEY	P	N	PR	MESOR +-SE	AMPLITUDE *	ACROPHASE (Ø) *
A 168 HOURS	<.001	4	6	117.75 12.08	0.98 (.92 1.05)	-237° (-199 -274)
B 84 HOURS	.018	4	11	117.56 12.09	0.59 (.24 3.56)	-83° (-2 -166)
C 24 HOURS	<.001	4	30	115.81 12.38	0.98 (.91 1.39)	-238° (-192 -280)

Figure 21. Blood pressure on four patients of Ignaz Zadek in the 1880s. Error ellipses around directed lines (representing by their length half the extent of rhythmic change (amplitude) and by their angle the timing of overall high values recurring in each cycle) are 95% confidence regions that indicate statistical significance at the 5% probability level when they do not overlap the pole (center of the graph). © Halberg.

Figure 22. "Flying blind" between office visits and the (false) platinum standard of a (chronobiologically-uninterpreted) 24-hour blood pressure trace does not allow the diagnosis of Vascular Variability Disorders, VVDs, which remain unrecognized by both those administering treatment and those receiving it. The latter can hence gain greatly from self-help in computer-aided surveillance by a website (see Figure 18). Pieter Breughel's painting "The Parable of the Blind Leading the Blind", which may be the appropriate description of the status quo, is reproduced by kind permission of the Fototeca della Soprintendenza of the BAS PSAE and of the Polo Museale of the City of Naples. © Halberg.

Figure 23A. Henri Marie Joseph Vallot (1854-1925) is described in Wikipedia as a French astronomer, geographer and Maecenas (i.e., philanthropist). He came from a rich family who never had financial problems; his wealth allowed him to cover a great deal of the cost of the construction of his observatory (refuge) on Mont Blanc, which he discovered in 1875. "Nonetheless, scientists never considered him as one of their own, even when his work in different fields -- botany, glaciology, construction, geology, photography, medicine, physiology, cartography, alpinism, meteorology, etc. -- was recognized as being of important scientific interest. We wish to show herein that he was indeed a pioneer of the study of clinical symptoms associated with sunspots (see Figure 23B)."

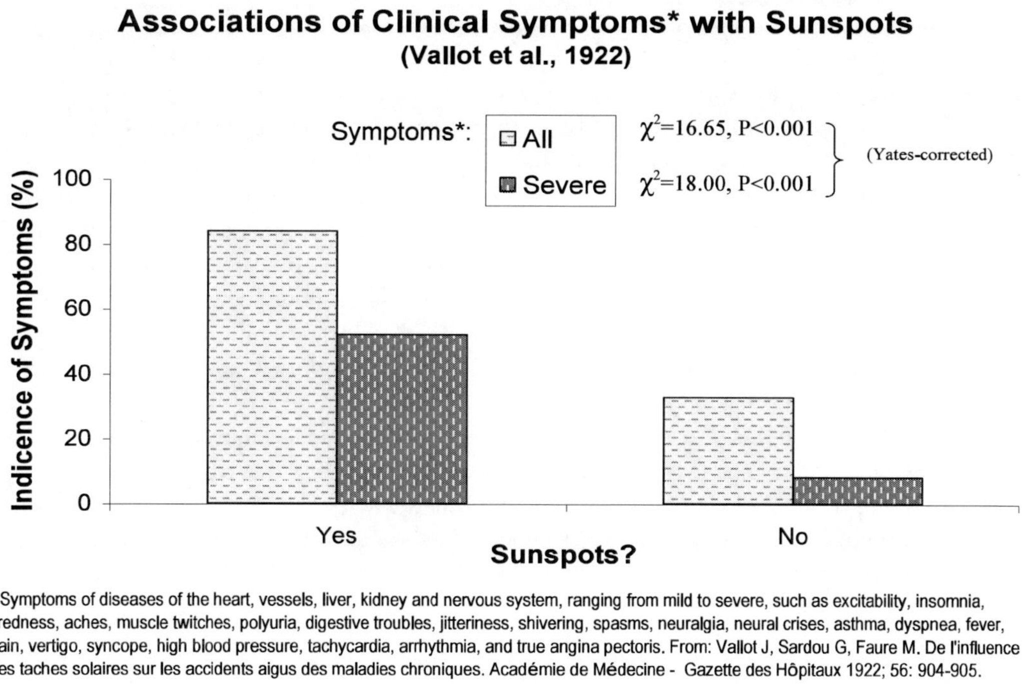

Figure 23B. Meta-analysis of what thoughtful and observant clinicians reported as an association of symptoms with solar activity over 80 years ago. Even mild symptoms, such as excitability, insomnia, tiredness, aches, muscle twitches, etc., have long been associated with sunspots (Vallot J, Sardou G, Faure M. De l'influence des taches solaires sur les accidents aigus des maladies chroniques. Académie de Médecine-Gazette des Hôpitaux 1922; 56: 904-905). We stand on the shoulders of many others who noted a lead in phase by a day of symptoms vs. sunspots.

We assume that these realistic problems, noted by Beaty for the USA, can be overcome in health care systems provided by the state or by others who do not wish to fly blind between office visits, i.e., most of the time, Figure 22.

We are reminded of mountaineers who risked their lives "because it's [the mountain] there". VVDs are here and there, and Joseph Vallot contributed more generally to the recognition that solar activity, reflected by the number of sunspots, can harm, Figure 23 [64].

GLOSSARY

(modified or amplified from Chronobiology glossary cf. Halberg F, Carandente F, Cornélissen G, Katinas GS. Glossary of chronobiology. Chronobiologia 1977; 4 [Suppl. 1], 189 pp) where more information ($^{\Delta}$) can be found.

Large ongoing international cooperative projects such as those revolving around chronobiologic self-help in health care often pose semantic problems that jeopardize the implementation of the work and lead to waste. This glossary should reduce, if not eliminate, semantic misunderstandings and thus contribute to the success of ongoing projects such as that on The BIOsphere and the COSmos, BIOCOS. Steps being implemented toward international cooperation require a selection of comparable if not unified reference standards.

Only thus can the definition of certain rhythm (cycle) characteristics become meaningful. By the same token, there is the need for using comparable analytical procedures that are generally applicable to systematically collected and stored data on blood pressure, heart rate and other time series. Broad international agreements can be reached toward the factual as well as semantic standardization of information capture, transfer, storage, analysis and updating from appropriate (data) bases: the characteristics of rhythms, estimated at different times and in different localities, will become amenable to a more facile and meaningful, direct comparison and integration. These are challenges and opportunities for the development of a serially updated individualized health form, card, booklet or equivalent, such as a cellphone, containing the information necessary for the precise early recognition of risk and thus for endeavors toward the prevention of diseases of individuals and (when data are pooled from many individuals) for learning about solar effects upon societal ills, for the development of countermeasures. In this glossary, the as-yet uninitiated reader may also find a stimulus toward gaining an interest in the time dimension necessary to reach a more dynamic understanding of the entire field of biology and broader science as it relates, beyond personal health, to society's ills and to our environment, in the atmosphere of the sun as a whole in which we happen to live.

ACROPHASE θ, ϕ, Φ

Measure of timing; the lag from a defined timepoint (acrophase$^{\Delta}$ reference) of the crest time in the function$^{\Delta}$ appropriately approximating a rhythm$^{\Delta}$; the phase$^{\Delta}$ angle of the crest, in relation to the specified reference timepoint, of a single best-fitting cosine (unless another approximating function is specified).

Units: angular measures: degrees, radians; time units: seconds, minutes, hours, days, months, years, decades, centuries etc.; or physiological episodic units: number of heart beats, respirations etc. Angular measures are directly applicable to any cycle$^{\Delta}$ length and hence are proposed for general use because of greater familiarity; degrees (with 360° equated to period of rhythm) are preferred over radians.

AMPLITUDE, A

Measure of one half the extent of rhythmic change in a cycle$^{\Delta}$ estimated by the sinusoidal (or other) function used to approximate the rhythm$^{\Delta}$, e.g., difference between maximum$^{\Delta}$ and MESOR$^{\Delta}$ of a best-fitting cosine.

Units: original physiological units, e.g., number of heart beats, mmHg in blood pressure, etc.

ANGULAR FREQUENCY, ω

Special case of frequency$^{\Delta}$ of a periodic process expressed in degrees or radians per unit of time obtained by equating one cycle$^{\Delta}$ to 2π, e.g., ω in equation

$$y(t) = M + A \cos(\omega t + \phi)$$

Used to approximate a rhythm$^{\Delta}$. Observe relation between angular frequency and frequency:

$$\omega = 2\pi/\tau = 2\pi f$$

Since frequency is the reciprocal of the period$^{\Delta}$:

$$f = 1/\tau$$

Note: equivalent to angular velocity, usually visualized on polar coordinates.

CHAT

Circadian hyper-amplitude-tension or circadian overswing, with circadian double amplitude exceeding the upper limit(s) of reference value(s) derived from clinically healthy peers matched by gender, age, and eventually ethnicity and geographic location.

CHRONODESM

Time-qualified reference interval, e.g., time-qualified prediction or tolerance interval.

CHRONOBIOLOGY

Computer-aided study in the biosphere of time structures, chronomes consisting of cycles, trends (that can be parts of cycles longer than a time series) and deterministic (and other) chaos (that can generate cycles).

CHRONOMICS

Computer-aided study of interacting time structures in the biosphere and in its environment.

CIRCADIAN

Relating to biologic variations or rhythms$^{\Delta}$ with a frequency$^{\Delta}$ of 1 cycle$^{\Delta}$ in 24 ± 4 h; circa (about, approximately) and dies (day or 24 h).

CIRCASEMISEPTAN

Half-weekly variation. Circasemiseptans characterize widely differing phenomena, such as the behavior on different lighting regimens of an enucleated giant green alga, or an aspect of the biochemistry of (anucleate) platelets and even sudden human death. Thus, in the last few decades in Canada, most sudden human cardiac deaths peak on Mondays, with a second peak on Thursdays. A 3.5-day cosine curve fits such data better than a 7-day cosine curve.

CIRCASEPTAN

About-weekly variation. Some human hormonal bioperiodicities, including rhythms$^{\Delta}$, follow a roughly weekly pattern, such as those in circulating cortisol. Circaseptans are also found to characterize death from a mouse malaria or the rejection of allografts of heart, pancreas or kidney in untreated rats. Human kidney transplant rejection episodes are also more likely to occur around the 7th, 14th, 21st and 28th days after operation, or near other multiples of 7 post-operative days.

CIRCATRIGINTAN

Variation, such as the human menstrual cycle, that approximates a month in duration; such bioperiodicities, including rhythms$^{\Delta}$, are also found before menarche and after menopause, and in men.

CONGRUENCE

Overlying or overlapping uncertainties (e.g., 95% confidence intervals) of two or more periods$^{\Delta}$ estimated in a time series$^{\Delta}$, given that the confidence interval of their amplitudes do not cover zero.

COSINOR

Statistical summary preferably with display of a biologic rhythm's$^{\Delta}$ amplitude$^{\Delta}$ and acrophase$^{\Delta}$ relations, on rectangular or polar coordinates; along the latter, by means of the length and the angle of a directed line, shown with a bivariate 95% or other statistical confidence$^{\Delta}$ region computed (at chosen trial period$^{\Delta}$) 1. to detect a rhythm$^{\Delta}$ (by a confidence$^{\Delta}$ region not overlapping zero, along rectangular coordinates, or the center of the plot, the pole, along polar coordinates) and 2. to estimate rhythm parameters$^{\Delta}$ with their uncertainties (confidence$^{\Delta}$ intervals).

Notes: among procedures for the analysis, mostly, of short time series$^{\Delta}$, three kinds of cosinor have been designed in an integrated routine, each appropriate to a different situation:

Single cosinor, cosinor-S procedure applicable to a single biologic time series$^{\Delta}$ (from an individual or group);

Group mean-cosinor, cosinor-G: a cosinor procedure applicable to data from two or more individuals for characterizing a rhythm$^{\Delta}$ in that particular group only;

Population mean-cosinor, cosinor-P: the original cosinor procedure applicable to parameter$^{\Delta}$ estimates from three or more biologic series for assessing the rhythm$^{\Delta}$ characteristics of a population. All three cosinors use a cosine function:

$$g(t) = M + A\cos(\omega t + \theta).$$

DEFICIENT HEART RATE VARIABILITY (DHRV)

A standard deviation of heart rate (determined around-the-clock for 7 days at 1-hour or shorter intervals) below the threshold of 7.5 beats/minute, a criterion to be further qualified for gender, age, and eventually ethnicity and geographic location.

DESYNCHRONIZATION

State of two or more previously synchronized rhythmic variables that have ceased to exhibit the same frequency$^{\Delta}$ and/ or the same acrophase$^{\Delta}$ relationships and show changing time relations.

ECPHASIA

Odd acrophase outside reference values of gender and age-matched peers.

ECFREQUENTIA

Odd frequency outside reference values of gender and age-matched peers.

ENTRAINMENT

Interaction between two or more organismic rhythms$^{\Delta}$ or the effect upon rhythm(s) of an (external) synchronizer$^{\Delta}$ resulting in identical frequencies among interactions or in frequencies constituting integral multiples of one another (frequency$^{\Delta}$ -- multiplication or demultiplication).

EXCESSIVE PULSE PRESSURE

Above the threshold of 60 mmHg (determined in a record of 7 days at 1-hour or shorter intervals), a criterion to be further qualified for gender and age.

FREE-RUNNING

Pertaining to continuance of bioperiodicity$^{\Delta}$ with a natural frequency$^{\Delta}$ usually at least slightly different from any known environmental schedule.

FREQUENCY, f

The number of occurrences of a given type of event in a given unit of time or the number of members in a population falling into a specified class.

Note: in the study of periodicity$^{\Delta}$ it is the number of cycles occurring per time unit, i.e., f is the reciprocal of the period (τ)

$$f = 1/\tau.$$

GLOCAL and GLOCALITY

Adjective and noun, respectively, beginning with the first syllable of GLObal and ending with the last syllable of loCAL, as "smog" is formed from SMoke and fOG. "Glocal" is proposed to designate, in principle and as method, an approach that is global and local both in time and in space. It is 1. global, both a. in time, insofar as it relates to the structure (or chronome) of a time series$^{\Delta}$ as a whole (in the longest available data series) and b. in space, insofar as it wishes to do so from the earth and other locations, such as the solar system, as a whole, as possible and reasonable and 2. local, again a. in time, insofar as it wishes to examine separately a set of intervals of different lengths and b. in space, namely separately from each of several terrestrial and other locations. As an example the incidence pattern of natality, morbidity or mortality can be studied in global and local statistics by spectra of entire time series and in intervals of each of the series of different length. In combination with spectral windows of an entire series, aligned gliding spectral windows, focusing on a given frequency region, and chronobiologic serial sections, focusing upon a single or few frequencies and their time course, are glocal procedures. The slogan "think globally, act locally" can thus be extended spatio-temporally.

INFRADIAN

Relating to certain biologic variations or rhythms$^{\Delta}$ with a frequency$^{\Delta}$ lower-than-circadian$^{\Delta}$.

LEAST-SQUARES METHOD

Estimation technique for determining quantities by minimizing the error (or residual) sum of squares. In a linear model, this method produces the best linear unbiased estimate (b.l.u.e.) in terms of variance. Note: two types of least-squares methods are considered: linear or nonlinear. The trial period is fixed in the former, allowed to vary in the latter.

MARKER RHYTHM

Rhythm$^{\Delta}$ of use in practical monitoring and, where appropriate, decision-making -- in applied or basic physiologic or pharmacologic work, in preventive health maintenance (prophylactic marker rhythm), risk monitoring (risk marker rhythm), for diagnostic purposes (diagnostic or screening marker rhythm), for timing therapy (chronotherapeutic marker rhythm) or for assessing therapeutic response (response marker rhythm) without any implication of necessarily causal relations between the process and its rhythmic marker.

MESOR, M

Rhythm-determined average of Midline Estimating Statistic of Rhythm, e.g., in the case of a single cosine approximation, the value midway between the highest and lowest values of function$^{\Delta}$ used to represent a rhythm$^{\Delta}$.

MESOR-HYPERTENSION, MH

For systolic and/or diastolic blood pressure, a transient$^{\Delta}$ or lasting elevation of the circadian$^{\Delta}$ (about-24-h) rhythm-adjusted mean (MESOR$^{\Delta}$, M) as validated statistically against the person's (patient's) own MESOR at another time in clinical health and/or against a peer reference standard.

PERIOD (τ)

Duration of one complete cycle$^{\Delta}$ in a rhythmic variation.

Note: biologic rhythms can be analyzed in terms of a spectrum$^{\Delta}$ with statistically significant components in several spectral$^{\Delta}$ domains. Period notation is customary within a given region or (e.g., circadian$^{\Delta}$) domain of the spectrum. Frequency$^{\Delta}$ (defined as the inverse of the period$^{\Delta}$ $f=1/\tau$) notation facilitates discussions of phenomena involving several broad spectral domains.

PREDICTION INTERVAL

A range of values expected to contain, on the average, a specified proportion of a population or of a distribution (of values) from an individual.

RESONANCE

Property of a system oscillating (or capable of oscillating) with some natural frequency$^{\Delta}$ (or rhythm$^{\Delta}$) to exhibit an increased amplitude$^{\Delta}$ (or to begin oscillating) when subjected to an external periodic influence or force with a frequency similar to that of the system, the amplitude of the resonant frequency increasing as the outside periodic influence approaches the natural frequency of the system.

RHYTHM

A periodic component of (biologic) time series$^{\Delta}$, demonstrated by inferential statistical means, preferably with objectively quantified characteristics (i.e., frequency$^{\Delta}$ f, acrophase$^{\Delta}$ [θ], amplitude$^{\Delta}$ A, MESOR$^{\Delta}$ M, and/or waveform$^{\Delta}$).

SYNCHRONIZER, Sy

Environmental periodicity determining the temporal placement of a given biologic rhythm$^{\Delta}$ along an appropriate time scale, by impelling the rhythm to assume synchronization$^{\Delta}$, i.e., its frequency$^{\Delta}$ or an integer multiple or submultiple of its frequency and a specifiable acrophase$^{\Delta}$ Note: also called zeitgeber, a time-giver, entraining agent (even though the environmental cycle does not give time), clue or cue. Adjectives, primary, dominant and secondary, describe relative roles played by different environmental synchronizers. In several strains of inbred mice fed ad libitum, the lighting regimen is the primary synchronizer of the blood eosinophil rhythm. Adjectives dominant and modifying also can be used to describe the effect of a given environmental factor in relation to a given rhythm. Under unusual circumstances a secondary synchronizer may become dominant. Thus in C_3H (Minnesota) mice subjected to a 50 percent restriction in dietary calories, the feeding time (of a diet restricted in calories) may be dominant over the lighting regimen. Moreover, under conditions of time-restricted access to food, synchronization by the lighting regiment may be largely though not fully overridden by meal timing. Thus, limited access time to food can be largely but not entirely dominant over lighting regimen with respect to the synchronization of the rhythm in telemetered intraperitoneal temperature. A secondary effect of the lighting regimen remains apparent as the result of interference among synchronizers. Finally, a rhythm can be influenced by secondary synchronizers and modifying factors, modulators, or, more generally, influencers.

ULTRADIAN

A variation with a frequency$^{\Delta}$ higher than 1 cycle in 20 h, i.e., with a period$^{\Delta}$ of less than 20 h. An example of an ultradian is the sleep-wake cycle of patients with narcolepsy, a sleep disorder in which patients fall asleep several times daily, e.g., with average periods of 1.7 h.

VASCULAR VARIABILITY ANOMALY (VVA):

An alteration of variability as compared to that of healthy peers, found in a 7-day record of measurements at hourly or shorter intervals analyzed as a whole, of blood pressure or heart rate. Examples are MESOR-hypertension; CHAT$^{\Delta}$ (circadian hyper-amplitude-tension); odd timing, or ecphasia; odd frequency$^{\Delta}$, or ecfrequentia; excessive pulse pressure and deficient heart rate variability.

VASCULAR VARIABILITY DISORDER (VVD)

A VVA replicated in at least three 7-day profiles at hourly or shorter intervals, each analyzed as a whole.

VASCULAR VARIABILITY SYNDROME (VVS)

Two or more kinds of VVDs present simultaneously in the same patient replicated in three 7-day/24-h records.

REFERENCES

[1] Halberg F, Cornelissen G, Otsuka K, Siegelova J, Fiser B, Dusek J, Homolka P, Sanchez de la Pena S, Singh RB, BIOCOS project. Extended consensus on need and means to detect vascular variability disorders (VVDs) and vascular variability syndromes (VVSs). *World Heart J.*, in press.

[2] Halberg F, Cornelissen G, Otsuka K, Watanabe Y, Katinas GS, Burioka N, Delyukov A, Gorgo Y, Zhao ZY, Weydahl A, Sothern RB, Siegelova J, Fiser B, Dusek J, Syutkina EV, Perfetto F, Tarquini R, Singh RB, Rhees B, Lofstrom D, Lofstrom P, Johnson PWC, Schwartzkopff O, International BIOCOS Study Group. Cross-spectrally coherent ~10.5- and 21-year biological and physical cycles, magnetic storms and myocardial infarctions. *Neuroendocrinol Lett* 2000; 21: 233-258.

[3] Halberg F, Cornelissen G, Sothern RB, Katinas GS, Schwartzkopff O, Otsuka K. Cycles tipping the scale between death and survival (= "life"). *Progress of Theoretical Physics* 2008; Suppl. 173: 153-181.

[4] Kamide Y. What human being cannot see can exist: a message from recent studies of solar-terrestrial relationships. *Biomed & Pharmacother* 2003; 17: 19s-23s.

[5] Otsuka K (Ed.) Proceedings, 1st International Symposium Workshop on Circadian Rhythms and Clinical Chronotherapy, 11 Nov 2000, Tokyo, Japan. *Biomed & Pharmacother* 2001; 55 (Suppl 1): 7s-190s.

[6] Otsuka K (Ed.) Proceedings, 2nd International Symposium Workshop on Circadian Rhythms and Clinical Chronotherapy, 17 Nov 2001, Tokyo, Japan. *Biomed & Pharmacother* 2002; 56 (Suppl 2): 231s-382s.

[7] Otsuka K (Ed.) Proceedings, 3rd International Symposium Workshop on Circadian Rhythms and Clinical Chronotherapy, 9 Nov 2002, Tokyo, Japan. *Biomed & Pharmacother* 2003; 57 (Suppl 1): 1s-198s.

[8] Otsuka K (Ed.) Proceedings, 4th International Symposium Workshop on Circadian Rhythms and Clinical Chronotherapy, 8 Nov 2003, Tokyo, Japan. *Biomed & Pharmacother* 2004; 58 (Suppl 1): S1-S188.

[9] Otsuka K (Ed.) Proceedings, 5th International Symposium Workshop on Circadian Rhythms and Clinical Chronotherapy, 6 Nov 2004, Tokyo, Japan. *Biomed. & Pharmacother* 2005; 59 (Suppl 1): S1-S261.

[10] Mikulecky M (Ed.) *The Moon and Living Matter.* Kosice, Slovakia, September 23-25, 1993. Bratislava: Slovak Medical Society; 1993. 97 pp.

[11] Mikulecky M (Ed.) *Sun, Moon and Living Matter.* Bratislava, Slovakia, June 28-July 1, 1994. Bratislava: Slovak Medical Society; 1994. 159 pp.

[12] Mikulecky M (Ed.). *Chronobiology and Its Roots in the Cosmos.* High Tatras, Slovakia, September 2-6, 1997. Bratislava: Slovak Medical Society; 1997. 287 pp.

[13] Halberg F, Kenner T, Fiser B, Siegelova J. (Eds.) *Proceedings, Noninvasive Methods in Cardiology*, Brno, Czech Republic, July 7-10, 2009. 402 p.

[14] Halberg F, Kenner T, Fiser B, Siegelova J. (Eds.) *Noninvasive Methods in Cardiology*, September 16-17, 2010, Brno, Czech Republic. Brno: Faculty of Medicine, Masaryk University. 164 p.

[15] Otsuka K, Cornelissen G, Norboo T, Takasugi E, Halberg F. Chronomics and "glocal" (combined global and local) assessment of human life. *Progress of Theoretical Physics* 2008; Suppl. 173: 134-152.

[16] Jozsa R, Halberg F, Cornelissen G, Zeman M, Kazsaki J, Csernus V, Katinas GS, Wendt HW, Schwartzkopff O, Stebelova K, Dulkova K, Chibisov SM, Engebretson M, Pan W, Bubenik GA, Nagy G, Herold M, Hardeland R, Hüther G, Pöggeler B, Tarquini R, Perfetto F, Salti R, Olah A, Csokas N, Delmore P, Otsuka K, Bakken EE, Allen J, Amory-Mazaudier C. Chronomics, neuroendocrine feedsidewards and the recording and consulting of nowcasts -- forecasts of geomagnetics. *Biomed & Pharmacother* 2005; 59 (Suppl 1): S24-S30.

[17] Chibisov SM, Cornelissen G, Halberg F. Magnetic storm effect on the circulation of rabbits. *Biomed. & Pharmacother* 2004; 58 (Suppl 1): S15-S19.

[18] Breus TK, Chibisov SM, Baevsky RM, Schebzuhov KV. *The chronostructure of heart biorhythms under effects of external environment factors.* Moscow: Russian University of People's Friendship, Polygraph Service; 2002. 231 pp.

[19] Otsuka K, Cornelissen G, Weydahl A, Holmeslet B, Hansen TL, Shinagawa M, Kubo Y, Nishimura Y, Omori K, Yano S, Halberg F. Geomagnetic disturbance associated with decrease in heart rate variability in a subarctic area. *Biomed & Pharmacother* 2001; 55 (Suppl 1): 51s-56s.

[20] Rosch PJ, Markov MS (Eds.). *Bioelectromagnetic Medicine.* New York: Marcel Dekker; 2004. 850 pp.

[21] Brown FA Jr. Response to pervasive geophysical factors and the biological clock problem. *Cold Spr Harb Symp quant Biol* 1960; 25: 57-71.

[22] Barnwell FH, Halberg F, Cornelissen G. Chronomics of an intertidal fiddler crab's locomotor activity under aquatically tideless laboratory conditions. *Proceedings, International Conference on the Frontiers of Biomedical Science: Chronobiology*, Chengdu, China, September 24-26, 2006, p. 89-92.

[23] West JB. Ibn al-Nafis, the pulmonary circulation, and the Islamic Golden Age. *J. Appl. Physiol.* 2008; 105: 1877-1880.

[24] Sarton G. *Introduction to the History of Science.* Vol. II: From Rabbi ben Ezra to Roger Bacon. Part II. Baltimore: Williams and Wilkins (for the Carnegie Institution of Washington); 1931. p. 1099-1101.

[25] Halberg F. Chronobiology: methodological problems. *Acta med. rom* 1980; 18: 399-440.

[26] Cornelissen G, Halberg F. Chronomedicine. In: Armitage P, Colton T. (Eds.) *Encyclopedia of Biostatistics,* 2nd ed. Chichester, UK: John Wiley and Sons Ltd; 2005. p. 796-812.

[27] Refinetti R, Cornelissen G, Halberg F. Procedures for numerical analysis of circadian rhythms. *Biological Rhythm Research* 2007; 38 (4): 275-325. *http://dx.doi.org/10.1080/09291010600903692*

[28] Gilbert W. *De magnete, magneticisque corporibus, et de magno magnete tellure; physiologia noua, plurimis et argumentis, et experimentis demonstrata* [On magnetism, magnetic bodies, and the great magnet Earth]. London: Short; 1600. 204 pp.

[29] Chapman S, Bartels J. *Geomagnetism.* 3rd ed. Oxford: Clarendon Press; 1962. 1049 pp.

[30] Roederer JG. Are magnetic storms hazardous to your health? *Eos, Transactions, American Geophysical Union* 1995; 76: 441, 444-445.

[31] Roederer JG. Tearing down disciplinary barriers. *Eos, Transactions, American Geophysical Union* 1985; 66: 681, 684-685.

[32] Gregori GP. Galaxy – Sun – Earth relations: The origin of the magnetic field and of the endogenous energy of the Earth, with implications for volcanism, geodynamics and climate control, and related items of concern for stars, planets, satellites, and other planetary objects. A discussion in a prologue and two parts. *Beiträge zur Geschichte der Geophysik und Kosmischen Physik*, Band 3, Heft 3, 2002, 471 pp. ISNN: 1615 - 2824 – NE: Gregori, Giovanni P. © Science Edition, Arbeitskreis Geschichte Geophysik / W. Schröder, Germany 2002.

[33] Halberg F, Breus TK, Cornelissen G, Bingham C, Hillman DC, Rigatuso J, Delmore P, Bakken E, International Womb-to-Tomb Chronome Initiative Group: Chronobiology in space. Keynote, 37th Ann. Mtg. Japan Soc. for Aerospace and Environmental Medicine, Nagoya, Japan, November 8-9, 1991. University of Minnesota/Medtronic *Chronobiology Seminar Series*, #1, December 1991, 21 pp. of text, 70 figures.

[34] Halberg F, Cornelissen G, Schack B, Wendt HW, Minne H, Sothern RB, Watanabe Y, Katinas G, Otsuka K, Bakken EE. Blood pressure self-surveillance for health also reflects 1.3-year Richardson solar wind variation: spin-off from chronomics. *Biomed & Pharmacother* 2003; 57 (Suppl 1): 58s-76s.

[35] Sigel F (Dreier W, Lerche D, Übers.; Göring H, Wissenschaftl. Red. der deutschsprachigen). *Schuld ist die Sonne*. Thun/Frankfurt am Main: Harri Deutsch; 1979. 215 pp.

[36] Halberg F, Cornelissen G, Otsuka K, Katinas G, Schwartzkopff O. Essays on chronomics spawned by transdisciplinary chronobiology: Witness in time: Earl Elmer Bakken. *Neuroendocrinol Lett* 2001; 22: 359-384.

[37] Halberg F, Cornelissen G, Katinas G, Tvildiani L, Gigolashvili M, Janashia K, Toba T, Revilla M, Regal P, Sothern RB, Wendt HW, Wang ZR, Zeman M, Jozsa R, Singh RB, Mitsutake G, Chibisov SM, Lee J, Holley D, Holte JE, Sonkowsky RP, Schwartzkopff O, Delmore P, Otsuka K, Bakken EE, Czaplicki J, International BIOCOS Group. Chronobiology's progress: season's appreciations 2004-2005. Time-, frequency-, phase-, variable-, individual-, age- and site-specific chronomics. *J. Appl.Biomed.* 2006; 4: 1-38. *http://www.zsf.jcu.cz/vyzkum/jab/4_1/halberg.pdf*

[38] Kovac M, Mikulecky M. Secular rhythms and Halberg's paraseasonality in the time occurrence of cerebral stroke. *Bratisl Lek Listy* 2005; 106 (2): 423-427.

[39] Mikulecky M, Florida PL. Daily birth numbers in Davao, Philippines, 1993-2003: Halberg's transyear stronger than year. Abstract, *26th Seminar, Man in His Terrestrial and Cosmic Environment*, Upice, Czech Republic, May 17-19, 2005.

[40] Kovac M, Mikulecky M. Time sequence of epileptic attacks from the point of view of possible lunisolar connections. *International Conference on the Frontiers of Biomedical Science: Chronobiology,* Chengdu, China, September 24-26, 2006, p. 175-179.

[41] Mikulecky M. Reanalysis of variability in south Brazil: Halberg's paraseasonality dominating again. *International Conference on the Frontiers of Biomedical Science: Chronobiology*, Chengdu, China, September 24-26, 2006, p. 187-188.

[42] Hamamatsu A, Cornelissen G, Otsuka Ku, Halberg F, Chibisov S (presenter). Linear-nonlinear rhythmometry documents a transyear and a cishalfyear in sudden cardiac death (ICD 10, code I46.1) in Tokyo. In: *Proceedings, International Symposium, Problems of ecological and physiological adaptation, People's Friendship University of*

Russia, Moscow, 30-31 Jan 2007. Moscow: People's Friendship University of Russia; 2007. p. 542-545.

[43] Halberg F, Cornelissen G, Otsuka K, Fiser B, Mitsutake G, Wendt HW, Johnson P, Gigolashvili M, Breus T, Sonkowsky R, Chibisov SM, Katinas G, Siegelova J, Dusek J, Singh RB, Berri BL, Schwartzkopff O. Incidence of sudden cardiac death, myocardial infarction and far- and near-transyears. *Biomed & Pharmacother* 2005; 59 (Suppl 1): S239-S261.

[44] Halberg F, Cornelissen G, Panksepp J, Otsuka K, Johnson D. Chronomics of autism and suicide. *Biomed & Pharmacother* 2005; 59 (Suppl 1): S100-S108.

[45] Cornelissen G, Halberg F. Chronomics of suicides and the solar wind. *Br. J. Psychiatry* 2006; 189: 567-568.

[46] Halberg F, Cornelissen G, Berk M, Dodd S, Henry M, Wetterberg L, Nolley E, Beaty L, Siegelova J, Fiser B, Wolff C, BIOCOS project. Solar signatures in Australian suicide incidence: gender differences in prominence of photic vs. nonphotic spectral components. In: Halberg F, Kenner T, Fiser B, Siegelova J. (Eds.) *Proceedings, Noninvasive Methods in Cardiology*, Brno, Czech Republic, October 4-7, 2008. p. 44-62. *http://web.fnusa.cz/files/kfdr2008/sbornik_2008.pdf*

[47] Stehr N, von Storch H. (Eds.) (Stehr B, Gamlin G, trans). *Eduard Brückner: the sources and consequences of climate change and climate variability in historical times.* Dordrecht/Boston: Kluwer Academic Publishers; 2000. 338 p.

[48] Stehr N, von Storch H. *Eduard Brückner's ideas -- relevant in his time and today.* Geesthacht: GKSS-Forschungszentrum Geesthacht GmbH; 2006. XXXV and 393 pp.

[49] Hoyt DV, Schatten KH. *The Role of the Sun in Climate Change.* New York/Oxford: Oxford University Press; 1997. 279 pp.

[50] Halberg F, Cornelissen G, Sothern RB, Czaplicki J, Schwartzkopff O. Thirty-five-year climatic cycle in heliogeophysics, psychophysiology, military politics, and economics. *Izvestiya, Atmospheric and Oceanic Physics* 2010; 46 (7): 844-864. (Back-translation from *Geophysical Processes and Biosphere* 2009; 8 [2]: 13-42.)

[51] Halberg Franz, Cornelissen G, Katinas G, Syutkina EV, Sothern RB, Zaslavskaya R, Halberg Francine, Watanabe Y, Schwartzkopff O, Otsuka K, Tarquini R, Perfetto P, Siegelova J. Transdisciplinary unifying implications of *circadian findings in the 1950s, J* Circadian Rhythms 2003; 1: 2. 61 pp. *www.JCircadianRhythms.com/content/pdf/1740-3391-2-3.pdf*

[52] Halberg F, Cornelissen G, Bernhardt K-H, Sampson M, Schwartzkopff O, Sonntag D. Egeson's (George's) transtridecadal weather cycling and sunspots. *Hist Geo Space Sci.* 2010; 1: 49-61.

[53] Cornelissen G, Halberg F, Sothern RB, Hillman DC, Siegelova J. Blood pressure, heart rate and melatonin cycles synchronization with the season, earth magnetism and solar flares. *Scripta med.* 2010; 83: 16-32.

[54] Halberg F, Cornelissen G, Schwartzkopff O. Quo vadis chronomics 2008: Measuring variability in us, among us and around us. In: Halberg F, Kenner T, Fiser B, Siegelova J. (Eds.) *Proceedings, Noninvasive Methods in Cardiology,* Brno, Czech Republic, October 4-7, 2008. p. 16-25. *http://web.fnusa.cz/files/kfdr2008/sbornik_2008.pdf*

[55] Halberg F, Cornelissen G, Katinas GS, Watanabe Y, Siegelova J. Cosmic inheritance rules: implications for health care and science. *Scripta med.* 2010; 83: 5-15.

[56] Wolff CL. The rotational spectrum of g-modes in the sun. *Astrophys. J.* 1983; 264: 667-676.

[57] Rieger E, Share GH, Forrest DJ, Kanbach G, Reppin C, Chupp EL. A 154-day periodicity in the occurrence of hard solar flares? *Nature* 1984; 312: 623-625.

[58] Halberg F, Engeli M, Hamburger C, Hillman D. Spectral resolution of low-frequency, small-amplitude rhythms in excreted 17-ketosteroid; probable androgen induced circaseptan desychronization. *Acta endocrinol* (Kbh) 1965; 50 (Suppl 103): 5-54.

[59] Cornelissen G, Halberg F, Pöllmann L, Pöllman B, Katinas GS, Minne H, Breus T, Sothern RB, Watanabe Y, Tarquini R, Perfetto F, Maggioni C, Wilson D, Gubin D, Otsuka K, Bakken EE. Circasemiannual chronomics: half-yearly biospheric changes in their own right and as a circannual waveform. *Biomed. & Pharmacother* 2003; 57 (Suppl 1): 45s-54s.

[60] Halberg F, Cornelissen G, Schwartzkopff O, Bakken EE. Cycles in the biosphere in the service of solar-terrestrial physics? In: Schröder W. (Ed.) *Case studies in physics and geophysics.* Bremen: Wilfried Schröder/Science Edition, 2006, p. 39-87. [Beiträge zur Geophysik und Kosmischen Physik/Journal for the History of Geophysics and Cosmical Physics, Special issue, 2006/2. ISSN 1615-2824]

[61] Halberg F, Cornelissen G, Grambsch P, McCraty R, Beaty L, Siegelova J, Homolka P, Hillman DC, Finley J, Thomas F, Kino T, Revilla M, Schwartzkopff O. Personalized chronobiologic cybercare; other chronomics' progress by transdisciplinary cycles' congruences: Season's Appreciations 2009. *J. Appl Biomed.* 2011; 9: 1-34. DOI 10.2478/v10136-009-0022-8.

[62] Cornelissen G, Grambsch P, Sothern RB, Katinas G, Otsuka K, Halberg F. Congruent biospheric and solar-terrestrial cycles. *J. Appl. Biomed.* 2011; 9: 1-40. DOI 10.2478/v10136-009-0023-7.

[63] Halberg F, Cornelissen G, Wilson D, Singh RB, De Meester F, Watanabe Y, Otsuka K, Khalilov E. Chronobiology and chronomics: detecting and applying the cycles of nature. *Biologist* 2009; 56 (4): 209-214.

[64] Vallot J, Sardou G, Faure M. De l'influence des taches solaires: sur les accidents aigus des maladies chroniques. *Gazette des Hôpitaux* 1922; 904-905.

[65] Chizhevsky [Tchijevsky] AL (de Smitt VP, trans and condensed). Physical factors of the historical process. *Cycles* 1971; 22: 11-27. *http://www.cyclesresearchinstitute.org /chizhevsky/chizhevsky1.pdf*

[66] Chizhevsky AL. Action de l'ionisation de l'atmosphère et de l'ionisation artificielle de l'air sur les organismes sains et les organismes malades. In: Piéry M. (Ed.) *Traité de Climatologie: Biologique et médicale.* Tome premier. Paris: Masson et Cie; 1934. p. 662-673.

[67] Ertel S. Space weather and revolutions: Chizhevsky's heliobiological claim scrutinized. *Studia Psychologica* 1996; 39: 3-22.

[68] Ertel S. Influenza pandemics and sunspots -- easing the controversy. *Naturwissenschaften* 1994; 82: 308-310.

[69] Halberg F, Smith HN, Cornelissen G, Delmore P, Schwartzkopff O, International BIOCOS Group. Hurdles to asepsis, universal literacy, and chronobiology -- all to be overcome. *Neuroendocrinol Lett.* 2000; 21: 145-160.

[70] Ahlgren A, Halberg F. Cycles of Nature: *An Introduction to Biological Rhythms.* Washington, DC: National Science Teachers Association; 1990. 87 pp. (Library of

Congress Catalog Card #89-063723; ISBN #0-87355-089-7. National Science Teachers Association, 1742 Connecticut Ave. N.W., Washington, DC 20009, www.nsta.org)

[71] Koop CE, Mosher R, Kun L, Geiling J, Grigg E, Long S, Macedonia C, Merrell RC, Satava R, Rosen JM. Future delivery of health care: cybercare. A distributed network-based health-care system. *IEEE Engineering in Medicine and Biology Mag* 2008; 29-38. doi 10.1109/MEMB.2008.929888

[72] Halberg F, Cornelissen G, Otsuka K, Siegelova J, Fiser B, Dusek J, Homolka P, Sanchez de la Pena S, Singh RB, BIOCOS project. Extended consensus on means and need to detect vascular variability disorders (VVDs) and *vascular variability syndromes (VVSs). Leibniz*-Online Nr. 5, 2009 (http://www2.hu-berlin.de/leibniz-sozietaet/journal/archiv_5_09.html). 35 pp; cf. Halberg F, Cornelissen G, Otsuka K, Siegelova J, Fiser B, Dusek J, Homolka P, Sanchez de la Pena S, Singh RB, BIOCOS project. Extended consensus on need and means to detect vascular variability disorders (VVDs) and vascular variability syndromes (VVSs). Geronto-Geriatrics: *Int. J. Gerontology-ChronomeGeriatrics* 2008; 11 (14): 119-146.

[73] Halberg F, Cornelissen G, Katinas G, Tvildiani L, Gigolashvili M, Janashia K, Toba T, Revilla M, Regal P, Sothern RB, Wendt HW, Wang ZR, Zeman M, Jozsa R, Singh RB, Mitsutake G, Chibisov SM, Lee J, Holley D, Holte JE, Sonkowsky RP, Schwartzkopff O, Delmore P, Otsuka K, Bakken EE, Czaplicki J, International BIOCOS Group. Chronobiology's progress: Part II, chronomics for an immediately applicable biomedicine. *J. Appl. Biomed.* 2006; 4: 73-86. *http://www.zsf.jcu.cz/vyzkum/jab/4_2/ halberg2.pdf*

[74] Cornelissen G, Halberg F, Hawkins D, Otsuka K, Henke W. Individual assessment of antihypertensive response by self-starting cumulative sums. *J. Medical Engineering and Technology* 1997; 21: 111-120.

[75] Adams C. Privacy requirements for low-cost medical systems. *Proceedings, International Conference on the Frontiers of Biomedical Science: Chronobiology*, Chengdu, China, September 24-26, 2006, p. 64-69.

[76] Beaty L. The status quo of chronobiologic 7-day/24-hour monitoring of blood pressure and heart rate. Geronto-Geriatrics: *Int. J. Gerontology-Chronome Geriatrics* 2008; 11 (14): 155.

[77] Beaty L. Web-enabled software to complement affordable automatic hardware: Phoenix website. Geronto-Geriatrics: *Int. J. Gerontology-Chronome Geriatrics* 2008; 11 (14): 156-158.

[78] Beaty L, Cornelissen G. Towards inexpensive ambulatory blood pressure variability monitors for home use: A first experiment using off-the-shelf devices as the basis for ABPMs. *http://www.phoenix.tc-ieee.org/0001_Bibliography/20081221beat_ FirstHomemade ABPM.pdf*

[79] Fossel M. Editor's Note (Halberg F, Cornelissen G, Halberg J, Fink H, Chen C-H, Otsuka K, Watanabe Y, Kumagai Y, Syutkina EV, Kawasaki T, Uezono K, Zhao ZY, Schwartzkopff O. Circadian Hyper-Amplitude-Tension, CHAT: a disease risk syndrome of anti-aging medicine. *J. Anti-Aging Med.* 1998; 1: 239-259). J Anti-Aging Med 1998; 1: 239.

[80] Bartter FC. Periodicity and medicine. In: Scheving LE, Halberg F, Pauly JE. (Eds.) *Chronobiology.* Tokyo: Igaku Shoin Ltd.; 1974. p. 6-13. On his patient whose blood pressure was diagnosed differently by two physicians who saw him at different times of

day: "By conventional standards, this patient is clearly normotensive every morning. Yet the BP determined each day at 6 in the afternoon provides especially convincing evidence that this patient is a hypertensive. ... My plea today [in 1974!] is that information contained in such curves [cosinor fits] become a routine minimal amount of information accepted for the description of a patient's blood pressure. The analysis of this information by cosinor should become a routine. It is essential that enough information be collected to allow objective characterization of a periodic phenomenon, to wit, an estimate of M [the time structure or chronome-adjusted mean, or MESOR] … an estimate of [the amplitude] A itself, and finally an estimate of acrophase, ϕ [a measure of timing]. In this way, a patient can be compared with himself at another time, or under another treatment, and the patient can be compared with a normal or with another patient."

[81] Bingham C, Arbogast B, Cornelissen Guillaume G, Lee JK, Halberg F. Inferential statistical methods for estimating and comparing cosinor parameters. *Chronobiologia* 1982; 9: 397-439.

[82] Hawkins DM. Self-starting CUSUM charts for location and scale. *The Statistician* 1987; 36: 299-315.

[83] Zadek I. *Die Messung des Blutdrucks am Menschen mittelst des Basch'chen Apparates.* Berlin, med. F., Diss., 25. Nov 1880. Berlin: Schumacher; 1880. 48 p.

[84] Halberg F, Schwartzkopff O, Cornelissen G, Hardeland R, Müller-Bohn T, Katinas G, Revilla MA, Beaty L, Otsuka K, Jozsa R, Zeman M, Csernus V, Hoogerwerf WA, Nagy G, Stebelova K, Olah A, Singh RB, Singh RK, Siegelova J, Dusek J, Fiser B, Czaplicki J, Kumagai Y, Chibisov SM, Frolov VA. Vaskuläres Variabilitäts-Syndrom (VVS) und andere Chronomik 2005-2007. In: Hardeland R. (Ed.) *Sonderdruck aus Abhandlungen der Leibniz-Sozietät der Wissenschaften*, Band 23: Facetten der Chronobiologie. Berlin: trafo verlag; 2008. p. 89-154.

[85] Janeway TC. *The clinical study of blood pressure*. New York: D. Appleton and Co.; 1904. 300 pp. "… it is essential that a record of the pressure be made at frequent intervals at some time previous [presumably to an examination], to establish the normal level and the extent of the periodic variations. When this is done, it may be possible to demonstrate changes of small extent, which, lacking this standard for comparison, would be considered within the limits of normal variation."

[86] Feinleib M, Rogot E, Sturrock PA. Solar activity and mortality in the United States. *Int. J. Epidemiol* 1975; 4: 227-229.

[87] Lipa BJ, Sturrock PA, Rogot E. Search for correlation between geomagnetic disturbances and mortality. *Nature* 1976; 259: 302-304.

[88] Halberg F, Bakken EE, Katinas GS, Cornelissen G, Zaslavskaya RM, Blank MA, Syutkina EV, Breus TK, Watanabe Y, Masalov A, Chibisov SM. Chronoastrobiology: Vernadsky's future science? Benefits from spectra of circadians and promise of a new transdisciplinary spectrum of near-matching cycles in and around us. *Proceedings, III International Conference, Civilization diseases in the spirit of V.I. Vernadsky*, People's Friendship University of Russia, Moscow, Oct. 10-12, 2005, p. 4-22.

[89] Düll B, Düll T. Erd- und Sonnenphysikalische Vorgänge in ihrer Bedeutung für Krankheits- und Todauslosung. *Nosokomeion* 1938; 9: 103-120.

[90] Stetson HT. *Sunspots in action*. New York: Ronald Press Co.; 1947. 252 pp.

[91] Dubrov AP (Sinclair FL, trans; Brown FA Jr, Ed). *The geomagnetic field and life: geomagnetobiology*. New York: Plenum Press, 1978: 318 pp.

[92] Halberg F, Cornelissen G, Otsuka K, Syutkina EV, Masalov A, Breus T, Viduetsky A, Grafe A, Schwartzkopff O. Chronoastrobiology: neonatal numerical counterparts to Schwabe's 10.5 and Hale's 21-year sunspot cycles. In memoriam Boris A. Nikityuk. *Int J Prenat Perinat Psychol Med* 2001; 13: 257-280.

[93] Cornelissen G, Halberg F, Wendt HW, Bingham C, Sothern RB, Haus E, Kleitman E, Kleitman N, Revilla MA, Revilla M Jr, Breus TK, Pimenov K, Grigoriev AE, Mitish MD, Yatsyk GV, Syutkina EV. Resonance of about-weekly human heart rate rhythm with solar activity change. *Biologia* (Bratislava) 1996; 51: 749-756.

[94] Halberg F, Cornelissen G, Sothern RB, Barnwell F, Cegielski N, Ilyia E, Siegelova J. The moon's and the genes' tides and double tides pulling the biosphere. In: Halberg F, Kenner T, Fiser B, Siegelova J (Eds.) *Noninvasive Methods in Cardiology*, September 16-17, 2010, Brno, Czech Republic. Brno: Faculty of Medicine, Masaryk University. p. 23-45.

[95] Cornelissen G, Frolov V, Chibisov S, Kharlitskaya E, Halberg F. Personalized chronome (time structure) complements genome detecting vascular variability disorders (VVDs), including MESOR-hypertension. *Vestnik Rossiyskogo Universiteta Droozhbi Narodov / Bulletin of People's Friendship University of Russia* 2010 (4): 16-25.

In: Cardiovascular Health and Chronomics
Editor: Ram B. Singh

ISBN: 978-1-62948-976-6

Chapter 5

CARDIAC SUDDEN DEATH OF ATHLETES: "ATHLETE'S HEART"

J. Fedacko*[1], D. Pella[1], R. B. Singh[2], B. Bohus[1], R. Rybar[1], J. Pella[1], P. Fedackova[1], K. Trejbal[1], T. Lopuchovsky[1], L. Merkovska[1], L. Jedlickova[1], M. Mraz[3], M. Babcak[4], I. Pirnikova[5] and J. Simonova[5]

[1]Third Dept. of Internal Medicine, Faculty of Medicine, P.J. Safarik University and L. Pasteur University Hospital Kosice, Slovakia
[2]Halberg Hospital and Research Institute, Moradabad, India
[3]Reiman Hospital, Presov, Slovakia
[4]First Pediatric Clinic, Faculty of Medicine, Safarik University Kosice, Slovakia
[5]First Dept. of Anaestesiology and Intensive Medicine, Faculty of Medicine, P.J. Safarik University and L. Pasteur University Hospital Kosice, Slovakia

ABSTRACT

Over the past several years there is a considerably interest among various experts and the lay public to find out the causes of sudden deaths in the trained athletes as well as in the clinical significance of cardiac symptoms as syncope and arrythmias. Catastrophic events in athletes are always unexpected, and although relatively uncommon, often achieve high visibility and convey a particularly devastating impact on the community. The most common causes of death in the young athletes are hypertrofic cardiomyopathy (need to be differentiated from "athlete's heart"), congenital coronary arteries anomalies and commotio cordis. Older athletes can also harbor occult cardiac disease, but the most common cause of death is atherosclerotic coronary artery disease (CAD). In view of these findings, the guidelines for prevention, management and rehabilitation were established, because the overall athlete population is in low risk dying for sudden cardiac death, but the risk exists.

* Correspondence author: Jan Fedacko, MD, PhD; 3rd Dept. of Internal Medicine, Faculty of Medicine, P.J. Safarik University and L. Pasteur University Hospital Kosice, Tr.SNP 1, Kosice, Slovakia. Tel/fax: +421-55-640-3861. Email: jan.fedacko@upjs.sk.

Keywords: Physical activity, games, hear health, stress and strain

INTRODUCTION

Athletic exertion; dynamic, aerobic or isometric (static, power) involves complex physiological, and morphological alterations of electrocardiographic manifestations depending upon on the level of physical exercise [2]. Physiological adaptation is an increase of ejection volume of the heart, arterio-venous oxygen differentiation, maximum oxygen consumption and reduction of resting heart rate and increase in heart rate variability. These conditions then result in the structural remodeling. There is an increase in left ventricular mass, enlargement of all four compartments and sometimes thickening of the left ventricular wall which occur in order to ensure all the cardiac function (systolic and diastolic) at high loads sports [8-13]. This physiological form of hypertrophy of the heart is called the "Athlete´s heart" which is generally regarded as a benign form of adaptation to continuous overload (training) and is not associated with increased cardiovascular morbidity. The increase in cardiac mass is different. In isometric workouts, it is the left ventricular wall thickening which is predominantly disproportional with left ventricular size. However, proportional hypertrophy may be associated with adequate left ventricular size which occurs predominantly during its dynamic workouts, especially during endurance exercise. Maximum increase in endurance exercises may be observed in cyclists, swimmers and skiers. This physiological cardiac hypertrophy is usually directly proportional to body surface and is less pronounced in women compared to man [8,11]. Among other physiological adaptations, changes in the electrocardiographic manifestations, particularly increased QRS voltage in the precordial leads, Q wave abnormalities during repolarization appear to be important [13].

RISK OF CARDIOVASCULAR DISEASES AMONG ATHLETES

Clinical and physiological differences between "Athlete´s heart" and pathological changes are critical for a trained athlete, because presence of the cardiovascular disease may cause disqualification of the profesional sportsman due to the increased risk of sudden cardiovascular death or progression of the disease among them [14]. On the other hand, the revaluation of diagnosis often leads to unnecessary restrictions on the sportsman, as well as psychological trauma on the athlete [2]. Adaptation of athlete´s heart easily imitates cardiovascular disease, especially when some dimensions protrude from the framework and standards mimic hypertrophic, dilated cardiomyopathy or arythmogenic right ventricular cardiomyopathy. It is therefore justified to know the alterations in different conditions, which are described in the table 1. However, it should be clear in mind that about 2% of athletes reach some imaginary intersection, and these athletes can not be clearly distinguished from physiological adaptation or from pathologies. These characteristics can be relatively well-recognizable by non-invasive method such as Doppler echocardiography [14].

CAUSES OF SUDDEN DEATH IN YOUNG ATHLETES (UNDER 35 YEARS)

Although the overall population of athletes is at a low risk, mainly congenital (usually undiagnosed) cardiovascular disease, are a major cause of sudden death in young athletes. Such deaths are relatively not very common in the general population, Apart from this, any cardiac disease, which can predispose sudden death in young non-athletes, may also be the cause of increased incidence of sudden deaths among athletes (Table. 2) [1,2-6,7,15]. Approximately 80% of deaths of the athletes were linked to the structural causes of cardiovascular disease [1,2,6,7].

Approximately one third of sudden death among athletes is due to hypertrophic cardiomyopathy (HCMP). HCMP is a relatively common genetic disorder (1:500) as characterized by asymmetric left ventricular hypertrophy without dilatation with heterogeneous clinical, morphological and genetic expression. It is possible that HCMP may be the leading cause of death in young athletes, majority of deaths occur during physical exertion. Sudden death due to ventricular tachyarrhythmias occur due to presence of an underlying electrically instability of the myocardial substrate [1,2,6,7,15]. This finding has been proven by histopathology, showing disorganisation of the myocardial structure and compensation with scares causing microvascular abnormalities resulting into flare myocardial ischemia (16). Athletes at high risk of HCMP are destined in the primary prevention for prophylactic implantable cardioverter-defibrillator [17].

Other common causes of death in athletes (about 20%) are congenital anomalies of coronary arteries. The most common is bad mouth of the left coronary artery from the right front Valsav sinus. [1,2,18]. Diagnosis of these anomalies should be borne in mind in young athletes with symptoms such as syncope and stenocardia pertaining to physical activity [18.19]. It is necessary to implement transesophageal echocardiography, angiography, CT or MRI to confirm the diagnosis. These patients do not show any abnormalities at 12 lead ECG, ergometry or ballast, since myocardial ischemia is episodic [18].

Other, but less common causes include premature coronary artery atherosclerosis, which is not very common in this population, and therefore remains often indistinguishable.

Table 1. Differential diagnosis condition of HCMP from "Athlete´s heart" [14]

HCMP		"Athlete´s heart"
+	Heterogeneous distribution of left ventricular hypertrophy	-
+	Left ventricular size less than 45 mm	-
-	The size of the left ventricle more than 55 mm	+
+	The enlarged left atrium	-
+	ECG changes	-
+	Abnormal filling of left ventricular	-
+	Female	-
-	Wall thinning associated with inefficiency	+
+	Family history of HCMP	-
-	VO2 max. more than 50ml/kg/min (More than 120% provided)	+

Table 2. Most common causes of sudden death in athletes [15]

	PRÍČINA	%
1	HCMP	36
2	Anomalies of the coronary arteries	17
3	Nedeterminované (perhaps HCMP)	10
4	Others	6
5	Coronary arteries disease	3
6	Prolaps of mitral valve	3
7	Arrhythmogenic right ventricular dysplasia	4
8	Dilatated CMP	3
9	Myocarditis	7
10	Aortal stenosis	3

Apart from this, dilated cardiomyopathy, arrhythmogenic (cardiomyopathy) of the right ventricular, are other rare coronary anomalies. Chronic myocarditis may be a undiagnosed cause of sudden deaths in athletes. Arrhythmogenic (cardiomyopathy) of the right ventricular is rare, familial disease associated with ventricular tachyarrhythmias or supraventricul tachyarrhythmias and sudden death. It is characterized by diffuse or segmental abnormalities of the right ventricle due to myocardial necrosis, fibrous or fatty tissue infiltration and coexisting myocarditis. These pathobiological changes may be associated with enlargement of the cardiac chambers and systolic dysfunction, T wave inversion and *epsilon* waves in right precordial leads. About 2% of deaths in young athletes are without a clear cause of death based on autopsies in the normal myocardial structure. These sudden deaths are probably associated with abnormalities such as disorders of ion channels such as long QT interval, Brugada syndrome and Wolff-Parkinson-White syndrome and disorders of the transmission system catecholamine polymorphic ventricular tachycardia, coronary vasospasm, and doping. [2,7,20-23].

Causes of Sudden Death in Athletes Older Than 35 Years

Older athletes may also suffer from hidden cardiovascular disease and die suddenly, abruptly and unexpectedly during sports shows by cyclists, marathon runners, rugby and squash, but the most common cause of sudden death in this age category is premature coronary artery atherosclerosis. Majority of the older athletes who die of sudden death due to coronary heart disease had well known risk factors, symptoms and possibly previous heart attack or other coronary intervention in their personal history. Finally, we should always exclude ischemic diseases such as HCMP and valvular heart disease among athletes. [4, 8].

Commotio Cordis

Sudden cardiac arrest can occur during sports activities, the abrupt, blunt striking the chest in the heart, with absence of cardiovascular disease. This condition is called commotio cordis, and is most common in baseball, softball, motoring but also in ice hockey (the puck), or other contact sports (boxing). It should be noted that many of these injuries occurs more frequently among recreational athletes, and most frequently in children (mean age 13 years), which allows configuration of the chest and facilitates the transfer of energy (shock) on the myocardium, where they apply two essential factors. The first is that the blow is directly located on the level of your heart, and the second is the time, respectively the phase in which power operates on the myocardium. The most fatal consequences of the blow due to the 20 ms window during repolarisation, just before the T wave, which makes the activation of potassium channels in the myocardium [24].

THE FREQUENCY AND DEMOGRAPHICS OF SUDDEN DEATH IN ATHLETES

Sudden death among athletes has become an important health problem in the western world. Sudden death occurs in approximately 1:200 000 per academic year and approximately 1:70 000 during the three-year career at the college in the United States [1,2]. In older athletes it is between 1:15 000-1:50 000 per year. Consequently the fact is that the persistent interest in the general public about these tragic events is in disproportion to the clinical significance in real terms. Despite the low prevalence of cardiovascular sudden death among athletes, it is often associated with social and emotional impact (with a maximum possible media coverage of the sudden death of a young athlete).

The most frequent cases have been reported in basketball, soccer, American football in the ratio 9:1 in favor of men. The ratio may be somewhat misleading, given the low participation of women in these sports. The most common cause of death for African-Americans is HCMP. Despite some limitations, these numbers cannot afford to underestimate the problem of sudden death. [1,2,6,25,26].

Table 3. World Health Organization guidelines were published in 1968, but are still applicable today [27]

1	The condition should be an important health problem.
2	There should be a treatment for the condition.
3	Facilities for diagnosis and treatment should be available.
4	There should be a latent stage of the disease.
5	There should be a test or examination for the condition.
6	The test should be acceptable to the population.
7	The natural history of the disease should be adequately understood
8	There should be an agreed policy on who to treat.
9	The total cost of finding a case should be economically balanced in relation to medical expenditure as a whole.
10	Case-finding should be a continuous process, not just a "once and for all" project.

Sudden death (SD) of children and adults ages 1 to 40 years has received increasing attention over the past decades. In this report, we provide an overview of this problem, consider several mass screening strategies to prevent these events, and critically appraise these strategies in light of the well known Wilson and Jungner criteria (Table 3).1 Finally, we present an alternative approach: optimal assessment of first-degree relatives of newly diagnosed patients with an inherited cardiac disease and young SD victims to identify those relatives at risk of sudden cardiac death (SCD) [27].

SD or its synonyms are generally defined as natural, unexpected death within 1 hour of the onset of symptoms [28] Four temporal elements should be considered in the use of this definition, that is, prodromes, onset of the terminal event, cardiac arrest, and biological death. The 1-hour definition refers to the period between onset of the terminal event, that is, acute changes in cardiovascular status, and cardiac arrest. Nonspecific prodromal symptoms, for example, chest pain, palpitations, or dyspnea, can be present during the days or weeks before a cardiac arrest. The biological legal death can occur days or weeks after the cardiac arrest, as patients can survive with irreversible brain damage and life support. In addition, when death occurs unwitnessed within 24 hours of being seen alive and functioning normally, this is also termed SD. In this report, the 1-hour definition is applied unless otherwise indicated.

To characterise the demographics and etiology of sudden cardiac death (SCD) in athletes referred to a tertiary cardiac pathology centre in the UK. A retrospective non-case controlled analysis [29], between 1996 and 2008, showed the hearts of 118 athletes referred for pathological assessment to ascertain the precise etiology of sudden cardiac death. The majority of athletes (n=113; 96%) were male and most (107; 91%) were amateurs participating predominantly in football, rugby and running. The mean (SD) age of death was 28 (12) years (range 7–59); 75% athletes were aged ≤35 years. Most deaths (81%) occurred during or immediately after exercise. Antecedent symptoms of cardiac disease were reported in 21 (18%) subjects, and 20 (17%) had a family history of premature cardiovascular disease and/or SCD. 25 (21%) athletes had relevant past medical history which included a known history of cardiac disease. Cardiomyopathy was the commonest cause of death and accounted for 62% of all the SCDs. A significantly high proportion of athletes (23%) exhibited a morphologically normal heart. Atherosclerotic coronary disease accounted for only 3% of cases and was confined to athletes aged >35 years. It is possible that SCD in sport is largely due to clinically silent cardiomyopathies or primary electrical disorders (morphologically normal heart). Antecedent symptoms and family history are absent in over 80% of cases, and therefore clinical screening with health questionnaires will fail to identify most athletes with potentially sinister cardiac disorders.

SCREENING FOR CARDIOVASCULAR ABNORMALITIES IN ATHLETES

Young trained athletes represent a unique challenge in terms of differential diagnosis of cardiovascular morbidity. Screening of athletes is done among pre-school children, during primary to high school and later during professional studies. The importance of screening is to diagnose and reduce the risk factors of sudden death unknown to these subjects [2,4,11]. It should be noted that even without the classical symptoms, athlete may have unforeseen cardiovascular disease [1,2,6,7]. Since the most common causes of sudden death are the structural and functional abnormalities, the basic diagnostic methods should include; a standard 12-lead ECG (QT interval, Brughada syndrome, WPW, LGL syndrome and other arrhythmias), 24-hour ECG Holter monitoring, echocardiography, MRI, CT, load ergometry, coronary angiography and, where appropriate, on an inclined plane test (HUT test). A family history of cardiovascular disease or sudden death may be quite important. Asymptomatic trained athletes without structural myocardial damage are often the object of arrhythmias on the basis of increased vagal tone and do not require invasive investigation or treatment [30,31].

Presence of ectopic complexes recorded by 24-hour ECG holter monitoring, including frequent ventricular premature complexes and supraventricular, nonsustained ventricular tachycardia not forecast imminent and most do not require treatment in athletes without structural myocardial involvement, but represent a differential diagnostic problem of differentiating these benign changes such as changes from the base of inflammation or infarction [4, 26, 31]. Despite all efforts in time, we are not able to diagnose asymptomatic (silent) cardiovascular disease in some cases [11].

In brief, highly trained, professional sportsman have a low incidence of sudden death from cardiovascular causes (athletes 35 years younger than 1:200.000, athletes older than 35 years 1:15.000-50.000). About one third of sudden death in young athletes are due to hypertrophic cardiomyopathy, as distinct from the heart "sport"" (Table 2). The second common most cause of death in athletes (about 20%) are congenital anomalies of coronary arteries. In children, a significant hazard is commotio cordis. In athletes older than 35 years, the premature coronary artery atherosclerosis appears to be the commonest cause of sudden death. The basic diagnostic methods should include: history (family + personal), 12-lead ECG, 24-hr. ECG Holter monitoring, echocardiography, MRI, CT, load ergometry, HUT test. Screening of athletes is expected to begin in early childhood, during the primary through high school to the age they become professionals.

A strong and constantly increasing interest to the general public (with a maximum possible media coverage) of sudden deaths of trained athletes from cardiovascular causes, often linked to social and emotional impact, are in disproportion to their low prevalence. The interest of the company persists because of erroneous myth that the trained athlete is the healthiest element of our society. The need for cardiological examination in at least all the top athletes should be mandatory. In higher age groups, sport serves as a healing method, and do not forget even for recreational sports, which should also not miss the need for preventive investigations.

So these subjects should properly investigated before advising exercises at specialized certified centers [32].

REFERENCES

[1] Maron BJ, Shirani J,Poliac LC, et al. Sudden death in young competitive athletes: Clinical, demographic, and pathological profiles. *JAMA* 276:199-204,1996.

[2] Maron BJ: Sudden death in young athletes. *N Engl. J. Med.* 349:1064-1075, 2003.

[3] Maron BJ, Mitten MJ, Quandt EK, et al. Competitive athletes with cardiovascular disease-the case of Nicholas Knapp. *N. Engl. J. Med.* 339:1632-1635, 1998.

[4] Estes NAM, Link MS, Cannom D, et al. Report of the NASPE Policy Conference on Arrhythmias and the Athlete. *J. Cardiovasc. Electrophysiol.* 12:1208-1219,2001.

[5] Maron BJ, Gohman TE, Kyle SB, et al. Clinical profile and spectrum of commotio cordis. *JAMA* 287:1142-1146, 2002

[6] Maron BJ, Carney KP, Lever HM, et al. Relationship of race to sudden death in competitive athletes with hypertrophic cardiomyopathy. *J. Am. Coll Cardiol.* 41:974-980, 2003.

[7] Corrado D, Basso C,Thiene G, et al. Spectrum of clinicopathologic manifestation of arryththmogenic right ventricular cardiomyopathy/dysplasia: A multicenter study. *J. Am. Coll. Cardiol*. 30:1512-1520, 1997.

[8] Pelliccia A, Culasso F, Di Paolo F, et al. Physiologic left ventricular cavity dilatation in elite athletes. *Ann. Intern. Med.* 130:23-31, 1999.

[9] Douglas PS, O´Toole ML, Katz SE, et al. Left ventricular hypertrophy in athletes. *Am. J. Cardiol.* 80:1384-1388, 1997

[10] Abernethy WB, Choo JK, Hutter AM Jr: Echocardiographic characteristics of professional football players. *J. Am. Coll Cardiol.* 41:280-284, 2003.

[11] Pluim BM, Zwinderman AH, van der Laarse A, van der Wall EE, The athlete´s heart: A meta-analysis of cardiac structure and function. *Circulation* 100:336-344, 1999.

[12] Pelliccia A, Maron BJ, de Luca R, et al. Remodeling of left ventricular hypetrophy in elite athletes after long-term deconditioning. *Circulation* 105:944-949, 2002.

[13] Pelliccia A, Maron BJ, Culasso F, et al. Clinical significance of abnormal electrocardiographic patterns in trained athletes. *Circulation* 102:278-284, 2000.

[14] Maron BJ, Pelliccia A, Spirito P:Cardiac disease in young trained athletes: Insights into methods for distinguishing athlete´s heart from structural heart disease with particular emphasis on hypertrophic cardiomyopathy. *Circulation* 91:1596-1601, 1995.

[15] Maron BJ, Thompson PD, Puffer JC, et al. Cardiovascular preparticipation screening of competitive athletes: A statement for health professionals from the Sudden Death Committee (Clinical Cardiology) and Congenital Cardiac Defects Committee (Cardiovascular Disease in the Young), American Heart Association. *Circulation* 94:850-856, 1996.

[16] Maron BJ, Hyperthrophic cardiomyopathy: A systematic review. *JAMA* 287:1308-1320, 2002.

[17] Maron BJ, Shen W-K, Link MS, et al., Efficacy of implantable cardioverter-defibrillators for the prevention of sudden death in patients with hypertrophic cardiomyopathy. *N. Engl. J.Med.* 342:365-373, 2000.

[18] Basso C, Maron BJ,Corrado D, Thiene G: Clinical profile of congenital coronary artery anomalies with origin from the wrong aortic sinus leading to sudden death in young competitive athletes. *J. Am. Coll Cardiol.* 35:1493-1501, 200.

[19] Goldschlager N, Epstein AE, Grub BP,et al. Etiologic considerations in the patient with syncope and an apparently normal heart. *Arch. Intern. Med.* 163:151-161,2003.

[20] Brugada J, Brugada R, Antzelevitch C,et al. Long-term follow-up of individuals with the electrocardiographic pattern of right bundle branch block and ST segmen elevation in precordial leads V1to V3. *Circulation* 105:73-78, 2002.

[21] Priori SG, Napolitano C, Memmi M, et al. Clinical and molecular characterization of patients with catecholaminergic polymorphic ventricular tachycardia. *Circulation* 106:69-74, 2002.

[22] Lange RA, Hillis LD: Cardiovascular complications of cocaine use. *N Engl. J. Med.* 345:351-358, 2001.

[23] Samenuk D, Link MS, Homoud MK,et al. Adverse cardiovascular events temporally associated with ma huang, an herbal source of ephedrine. *Mayo Clin. Proc.* 77:12-16, 2002.

[24] Link MS, Wang PJ, Pandian NG,et al. An experimental model of sudden death due to low-energy chest wall impact (commotio cordis).*N. Engl J. Med.* 338:1805-1811,1998

[25] Corrado D, Basso C, Rizzoli G, et al. Does sports activity enhance the risk of sudden death in adolescents and youngh adults? *J. Am. Coll Cardiol.* 42:1959-1963, 2003.

[26] Biffi A, Pelliccia A, Verdile L,et al. Long-term clinical significanceof frequent and complex ventricular tachyarrhythmias in trained athletes. *J. Am. Coll Cardiol.* 40:446-452, 2002.

[27] Wilson JMG, Jungner G. Principles and Practice of Screening for Disease. *WHO Chronicle* 1968;22(11):473.

[28] van der Werf C, van Langen IM, Wilde AAM. Sudden Death in the Young;What Do We Know About It and How to Prevent? *Circulation: Arrhythmia and Electrophysiology.* 2010;3:96-104.

[29] De Noronha SV, Sharma S, Papadakis M, Desai S,Whyte G,Sheppard MN. Aetiology of sudden cardiac death in athletes in the United Kingdom: a pathological study. *Heart* 2009;95:1409-1414.

[30] Kiško A, Kmec J, Eliašová A, Dernarová Ľ, Uherová Z, Hornáková A, Shyp A, Horlenko O, Kiško N, Svystak V. Strategy of Cardiovascular Testing in Young athletes for the Prevention of cardiac omplications. *Medicina sportiva* 2008; 14: The 15-th Sports medicine Balkan Congress (Abstract).

[31] Kiško A, Kmec J, Eliašová A, Dernarová Ľ, Uherová Z, Hornáková A, Shyp A, Horlenko O, Kiško N, Svystak V. Protocol of cardiac examination in Young athletes for the Prevention of sports- related cardiovascular complications. Archivos de medicina del deporte 2008; XXV (6) Abstract: 482.

[32] Kiško A, Kmec J, Eliašová A, Derňarová Ľ. Is there a way to prevent cardiovasular complications and fatalities in athletes population? *Humanum* 2009, 3: 417- 421.

In: Cardiovascular Health and Chronomics
Editor: Ram B. Singh

ISBN: 978-1-62948-976-6

Chapter 6

THE ROLE OF IMMUNOGENICITY IN CARDIOVASCULAR DISEASE

***Michael Jan*[1,2,3,#], *Anthony T. Virtue*[1,2,#], *Meghanaben Pansuria*[1,2,#], *Jingshan Liu*[1,2,#], *Xinyu Xiong*[1,2,#], *Pu Fang*[1,2,#], *Shu Meng*[1,2], *Hong Wang*[1,2,3] and *Xiao-Feng Yang*[1,2,*#]**

[1]Department of Pharmacology,
[2]Independence Blue Cross Cardiovascular Research Center
[3]Sol Sherry Thrombosis Research Center
Temple University School of Medicine, Philadelphia, PA, US

ABSTRACT

Recently, many of the complexities associated with cardiovascular diseases (CVD) have been unlocked. However, despite these breakthroughs, CVD and its related complications are the leading contributors of morbidity and mortality worldwide, which indicates the shortcomings of current treatment regimens and the need for continued research. Published data within the field clearly indicates that CVD are built on inflammation and autoimmune platforms, though a strong, fundamental understanding of the mechanisms remains elusive. Areas such as the mechanisms underlying increased immunogenicity of self-proteins in the cardiovascular system, the roles of immunogenic auto-antigens in eliciting inflammatory autoimmune responses, and the immunosuppressive mechanisms involved in controlling inflammatory and autoimmune cardiovascular diseases remain to be well-understood. We will delve into these topics and the advancements made within the field in this review. Specifically, we will concentrate on the innate and adaptive immune responses mediating immunogenicity; the mechanisms of inflammation and autoimmunity in atherogenesis; the mechanisms of inflammation and autoimmunity in diabetic atherosclerosis; immunogenicity and stem cell therapy; as well as immunogenicity and immunosuppression. In depth examination

* Corresponding Author: Xiao-Feng Yang, MD, PhD, Department of Pharmacology, Temple University School of Medicine, 3420 North Broad Street, MRB Room 325. Philadelphia, PA 19140. USA. Tel: 1-215-707-5985. Fax: 1-215-707-7068. Email: xfyang@temple.edu.
# MJ, ATV, MP, JL, XX, and PF Contributed Equally to This Work.

and comprehension of these topics will provide insight into the recent progress of the field and bring to the forefront potentially novel therapeutic avenues.

Keywords: Immunogenicity, inflammation, autoimmunity, atherosclerosis, diabetes, stem cell therapy, immunosuppression

ABBREVIATIONS

AGE – advanced glycation endproduct
APC – antigen presenting cell
Apo – apolipoprotein
BMC – bone marrow mononuclear cell
CCL – CC chemokine ligand
CCR – CC chemokine receptor
Cuox–LDL – copper-oxidized low-density lipoprotein
CVD – cardiovascular disease
CX3CL – CX3C chemokine ligand
CX3CR – CX3C chemokine receptor
CXCL – CXC chemokine ligand
CXCR – CXC chemokine receptor
DC – dendritic cell
DM – diabetes mellitus
EC – endothelial cell
EPC – endothelial progenitor cell
ERK – extracellular signal-regulated kinase
ESC – embryonic stem cell
ET – endothelin
FFA – free fatty acid
GM-CSF – granulocyte macrophage colony-stimulating factor
HLA – human leukocyte antigens
HO – heme oxygenase
HSC – hematopoietic stem cell
IFN – interferon
IKK-β – inhibitor of NF-κB kinase subunit β
IL – interleukin
IR – insulin resistance
IRS – insulin receptor substrate
JNK – c-Jun N-terminal kinase
LDL – low-density liproprotein
LO – lipoxygenase
MΦ – macrophage
Mac-1 – macrophage-1 antigen
MAPK – mitogen-activated protein kinase
MC – monocyte
MDA–LDL – malondialdehye-modified low-density lipoprotein
MHC – major histocompatibility complex
mHC – minor histocompatibility complex
MI – myocardial infarction
MIF – migration inhibitory factor
MMP – matrix metalloproteinase
MSC – mesenchymal stem cell
NK – natural killer
NOD – non-obese diabetic
oxLDL – oxidized low-density lipoprotein
PAMP – pathogen-associated molecular pattern
PI3K – phosphoinositide 3-kinase
PPAR – peroxisome proliferator-activated receptor
PRR – pattern recognition receptor
PSGL – P-selectin glycoprotein ligand
RAGE – receptor for advanced glycation endproduct
ROS – reactive oxygen species
SEREX – serological analysis of recombinant tumor cDNA expression libraries
sHSP – soluble heat shock protein
SkMb – skeletal myoblast
SLE – systemic lupus erythematosus
SMC – smooth muscle cell
SR – scavenger receptor
T1DM – type 1 diabetes mellitus
Tfh – T follicular helper cells
Th – T-helper
Th1 – type 1 T-helper
Th2 – type 2 T-helper
TIMP – tissue inhibitor of metalloproteinases
TLR – toll-like receptor
TNF – tumor necrosis factor
Treg – regulatory T-cell
VCAM – vascular cell adhesion molecule
VLA – very late antigen

I. Introduction

A cardiovascular disease with an onset characterized by endothelial injury and lipid dysregulation, atherosclerosis was recently discovered to be driven by an inflammatory component [1,2]. Shortly thereafter, research demonstrated that both the innate and adaptive immune systems contributed to the pathogenesis of atherosclerosis. Research pursuing this breakthrough yielded fruitful results providing cellular mechanisms and new therapeutic targets. In particular, the adaptive immune system was found to contribute through a complex autoimmune process [3,4]. This finding led to development in therapeutic avenues such as vaccines [4-6]. When considering atherosclerosis as a disease it is vital to remember that lipid dysregulation, inflammation, and autoimmunity are not contending viewpoints. Rather, atherosclerosis should be thought of as a multi-faceted disease and its complexity indicates a necessity to consider multiple aspects when designing therapeutics. In this chapter, we will explore the immunogenicity of cardiovascular diseases (CVD) to better elucidate the mechanisms behind these diseases. This could provide inroads for the pioneering effort to target disease development instead of merely treating symptoms. This paper will first review the mechanisms critical to modulating immunogenicity by the adaptive and innate immune systems and then explore how these mechanisms contribute to CVD development. Finally, therapeutic strategies involving stem cell therapy, common suppressive drugs, and future therapeutic directions will be discussed to conclude this review. The progress from a variety of scientific fields will be highlighted illustrating the complexity of CVD and the requirement of multi-discipline collaboration to illuminate disease pathology and development of therapeutic strategies.

II. Innate and Adaptive Immune Responses and Immunogenicity

A concise overview of relevant immunological effectors and their influence on the adaptive and innate immune system will now be assessed to lay the foundation for further immunogenicity discussions. Consisting of a network of cells, tissues, and organs, the immune system protects its host against agents perceived as pathogens. In order to accomplish this function, the immune system must be capable of differentiating between self and non-self. Following the recognition of non-self agents, the system must also be competent in the elimination of these agents. The immune responses required for recognition and elimination can be categorized into two major classes, innate or adaptive. An innate immune response can occur instantaneously or evolve over several hours, the duration often dependent on the actions taken. The response could simply involve the recognition and removal of a foreign agent, or it may entail the recruitment and activation of other immune effectors. If a foreign agent endures the innate immune response, the adaptive immune system will be activated. Unlike the timely action of the innate system, the adaptive immune response requires several days to respond to the foreign agents [7]. The components of the adaptive immune system vary in activity with antigen-specific lymphocytes targeting unique pathogens for removal while memory cells safeguard from future infection by the same pathogen. It is important to recognize that the differences in duration and longevity between the two systems can be attributed to the unique mechanisms of each [7,8].

II.A. Activation of the Innate Immune Response

II.A.1. Innate Immune Receptors

The actions taken by the innate immune system are dependent on the specific pathogen present and the cellular receptors activated. Despite lacking the remarkable specificity of the adaptive immune system, the innate immune system is able to recognize numerous cellular patterns common among microbes and toxins through unique receptors referred to as pattern recognition receptors (PRRs) [7,8]. One such receptor, Toll-like receptor 4 (TLR4), is capable of recognizing lipopolysaccharide, which is found in the cell walls of gram negative bacteria. Similarly, TLR2 can recognize the proteoglycans of gram positive bacteria [8]. In addition to PRRs, activated phagocytic receptors can initiate the destruction of foreign agents while chemotatic receptors can mediate cytokine release causing the recruitment of effectors capable of aiding in pathogen removal. Importantly, the activation of any of these receptors not only stimulates an innate immune response but also induces the adaptive immune system [7].

II.A.2. Role of Inflammasomes in Innate Immunity

Cellular receptors, effectors, and signaling molecules work in concert to facilitate innate immune responses. Aforementioned, PRRs have the ability to identify and associate with unique pathogen-associated molecular patterns (PAMPs), leading to distinct innate immune responses [9]. One particular element, the inflammasome, vitally participates in these responses by secreting pro-inflammatory cytokines [8]. Once initiated, the assembly of this enzyme complex is dependent on individual component expression that fluctuates in a tissue-dependent fashion. To address this, a three-tier model was constructed to categorize the "inflammation privilege" of various tissues. First tier tissues constitutively express all of the inflammasome components, while the second tier expresses all but one component and the third tier lacks more than one element. The lack in constitutive expression is significant because upregulation of that particular component is required prior to inflammasome assembly [10]. Therefore, the tiers can be used to delineate the ability of specific tissues to mount acute inflammatory responses. The importance of inflammasomes in the innate immune response is highlighted by the activity of adjuvants. An adjuvant will often be co-administered with a drug or vaccine to generate a stronger immune response that enhances the potency and longevity of the drug or vaccine [11,12]. The adjuvant alum was utilized for nearly eighty years without complete understanding of its mechanism. It was eventually determined that the enhanced immune response to alum was the result of the activation of the NALP3 inflammasome [13].

II.A.3. Inflammation

The hallmarks of inflammation include swelling, redness, pain, and fever, which are all symptoms of innate immune system activity [7,14]. Swelling is the result of inflammation that occurs in response to the release of cytokines and other mediators from activated macrophages (MΦs) and other resident effectors. Functionally, swelling attracts additional effector molecules like neutrophils that are capable of eliminating perceived pathogens, acts as a physical barrier against the spread of infection, and promotes the repair of injured tissue [7]. The other hallmarks result from innate immune effects on the vasculature. Corresponding with redness and fever, an increase in vessel diameter results in augmented blood flow allowing for the rapid accumulation of inflammatory mediators while swelling and pain can be attributed to increased vascular permeability leading to exudate accumulation [7].

II.B. Innate Immune Response Effectors

II.B.1. Phagocytes

Phagocytosis is a mechanism by which an effector can eliminate a pathogen. Cells like neutrophils, MΦs, and monocytes (MCs) engulf pathogens, internalizing them within vesicles. These vesicles are then directed towards lysosomes or other enzymes that degrade the pathogens. This process is based on receptor-mediated endocytosis and can be triggered by PAMP-PRR or Fc-antibody binding [8]. Other cells, termed natural killer (NK) cells, express Fc receptors and are activated by antibody-bound pathogens. This activation leads to the subsequent destruction of the pathogen and is termed antibody-dependent cellular cytotoxicity. The regulation of NK cells is critical for normal immune functioning and is dependent on the presence of activating and inhibitory receptors. Inhibitory NK cell receptors recognize major histocompatibility complex (MHC) class I alleles presented on all nucleated cells. Following infection, a cell's MHC class I expression is diminished leading to destruction by NK cells. In particular, NK cells are potent eliminators of tumor cells and virally-infected cells [8,15].

II.B.2. Cytokines

Cytokines play a major role in mediating the innate immune response. Both resident cells and recruited effectors are capable of releasing cytokines, which signal a myriad of effects. The diversity in roles, targets, and origins of cytokines has made sub-categorization difficult and because of this there is some debate as to whether some cytokines should be labeled as hormones and vice versa. Nevertheless, functional classifications prove useful when considering cytokines within specific contexts – e.g., pro-inflammatory vs. anti-inflammatory or type 1 vs. type 2 (pro-Th1 vs. pro-Th2, *see discussion of adaptive immunity below*). Cytokines act on transmembrane receptors located on target cells that typically activate the JAK-STAT pathway. Thus it can be insinuated that cytokines achieve their cellular effect via induction of new gene transcription and new protein synthesis [8,16].

II.B.3. Complement Cascade

The complement system consists of more than twenty-five proteins working in concert to form a biochemical cascade. Primarily consisting of protease zymogens, the activation of these proteins is cleavage-dependent. Thus, activation of one protein typically results in the successive cleavage of the next zymogen within the cascade. The complement system is responsible for the recruitment of effectors by inducing inflammation, identifying pathogens for removal, and eliminating certain foreign agents. These functions are carried out through three distinct, activation-dependent pathways utilizing different effectors. Capable of identifying and eliminating pathogens, the classical and lectin pathways are distinctly initiated either by antigen-antibody complexes or bacterial cell wall mannans, respectively. Meanwhile, the alternative pathway is activated spontaneously by microbial components [8,17-19].

II.C. Overview of the Adaptive Immune Response

An adaptive immune response is the result of a pathogen eluding elimination by the innate immune system. The expression of MHC/antigen complexes, T-/B-cell antigen receptors, and co-stimulatory molecules on the surface of innate immune cells lead to the activation of adaptive immune lymphocyte effectors. Consisting of dendritic cells, B-cells, and MΦs, professional antigen presenting cells (APCs) activate lymphocyte effectors through antigen epitope presentation (Figure 1). B-cells are activated similarly although the presence of $CD4^+$ T-helper (Th) cells is also required. These Th cells can be further classified into several subsets based on their cytokine secretion. Two particular subsets, Th-1 and Th-2, were the original two groups and are subsequent progenitors of $CD4^+$ cells. Th-1 cells are generated in the presence of interferon-γ (IFN-γ) and interleukin-12 (IL-12), while IL-4 induces Th-2 cells.

Additionally, Th-1 cells produce IFN-γ and IL-2, whereas Th-2 cells secrete IL-4, IL-5, IL-10, and IL-13, which promote B-cell activation leading to humoral immunity [3,20]. Furthermore, these two Th subsets maintain the ability to regulate one another. Upon the discovery of IL-9-producing and IL-17-producing cells, the Th9 and Th17 cell subsets were created [21,22]. The Th9 subset promotes T-cell proliferation while the Th17 subset mediates defense against bacteria as well as a variety of autoimmune responses [21,22].

Another subset, T follicular helper cells (Tfh), was observed to target B-cell follicles where they induce B-cell antibody production [23]. All of the aforementioned effectors of the adaptive immune response function with a high degree of specificity to eliminate foreign pathogens and create immunological memory. The extent of B- and T-cell receptors, 10^{18} and 10^{14}, respectively, gives rise to the adaptive immune system's extraordinary specificity [20]. Despite this, occasionally an adaptive immune response is inadvertently mounted against self-antigens.

II.D. Pathological Consequences of Unregulated Immune Responses

The complexity of the innate and adaptive immune systems and the need for finely-tuned modulation perpetuates the opportunity for harmful effects if a malfunction in regulation occurs. Despite the beneficial need for acute inflammation to remove foreign agents, chronic inflammation is almost always detrimental [24]. Inadvertent self-targeting and tissue destruction can occur in response to overproduction of inflammatory mediators, reactive oxygen species (ROS), and complement cascade proteins. This scenario occurs in paroxysmal nocturnal hemoglobinuria where healthy red blood cells are eliminated. This ailment is the result of complement cascade hyperactivity due to a deficiency in decay-accelerating factor and CD59. Normally these factors function to protect against complement system self-activation. Similarly, hereditary angioneurotic edema occurs as a result of C1-inhibitor deficiency. This defect leads to fluid accumulation in tissues as a result of excessive secretion of vasoactive mediators. Both of these examples highlight the vital importance of self-recognition by the immune system.

A malfunction or loss of self-recognition can have life-threatening consequences as self-antigens are marked for destruction by the adaptive immune system. Atherosclerosis, systemic lupus erythematosus (SLE), multiple sclerosis, rheumatoid arthritis, and type 1 diabetes mellitus (T1DM) all contain an autoimmune component where self-antibodies are synthesized.

Examination of autoimmune diseases in twins reveals an underlying genetic connection to disease progression. However, genetic factors do not solely affect autoimmunity; environmental factors have also been shown to have an impact [25,26]. The detailed mechanisms remain to be fully comprehended though, due to the complexity and difficulty associated with studying environmentally-induced autoimmune diseases. Despite this, it is known that environmental effectors such as drugs and medications can alter DNA methylation patterns leading to SLE-like autoimmune disease. This finding along with other similar results suggests a possible epigenetic contribution to autoimmunity.

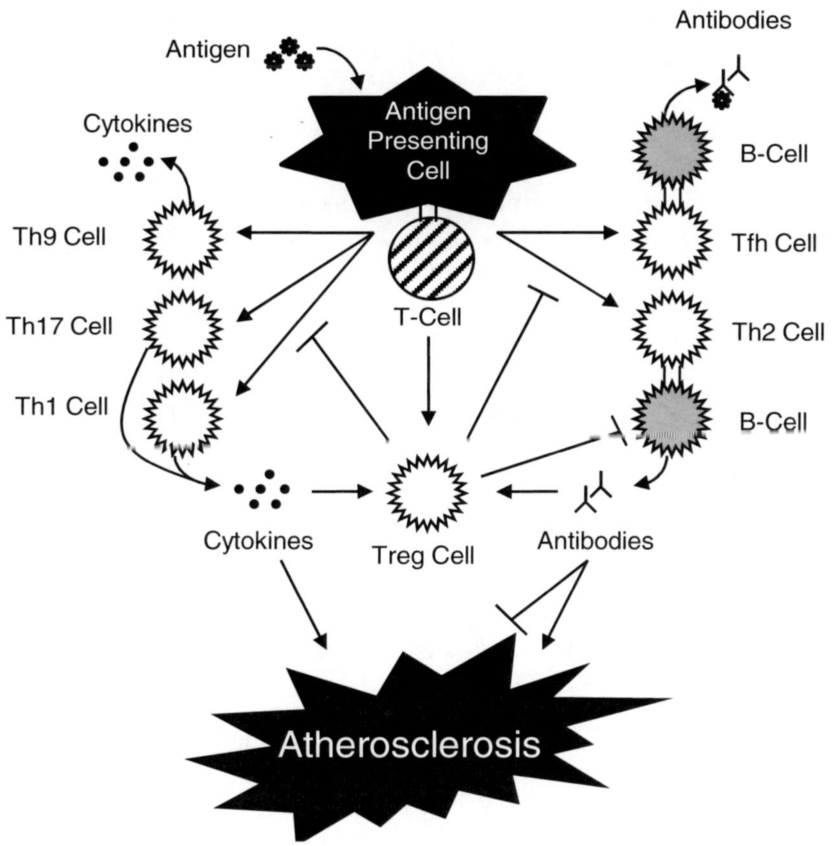

Figure 1. Activation of Adaptive Immune Responses in Atherosclerosis. This schematic depicts the roles of T-cells and B-cells in mediating immune responses to antigens in atherosclerosis.

III. Role of Inflammation and Autoimmunity in Atherogenesis

Atherosclerosis, characterized by the accumulation of plaques in large- and medium-sized arteries, frequently progresses to more severe complications like myocardial infarction (MI) or cerebrovascular accident. Atherosclerotic plaques are biologically complex consisting of altered lipids, a necrotic core, calcified regions, inflamed smooth muscle cells (SMCs), activated and dysfunctional endothelial cells (ECs), leukocytes, and foam cells [1]. The majority of atherosclerosis instances can be attributed to traditional risk factors such as age, hypertension, diabetes mellitus (DM), hyperlipidemia, sex, hyperglycemia, obesity, and cigarette smoking [27,28]. However, nontraditional risk factors like hyperhomocysteinemia and bacterial infection have also been characterized [29-32]. Whether traditional or nontraditional, atherosclerosis risk factors sustain the ability to influence inflammatory and/or autoimmune responses. These findings provide evidence of the link between immunity and atherogenesis.

III.A. Innate Immune Response Effectors and Atherogenesis

III.A.1. Monocytes and Macrophages

Contained within the heterogeneous atherosclerotic plaques are immune effectors such as MΦs, T-cells and a few B-cells [33]. Originating from MCs, MΦs partake in multiple functions such as pro-inflammatory cytokine secretion, vascular cell wall remodeling, and lipid retention. Additionally, they express immune modulating TLRs, PRRs, and scavenger receptors (SRs). The receptor expression and cellular roles of MΦs serve to bridge the gap between innate and adaptive immune interaction during atherosclerosis [34]. It is well-characterized that MΦs can become foam cells, an early indicator of atherosclerosis, as a result of lipid accumulation [35]. Also there is evidence that suggests MΦ apoptosis can affect atherogenesis. In early lesions, induction of MΦ apoptosis hinders atherogenesis, whereas impediment of MΦ apoptosis in later stages may contribute to secondary necrosis. This leads to increased pro-inflammatory responses and additional apoptotic signals for SMCs, ECs, and leukocytes within the plaques [34]. MCs are also capable of developing into dendritic cells (DCs), which may enhance naive lymphocyte recruitment into atherosclerotic vessels with the production of CC chemokine ligand 19 (CCL19) and CCL21 [36].

III.A.2. Mast Cells and Neutrophils

Mast cells and neutrophils are effectors of the innate immune system responsible for the degradation of foreign agents by lytic enzymes. This enzymatic activity may also be relevant in atherosclerosis development. Mast cells are believed to contribute to atherosclerotic plaque destabilization through the enzymatic activity of proteases such as tryptase and chymase. Their contribution is further supported by the concentrated presence of mast cells in locations of plaque rupture. Furthermore, the activation of perivascular mast cells correlates with intraplaque hemorrhage, MΦ and EC apoptosis, vascular leakage, and CXC chemokine receptor 2 (CXCR2)- and very late antigen 4 (VLA-4)-mediated recruitment of leukocytes to the plaque [37]. Mast cells also maintain the capability to affect lipid metabolism through apolipoprotein E (ApoE)- and ApoA-II-dependent cholesterol efflux [38]. Aside from mast cells, the short-lived phagocytic neutrophils possess biologically active molecules like myeloperoxidase and proteinases. When neutrophil concentration is increased in plaques it perpetuates a pro-inflammatory phenotype and apoptosis. This behavior indicates a pro-inflammatory function for neutrophils in atherosclerosis [39].

III.A.3. Platelets as Nontraditional Effectors

Platelets represent an interesting regulatory nexus for various physiological and pathological responses. Despite canonically performing protective roles, activated platelets expressing P-selectin are detected at different stages of atherosclerosis and transiently interact with the endothelium of atherosclerotic carotid arteries of $ApoE^{-/-}$ mice *in vivo*. This interaction results in the immobilization of platelet-derived CCL5 and CXC chemokine ligand 4 (CXCL4) on atherosclerotic endothelium [40,41]. Additionally, it is known that platelet glycoproteins GPIIb/IIIa, GPIb, and endothelial von Willebrand factor are at least partially responsible for these platelet-endothelial interactions. Immobilized platelets are also capable of interacting with leukocytes through P-selectin/P-selectin glycoprotein ligand (PSGL)-1 interactions that activate macrophage-1 antigen (Mac-1) and VLA-4 integrins and may facilitate firm MC adhesion [40]. Another function of platelets is to initiate the rolling of DCs through PSGL-1-dependent interactions between Mac-1 and junctional adhesion molecule C in injured carotid arteries [42]. They are also required for CX_3C chemokine ligand 1 (CX3CL1)-induced leukocyte adhesion at high shear rates. Both soluble and membrane-bound CX3CL1 trigger P-selectin expression on adherent platelets, which facilitates the local accumulation of leukocytes under arterial shear stress [43]. Overall, platelets act as bridges between inflamed endothelium and circulating blood cells as well as recruiters of leukocytes to inflamed atherosclerotic endothelium.

III.B. Cytokines and Atherogenesis

With the ability to modulate immune responses, cytokines can predictably affect atherogenesis through mediation of endothelial permeability, expression of adhesion molecules, SR, lipid metabolism, as well as proliferation and migration of SMCs and ECs. Atherosclerosis-influencing cytokines can be generated by ECs, MΦs, mast cells, and platelets. ECs release IL-3, stem cell factor, granulocyte macrophage colony-stimulating factor (GM-CSF), granulocyte colony-stimulating factor, and macrophage colony-stimulating factor, which all can affect hematopoietic cell proliferation [44]. Meanwhile, MΦs can secrete both pro- and anti-inflammatory cytokines when present in a hyperlipidemic state as seen in atherosclerosis. Mast cells, on the other hand, produce cytokines IL-6 and IFN-γ, which are critical for mast cell-dependent acceleration of atherosclerosis. Finally, platelets yield the pro-atherogenic mediators IL-1β, CD40L, and CXCL4 when activated. It should be noted that cytokines can also influence extracellular matrix composition by varying the expression of matrix metalloproteinases (MMP) -1, -3, -8, -9, and -12 as well as tissue inhibitors of metalloproteinases (TIMP) -1, -2, and -3 [45].

III.B.1. Pro-Inflammatory Cytokines

As a consequence of their immunological nature, pro-inflammatory cytokines typically contribute to atherogenesis. An example is TNF-α, which has been discovered to promote fatty streak formation, a characteristic of early atherosclerosis [46,47]. It is also understood that the cytokine IL-1 promotes leukocyte adhesion to ECs and leukocyte transmigration across the endothelium. Moreover, IL-1 functions as a local autocrine and paracrine stimulator of other cytokines [44]. IL-2 is known to have pro-atherogenic effects as well, which can be inhibited with IL-2 specific antibodies [48]. Similarly, IL-18 augments atherosclerosis progression via up-regulation of IFN-γ, which participates in all phases of atherogenesis [49]. When secreted by aortic plasmacytoid DCs, IFN-α has been demonstrated to enhance the expression of TNF-related apoptosis-inducing ligand on $CD4^+$ T-cells and increase their cytolytic capacity toward SMCs [50]. In addition, IFN-α stimulates TLR4 expression on APCs thereby increasing pathogen sensitivity. Furthermore, IFN-α intensifies TNF-α, IL-12, and MMP-9 production, thus leading to further plaque destabilization [51]. Found within atherosclerotic plaques, TLR1, TLR2, TLR4, and TLR5 can influence cytokine expression following activation. In fact, it has been confirmed that atherosclerosis-associated oxidized low-density lipoprotein (oxLDL) can modulate TLR4 expression and therefore affect cytokine expression.

III.B.2. Anti-Inflammatory Cytokines

As opposed to pro-inflammatory cytokines, it can be generalized that anti-inflammatory cytokines typically have anti-atherogenic effects. This idea has been supported by experimental conclusions. The anti-inflammatory cytokine IL-10 is generated by MCs, Th2 cells, B-cells, and MΦs. Its presence reduces vascular inflammation and inhibits lesion progression by inducing an anti-inflammatory MΦ phenotype [52]. Additionally, IL-10 maintains the delicate balance between Th1 and Th2 responses. If IL-10 expression is knocked down in mice, a phenotype characterized by T-cell accumulation, increased IFN-γ production and limited collagen in atherosclerotic vessels is witnessed [44]. IL-33, another anti-inflammatory cytokine, also provides beneficial effects. Experimentally, IL-33 has been demonstrated to significantly reduce atherosclerosis in $ApoE^{-/-}$ mice. This physiological result can be attributed to increases in IL-4, IL-5, and IL-13 production, reduction of pro-inflammatory cytokines, and diminished IFN-γ secretion. These alterations lead to the alteration of the Th1 response to a Th2 response [53]. In addition, it is hypothesized that IL-33 can counteract harmful oxLDL through IL-5-dependent secretion of oxLDL antibodies [53]. In addition, there is supporting data for the beneficial effects of TGF-β in atherosclerosis by modulating SMC and EC phenotypes, as well as Th1 functions.

III.B.3. Chemokines

Chemokines are a subdivision of cytokines possessing the ability to influence chemotaxis. In correlation to atherosclerosis they play an important function in directing leukocytes into atherosclerosis-prone vessels. The first chemokine to be associated with atherosclerosis, CCL2 and its receptor CC chemokine receptor 2 (CCR2), integrally participate in MC recruitment from the bone marrow into the arterial wall [54 55].

Likewise, platelets facilitate MC and T-cell recruitment into the endothelium through the secretion of pro-inflammatory chemokines CCL5 and CXCL4 [40]. The expression of

CXCL4 can perpetuate atherosclerosis by potentially activating ECs through NF-κB activation, inducing E-selectin expression, and advocating oxLDL binding to ECs [56]. Another chemokine with implications in atherosclerosis development is CXCL8. It can promote both migration and proliferation of SMCs and ECs as well as affect neovascularization. Oppositely, the chemokine CXCL1 initiates MC arrest by increasing production of VLA-4 integrin [57]. In low stress regions, CCL2, CXCL10, CXCL1 and CX3CL1 are present with CX3CL1 being solely observed in areas with thinner fibrous caps and larger necrotic cores [58].

In atherosclerotic plaques, CX3CL1 adopts a transmembrane form that can attract CX_3C chemokine receptor 1 (CX3CR1) positive NK cells, subsets of T-cells, and MCs [59]. A different chemokine, CXCL10, can regulate the balance between the effector and regulatory arms of the immune system in aortas by reducing regulatory T-cell (Treg) expression. Chemokines also act as potent mitogenic and chemotactic factors for SMCs [60]. In research conducted with migration inhibitory factor ($MIF^{-/-}$) and low-density liproprotein receptor ($Ldlr^{-/-}$), dual knockouts indicate a correlation of MIF with atherosclerosis.

This link is found in the regulation of lipid deposition, protease expression, and intimal thickening [61]. Additionally, when MIF binds to the CXCR2 and CD74 receptor complex, elevated MC targeting of atherosclerotic endothelium was observed [62].

III.C. Atherogenesis and Inflammatory Mediators

Once an innate immune response has been mounted and inflammation has been induced, a number of pro-inflammatory mediators are expressed that can impact the development of atherosclerosis. Leukotrienes, inflammatory lipid mediators, are a component of the 5-lipoxygenase (5-LO) pathway with pro-atherogenic aptitude [63]. Furthermore, plaque instability is augmented with the up-regulation of 5-LO and leukotriene A4. In response to LDL oxidation another inflammatory mediator, heme oxygenase (HO)-1, is induced leading to a reduction in MC chemotaxis [64]. Therapeutically, HO-1 has demonstrated protection from atherosclerosis in an antioxidant-dependent manner [65].

This result is substantiated by the fact that excessive generation of ROS can induce atherosclerosis through the activation of the vascular endothelium and components of the immune system. One of the ways a superoxide can affect atherogenesis is through the activation of SMC mitogenic signaling pathways [66]. Furthermore, NADPH-induced superoxide has been shown to enhance platelet recruitment [67]. oxLDL is tightly linked to atherosclerosis and can influence the migration of MCs to the aortic wall by alternating from CCR2 to CX3CR1 expression through a peroxisome proliferator-activated receptor-γ (PPAR-γ)-dependent pathway [68].

III.D. Atherogenesis and the Adaptive Immune Response

III.D.1. T-Cells

Experimentation in T- and B-cell deficient hypercholesterolemic mice has been conducted to confirm the role of adaptive immunity in atherogenesis. It was observed in these mice that disease progression was significantly attenuated [69-72]. Furthermore, lesion formation was resurrected when $CD4^+$ T-cells were transferred into these deficient mice [73]. Aforementioned, atherosclerotic plaques are comprised of numerous cells that secrete IFN-γ, IL-12, IL-15, IL-18, and TNF with only a few cells that produce IL-4 [74-76]. This finding implies a role for Th1 cells and not Th2 cells in promoting atherosclerosis. This notion was confirmed in studies where lesion reduction was witnessed when mediators contributing to T-cell differentiation into Th1 cells were suppressed [77-82]. Moreover, administration of Th1 differentiating IFN-γ expectantly induces atherosclerosis in mice whereas a reduction in atherosclerosis incidence was observed when Th1 cells were pharmacologically inhibited [83,84]. Contrary to Th1 cells, Th2 cells reveal anti-atherosclerotic effects in animal studies. Mice genetically altered to yield Th2 immune responses (as opposed to Th1 immune responses) showed reduced fatty streak formation. This finding could be reversed by inhibiting T-cell differentiation into Th2 cells [85,86]. Additionally, over-expression in hypercholesterolemic mice of the Th2 cytokine IL-10 led to reduced lesion size by 50% [87]. Conversely, a deficiency in the IL-5, another Th2 cytokine, resulted in an increase in atherosclerosis [88]. Moreover, studies involving over-expression of the Th2-inducing cytokine IL-4 show inconsistent results [89,90]. Thus, it can be concluded that Th1 cells have a clear pro-atherogenic role, whereas Th2 cells have a more complex role in atherosclerosis that is yet to be fully elucidated.

Th1 cytokines primarily encourage atherosclerosis by perpetuating the inflammatory mechanisms discussed previously (Figure 1). Located within atherosclerotic plaques, the pro-inflammatory cytokines IL-12, IL-18, IFN-γ, and TNF, induce Th1 differentiation resulting in further cytokine production and activate both MΦs and vascular cells. Individually, IFN-γ is capable of activating MΦs while simultaneously inhibiting EC and SMC [91,92]. These actions result in the secretion of pro-inflammatory cytokines, pro-thrombotic factors, and vasoactive mediators. Meanwhile, TNF can stimulate the NF-κB pathway in vascular cells furthering inflammation and causing the production of ROS, proteolytic enzymes, and pro-thrombotic factors [93-95]. The activation of these two pathways provides strong evidence that Th1 adaptive immune responses participate in atherosclerosis. It should be noted that a correlation between atherosclerosis and other pro-atherosclerotic pathways associated with Th2, Th9, Th17, Tfh, and $CD8^+$ T-cells, has yet to be established [4,96]. However, it is believed that these cells participate in modulating pro- and anti-atherosclerotic pathways as a result of their common lineage. This implies that the nuances governing their differentiation ultimately affect pro- or anti-atherosclerotic responses. Tfh cells maintain the ability to regulate B-cell response and antibody production while Th2 cells modulate Th1 cell differentiation and activity. In addition, Th17 cells can modulate immunosuppressive Tregs [22,96] and secrete IL-17, which has been indicated in affecting autoimmune responses in atherosclerosis (Figure 1) [52,97,98].

III.D.2. Tregs

$CD4^+CD25^{high}$ Tregs are capable of suppressing both the innate and adaptive immune systems [99]. Expectedly, the loss of Treg function has been found to perpetuate autoimmune responses (Figure 1) [100,101]. The identification of a Treg-specific apoptotic pathway that relies on IL-2 participation provides evidence that Treg apoptotic/survival pathways can serve as therapeutic targets in Treg-based immunotherapy. It has also been demonstrated that inhibition of Treg results in accelerated vascular inflammation. This heightened inflammation can understandably contribute to atherosclerosis exacerbation [102-104]. For activation, Tregs depend on T-cell receptor activity and co-stimulation similar to effector T-cells. Tregs and effector T-cells differ however, in the fact that a distinguished pathway that exclusively regulates Tregs has yet to be identified [105]. Superagonistic CD28 antibody was attributed with preferentially-activating Tregs in animal models but phase I clinical trials identified a lack of specificity that triggered a cytokine storm [106-108]. Despite this, direct Treg transplant has generated positive results [109].

This line of therapy could prove useful in treating atherosclerosis where plaques contain a depleted number of Tregs (1%–5% of all T-cells versus normally 25% of all T-cells) as well as other T-cell dysfunction ailments [110]. It is important to note when exploring Treg therapies that excessive T-cell suppression can compromise cellular defense systems. The development of Treg-based atherosclerosis therapies should ideally target specific autoantigens both involved in atherogenesis and located in atherosclerotic plaques.

III.D.3. B-Cells

In addition to T-cell mediated cellular responses, humoral responses modulated by B-cells are also part of the adaptive immune response. Experiments with T- and B-cell deficient mice revealed a pro-atherogenic function of these mediators [69-72]. Data exists where splenectomy and consequent loss of some T- and B-cells led to increased atherosclerosis. Furthermore, this finding could be reversed through adoptive transfer of B-cells [111-113]. These conflicting results demonstrate the complex nature of B-cell activity. It is postulated that the principal role of B-cells in atherosclerosis is the production of antibodies, which may be either pro- or anti-atherogenic [3,96]. Moreover, B-cells contain the ability to secrete cytokines and serve as APCs. This capacity suggests the ability of B-cells to modulate T-cell adaptive immune response; however, this function has not been adequately examined in atherosclerosis. A limiting factor in fully comprehending B-cell mediated cellular responses is the inability to completely characterize B-cell subsets [114,115]. Discovered in mice, specific B-cell subsets have yet to be concretely identified in humans [115]. However, it is postulated, that a B-cell subset termed B1-cells produce natural antibodies of the IgM subtype in a T-cell and antigen independent manner [116]. The B1-cells function as a first line of defense against foreign pathogens with their primary location in the peritoneal cavity [114,116]. Interestingly, oxLDL-specific IgM antibodies can be detected in the circulation of individuals with minimal levels of atherosclerosis [117].

III.D.4. oxLDL as an Autoantigen

$CD4^+$ T-cells located within atherosclerotic plaques exhibit specificity for exogenous and endogenous antigens such as oxLDL and heat shock protein 60 (HSP60) [75,118-120]. A well characterized biomarker, oxLDL is the preeminent autoantigen associated with atherosclerosis. [121]. Atherogenesis is first initiated as LDL particles embedded within the arterial intima become oxidized leading to the accumulation of fatty streaks [122-124]. The oxLDL is then taken up by APCs via the SR pathway. At this point the oxLDL molecules undergo proteolytic processing and then bind to MHC class II molecules for presentation to T-cells [3]. Since native LDL components do not elicit a T-cell response it is postulated that oxidation of LDL is responsible for the loss of immunogenic tolerance. Compounding the issue, hypercholesterolemia is thought to impair DC mobility causing diminished immunogenic tolerance due to inhibition of clearance functions [4,125-127]. Furthermore, the existence of oxLDL specific antibodies provides support that an adaptive immune response to oxLDL occurs during atherosclerosis [119]. This response is believed to be a balance between pathological autoimmunity and protective immunity [4].

III.D.5. HSP as an Autoantigen

Primarily functioning as chaperones, HSPs are highly conserved and abundantly expressed. Despite their conservation they have also been implicated in autoimmune atherosclerosis [128]. The notion that HSPs may have negative health implications developed from observations of elevated expression in patients with systemic hypertension, coronary artery disease, carotid atherosclerosis, MI, and ischemia [6].

Additionally, high levels of antibodies against mycobacterial HSP65 independently predict stroke, MI, and cardiovascular death [129]. The cross-reactivity of HSPs with other bacterial HSPs (e.g., *Mycobacteria*, *Chlamydia*) and the detection of bacterial HSPs by antibodies suggest that HSPs may function as autoantigens [6].

In fact, *Chlamydia pneumoniae* have HSPs discovered to be located in atherosclerotic plaques [6,29]. Meanwhile, other infectious agents such as herpes simplex and cytomegalovirus have also been found in atherosclerotic lesions [20]. Even though these infections do not directly contribute to atherosclerosis, their presence can have indirect effects by activating PRRs leading to immune responses. This indirect effect is coined "molecular mimicry" [130].

Comprehensive evidence of this can be seen with the activation of TLR9 located on DCs in atherosclerotic plaques. This activation results in an autoimmune T-cell attack on vascular SMCs [50] and clearly reveals how an unassociated pathogen can generate a pro-atherogenic autoimmune response.

III.E. Atherogenesis Resulting from Other Pathologies

As the connections between atherosclerosis, inflammation, and autoimmunity are unearthed it can be postulated that chronic inflammation and other autoimmune diseases will have an impact on atherosclerosis progression. This hypothesis is supported by an increase in atherosclerosis progression seen in the presence of rheumatoid arthritis [131], SLE, systemic sclerosis [132], Wegener's granulomatosis [133], and antiphospholipid syndrome [134].

In addition, diabetic mice develop atherosclerosis at an accelerated rate with diabetes, glucose, and products of glucose metabolism believed to induce atherosclerosis under hyperlipidemic conditions. Furthermore, the symptoms associated with a pre-diabetic state – abdominal obesity, elevated LDL cholesterol, and increased blood pressure – are all positively correlated with CVD. Finally, experimental mice expressing human aldose reductase, which leads to elevated glucose levels, develop atherosclerosis at an accelerated rate compared with controls [135].

IV. Inflammation and Autoimmunity in Diabetic Atherosclerosis

DM is defined by elevated blood glucose levels due to either insufficient pancreatic insulin production or insufficient insulin use. T1DM is an absolute loss of insulin secretion from autoimmune-mediated destruction of pancreatic β-cells. On the other hand, type 2 DM (T2DM) is a relative insulin deficiency occurring because of defective β-cell insulin secretion due to peripheral insulin resistance (IR) (Figure 2) [136]. Diabetes is a growing pandemic currently afflicting more than 220 million people worldwide and predicted to ail twice as many individuals by 2030 [137]. These overwhelming values indicate the necessity of determining the mechanism of IR so novel therapeutic targets can be identified.

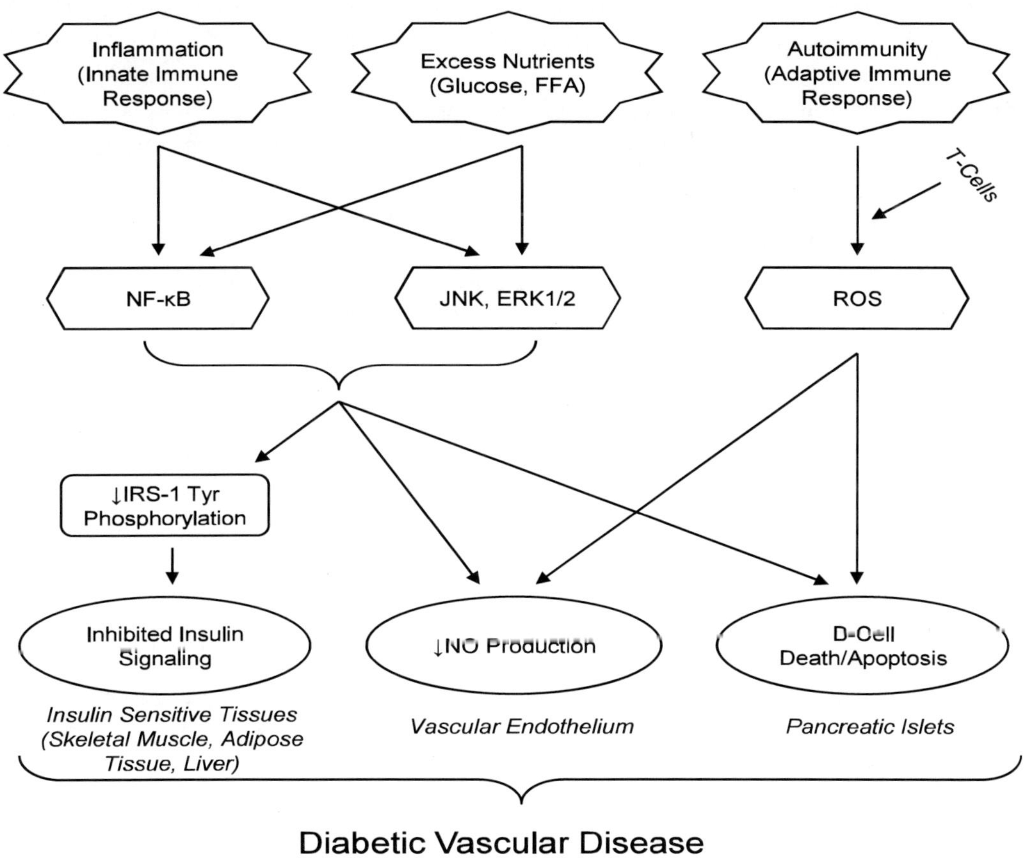

Figure 2. The Immunopathology of Type 2 Diabetes Mellitus. This flow chart shows how both innate and adaptive immune responses contribute to T2DM pathology.

IV.A. Insulin Resistance

Insulin is a mediator of fuel homeostasis stimulating glucose uptake and suppressing the release of stored lipid from adipose tissue. Publications have positively linked the advancement of IR with hepatic steatosis, abdominal obesity, and chronic subclinical inflammation [138-142]. This suggests that elevated cytokine levels may contribute to peripheral IR as well as decreasing β-cell function and mass. Pro-inflammatory cytokine levels can affect susceptible tissue like blood, brain, adipocytes, endothelium, liver, muscle, pancreatic islet cells, and muscle, which are used as diagnostic markers for T2DM and metabolic syndrome [138,140]. Moreover, inflammatory cytokines and free fatty acids (FFA) can spur IR and activate IKK-β (inhibitor of NF-κB kinase subunit β) in the presence of inflammatory stimuli. IKK-β is capable of activating NF-κB and inhibiting insulin signaling through the restraint of insulin receptor substrate (IRS) tyrosine phosphorylation. Furthermore, suppression of IKK-β protects against the deleterious effects of TNF-α and FFA on insulin signaling. Studies show that overexpression of a constitutively active version of IKK-β in the liver of rodents causes IR and diabetes in liver and muscle [143-148]. Interestingly, obesity or a high fat diet has been demonstrated to cause activation of NF-κB and its pro-inflammatory targets in the liver. In addition, an obese or diabetic state can lead to elevated levels of c-Jun N-terminal kinase (JNK) and extracellular signal-regulated kinase (ERK) activity in several tissues under metabolic and inflammatory stress [149-153]. The above findings indicate that prolonged exposure to lipids and other metabolic fuels can affect IR development through the alteration of cytokine production within the liver, adipose tissue, and infiltrating immune cells.

IV.B. β-Cell Destruction

The graded serum levels of acute phase reactants (e.g., CRP, sialic acid) and their cytokine mediators (e.g., TNF-α, IL-1, IFN-γ) in correlation with diabetes, obesity, metabolic syndrome, hypertension, and atherosclerosis development suggests a role in disease pathogenesis [140,154-156]. A similar mechanism may occur in the pancreatic islet cells of people with T2DM [157,158]. While T1DM is immune-mediated and based upon pancreatic islet inflammation [159,160], T2DM is defined by inadequate insulin production in the face of IR [161]. With insulin secretion dependent on pancreatic β-cell function and mass, the pancreas is remarkably capable of adapting to conditions that require elevated insulin production such as obesity, pregnancy, and growth hormone excess, by augmenting its β-cell mass [162]. However, in T2DM, β-cell function declines with disease progression [14,162] and β-cell dysfunction was found to be the major contributor in the transition from normoglycemia to DM in an experimental population of Pima Indians [163]. Despite etiological and pathogenic differences between T1DM and T2DM, pancreatic β-cell demise by apoptosis is a common cellular denominator. Eventually, in both situations, β-cell destruction occurs leading to clinical manifestations of absolute or relative insulin deficiency [164]. As hyperglycemia develops, additional processes like islet inflammation, O-linked glycosylation, and amyloid deposition accelerate β-cell demise. This results in a deterioration of β-cell function and a loss of controlled glucose stimulated insulin secretion [162,165]. The chronic hyperglycemia then leads to ROS production, which induces pancreatic β-cell oxidative damage. Additionally, both ROS and IL-1β activate NF-κB, a critical mediator of inflammatory responses. Apoptosis in the presence of pro-inflammatory cytokines like IL-1β and TNF-α can provoke additional immune responses. Therefore, the chronic increase in inflammatory mediators observed in T2DM and the elevated secretion of TNF-α, IL-6, and IL-1β by adipose tissue [166] could have a significant effect on noninsulin-sensitive cells.

IV.C. Diabetic Atherosclerosis

Atherosclerosis is a frequent complication of DM but despite strong epidemiologic evidence of their correlation, the mechanism connecting the two has yet to be determined. Postulations made from early studies suggest that DM exacerbates many traditional risk factors for atherosclerosis [167] and in fact, DM patients often exhibit elevated levels of oxLDL, hyperinsulinemia, oxidative stress, hyperglycemia, and increased FFA, which all are associated with endothelial dysfunction [167-172]. Another possible contributor to the accelerated atherosclerosis development is increased LDL, reduced high-density lipoprotein, and increased triglycerides connected to diabetic dyslipidemia [167]. Furthermore, individuals with DM have hypercoagulability, impaired fibrinolysis and platelet hyperaggregability, which all augment thrombogenicity and contribute to atherosclerosis [173-175]. Another supporting piece of evidence is an oxidative stress mechanism unique to DM with atherosclerosis effects. It involves the formation of advanced glycation endproducts (AGE) with altered structure and function. The receptor for AGE (RAGE) is expressed by ECs, SMCs, MCs, and lymphocytes with enhanced expression in atherosclerotic lesions. The administration of soluble recombinant RAGE neutralizes AGE, thereby reducing NF-κB induction, vascular cell adhesion molecule 1 (VCAM-1), and tissue factor expression. This in turn causes a reduction in atherosclerotic lesion burden [176]. The AGE-RAGE pathway is known to elevate oxidative stress, endothelial dysfunction, expression of adhesion molecules, as well as the generation of inflammatory cytokines and tissue factor. The modulation of these cellular factors can ultimately effect atherosclerosis progression [177].

IV.D. Unifying Immune Responses of DM and CVD

The role of the immune system in modulating diabetic atherosclerosis has revealed some mechanistic associations. S100A9$^+$ MΦs are pro-inflammatory biomarkers for cardiovascular events and are recruited to atherosclerotic lesions in T1DM during plaque disruption [178]. In addition, T1DM conditions can induce S100A9$^+$ MΦ differentiation. Meanwhile, other findings indicate that high concentrations of glucose can induce MΦ CD36 expression [179]. Significantly, CD36 is a MΦ SR that mediates oxLDL uptake and subsequent foam cell development. Moreover, studies with CD36 have demonstrated that oxLDL promotes assembly of a TLR4/TLR6 heterodimer [180]. These conclusions provide a direct mechanistic relationship between DM and atherosclerosis with immune mediation.

IV.E. Endothelial Dysfunction as an Ancestor for DM and CVD

Development of CVD is the principal complication in T2DM but it can precede its development as well [181]. This intertwined relationship lends itself to the idea of mutual ancestors. While IR is an accepted factor in the pathogenesis of T2DM its link to CVD has not been well defined [182]. However, inflammation is a common tie being a precursor of CVD, associated with IR, and preceding the progression of T2DM [156,183]. Inflammatory mediators are known to function pathogenically by enhancing systemic endothelial dysfunction, leading to hindered NO and vascular hemostatic agent production [184]. Imbalance of the vasodilatory agent NO and the vasoconstrictor endothelin-1 (ET-1) can further contribute to endothelial dysfunction [185]. Noteworthy is the fact insulin can stimulate production of NO from the endothelium and cause vasodilation through the binding and activation of the insulin receptor. This mechanism is similar to metabolic insulin signaling in skeletal muscle and adipose tissue [186-188]. Once the insulin receptor is bound, tyrosine phosphorylation of IRS occurs resulting in activation of both the phosphoinositide 3-kinase (PI3K)-Akt pathway and the mitogen-activated protein kinase (MAPK) pathway [189,190]. A key feature of IR is the alteration of the PI3K-Akt pathway, though other insulin-signaling branches, including Ras/MAPK-dependent pathways, remain unmediated [191]. As a result, hyperinsulinemia will lead to an imbalance between PI3K- and MAPK-dependent pathways. This imbalance leads to elevated vascular inflammation via enhanced expression of VCAM-1, intracellular adhesion molecule 1, E-selectin, monocyte chemotactic protein 1, and IL-6. Additionally, the imbalance will lead to a reduction in available NO in the endothelium [191,192]. This results in diminished eNOS activity, decreased NO generation, and increased ET-1 secretion, a characteristic of endothelial dysfunction. The elevated levels of ET-1 increase serine phosphorylation of IRS-1, leading to a reduction of PI3K activity in vascular SMCs and hindering insulin-stimulated translocation of glucose transporter type 4 in adipocytes [193]. It can be concluded that IR-mediated reduction of NO in the endothelium contributes to accelerated atherosclerosis through the regulation of vasoconstriction, inflammation, and thrombosis [194,195]. The categorization of endothelial dysfunction as a T2DM precursor may provide additional vantages for diabetes prevention and treatment [196,197]. Moreover, therapies targeting endothelial dysfunction may prove vital in preventing IR development and therefore T2DM [198-200].

IV.F. Therapeutic Strategies

IR is a leading factor in T2DM and is the central component of metabolic syndrome [163,201]. Recent studies suggest that pro-inflammatory cytokines, adipocytokines, and transcription factors are involved in the pathogenesis of IR. This claim is supported by clinical observations where patients with metabolic syndrome and T2DM had a positive correlation between IR and chronic inflammation. Based on these data, methods to attenuate IR by inflammation modulation were investigated. The insulin sensitizer thiazolidinedione exerts its effects through PPAR-γ by enhancing the effects of insulin in metabolic tissues and by directly decreasing peripheral IR [187,202]. Also, the receptor of interest is expressed in all major cell types involved in vascular lesions – MCs, MΦs, ECs and vascular SMCs. Additionally, the PPAR-γ agonist glitazone has numerous beneficial effects like improvement of glycemic control, decreased levels of circulating FFA, and anti-inflammatory effects on EC and leukocyte function [203,204]. Other beneficial anti-atherogenic effects have also been reported in several studies using PPAR-γ ligands [205,206]. Potent inhibition of inflammation, blockade of MΦs differentiation, and cytokine secretion [207] as well as inhibition of vascular SMC proliferation and migration have been shown [208]. In addition, treatment with PPAR-γ agonists in T2DM patients reduces plasma levels of inflammatory biomarkers that are predictive of CVD [209]. These results encourage and suggest that anti-inflammatory agents can improve tissue susceptibility to insulin and are promising for the treatment of metabolic syndrome and T2DM.

V. Immunogenicity and Stem Cell Therapy

Blood transfusions can be considered the first primitive form of cell therapy. Since then the concept of cell therapy has been refined to include only essential components needed to restore function or repair injury. These components may undergo modifications altering their functionality or enhancing their versatility. This ability to control cell functionality is one reason why cell therapy is an attractive avenue. Other reasons include the relative abundance of cells from countless donors and the inclusion of cellular features that are advantageous for targeting. Moreover, specific cell types demonstrate tremendous versatility and can be genetically modified to meet specific therapeutic demands [210]. At the forefront, stem cells contain immense cell therapeutic potential with their self-renewing ability and capacity to differentiate into more specialized cell types. We will outline key therapeutic concerns and discuss some progress in the stem cell therapeutic field.

V.A. Immunogenicity of Stem Cells

V.A.1. Antigen Matching

The idea of stem cell therapy in CVD is promising though apprehension regarding inflammation and immune responses needs to be addressed [211,212]. Allogeneic cells, tissues, and organs regularly elicit host immune responses. The host immune system "recognizes" certain protein antigens as foreign, leading to graft destruction. Due to their participation in tissue compatibility these molecules are named alloantigens or histocompatibility antigens. In allogeneic transplantations three distinct classes of antigens exist. The first class consists of MHC antigens, also termed human leukocyte antigens (HLA), which are the major contributors of tissue compatibility. The other two groups, minor histocompatibility complex (mHC) antigens and ABO blood group antigens, play a relatively minor role in tissue compatibility compared to MHC antigens [213,214]. Host T-cell recognition of MHC antigens from the donor is the primary cause for the rejection of an allogeneic transplant. To alleviate this issue, HLA matching between the donor and the recipient is conducted but is extremely difficult. A high number of polymorphisms results in hundreds of alleles for some HLA genes [215]. Also, the combination of the alleles can reach in the billions due to the two sets of HLA genes with six alleles per set [216]. This variability makes it virtually impossible to find the perfect match in genetically unrelated individuals.

V.A.2. Stem Cell Antigen Expression

In order to determine the feasibility of stem cell therapy it is critical to evaluate the MHC antigen cell surface expression to alleviate the risk of graft reject. Utilizing flow cytometry, it was characterized that undifferentiated human embryonic stem cells (ESCs) have very low MHC class I antigen expression [217,218]. Inciting human ESCs to differentiate into embryonic bodies or teratomas increases MHC class I antigen expression 2-4 fold and 8-10 fold, respectively [218]. Further, supporting studies in mice revealed that MHC class I antigens are not expressed in undifferentiated mouse ESCs and are only present at low levels following prolonged differentiation [219]. It was also discovered that epitope processing by immunoproteasomes is accelerated in the presence of IFN-γ [220]. This suggests that increased immunogenicity of stem cells may be related to stimuli such as stress, cytokines, and drugs that hasten self-antigen epitope processing [221]. These findings suggest ESC therapy of CVD could be feasible; however, elevated levels of IFN-γ and similar cytokines could result in amplified MHC expression and therefore needs to be characterized.

V.A.3. Immunogenicity of Self Antigens

Aside from alloantigens it is also necessary to comprehend the mechanism of how non-mutated self-protein antigens obtain immunogenicity and elicit immune responses [222]. The most common mechanism for non-mutated self antigen immunogenicity is the abnormal overexpression of antigens in tumors [223]. This elevated expression can be a product of either transcriptional or post-transcriptional mechanisms and can be identified using serological analysis of recombinant tumor cDNA expression libraries (SEREX). This technique involves the screening of patient's sera [224] for the presence of autoantigens [225]. Additionally, a stimulation-responsive splicing model for autoantigen and self-tumor antigen selection has been established [221]. Utilization of several strategies [226-231] has been employed to identify immunogenic antigen alleles like modified SEREX [224], database-mining [232], and immunogenic isoform mapping. Despite investigation, the detailed antigen repertoire of normal stem cells remains to be well-defined. Gaining insight into these autoantigens and self-tumor antigens would prove vital in designing more effective cell replacement therapeutics.

V.B. Stem Cell Subset Selection

V.B.1. Embryonic Stem Cells

An important consideration when developing cell therapies is the selection of cell subsets. Each has its advantages and disadvantages such as ESCs that maintain the flexibility to differentiate into any cell type within the adult body. It could also be rationalized that the reduction in MHC I expression of ESCs would reduce the likelihood of tissue rejection. However, the rates of tissue rejection in recipient $MHC^{-/-}$ mice are comparable to wild-type control groups [233]. This suggests that selection of ESCs to limit immunological rejection may not be practical. Additionally, ESCs show a propensity to develop into teratomas when injected *in vivo* [234,235]. In order to alleviate this risk, *in vitro* differentiated ESCs could be selected for therapy, though the gain of specificity and control comes at the expense of flexibility and a gain in immunogenicity [236,237]. Further elucidation of ESC differentiation mechanisms could prove fruitful in fully understanding ESC differentiation.

V.B.2. Adult Stem Cells

Over the past decade the majority of CVD stem cell therapy has focused on adult stem cells. The utilization of adult stem cell subsets allows for the circumvention of graft rejection through autologous harvest. Another benefit of adult stem cells is the avoidance of *in vitro* expansion before use in transplantation. This alleviates the possibility of abnormal growth under culturing conditions. Thus far adult stem cell therapy has focused on the use of skeletal myoblasts (SkMbs), bone marrow mononuclear cells (BMCs), and endothelial progenitor cells (EPCs).

V.B.3. Skeletal Myoblasts

SkMbs were the first adult cell type to be transplanted into cardiovascular tissue. They were successfully transferred into recipients with injured hearts [238] and shown to restore myocardial function in animal models [239]. However, the mechanisms leading to improved heart function are still unknown. Working with SkMbs is advantageous due to their resistance to ischemia and their ability to differentiate into myotubes, which can potentially supplement injured myocardium. However, it should be noted that these myotubes are incapable of electrically integrating with the myocardium and are associated with arrhythmogenesis after transplantation [240].

V.B.4. Bone Marrow Mononuclear Cells

BMCs consist of hematopoietic stem cells (HSCs), mesenchymal stem cells (MSCs), and EPCs, all of which can independently improve cardiac function. In fact, a unique subset of BMCs was found to differentiate into cardiomyocytes in mouse models [241]. However, from a therapeutic standpoint this finding was irrelevant due to the exceptionally low rate of differentiation [242]. Additional investigations in animal models also revealed the inability of HSCs to differentiate into cardiomyocytes. This was further supported in clinical studies where the effect of BMCs on cardiac function had little or no improvement [243]. On the other hand MSCs, unique from other BMCs, demonstrated the ability to differentiate into cardiomyocytes under stringent *in vitro* conditions [244,245]. This capacity was also observed *in vivo* though at an extremely low rate. A potential advantage for the use of MSCs in therapy is their relatively low immunogenicity compared with other stem cells [246]. Despite these findings there is little concrete evidence that BMCs can differentiate into cardiomyocytes at a physiological rate *in vivo*.

V.B.5. "Cardiac Stem Cells"

Recent publications have characterized a subset of endogenous cardiac progenitor cells that are capable of differentiating into various cardiac cells including cardiomyocytes [247,248]. These epicardial resident progenitor cells had previously been shown to generate SMC, EC, and fibroblasts; however, until recently their ability to produce cardiomyocytes had not been demonstrated. The discovery of this capability should provide new avenues for cardiovascular stem cell therapy. The use of endogenous progenitors is advantageous because it will allow the circumvention of possible graft rejection. Also, the ability to manipulate these "cardiac stem cells" may provide a novel approach to the restoration of depleted cardiomyocytes following ischemia or MI [249].

V.C. Stem Cell Therapy

V.C.1. Vascular Therapy

Endothelial injury and dysfunction are common contributors to vascular diseases. A promising candidate for use in vascular cell therapy, EPCs are capable of combating this complication. After being discovered in 1997 by Asahara et al. studies involving EPCs have yielded promising results in the cell therapy field [250]. It was indicated that an age-related reduction in progenitor cells contributes to the development of atherosclerosis [251]. This idea was experimentally verified by transferring EPCs from young non-atherosclerotic mice to older hypercholesterolemic mice, which led to atherosclerosis development inhibition via mechanisms of vascular repair and inflammation attenuation [251,252]. Despite the success of this model, additional experimentation with other subsets has led to differentiation into cell types that contribute to atherosclerosis development [253]. It should also be noted that while stem cell therapy in the treatment of atherosclerosis demonstrates important scientific principles it may be an unnecessarily complicated approach [254,255].

V.C.2. Cardiac Therapy

A focal point of cardiovascular stem cell therapy is the regeneration of cardiac myocytes following ischemia or MI. This is important due to the inadequate regenerative capacity of the human cardiac muscle in compensating for the substantial loss of the myocardium by MI or other CVD. While SkMbs in mammals are capable of sufficiently regenerating after widespread or repeated injury [256,257], mammalian cardiomyocytes rarely divide either before or after injury [258,259]. However, it should be noted that differentiation of cardiomyocytes can be accelerated through the over-expression of specific transgenic genes in mouse models [259]. The identification of an ideal stem cell subset for cardiac therapy is critical for successful therapy. The cell subset should be immunologically innate, capable of reconstituting the infarct zone with cardiomyocytes, pro-angiogenic, able to repair damaged vasculature, and electrically integrable with surviving myocardium.

VI. Immunogenicity and Immunosuppression

Thus far, the content of this review has concentrated on the role of the immune system in mediating CVD. The final section will focus on the treatment and neutralization of undesired immune responses. Controlled suppression of the immune system is a delicate endeavor requiring precision to prevent unwanted effects. Original immunosuppressants like corticosteroids have been used to effectively suppress the innate immune response but lacked specificity leading to the consequent stemming of the adaptive immune response. Drugs were then developed with more specific activity capable of targeting distinct mediators of the adaptive immune system. However, it is important to consider that any immunosuppression, specific or broad, inhibits not only the harmful activity but the protective mechanisms of the immune system as well. An ideal immunosuppressive agent would effectively mitigate the harmful mechanisms of the immune system with little or no effect on the protective effects. Agent specificity is vital in accomplishing this. Fortunately, the adaptive immune system relies heavily on specificity for its regular function. Therefore, the use of antibodies for therapeutic inhibition of immune responses has been heavily investigated.

VI.A. Anti-Inflammatory Drugs

The activation of the adaptive immune system following innate immune activity has several therapeutic implications. It explains why the co-administration of adjuvants optimizes the potency and longevity of vaccines and it elucidates why the administration of innate immunosuppressants can inherently eliminate the adaptive immune response. This is exemplified by corticosteroids, which have potent anti-inflammatory effects due to diminished IL-1, IL-3, IL-4, IL-5, IL-8, TNF-α, and GM-CSF expression [7]. Since corticosteroids bind to intracellular receptors in almost every cell type, their effect is broad with both beneficial and toxic results. Fortunately, this negative effect can be mitigated through combinational therapy allowing for lower therapeutic doses, thereby accentuating the favorable effects while diminishing the toxic effects. Corticosteroids are often co-administered with cytotoxic drugs like azathioprine and cyclophosphamide, a class of immunosuppressive drugs that interferes with DNA synthesis. In small doses these drugs target rapidly dividing cells like lymphocytes, thus suppressing the adaptive immune response. Another class of immunosuppressive drugs target T-cell proliferation through the inhibition of IL-2 induction [7]. Even though this class of agents is effective, its insufficient specificity makes standard use unreasonable. In fact, the harmful side effects restrict use to extreme circumstances like organ transplant, graft rejection, and various autoimmune diseases.

VI.B. Immunosuppressive Antibodies

Immunosuppressive antibodies have been employed in two capacities to ensure a high level of specificity. Depleting antibodies, the first subset, associate with lymphocytes by forming a complex that is later eliminated by NK cells. The second group is blocking antibodies, which inhibit the activity of the target protein but do not cause the protein's destruction [8].

VI.B.1. Receptor Antibodies

Certain antibodies elicit their physiological effect through interactions with receptor complexes. Exemplifying this interaction, anti-CD3-specific antibodies act on the CD3 antibody/T-cell receptor complex to inhibit immunological responses. These antibodies have been administrated therapeutically to prevent short-term allograft rejection while their use in non-obese diabetic (NOD) mice resulted in the long-term remission of autoimmune diabetes [260]. Furthermore, the use of anti-CD3 antibodies in $Ldlr^{-/-}$ mice on a high cholesterol diet led to the inhibition of additional plaque formation as well as the halted progression of previously established lesions. Interestingly, an increase in both anti-inflammatory cytokine TGF-β in lymph node cells and the expression of the Treg marker Foxp3 in spleen cells was reported. This finding suggests that short-term treatment with an anti-CD3 antibody leads to the induction of a Treg phenotype, known to aid in the restoration of self-tolerance in a mouse model of atherosclerosis [261].

VI.B.2. Co-Stimulatory Signal Antibodies

Required for T-cell activation, co-stimulatory signals can be considered targets for immunosuppressive therapy. CD40 and its ligand CD40L have been demonstrated to be co-expressed on ECs, MΦs, SMCs, and T-cells present in atherosclerotic lesions. The inhibition of CD40L-CD40 signaling with antibodies results in the reduction of experimental autoimmune diseases like collagen-induced arthritis, lupus nephritis, acute or chronic graft-versus-host disease, multiple sclerosis, and thyroiditis. Additionally, the use of an anti-CD40L antibody in $Ldlr^{-/-}$ mice on a high cholesterol diet led to a decline in aortic atherosclerotic lesion area. Furthermore, a reduction in lesion lipid content and the presence of MΦs and lymphocytes was also reported [262]. In spite of these findings, it should be stated that treatment in $Ldlr^{-/-}$ or $ApoE^{-/-}$ mice with established atherosclerosis did not cause atherosclerotic lesion decline. It did, however, spur a change in lesion composition resulting in a reduction of lipids and MΦs while increasing the content of collagen and SMCs [90,263]. Ultimately, these modifications contributed to a more stable plaque phenotype. Although therapeutics involving CD40 and its ligand were not further pursued due to the potential risk of thrombotic complications caused by CD40L expression on platelets, it provides a useful example of how co-stimulatory signal antibodies can be used in immunosuppressive therapy.

VI.C. Tolerance

The ability to restore immune tolerance is a major goal in immunotherapy – particularly in regard to autoimmune diseases. With an early onset, most atherosclerosis therapies are directed at reducing progression by limiting innate and adaptive immune responses instead of prevention. Oral or intranasal immunization is a common method utilized to develop tolerance [264]. Administration of autoantigens by these routes has been shown to reduce organ-specific inflammation and disease expression in several models of autoimmunity. Mucosal antigens are capable of restoring tolerance through the induction of antigen-specific T-cell immune responses characterized by the secretion of IL-4, IL-10, and TGF-β [265]. Tolerance is gained mechanistically through antigen site-directed bystander suppression by the T-cell response effectors [266,267]. Mucosal tolerance treatments have already shown their validity in the treatment of HSPs and oxLDL in murine atherosclerosis models. The success of mucosal tolerance treatments has led to human clinical trials [265].

VI.C.1. HSPs

Thought to contribute to atherogenesis, HSPs are expressed by ECs under duress typically associated with classic atherosclerosis risk factors such as high blood pressure, smoking, DM, and chemically altered lipoproteins [268,269]. These risk factors may lead to EC activation and death therefore resulting in the release of HSPs into intercellular spaces.

These soluble HSPs (sHSPs) are pro-atherogenic in nature, causing the secretion of pro-inflammatory cytokines by MΦs and the expression of adhesion molecules by ECs [270]. In addition, MΦs present antigens to B-cells, which produce HSP autoantibodies. These autoantibodies interact with HSPs expressed on activated ECs further exacerbating the situation. The elevated expressions of both HSPs and HSP autoantibodies have been positively correlated with atherosclerosis [269,271,272]. Interestingly, the nasal and oral administrations of HSP65 in Ldlr$^{-/-}$ mice seem to cause an immunization effect, resulting in a reduction of atherosclerotic lesions induced by a high cholesterol diet [273,274]. Maron et al. observed a decrease in the proliferation of splenocytes and a decline in the total anti-HSP65 IgG titers with both administration routes. Meanwhile, nasal administration also resulted in decreased MΦ and T-cell lesion numbers, reduced expression of IFN-γ, increased IL-10 expression, and increased Th2 dependent anti-HSP65 IgG1 titers [274]. Additional research was conducted with full length HSP60 or a 16aa HSP60 fragment (253 to 268) in Ldlr$^{-/-}$ mice that were subsequently fed a Western type diet [275]. These oral treatments both resulted in an 80% reduction of lesion formation in carotid arteries following collar placement and a 27% reduction in plaque area at the aortic root of HSP60-treated mice. Furthermore, augmented populations of $CD4^+CD25^+Foxp3^+$ Tregs in lymphoid organs and blood were found. *In vitro* follow-up revealed that HSP60 treatment of lymph node cells from HSP60-tolerized mice secreted additional levels of IL-10 and TGF-β. Meanwhile, endogenous anti-HSP60 antibody levels were unaffected.

VI.C.2. oxLDL

Although the induction of immune responses against oxLDL by active immunization strategies can mediate atheroprotection as discussed above, pro-atherogenic responses specific for oxLDL have been suggested to exist as well. For example, transfer of $CD4^+$ T-cells from mice immunized with malondialdehyde-modified LDL (MDA-LDL) induced significantly more atherosclerosis in recipient ApoE$^{-/-}$ SCID mice than T-cells from control mice. This response is thought to be the result of IFN-γ-secreting Th1 cells that are believed to mediate pro-inflammatory effects in lesions. To resolve this, an oral tolerance strategy using human copper-oxidized LDL (Cuox-LDL) and MDA-LDL was employed. Interestingly, only Ldlr$^{-/-}$ mice that received Cuox-LDL through oral administration developed significantly less atherosclerosis, while MDA-LDL feeding did not have an effect. Meanwhile, oral administration of Cuox-LDL led to a significant increase in TGF-β production and levels of $CD4^+CD25^+Foxp3^+$ cells in the spleen, mesenteric lymph nodes, and in lymph node cells of tolerized mice [276]. No effects were observed in antibody titers to Cuox-LDL. Thus, similar to mucosal tolerance approaches with HSP60/65, the induction of Treg populations is associated with the decrease in atherogenesis, though direct immunological suppression of endogenous responses has not been well documented.

CONCLUSION

In this review, we thoroughly examined the role of immunogenicity in the development, participation, and exacerbation of CVD. A brief overview of immunogenicity and the innate and adaptive responses were discussed to provide the necessary foundation required for the comprehension of the concepts discussed. The recently established involvement of inflammatory/autoimmune mechanisms in atherogenesis has spurred the attention of researchers from a variety of scientific fields leading to an escalation of knowledge and novel findings. These advances will prove valuable in other fields that share etiology with atherogenesis such as diabetes. In addition, the elucidation of immunogenicity in atherosclerogenesis has led to the investigation of novel therapies. Potential treatment with stem cells has come to the forefront and requires judicious review due to their impending ability to trigger immune responses. Despite this possible complication, stem cells offer vast potential in future therapy and for the understanding of unknown mechanisms of pathophysiology. The examination of selective immune suppression concluded this review, providing possible insight into the immunogenic mechanisms that govern atherogenesis and the development of prospective therapeutics. Undeniable are the substantial advances that have been made in the field of CVD as the scientific disciplines of vascular biology, immunology, epidemiology, and bioinformatics become interpolated in the quest for a more comprehensive knowledge of CVD.

References

[1] Ross R. Atherosclerosis--an inflammatory disease. *N. Engl. J. Med.* 1999;340:115-126.

[2] Ross R, Harker L. Hyperlipidemia and atherosclerosis. *Science*. 1976;193:1094-1100.

[3] Hansson GK, Libby P. The immune response in atherosclerosis: a double-edged sword. *Nat. Rev. Immunol.* 2006;6:508-519.

[4] Nilsson J, Hansson GK. Autoimmunity in atherosclerosis: a protective response losing control? *J. Intern. Med.* 2008;263:464-478.

[5] Jan M, Meng S, Chen NC, Mai J, Wang H, Yang XF. Inflammatory and autoimmune reactions in atherosclerosis and vaccine design informatics. *J. Biomed Biotechnol.* 2010;2010:459798.

[6] Riley E, Dasari V, Frishman WH, Sperber K. Vaccines in development to prevent and treat atherosclerotic disease. *Cardiol. Rev.* 2008;16:288-300.

[7] Janeway CA, Travers P, Walport M, Shlomchik MJ. *Immunobiology* (ed 6th). New York: Garland Science Publishing; 2005.

[8] Chaplin DD. Overview of the immune response. *J. Allergy Clin. Immunol.* 2010;125:S3-23.

[9] Medzhitov R. Recognition of microorganisms and activation of the immune response. *Nature.* 2007;449:819-826.

[10] Yin Y, Yan Y, Jiang X, et al. Inflammasomes are differentially expressed in cardiovascular and other tissues. *Int. J. Immunopathol Pharmacol.* 2009;22:311-322.

[11] Reed SG, Bertholet S, Coler RN, Friede M. New horizons in adjuvants for vaccine development. *Trends Immunol.* 2009;30:23-32.

[12] Wack A, Rappuoli R. Vaccinology at the beginning of the 21st century. *Curr. Opin. Immunol.* 2005;17:411-418.

[13] Eisenbarth SC, Colegio OR, O'Connor W, Sutterwala FS, Flavell RA. Crucial role for the Nalp3 inflammasome in the immunostimulatory properties of aluminium adjuvants. *Nature.* 2008;453:1122-1126.

[14] Wellen KE, Hotamisligil GS. Inflammation, stress, and diabetes. *J. Clin. Invest.* 2005;115:1111-1119.

[15] Jonsson AH, Yokoyama WM. Natural killer cell tolerance licensing and other mechanisms. *Adv. Immunol.* 2009;101:27-79.

[16] Fonacier LS, Dreskin SC, Leung DY. Allergic skin diseases. *J. Allergy Clin. Immunol.* 2010;125:S138-149.

[17] Botto M, Kirschfink M, Macor P, Pickering MC, Wurzner R, Tedesco F. Complement in human diseases: Lessons from complement deficiencies. *Mol. Immunol.* 2009;46:2774-2783.

[18] Carroll MC. The complement system in regulation of adaptive immunity. *Nat. Immunol.* 2004;5:981-986.

[19] Longhi MP, Harris CL, Morgan BP, Gallimore A. Holding T cells in check--a new role for complement regulators? *Trends Immunol.* 2006;27:102-108.

[20] Binder CJ, Chang MK, Shaw PX, et al. Innate and acquired immunity in atherogenesis. *Nat. Med.* 2002;8:1218-1226.

[21] Veldhoen M, Uyttenhove C, van Snick J, et al. Transforming growth factor-beta 'reprograms' the differentiation of T helper 2 cells and promotes an interleukin 9-producing subset. *Nat. Immunol.* 2008;9:1341-1346.

[22] Weaver CT, Hatton RD. Interplay between the TH17 and TReg cell lineages: a (co-)evolutionary perspective. *Nat. Rev. Immunol.* 2009;9:883-889.

[23] King C. New insights into the differentiation and function of T follicular helper cells. *Nat. Rev. Immunol.* 2009;9:757-766.

[24] Gabay C, Kushner I. Acute-phase proteins and other systemic responses to inflammation. *N. Engl. J. Med.* 1999;340:448-454.

[25] Doevendans PA, Jukema W, Spiering W, Defesche JC, Kastelein JJ. Molecular genetics and gene expression in atherosclerosis. *Int. J. Cardiol.* 2001;80:161-172.

[26] Hewagama A, Richardson B. The genetics and epigenetics of autoimmune diseases. *J. Autoimmun.* 2009;33:3-11.

[27] Berenson GS, Srinivasan SR, Bao W, Newman WP, 3rd, Tracy RE, Wattigney WA. Association between multiple cardiovascular risk factors and atherosclerosis in children and young adults. The Bogalusa Heart Study. *N. Engl. J. Med.* 1998;338:1650-1656.

[28] Tegos TJ, Kalodiki E, Sabetai MM, Nicolaides AN. The genesis of atherosclerosis and risk factors: a review. *Angiology*. 2001;52:89-98.

[29] Campbell LA, Kuo CC. Chlamydia pneumoniae--an infectious risk factor for atherosclerosis? *Nat. Rev. Microbiol.* 2004;2:23-32.

[30] Maron BA, Loscalzo J. The treatment of hyperhomocysteinemia. *Annu. Rev. Med.* 2009;60:39-54.

[31] Wang H, Jiang X, Yang F, et al. Hyperhomocysteinemia accelerates atherosclerosis in cystathionine beta-synthase and apolipoprotein E double knock-out mice with and without dietary perturbation. *Blood.* 2003;101:3901-3907.

[32] Zhang D, Jiang X, Fang P, et al. Hyperhomocysteinemia promotes inflammatory monocyte generation and accelerates atherosclerosis in transgenic cystathionine beta-synthase-deficient mice. *Circulation.* 2009;120:1893-1902.

[33] Jonasson L, Holm J, Skalli O, Bondjers G, Hansson GK. Regional accumulations of T cells, macrophages, and smooth muscle cells in the human atherosclerotic plaque. *Arteriosclerosis.* 1986;6:131-138.

[34] Tabas I. Consequences and therapeutic implications of macrophage apoptosis in atherosclerosis: the importance of lesion stage and phagocytic efficiency. Arterioscler *Thromb Vasc Biol.* 2005;25:2255-2264.

[35] Galkina E, Ley K. Immune and inflammatory mechanisms of atherosclerosis (*). *Annu. Rev. Immunol.* 2009;27:165-197.

[36] Erbel C, Sato K, Meyer FB, et al. Functional profile of activated dendritic cells in unstable atherosclerotic plaque. *Basic Res. Cardiol.* 2007;102:123-132.

[37] Bot I, de Jager SC, Zernecke A, et al. Perivascular mast cells promote atherogenesis and induce plaque destabilization in apolipoprotein E-deficient mice. *Circulation.* 2007;115:2516-2525.

[38] Lee M, Calabresi L, Chiesa G, Franceschini G, Kovanen PT. Mast cell chymase degrades apoE and apoA-II in apoA-I-knockout mouse plasma and reduces its ability to promote cellular cholesterol efflux. *Arterioscler. Thromb. Vasc. Biol.* 2002;22:1475-1481.

[39] Zernecke A, Bot I, Djalali-Talab Y, et al. Protective role of CXC receptor 4/CXC ligand 12 unveils the importance of neutrophils in atherosclerosis. *Circ. Res.* 2008;102:209-217.

[40] Huo Y, Schober A, Forlow SB, et al. Circulating activated platelets exacerbate atherosclerosis in mice deficient in apolipoprotein E. *Nat. Med.* 2003;9:61-67.

[41] Schober A, Manka D, von Hundelshausen P, et al. Deposition of platelet RANTES triggering monocyte recruitment requires P-selectin and is involved in neointima formation after arterial injury. *Circulation.* 2002;106:1523-1529.

[42] Langer HF, Daub K, Braun G, et al. Platelets recruit human dendritic cells via Mac-1/JAM-C interaction and modulate dendritic cell function in vitro. Arterioscler *Thromb. Vasc. Biol.* 2007;27:1463-1470.

[43] Schulz C, Schafer A, Stolla M, et al. Chemokine fractalkine mediates leukocyte recruitment to inflammatory endothelial cells in flowing whole blood: a critical role for P-selectin expressed on activated platelets. *Circulation.* 2007;116:764-773.

[44] Tedgui A, Mallat Z. Cytokines in atherosclerosis: pathogenic and regulatory pathways. *Physiol. Rev.* 2006;86:515-581.

[45] Newby AC. Dual role of matrix metalloproteinases (matrixins) in intimal thickening and atherosclerotic plaque rupture. *Physiol. Rev.* 2005;85:1-31.

[46] Canault M, Peiretti F, Poggi M, et al. Progression of atherosclerosis in ApoE-deficient mice that express distinct molecular forms of TNF-alpha. *J. Pathol.* 2008;214:574-583.

[47] Xiao N, Yin M, Zhang L, et al. Tumor necrosis factor-alpha deficiency retards early fatty-streak lesion by influencing the expression of inflammatory factors in apoE-null mice. *Mol. Genet Metab.* 2009;96:239-244.

[48] Upadhya S, Mooteri S, Peckham N, Pai RG. Atherogenic effect of interleukin-2 and antiatherogenic effect of interleukin-2 antibody in apo-E-deficient mice. *Angiology.* 2004;55:289-294.

[49] Whitman SC, Ravisankar P, Daugherty A. Interleukin-18 enhances atherosclerosis in apolipoprotein E(-/-) mice through release of interferon-gamma. *Circ. Res.* 2002;90:E34-38.

[50] Niessner A, Sato K, Chaikof EL, Colmegna I, Goronzy JJ, Weyand CM. Pathogen-sensing plasmacytoid dendritic cells stimulate cytotoxic T-cell function in the atherosclerotic plaque through interferon-alpha. *Circulation.* 2006;114:2482-2489.

[51] Niessner A, Shin MS, Pryshchep O, Goronzy JJ, Chaikof EL, Weyand CM. Synergistic proinflammatory effects of the antiviral cytokine interferon-alpha and Toll-like receptor 4 ligands in the atherosclerotic plaque. *Circulation.* 2007;116:2043-2052.

[52] Taleb S, Romain M, Ramkhelawon B, et al. Loss of SOCS3 expression in T cells reveals a regulatory role for interleukin-17 in atherosclerosis. *J. Exp. Med.* 2009;206:2067-2077.

[53] Miller AM, Xu D, Asquith DL, et al. IL-33 reduces the development of atherosclerosis. *J. Exp. Med.* 2008;205:339-346.

[54] Serbina NV, Pamer E.G. Monocyte emigration from bone marrow during bacterial infection requires signals mediated by chemokine receptor CCR2. *Nat. Immunol.* 2006;7:311-317.

[55] Tsou CL, Peters W, Si Y, et al. Critical roles for CCR2 and MCP-3 in monocyte mobilization from bone marrow and recruitment to inflammatory sites. *J. Clin. Invest.* 2007;117:902-909.

[56] Yu G, Rux AH, Ma P, Bdeir K, Sachais BS. Endothelial expression of E-selectin is induced by the platelet-specific chemokine platelet factor 4 through LRP in an NF-kappaB-dependent manner. *Blood.* 2005;105:3545-3551.

[57] Huo Y, Weber C, Forlow SB, et al. The chemokine KC, but not monocyte chemoattractant protein-1, triggers monocyte arrest on early atherosclerotic endothelium. *J. Clin. Invest.* 2001;108:1307-1314.

[58] Cheng C, Tempel D, van Haperen R, et al. Shear stress-induced changes in atherosclerotic plaque composition are modulated by chemokines. *J. Clin. Invest.* 2007;117:616-626.

[59] Kraaijeveld AO, de Jager SC, van Berkel TJ, Biessen EA, Jukema JW. Chemokines and atherosclerotic plaque progression: towards therapeutic targeting? *Curr. Pharm. Des.* 2007;13:1039-1052.

[60] Heller EA, Liu E, Tager AM, et al. Chemokine CXCL10 promotes atherogenesis by modulating the local balance of effector and regulatory T cells. *Circulation.* 2006;113:2301-2312.

[61] Pan JH, Sukhova GK, Yang JT, et al. Macrophage migration inhibitory factor deficiency impairs atherosclerosis in low-density lipoprotein receptor-deficient mice. *Circulation.* 2004;109:3149-3153.

[62] Bernhagen J, Krohn R, Lue H, et al. MIF is a noncognate ligand of CXC chemokine receptors in inflammatory and atherogenic cell recruitment. *Nat. Med.* 2007;13:587-596.

[63] Radmark O, Samuelsson B. 5-lipoxygenase: regulation and possible involvement in atherosclerosis. *Prostaglandins Other Lipid Mediat.* 2007;83:162-174.

[64] Ishikawa K, Navab M, Leitinger N, Fogelman AM, Lusis AJ. Induction of heme oxygenase-1 inhibits the monocyte transmigration induced by mildly oxidized LDL. *J. Clin. Invest.* 1997;100:1209-1216.

[65] Wu BJ, Kathir K, Witting PK, et al. Antioxidants protect from atherosclerosis by a heme oxygenase-1 pathway that is independent of free radical scavenging. *J. Exp. Med.* 2006;203:1117-1127.

[66] Vendrov AE, Hakim ZS, Madamanchi NR, Rojas M, Madamanchi C, Runge MS. Atherosclerosis is attenuated by limiting superoxide generation in both macrophages and vessel wall cells. *Arterioscler Thromb Vasc Biol.* 2007;27:2714-2721.

[67] Freedman JE. Oxidative stress and platelets. *Arterioscler .Thromb. Vasc. Biol.* 2008;28:s11-16.

[68] Barlic J, Zhang Y, Foley JF, Murphy PM. Oxidized lipid-driven chemokine receptor switch, CCR2 to CX3CR1, mediates adhesion of human macrophages to coronary artery smooth muscle cells through a peroxisome proliferator-activated receptor gamma-dependent pathway. *Circulation.* 2006;114:807-819.

[69] Dansky HM, Charlton SA, Harper MM, Smith JD. T and B lymphocytes play a minor role in atherosclerotic plaque formation in the apolipoprotein E-deficient mouse. *Proc. Natl .Acad. Sci. USA*. 1997;94:4642-4646.

[70] Daugherty A, Pure E, Delfel-Butteiger D, et al. The effects of total lymphocyte deficiency on the extent of atherosclerosis in apolipoprotein E-/- mice. *J. Clin. Invest.* 1997;100:1575-1580.

[71] Reardon CA, Blachowicz L, White T, et al. Effect of immune deficiency on lipoproteins and atherosclerosis in male apolipoprotein E-deficient mice. *Arterioscler Thromb Vasc Biol.* 2001;21:1011-1016.

[72] Song L, Leung C, Schindler C. Lymphocytes are important in early atherosclerosis. *J. Clin. Invest.* 2001;108:251-259.

[73] Zhou X, Nicoletti A, Elhage R, Hansson GK. Transfer of CD4(+) T cells aggravates atherosclerosis in immunodeficient apolipoprotein E knockout mice. *Circulation.* 2000;102:2919-2922.

[74] Frostegard J, Ulfgren AK, Nyberg P, et al. Cytokine expression in advanced human atherosclerotic plaques: dominance of pro-inflammatory (Th1) and macrophage-stimulating cytokines. *Atherosclerosis.* 1999;145:33-43.

[75] Stemme S, Faber B, Holm J, Wiklund O, Witztum JL, Hansson GK. T lymphocytes from human atherosclerotic plaques recognize oxidized low density lipoprotein. *Proc. Natl. Acad. Sci. USA*. 1995;92:3893-3897.

[76] Uyemura K, Demer LL, Castle SC, et al. Cross-regulatory roles of interleukin (IL)-12 and IL-10 in atherosclerosis. *J. Clin. Invest.* 1996;97:2130-2138.

[77] Branen L, Hovgaard L, Nitulescu M, Bengtsson E, Nilsson J, Jovinge S. Inhibition of tumor necrosis factor-alpha reduces atherosclerosis in apolipoprotein E knockout mice. *Arterioscler Thromb Vasc. Biol.* 2004;24:2137-2142.

[78] Buono C, Binder CJ, Stavrakis G, Witztum JL, Glimcher LH, Lichtman AH. T-bet deficiency reduces atherosclerosis and alters plaque antigen-specific immune responses. *Proc. Natl. Acad. Sci. USA*. 2005;102:1596-1601.

[79] Buono C, Come CE, Stavrakis G, Maguire GF, Connelly PW, Lichtman AH. Influence of interferon-gamma on the extent and phenotype of diet-induced atherosclerosis in the LDLR-deficient mouse. *Arterioscler Thromb Vasc Biol.* 2003;23:454-460.

[80] Davenport P, Tipping PG. The role of interleukin-4 and interleukin-12 in the progression of atherosclerosis in apolipoprotein E-deficient mice. *Am. J. Pathol.* 2003;163:1117-1125.

[81] Elhage R, Jawien J, Rudling M, et al. Reduced atherosclerosis in interleukin-18 deficient apolipoprotein E-knockout mice. *Cardiovasc. Res.* 2003;59:234-240.

[82] Gupta S, Pablo AM, Jiang X, Wang N, Tall AR, Schindler C. IFN-gamma potentiates atherosclerosis in ApoE knock-out mice. *J. Clin. Invest.* 1997;99:2752-2761.

[83] Laurat E, Poirier B, Tupin E, et al. In vivo downregulation of T helper cell 1 immune responses reduces atherogenesis in apolipoprotein E-knockout mice. *Circulation.* 2001;104:197-202.

[84] Whitman SC, Ravisankar P, Elam H, Daugherty A. Exogenous interferon-gamma enhances atherosclerosis in apolipoprotein E-/- mice. *Am. J. Pathol.* 2000;157:1819-1824.

[85] Huber SA, Sakkinen P, David C, Newell MK, Tracy RP. T helper-cell phenotype regulates atherosclerosis in mice under conditions of mild hypercholesterolemia. *Circulation.* 2001;103:2610-2616.

[86] Paigen B, Morrow A, Brandon C, Mitchell D, Holmes P. Variation in susceptibility to atherosclerosis among inbred strains of mice. *Atherosclerosis.* 1985;57:65-73.

[87] Pinderski LJ, Fischbein MP, Subbanagounder G, et al. Overexpression of interleukin-10 by activated T lymphocytes inhibits atherosclerosis in LDL receptor-deficient Mice by altering lymphocyte and macrophage phenotypes. *Circ. Res.* 2002;90:1064-1071.

[88] Binder CJ, Hartvigsen K, Chang MK, et al. IL-5 links adaptive and natural immunity specific for epitopes of oxidized LDL and protects from atherosclerosis. *J. Clin. Invest.* 2004;114:427-437.

[89] King VL, Szilvassy SJ, Daugherty A. Interleukin-4 deficiency decreases atherosclerotic lesion formation in a site-specific manner in female LDL receptor-/- mice. *Arterioscler Thromb. Vasc. Biol.* 2002;22:456-461.

[90] Lutgens E, Cleutjens KB, Heeneman S, Koteliansky VE, Burkly LC, Daemen MJ. Both early and delayed anti-CD40L antibody treatment induces a stable plaque phenotype. *Proc. Natl. Acad. Sci. USA.* 2000;97:7464-7469.

[91] Friesel R, Komoriya A, Maciag T. Inhibition of endothelial cell proliferation by gamma-interferon. *J. Cell Biol.* 1987;104:689-696.

[92] Hansson GK, Hellstrand M, Rymo L, Rubbia L, Gabbiani G. Interferon gamma inhibits both proliferation and expression of differentiation-specific alpha-smooth muscle actin in arterial smooth muscle cells. *J. Exp. Med.* 1989;170:1595-1608.

[93] Lee E, Vaughan DE, Parikh SH, et al. Regulation of matrix metalloproteinases and plasminogen activator inhibitor-1 synthesis by plasminogen in cultured human vascular smooth muscle cells. *Circ. Res.* 1996;78:44-49.

[94] Saren P, Welgus HG, Kovanen PT. TNF-alpha and IL-1beta selectively induce expression of 92-kDa gelatinase by human macrophages. *J. Immunol.* 1996;157:4159-4165.

[95] van Hinsbergh VW, van den Berg EA, Fiers W, Dooijewaard G. Tumor necrosis factor induces the production of urokinase-type plasminogen activator by human endothelial cells. *Blood.* 1990;75:1991-1998.

[96] Andersson J, Libby P, Hansson GK. Adaptive immunity and atherosclerosis. Clin Immunol. 2009;134:33-46.

[97] Eid RE, Rao DA, Zhou J, et al. Interleukin-17 and interferon-gamma are produced concomitantly by human coronary artery-infiltrating T cells and act synergistically on vascular smooth muscle cells. *Circulation.* 2009;119:1424-1432.

[98] Xie JJ, Wang J, Tang TT, et al. The Th17/Treg functional imbalance during atherogenesis in ApoE(-/-) mice Th17/Treg imbalance in atherogenesis. *Cytokine.* 2009.

[99] Yang XF, Yin Y, Wang H. Vascular Inflammation and Atherogenesis Are Activated Via Receptors for PAMPs and Suppressed By Regulatory T Cells. *Drug Discov. Today Ther Strate.g.* 2008;5:125-142.

[100] von Herrath MG, Harrison LC. Antigen-induced regulatory T cells in autoimmunity. *Nat. Rev. Immunol.* 2003;3:223-232.

[101] Yan Y, Chen Y, Yang F, et al. HLA-A2.1-restricted T cells react to SEREX-defined tumor antigen CML66L and are suppressed by CD4+CD25+ regulatory T cells. *Int. J. Immunopathol Pharmacol.* 2007;20:75-89.

[102] Mallat Z, Ait-Oufella H, Tedgui A. Regulatory T-cell immunity in atherosclerosis. *Trends Cardiovasc Med.* 2007;17:113-118.

[103] Xiong Z, Song J, Yan Y, et al. Higher expression of Bax in regulatory T cells increases vascular inflammation. *Front Biosci.* 2008;13:7143-7155.

[104] Xiong Z, Yan Y, Song J, et al. Expression of TCTP antisense in CD25(high) regulatory T cells aggravates cuff-injured vascular inflammation. *Atherosclerosis.* 2009;203:401-408.

[105] Riley JL, June CH. The CD28 family: a T-cell rheostat for therapeutic control of T-cell activation. *Blood.* 2005;105:13-21.

[106] Beyersdorf N, Gaupp S, Balbach K, et al. Selective targeting of regulatory T cells with CD28 superagonists allows effective therapy of experimental autoimmune encephalomyelitis. *J. Exp. Med.* 2005;202:445-455.

[107] Lin T, Hu J, Wang D, Stocco DM. Interferon-gamma inhibits the steroidogenic acute regulatory protein messenger ribonucleic acid expression and protein levels in primary cultures of rat Leydig cells. *Endocrinology.* 1998;139:2217-2222.

[108] Suntharalingam G, Perry MR, Ward S, et al. Cytokine storm in a phase 1 trial of the anti-CD28 monoclonal antibody TGN1412. *N. Engl. J. Med.* 2006;355:1018-1028.

[109] Riley JL, June CH, Blazar BR. Human T regulatory cell therapy: take a billion or so and call me in the morning. *Immunity.* 2009;30:656-665.

[110] de Boer OJ, van der Meer JJ, Teeling P, van der Loos CM, van der Wal AC. Low numbers of FOXP3 positive regulatory T cells are present in all developmental stages of human atherosclerotic lesions. *PLoS One.* 2007;2:e779.

[111] Caligiuri G, Nicoletti A, Poirier B, Hansson GK. Protective immunity against atherosclerosis carried by B cells of hypercholesterolemic mice. *J. Clin. Invest.* 2002;109:745-753.

[112] Major AS, Fazio S, Linton MF. B-lymphocyte deficiency increases atherosclerosis in LDL receptor-null mice. *Arterioscler. Thromb Vasc. Biol.* 2002;22:1892-1898.

[113] Witztum JL. Splenic immunity and atherosclerosis: a glimpse into a novel paradigm? *J. Clin. Invest.* 2002;109:721-724.

[114] van Leeuwen M, Damoiseaux J, Duijvestijn A, Tervaert JW. The therapeutic potential of targeting B cells and anti-oxLDL antibodies in atherosclerosis. *Autoimmun Rev.* 2009;9:53-57.

[115] Yanaba K, Bouaziz JD, Matsushita T, Magro CM, St Clair EW, Tedder TF. B-lymphocyte contributions to human autoimmune disease. *Immunol. Rev.* 2008;223:284-299.

[116] Binder CJ, Silverman GJ. Natural antibodies and the autoimmunity of atherosclerosis. *Springer Semin Immunopathol.* 2005;26:385-404.

[117] Shaw PX, Horkko S, Tsimikas S, et al. Human-derived anti-oxidized LDL autoantibody blocks uptake of oxidized LDL by macrophages and localizes to atherosclerotic lesions in vivo. *Arterioscler. Thromb Vasc Biol.* 2001;21:1333-1339.
[118] de Boer OJ, van der Wal AC, Houtkamp MA, Ossewaarde JM, Teeling P, Becker AE. Unstable atherosclerotic plaques contain T-cells that respond to Chlamydia pneumoniae. *Cardiovasc. Res.* 2000;48:402-408.
[119] Palinski W, Rosenfeld ME, Yla-Herttuala S, et al. Low density lipoprotein undergoes oxidative modification in vivo. *Proc. Natl. Acad. Sci. USA.* 1989;86:1372-1376.
[120] Paulsson G, Zhou X, Tornquist E, Hansson GK. Oligoclonal T cell expansions in atherosclerotic lesions of apolipoprotein E-deficient mice. *Arterioscler Thromb Vasc Biol.* 2000;20:10-17.
[121] Witztum JL, Steinberg D. Role of oxidized low density lipoprotein in atherogenesis. *J. Clin. Invest.* 1991;88:1785-1792.
[122] Rader DJ, Daugherty A. Translating molecular discoveries into new therapies for atherosclerosis. Nature. 2008;451:904-913.
[123] Skalen K, Gustafsson M, Rydberg EK, et al. Subendothelial retention of atherogenic lipoproteins in early atherosclerosis. *Nature.* 2002;417:750-754.
[124] Tabas I, Williams KJ, Boren J. Subendothelial lipoprotein retention as the initiating process in atherosclerosis: update and therapeutic implications. *Circulation.* 2007;116:1832-1844.
[125] Ait-Oufella H, Kinugawa K, Zoll J, et al. Lactadherin deficiency leads to apoptotic cell accumulation and accelerated atherosclerosis in mice. *Circulation.* 2007;115:2168-2177.
[126] Angeli V, Llodra J, Rong JX, et al. Dyslipidemia associated with atherosclerotic disease systemically alters dendritic cell mobilization. *Immunity.* 2004;21:561-574.
[127] Weyand CM, Goronzy JJ. Medium- and large-vessel vasculitis. *N. Engl. J. Med.* 2003;349:160-169.
[128] Gupta M, Vavasis C, Frishman WH. Heat shock proteins in cardiovascular disease a new therapeutic target. *Cardiol. Rev.* 2004;12:26-30.
[129] Veres A, Fust G, Smieja M, et al. Relationship of anti-60 kDa heat shock protein and anti-cholesterol antibodies to cardiovascular events. *Circulation.* 2002;106:2775-2780.
[130] Bachmaier K, Penninger JM. Chlamydia and antigenic mimicry. *Curr. Top Microbiol. Immunol.* 2005;296:153-163.
[131] Szekanecz Z, Kerekes G, Der H, et al. Accelerated atherosclerosis in rheumatoid arthritis. *Ann. N Y Acad Sci.* 2007;1108:349-358.
[132] Sherer Y, Cerinic MM, Bartoli F, et al. Early atherosclerosis and autoantibodies to heat-shock proteins and oxidized LDL in systemic sclerosis. *Ann. N Y Acad. Sci.* 2007;1108:259-267.
[133] de Leeuw K, Kallenberg C, Bijl M. Accelerated atherosclerosis in patients with systemic autoimmune diseases. *Ann. N Y Acad. Sci.* 2005;1051:362-371.
[134] Belizna CC, Richard V, Thuillez C, Levesque H, Shoenfeld Y. Insights into atherosclerosis therapy in antiphospholipid syndrome. *Autoimmun. Rev.* 2007;7:46-51.
[135] Vikramadithyan RK, Hu Y, Noh HL, et al. Human aldose reductase expression accelerates diabetic atherosclerosis in transgenic mice. *J. Clin. Invest.* 2005;115:2434-2443.

[136] Stumvoll M. Type 2 diabetes : principles of pathogensis and therapy. *Lancet.* 2005;365:1333-1346.
[137] Diabetes. World Health Organization; 2009 Nov.
[138] Fernandez-Real JM, Ricart W. Insulin resistance and chronic cardiovascular inflammatory syndrome. *Endocr. Rev.* 2003;24:278-301.
[139] Festa A, D'Agostino R, Jr., Howard G, Mykkanen L, Tracy RP, Haffner SM. Chronic subclinical inflammation as part of the insulin resistance syndrome: the Insulin Resistance Atherosclerosis Study (IRAS). *Circulation.* 2000;102:42-47.
[140] Pickup JC. Inflammation and activated innate immunity in the pathogenesis of type 2 diabetes. *Diabetes Care.* 2004;27:813-823.
[141] Pickup JC, Crook MA. Is type II diabetes mellitus a disease of the innate immune system? *Diabetologia.* 1998;41:1241-1248.
[142] Pickup JC, Mattock MB, Chusney GD, Burt D. NIDDM as a disease of the innate immune system: association of acute-phase reactants and interleukin-6 with metabolic syndrome X. *Diabetologia.* 1997;40:1286-1292.
[143] Arkan MC, Hevener AL, Greten FR, et al. IKK-beta links inflammation to obesity-induced insulin resistance. *Nat. Med.* 2005;11:191-198.
[144] Cai D, Yuan M, Frantz DF, et al. Local and systemic insulin resistance resulting from hepatic activation of IKK-beta and NF-kappaB. *Nat. Med.* 2005;11:183-190.
[145] Gual P, Gremeaux T, Gonzalez T, Le Marchand-Brustel Y, Tanti JF. MAP kinases and mTOR mediate insulin-induced phosphorylation of insulin receptor substrate-1 on serine residues 307, 612 and 632. *Diabetologia.* 2003;46:1532-1542.
[146] Rohl M, Pasparakis M, Baudler S, et al. Conditional disruption of IkappaB kinase 2 fails to prevent obesity-induced insulin resistance. *J. Clin. Invest.* 2004;113:474-481.
[147] Wunderlich FT, Luedde T, Singer S, et al. Hepatic NF-kappa B essential modulator deficiency prevents obesity-induced insulin resistance but synergizes with high-fat feeding in tumorigenesis. *Proc. Natl. Acad. Sci. USA.* 2008;105:1297-1302.
[148] Yuan M, Konstantopoulos N, Lee J, et al. Reversal of obesity- and diet-induced insulin resistance with salicylates or targeted disruption of Ikkbeta. *Science.* 2001;293:1673-1677.
[149] Hotamisligil GS, Peraldi P, Budavari A, Ellis R, White MF, Spiegelman BM. IRS-1-mediated inhibition of insulin receptor tyrosine kinase activity in TNF-alpha- and obesity-induced insulin resistance. *Science.* 1996;271:665-668.
[150] Kanety H, Feinstein R, Papa MZ, Hemi R, Karasik A. Tumor necrosis factor alpha-induced phosphorylation of insulin receptor substrate-1 (IRS-1). Possible mechanism for suppression of insulin-stimulated tyrosine phosphorylation of IRS-1. *J. Biol. Chem.* 1995;270:23780-23784.
[151] Kwon G, Xu G, Marshall CA, McDaniel ML. Tumor necrosis factor alpha-induced pancreatic beta-cell insulin resistance is mediated by nitric oxide and prevented by 15-deoxy-Delta12,14-prostaglandin J2 and aminoguanidine. A role for peroxisome proliferator-activated receptor gamma activation and inos expression. *J. Biol. Chem.* 1999;274:18702-18708.
[152] Leahy JL. Pathogenesis of type 2 diabetes mellitus. *Arch. Med. Res.* 2005;36:197-209.
[153] Sethi JK, Hotamisligil GS. The role of TNF alpha in adipocyte metabolism. *Semin Cell Dev Biol.* 1999;10:19-29.

[154] Duncan BB, Schmidt MI. The epidemiology of low-grade chronic systemic inflammation and type 2 diabetes. *Diabetes Technol. Ther.* 2006;8:7-17.

[155] Fernandez-Real JM, Pickup JC. Innate immunity, insulin resistance and type 2 diabetes. Trends *Endocrinol. Metab.* 2008;19:10-16.

[156] Schmidt MI, Duncan BB, Sharrett AR, et al. Markers of inflammation and prediction of diabetes mellitus in adults (Atherosclerosis Risk in Communities study): a cohort study. *Lancet.* 1999;353:1649-1652.

[157] Dandona P, Aljada A. A rational approach to pathogenesis and treatment of type 2 diabetes mellitus, insulin resistance, inflammation, and atherosclerosis. *Am. J. Cardiol.* 2002;90:27G-33G.

[158] Marette A. Mediators of cytokine-induced insulin resistance in obesity and other inflammatory settings. *Curr. Opin. Clin. Nutr .Metab Care.* 2002;5:377-383.

[159] Eizirik DL, Mandrup-Poulsen T. A choice of death--the signal-transduction of immune-mediated beta-cell apoptosis. *Diabetologia.* 2001;44:2115-2133.

[160] Winer S, Tsui H, Lau A, et al. Autoimmune islet destruction in spontaneous type 1 diabetes is not beta-cell exclusive. *Nat. Med.* 2003;9:198-205.

[161] Savage DB, Petersen KF, Shulman GI. Mechanisms of insulin resistance in humans and possible links with inflammation. *Hypertension.* 2005;45:828-833.

[162] Butler AE, Janson J, Bonner-Weir S, Ritzel R, Rizza RA, Butler PC. Beta-cell deficit and increased beta-cell apoptosis in humans with type 2 diabetes. *Diabetes.* 2003;52:102-110.

[163] Weyer C, Bogardus C, Mott DM, Pratley RE. The natural history of insulin secretory dysfunction and insulin resistance in the pathogenesis of type 2 diabetes mellitus. *J. Clin. Invest.* 1999;104:787-794.

[164] Taylor SI, Accili D, Imai Y. Insulin resistance or insulin deficiency. Which is the primary cause of NIDDM? *Diabetes.* 1994;43:735-740.

[165] Clark A, Wells CA, Buley ID, et al. Islet amyloid, increased A-cells, reduced B-cells and exocrine fibrosis: quantitative changes in the pancreas in type 2 diabetes. *Diabetes Res.* 1988;9:151-159.

[166] Prentki M, Nolan CJ. Islet beta cell failure in type 2 diabetes. J Clin Invest. 2006;116:1802-1812.

[167] Hurst RT, Lee RW. Increased incidence of coronary atherosclerosis in type 2 diabetes mellitus: mechanisms and management. *Ann. Intern. Med.* 2003;139:824-834.

[168] Anderson TJ, Meredith IT, Charbonneau F, et al. Endothelium-dependent coronary vasomotion relates to the susceptibility of LDL to oxidation in humans. *Circulation.* 1996;93:1647-1650.

[169] Despres JP, Lamarche B, Mauriege P, et al. Hyperinsulinemia as an independent risk factor for ischemic heart disease. *N. Engl. J. Med.* 1996;334:952-957.

[170] Guzik TJ, Mussa S, Gastaldi D, et al. Mechanisms of increased vascular superoxide production in human diabetes mellitus: role of NAD(P)H oxidase and endothelial nitric oxide synthase. *Circulation.* 2002;105:1656-1662.

[171] Steinberg HO, Tarshoby M, Monestel R, et al. Elevated circulating free fatty acid levels impair endothelium-dependent vasodilation. *J. Clin. Invest.* 1997;100:1230-1239.

[172] Williams SB, Goldfine AB, Timimi FK, et al. Acute hyperglycemia attenuates endothelium-dependent vasodilation in humans in vivo. *Circulation.* 1998;97:1695-1701.

[173] Davi G, Catalano I, Averna M, et al. Thromboxane biosynthesis and platelet function in type II diabetes mellitus. *N. Engl. J. Med.* 1990;322:1769-1774.

[174] Sobel BE, Woodcock-Mitchell J, Schneider DJ, Holt RE, Marutsuka K, Gold H. Increased plasminogen activator inhibitor type 1 in coronary artery atherectomy specimens from type 2 diabetic compared with nondiabetic patients: a potential factor predisposing to thrombosis and its persistence. *Circulation.* 1998;97:2213-2221.

[175] Thogersen AM, Jansson JH, Boman K, et al. High plasminogen activator inhibitor and tissue plasminogen activator levels in plasma precede a first acute myocardial infarction in both men and women: evidence for the fibrinolytic system as an independent primary risk factor. *Circulation.* 1998;98:2241-2247.

[176] Park L, Raman KG, Lee KJ, et al. Suppression of accelerated diabetic atherosclerosis by the soluble receptor for advanced glycation endproducts. *Nat. Med.* 1998;4:1025-1031.

[177] Basta G, Schmidt AM, De Caterina R. Advanced glycation end products and vascular inflammation: implications for accelerated atherosclerosis in diabetes. *Cardiovasc. Res.* 2004;63:582-592.

[178] Johansson F, Kramer F, Barnhart S, et al. Type 1 diabetes promotes disruption of advanced atherosclerotic lesions in LDL receptor-deficient mice. *Proc. Natl. Acad. Sci. USA*. 2008;105:2082-2087.

[179] Griffin E, Re A, Hamel N, et al. A link between diabetes and atherosclerosis: Glucose regulates expression of CD36 at the level of translation. *Nat. Med.* 2001;7:840-846.

[180] Stewart CR, Stuart LM, Wilkinson K, et al. CD36 ligands promote sterile inflammation through assembly of a Toll-like receptor 4 and 6 heterodimer. *Nat. Immunol.* 2010;11:155-161.

[181] Hu FB, Stampfer MJ, Haffner SM, Solomon CG, Willett WC, Manson JE. Elevated risk of cardiovascular disease prior to clinical diagnosis of type 2 diabetes. *Diabetes Care.* 2002;25:1129-1134.

[182] Ruige JB, Assendelft WJ, Dekker JM, Kostense PJ, Heine RJ, Bouter LM. Insulin and risk of cardiovascular disease: a meta-analysis. *Circulation.* 1998;97:996-1001.

[183] Pradhan AD, Manson JE, Rifai N, Buring JE, Ridker PM. C-reactive protein, interleukin 6, and risk of developing type 2 diabetes mellitus. *Jama.* 2001;286:327-334.

[184] Kim F, Tysseling KA, Rice J, et al. Free fatty acid impairment of nitric oxide production in endothelial cells is mediated by IKKbeta. *Arterioscler. Thromb. Vasc. Biol.* 2005;25:989-994.

[185] Pirola L, Johnston AM, Van Obberghen E. Modulation of insulin action. *Diabetologia.* 2004;47:170-184.

[186] Huang PL. eNOS, metabolic syndrome and cardiovascular disease. *Trends Endocrinol. Metab.* 2009;20:295-302.

[187] Kim JA, Montagnani M, Koh KK, Quon MJ. Reciprocal relationships between insulin resistance and endothelial dysfunction: molecular and pathophysiological mechanisms. *Circulation.* 2006;113:1888-1904.

[188] Semenkovich CF. Insulin resistance and atherosclerosis. *J. Clin. Invest.* 2006;116:1813-1822.

[189] Montagnani M, Golovchenko I, Kim I, et al. Inhibition of phosphatidylinositol 3-kinase enhances mitogenic actions of insulin in endothelial cells. *J. Biol. Chem.* 2002;277:1794-1799.

[190] Montagnani M, Quon MJ. Insulin action in vascular endothelium: potential mechanisms linking insulin resistance with hypertension. *Diabetes Obes Metab.* 2000;2:285-292.

[191] Jiang ZY, Lin YW, Clemont A, et al. Characterization of selective resistance to insulin signaling in the vasculature of obese Zucker (fa/fa) rats. *J. Clin. Invest.* 1999;104:447-457.

[192] Cusi K, Maezono K, Osman A, et al. Insulin resistance differentially affects the PI 3-kinase- and MAP kinase-mediated signaling in human muscle. *J. Clin. Invest.* 2000;105:311-320.

[193] Jiang ZY, Zhou QL, Chatterjee A, et al. Endothelin-1 modulates insulin signaling through phosphatidylinositol 3-kinase pathway in vascular smooth muscle cells. *Diabetes.* 1999;48:1120-1130.

[194] Deanfield JE, Halcox JP, Rabelink TJ. Endothelial function and dysfunction: testing and clinical relevance. *Circulation.* 2007;115:1285-1295.

[195] Widlansky ME, Gokce N, Keaney JF, Jr., Vita JA. The clinical implications of endothelial dysfunction. *J. Am. Coll Cardiol.* 2003;42:1149-1160.

[196] Caballero AE. Endothelial dysfunction in obesity and insulin resistance: a road to diabetes and heart disease. *Obes. Res.* 2003;11:1278-1289.

[197] Meigs JB, Hu FB, Rifai N, Manson JE. Biomarkers of endothelial dysfunction and risk of type 2 diabetes mellitus. *Jama.* 2004;291:1978-1986.

[198] Halcox JP, Schenke WH, Zalos G, et al. Prognostic value of coronary vascular endothelial dysfunction. *Circulation.* 2002;106:653-658.

[199] Hingorani AD, Cross J, Kharbanda RK, et al. Acute systemic inflammation impairs endothelium-dependent dilatation in humans. *Circulation.* 2000;102:994-999.

[200] Yudkin JS, Stehouwer CD, Emeis JJ, Coppack SW. C-reactive protein in healthy subjects: associations with obesity, insulin resistance, and endothelial dysfunction: a potential role for cytokines originating from adipose tissue? *Arterioscler. Thromb. Vasc. Biol.* 1999;19:972-978.

[201] Grundy SM. Obesity, metabolic syndrome, and cardiovascular disease. *J. Clin. Endocrinol. Metab.* 2004;89:2595-2600.

[202] Kota BP, Huang TH, Roufogalis BD. An overview on biological mechanisms of PPARs. *Pharmacol. Res.* 2005;51:85-94.

[203] Berger JP, Akiyama TE, Meinke PT. PPARs: therapeutic targets for metabolic disease. *Trends Pharmacol. Sci.* 2005;26:244-251.

[204] Michalik L, Auwerx J, Berger JP, et al. International Union of Pharmacology. LXI. Peroxisome proliferator-activated receptors. *Pharmacol.Rev.* 2006;58:726-741.

[205] Collins AR, Meehan WP, Kintscher U, et al. Troglitazone inhibits formation of early atherosclerotic lesions in diabetic and nondiabetic low density lipoprotein receptor-deficient mice. *Arterioscler Thromb. Vasc. Biol.* 2001;21:365-371.

[206] Ricote M, Li AC, Willson TM, Kelly CJ, Glass CK. The peroxisome proliferator-activated receptor-gamma is a negative regulator of macrophage activation. *Nature.* 1998;391:79-82.

[207] Marx N, Kehrle B, Kohlhammer K, et al. PPAR activators as antiinflammatory mediators in human T lymphocytes: implications for atherosclerosis and transplantation-associated arteriosclerosis. *Circ. Res.* 2002;90:703-710.

[208] Hartge MM, Unger T, Kintscher U. The endothelium and vascular inflammation in diabetes. *Diab. Vasc. Dis. Res.* 2007;4:84-88.

[209] Libby P, Ridker PM, Maseri A. Inflammation and atherosclerosis. *Circulation.* 2002;105:1135-1143.

[210] Gulati R, Simari RD. Defining the potential for cell therapy for vascular disease using animal models. *Dis. Model. Mech.* 2009;2:130-137.

[211] Bradley JA, Bolton EM, Pedersen RA. Stem cell medicine encounters the immune system. *Nat. Rev. Immunol.* 2002;2:859-871.

[212] Yang XF. Immunology of stem cells and cancer stem cells. *Cell Mol Immunol.* 2007;4:161-171.

[213] Roopenian D, Choi EY, Brown A. The immunogenomics of minor histocompatibility antigens. *Immunol Rev.* 2002;190:86-94.

[214] Watkins WM. The ABO blood group system: historical background. *Transfus Med.* 2001;11:243-265.

[215] Schreuder GM, Hurley CK, Marsh SG, et al. The HLA Dictionary 2001: a summary of HLA-A, -B, -C, -DRB1/3/4/5 and -DQB1 alleles and their association with serologically defined HLA-A, -B, -C, -DR and -DQ antigens. *Eur. J. Immunogenet.* 2001;28:565-596.

[216] Rubinstein P. HLA matching for bone marrow transplantation--how much is enough? *N. Engl. J. Med.* 2001;345:1842-1844.

[217] Draper JS, Pigott C, Thomson JA, Andrews PW. Surface antigens of human embryonic stem cells: changes upon differentiation in culture. *J. Anat.* 2002;200:249-258.

[218] Drukker M, Katz G, Urbach A, et al. Characterization of the expression of MHC proteins in human embryonic stem cells. *Proc. Natl. Acad. Sci. USA.* 2002;99:9864-9869.

[219] Tian L, Catt JW, O'Neill C, King NJ. Expression of immunoglobulin superfamily cell adhesion molecules on murine embryonic stem cells. *Biol. Reprod.* 1997;57:561-568.

[220] Yang XF, Mirkovic D, Zhang S, et al. Processing sites are different in the generation of HLA-A2.1-restricted, T cell reactive tumor antigen epitopes and viral epitopes. *Int. J. Immunopathol Pharmacol.* 2006;19:853-870.

[221] Yang F, Chen IH, Xiong Z, Yan Y, Wang H, Yang XF. Model of stimulation-responsive splicing and strategies in identification of immunogenic isoforms of tumor antigens and autoantigens. *Clin. Immunol.* 2006;121:121-133.

[222] Yang F, Yang XF. New concepts in tumor antigens: their significance in future immunotherapies for tumors. Cell Mol Immunol. 2005;2:331-341.

[223] Zinkernagel RM, Hengartner H. Regulation of the immune response by antigen. *Science.* 2001;293:251-253.

[224] Sahin U, Tureci O, Schmitt H, et al. Human neoplasms elicit multiple specific immune responses in the autologous host. *Proc. Natl. Acad. Sci. USA.* 1995;92:11810-11813.

[225] Chen YT. Identification of human tumor antigens by serological expression cloning: an online review on SEREX. *Cancer Immunity;* 2004.

[226] Xiong Z, Yan Y, Liu E, et al. Novel tumor antigens elicit anti-tumor humoral immune reactions in a subset of patients with polycythemia vera. *Clin. Immunol.* 2007;122:279-287.

[227] Wu CJ, Yang XF, McLaughlin S, et al. Detection of a potent humoral response associated with immune-induced remission of chronic myelogenous leukemia. *J. Clin. Invest.* 2000;106:705-714.

[228] Xiong Z, Liu E, Yan Y, et al. An unconventional antigen translated by a novel internal ribosome entry site elicits antitumor humoral immune reactions. *J. Immunol.* 2006;177:4907-4916.

[229] Xiong Z, Liu E, Yan Y, et al. A novel unconventional antigen MPD5 elicits anti-tumor humoral immune responses in a subset of patients with polycythemia vera. *Int. J. Immunopathol. Pharmacol.* 2007;20:373-380.

[230] Yang XF, Wu CJ, Chen L, et al. CML28 is a broadly immunogenic antigen, which is overexpressed in tumor cells. *Cancer Res.* 2002;62:5517-5522.

[231] Yang XF, Wu CJ, McLaughlin S, et al. CML66, a broadly immunogenic tumor antigen, elicits a humoral immune response associated with remission of chronic myelogenous leukemia. *Proc. Natl. Acad. Sci. USA.* 2001;98:7492-7497.

[232] Ng B, Yang F, Huston DP, et al. Increased noncanonical splicing of autoantigen transcripts provides the structural basis for expression of untolerized epitopes. *J. Allergy Clin. Immunol.* 2004;114:1463-1470.

[233] Grusby MJ, Auchincloss H, Jr., Lee R, et al. Mice lacking major histocompatibility complex class I and class II molecules. *Proc. Natl. Acad. Sci. USA.* 1993;90:3913-3917.

[234] Laflamme MA, Murry CE. Regenerating the heart. *Nat Biotechnol.* 2005;23:845-856.

[235] Nussbaum J, Minami E, Laflamme MA, et al. Transplantation of undifferentiated murine embryonic stem cells in the heart: teratoma formation and immune response. *Faseb J.* 2007;21:1345-1357.

[236] Behfar A, Perez-Terzic C, Faustino RS, et al. Cardiopoietic programming of embryonic stem cells for tumor-free heart repair. *J. Exp. Med.* 2007;204:405-420.

[237] Tomescot A, Leschik J, Bellamy V, et al. Differentiation in vivo of cardiac committed human embryonic stem cells in postmyocardial infarcted rats. *Stem Cells.* 2007;25:2200-2205.

[238] Chiu RC, Zibaitis A, Kao RL. Cellular cardiomyoplasty: myocardial regeneration with satellite cell implantation. *Ann. Thorac. Surg.* 1995;60:12-18.

[239] Taylor DA, Atkins BZ, Hungspreugs P, et al. Regenerating functional myocardium: improved performance after skeletal myoblast transplantation. *Nat. Med.* 1998;4:929-933.

[240] Leobon B, Garcin I, Menasche P, Vilquin JT, Audinat E, Charpak S. Myoblasts transplanted into rat infarcted myocardium are functionally isolated from their host. *Proc. Natl. Acad. Sci. USA.* 2003;100:7808-7811.

[241] Leri A, Kajstura J, Anversa P. Cardiac stem cells and mechanisms of myocardial regeneration. *Physiol. Rev.* 2005;85:1373-1416.

[242] Jackson KA, Majka SM, Wang H, et al. Regeneration of ischemic cardiac muscle and vascular endothelium by adult stem cells. *J. Clin. Invest.* 2001;107:1395-1402.

[243] Abdel-Latif A, Bolli R, Tleyjeh IM, et al. Adult bone marrow-derived cells for cardiac repair: a systematic review and meta-analysis. *Arch. Intern. Med.* 2007;167:989-997.

[244] Caplan AI, Dennis JE. Mesenchymal stem cells as trophic mediators. *J. Cell Biochem.* 2006;98:1076-1084.

[245] Wollert KC, Drexler H. Clinical applications of stem cells for the heart. *Circ. Res.* 2005;96:151-163.

[246] Amado LC, Saliaris AP, Schuleri KH, et al. Cardiac repair with intramyocardial injection of allogeneic mesenchymal stem cells after myocardial infarction. *Proc. Natl. Acad. Sci. USA*. 2005;102:11474-11479.

[247] Cai CL, Martin JC, Sun Y, et al. A myocardial lineage derives from Tbx18 epicardial cells. *Nature.* 2008;454:104-108.

[248] Zhou B, Ma Q, Rajagopal S, et al. Epicardial progenitors contribute to the cardiomyocyte lineage in the developing heart. *Nature.* 2008;454:109-113.

[249] Kubo H, Jaleel N, Kumarapeli A, et al. Increased cardiac myocyte progenitors in failing human hearts. *Circulation*. 2008;118:649-657.

[250] Asahara T, Murohara T, Sullivan A, et al. Isolation of putative progenitor endothelial cells for angiogenesis. *Science*. 1997;275:964-967.

[251] Rauscher FM, Goldschmidt-Clermont PJ, Davis BH, et al. Aging, progenitor cell exhaustion, and atherosclerosis. *Circulation*. 2003;108:457-463.

[252] Hibbert B, Olsen S, O'Brien E. Involvement of progenitor cells in vascular repair. *Trends Cardiovasc Med.* 2003;13:322-326.

[253] Sata M, Saiura A, Kunisato A, et al. Hematopoietic stem cells differentiate into vascular cells that participate in the pathogenesis of atherosclerosis. *Nat Med.* 2002;8:403-409.

[254] Libby P. Bone marrow: a fountain of vascular youth? *Circulation*. 2003;108:378-379.

[255] SoRelle R. Aging of progenitor cells contributes to atherosclerosis. *Circulation.* 2003;108:e9006-9007.

[256] Shi X, Garry DJ. Muscle stem cells in development, regeneration, and disease. *Genes Dev*. 2006;20:1692-1708.

[257] Wagers AJ, Conboy IM. Cellular and molecular signatures of muscle regeneration: current concepts and controversies in adult myogenesis. *Cell.* 2005;122:659-667.

[258] Ahuja P, Sdek P, MacLellan WR. Cardiac myocyte cell cycle control in development, disease, and regeneration. *Physiol. Rev*. 2007;87:521-544.

[259] Rubart M, Field LJ. Cardiac regeneration: repopulating the heart. *Annu. Rev. Physiol.* 2006;68:29-49.

[260] Chatenoud L, Thervet E, Primo J, Bach JF. Anti-CD3 antibody induces long-term remission of overt autoimmunity in nonobese diabetic mice. *Proc. Natl Acad. Sci. USA*. 1994;91:123-127.

[261] Steffens S, Burger F, Pelli G, et al. Short-term treatment with anti-CD3 antibody reduces the development and progression of atherosclerosis in mice. *Circulation.* 2006;114:1977-1984.

[262] Mach F, Schonbeck U, Sukhova GK, Atkinson E, Libby P. Reduction of atherosclerosis in mice by inhibition of CD40 signalling. *Nature.* 1998;394:200-203.

[263] Schonbeck U, Sukhova GK, Shimizu K, Mach F, Libby P. Inhibition of CD40 signaling limits evolution of established atherosclerosis in mice. *Proc. Natl. Acad. Sci. USA.* 2000;97:7458-7463.

[264] Holmgren J, Czerkinsky C, Eriksson K, Mharandi A. Mucosal immunisation and adjuvants: a brief overview of recent advances and challenges. *Vaccine.* 2003;21 Suppl 2:S89-95.
[265] Faria AM, Weiner HL. Oral tolerance: mechanisms and therapeutic applications. *Adv. Immunol.* 1999;73:153-264.
[266] Chen Y, Kuchroo VK, Inobe J, Hafler DA, Weiner HL. Regulatory T cell clones induced by oral tolerance: suppression of autoimmune encephalomyelitis. *Science.* 1994;265:1237-1240.
[267] Miller A, Lider O, Weiner HL. Antigen-driven bystander suppression after oral administration of antigens. *J. Exp. Med.* 1991;174:791-798.
[268] Amberger A, Maczek C, Jurgens G, et al. Co-expression of ICAM-1, VCAM-1, ELAM-1 and Hsp60 in human arterial and venous endothelial cells in response to cytokines and oxidized low-density lipoproteins. *Cell Stress Chaperones.* 1997;2:94-103.
[269] Wick G, Knoflach M, Kind M, Henderson B, Bernhard D. Heat shock proteins and stress in atherosclerosis. *Autoimmun. Rev.* 2004;3 Suppl 1:S30-31.
[270] Kol A, Bourcier T, Lichtman AH, Libby P. Chlamydial and human heat shock protein 60s activate human vascular endothelium, smooth muscle cells, and macrophages. *J. Clin. Invest.* 1999;103:571-577.
[271] Schett G, Xu Q, Amberger A, et al. Autoantibodies against heat shock protein 60 mediate endothelial cytotoxicity. *J. Clin. Invest.* 1995;96:2569-2577.
[272] Xu Q, Schett G, Perschinka H, et al. Serum soluble heat shock protein 60 is elevated in subjects with atherosclerosis in a general population. *Circulation.* 2000;102:14-20.
[273] Harats D, Yacov N, Gilburd B, Shoenfeld Y, George J. Oral tolerance with heat shock protein 65 attenuates Mycobacterium tuberculosis-induced and high-fat-diet-driven atherosclerotic lesions. *J. Am. Coll Cardiol.* 2002;40:1333-1338.
[274] Maron R, Sukhova G, Faria AM, et al. Mucosal administration of heat shock protein-65 decreases atherosclerosis and inflammation in aortic arch of low-density lipoprotein receptor-deficient mice. *Circulation.* 2002;106:1708-1715.
[275] van Puijvelde GH, van Es T, van Wanrooij EJ, et al. Induction of oral tolerance to HSP60 or an HSP60-peptide activates T cell regulation and reduces atherosclerosis. *Arterioscler Thromb Vasc Biol.* 2007;27:2677-2683.
[276] van Puijvelde GH, Hauer AD, de Vos P, et al. Induction of oral tolerance to oxidized low-density lipoprotein ameliorates atherosclerosis. *Circulation.* 2006;114:1968-1976.

In: Cardiovascular Health and Chronomics
Editor: Ram B. Singh
ISBN: 978-1-62948-976-6

Chapter 7

EVALUATING THE FORM OF NONSINUSOIDAL VARIATIONS

George S. Katinas[1], Mikhail V. Dementyev[2], Franz Halberg[1], Patricia Grambsch[1], Anatoly V. Sorokin[2] and Germaine Cornélissen[1]

[1]Halberg Chronobiology Center, University of Minnesota, Minneapolis, MN, US
[2]Chelyabinsk State Medical Academy, Chelyabinsk, Russia

PREMISES

Structures in time series can be viewed as consisting of chaos, trends and cycles (chronomes). Since Fourier's time, cycles have been harmonically resolved; since Gauss', the fitting of sinusoids can be carried out on unequidistant data. But the resolution of dynamics must also consider methods that do not depend on sinusoids. To the extent that biological processes are dissipative, they are subject to various changing nonlinear influences; hence they are not necessarily stable.

In principle the dynamics cannot always be sinusoidal, although cosine fitting is a useful approximation of the recurrent aspects of the given series.

STATUS QUO

An important approximation of geological, biological and other processes by the sinusoids (co-sinusoids) was suggested and introduced worldwide by F. Halberg [1]. Its usage in iteration procedures reveals hidden periodicities and evaluates the estimations of their occurrence. The estimation of the uncertainties of the parameters of such functions is also implemented (microscopic approach [2]).

For many decades the cosinor has played an exceedingly important role for the recognition of a time structure (chronome) of natural geo- and biospheric, and social processes [3-8].

Although, theoretically, the shape of any curve might be approximated (described) by using the sum of many sinusoidal harmonics, in practice many components must be included in the model if one is likely to avoid discernible distortions of the real profile of the signal.

AIM

The goal of this work was to elaborate a method allowing us: 1. to reveal the hidden periodicities having any form of their waves, 2. to restore their form not considering them as a sum of trigonometric functions, and 3. to evaluate the statistical significance of the profile's characteristics (parameters).

ALGORITHM

The main principles of the algorithm are based on combinations of the method of superposed epochs [9] and the filtering of oscillations after main principles described by Savitzky and Golay [10]. As the first step, the original data are folded (stacked) according both to the period length and to the reference point chosen (plexogram [2, 11, 12]). The best fitted period for plexogram plotting should be found first by means of iterations of the period length, and computing the minimal sum of residual squares for every trial period. Figure 1 shows an example of folding the records of systolic (S) blood pressure (BP) obtained by automatic monitoring during 84 hours with 30-minute intervals. The period length was chosen as 24 hours, and the phase reference point as 00:00. In the chronogram (top), measurements related to different days are shown in different colors and connected by lines. In the plexogram, they are not connected but the colors remain the same. By the naked eye, two elevations of the SBP might be noted before 9 a.m. and after 6 p.m., but both noise and high-frequency oscillations mask the useful signal. This macroscopic approach [2] is too rough, and still cannot characterize any parameters of the profile of the oscillation (SBP in this case).

In the second step, high-frequency oscillations (noise included) are filtered out. By changing and setting parameters of the filter, it is possible to keep only such details of the profile which correspond to the chosen instantaneous (momentary) frequencies. It should be stressed that the form of such oscillation is not necessarily sinusoidal. Even the straight line can be successfully approximated by means of this filter (the third and fourth coefficients would become equal to zero).

Figure 2 demonstrates different modes of application of the filter to the plexogram. Black dots connected with thin lines are the same original measurements that were shown as not connected in the lower section of Figure 1.

By changing the parameters of the filter, it is possible to approximate the process keeping only the oscillation having the longest period (equal to the length of the whole cycle) or to present its superposition with the shorter ones. Which of the oscillations remain, and which of them are filtered out, depends on the length of the gliding interval.

Note that even the 24-hour component is not sinusoidal: the curve is usually non-symmetrical, the ascending slope is shorter than the descending one.

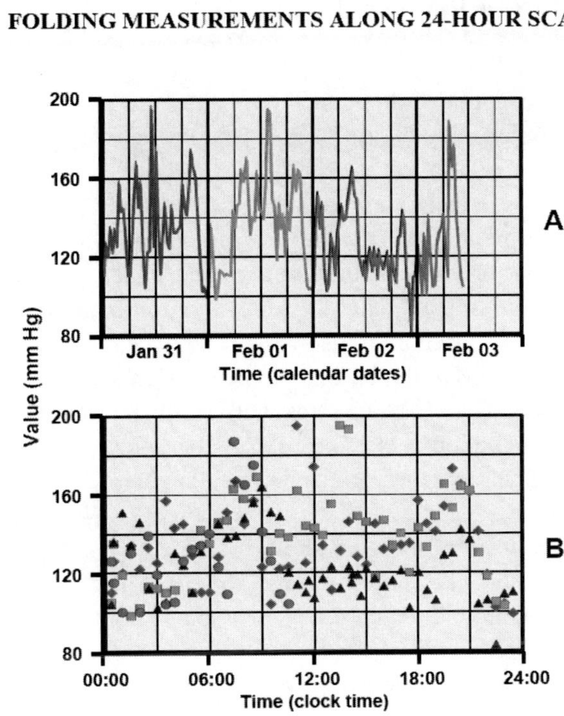

Figure 1. Plexogram composition. A - Chronogram of systolic blood pressure (BP) during 84 hours. Measurements obtained during consecutive days are drawn in different colors. B - Superposition of measurements according to clock time. Observations made on different days are shown by different symbols but colors are kept the same. Data started on March 31, 1998 from ~10-year long series of automatic BP monitoring by GSK, a man 72 years old at start of half-hourly measurements for over 10 years (1997-2009).

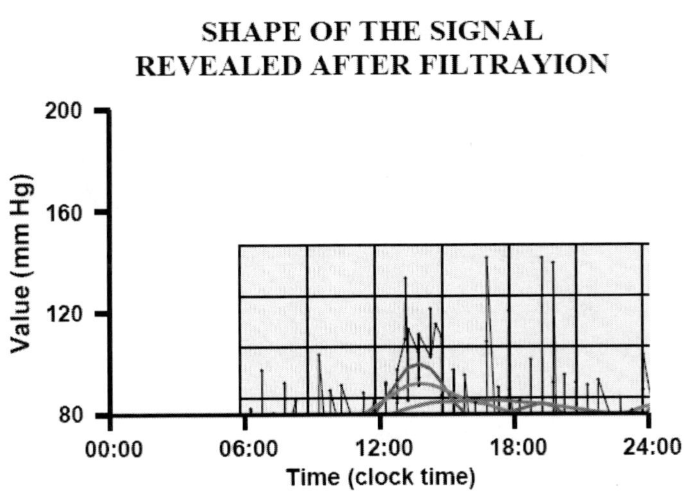

Figure 2. Filtration of noise and high-frequency oscillations. Black dots connected by thin lines: original data shown in Figure 1B. Colored lines: signal after filtering noise and high-frequency oscillations: Blue, 24-hour and longer; green, 12-hour and longer; red, 8-hour and longer spectral components are kept. The same data as those in Figure 1 are used.

DESCRIPTION OF THE SIGNAL'S SHAPE AND I7TS ANALYSIS

In the third step, the application of different computational and statistical analytical procedures provides a microscopic approach [2], and allows us to find out next the characteristics of the signal:

1: mid-level, 2: maximal and minimal values of original data, 3: clock time of their occurrence, 4: swing of original data, 5: values of peak and pit (maximal and minimal values of the approximated curve), 6: phase of their occurrence (clock time if period is equal to 24 hours), 7: swing of the approximated data, 8: values of additional elevations and depressions located between peak and pit, 9: their statistical significance, 10: phase of their occurrence, 11: confidence limits (corridors) at the chosen statistical significance of approximated values for the mid-level, for abscissas and ordinates as well as for the timing of peak, pit, elevations and depressions, 12: determination coefficient, 13: evaluation of quality of the approximation as a whole, 14: speed of the process at various phases of cycle, 15: acceleration of the process at various phases of cycle.

Figure 3 demonstrates some of these results.

1 – The term MID-LEVEL, preferred by GSK (or mid-value), should be used as generalization for the value around which deviations come about. Depending on the approach the mid-value may be computed as any mean (arithmetic, geometrical or harmonic), or as a median; for trigonometrical (e.g., cosine) approximation the term MESOR (Midline Estimating Statistic of Rhythm) was introduced [1]. Since the method proposed in this paper for estimating the signal's shape is not based on the cosine approximation, the term MESOR in principle cannot be used.

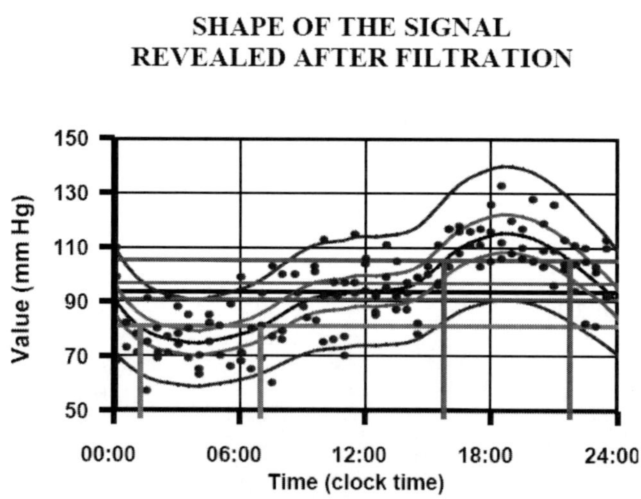

Figure 3. Some characteristics of waveform computed after filtration of the original data. Blue dots: original data folded according to 24 hours. Lines: blue: 95% confidence corridor (CC) of data, black curve: approximation, congruent green: its 95% CC, horizontal black: mid-level, parallel brown: its 95% CC, horizontal red and violet: cut-off lines for peak and pit areas corresponding to P=0.001, vertical red and violet: limits of the peak and pit areas. Data: diastolic blood pressure on Jan 14 – 16, 2008 from ~10-year long series of automatic BP monitoring started on March 31, 1998 by GSK, a man 72 years old at start of half-hourly measurements for over 10 years (1997-2009). 8-hour and longer spectral components are kept.

The term Mean Value shows the method of computation, and can be related to any average independent of its specificity. The mid-level in the program "form" is computed as the arithmetic mean.

4, 7 – Swing of oscillation means the difference between maximum and minimum of data set (either original or approximated). Reciprocally the notion double amplitude is used for trigonometrical (e.g., cosine) approximation.

5 – Peak and Pit of the signal mean the highest and the lowest points of approximation. Both the peak and the pit are unique for the process; its profile can have only one peak and only one pit. Therefore these terms cannot be used in plural (peaks and pits) being employed for a single curve only.

Sometimes the term crest is used to denote the highest value; however, it is less precise than "peak": a crest would be preferable to describe any elongated object, e.g., for the sequence of highest points at a 3D presentation (compare "the peak Everest" and the "mountain crest Himalayas"). For the same reason the term "pit" should be used, rather than 'trough".

6 –The phases of peak and pit show their position in the cycle (in relation to any reference time point). They can be expressed in radians, degrees of the cycle or as the clock time if the length of the cycle is equal to 24 hours. The parallels at the cosine approximation are the acrophase for a single cosine curve and orthophase for the superposition of several cosine harmonics.

8 – Elevations and Depressions of the process are the intermediate climb-ups and climb-downs between peak and pit. They can also be called a hills and delves. If they exist, the increasing and/or decreasing stages of the process are not monotonous, and some extrema appear. They should be taken into account if the values of ordinates of neighboring extrema differ statistically significantly. If no statistical significance is revealed, the terms increases and decreases should be used.

12 – The determination coefficient shows the power of the detected signal (often called the useful signal, described by the applied model) in relation to common power of the signal which also includes both spectral components and noise (filtered out during the approximation). It is a general statistical term which when applied to the cosine approximation and expressed in percent is called percentage rhythm [2].

13 – Evaluation of approximation as a whole (its goodness or lack of fit) shows the probability of rejecting the model which was used for description of the signal profile. It is computed as the F-criterion when comparing the original and approximated ordinates (Excel Main Menu – Service – Data Analysis – Regression).

14, 15 – Modeling the curve profile permits computing the first and second finite differences corresponding to the first and second derivatives of a process. They show speed and acceleration of the process at different phases of the cycle (Figure 3). In mechanical processes, acceleration changes when any outside force is applied to a moving body. If no force is applied, speed remains constant and the computed acceleration is equal to zero. If any force increases speed, acceleration becomes positive; in the opposite case, it is negative. In relation to biological processes, e.g., blood pressure, the appearance of positive acceleration can give evidence that any physiological mechanism begins stimulating this process. If acceleration becomes negative, it may mean that for some reasons, the process is hampered. Of course an analysis of speed and acceleration of physiological processes cannot clarify concrete branches, chains, loops and/or their interactions as elements of complicated and sophisticated feedsidewards [13], but even so, it becomes possible to choose the time when medical intervention could be most useful.

The possibility to compute speed and acceleration after approximation of the waveform of a non-sinusoidal signal was checked in the modeling examples. The results of a saw-shaped signal follow. In Figure 4, this signal is shown by a blue line. Abscissa of the graph is presented in degrees of the signal length. The original signal begins as a straight line from 0 ordinate at 0 degree of the cycle, reaches 200 at 324 degrees and returns to 0 at 360 degrees. It was approximated in two ways: by 4 cosine harmonics (red line, left) and by the "FORM" program (green line, right).

Quantitative and qualitative comparisons of approximations are shown in Table 1: most distortions (expressed as percentage of parameters estimated – in parentheses) are less when the algorithm "Form" was applied for approximation. A very demonstrative advantage of the Form-algorithm as compared with the trigonometric approximation may be shown by means of acceleration curves (Figure 5). Splashes coinciding with extrema and no wobbling along the straight part of the signal can be found if the first approach was applied, and weaker deviations are present at the extrema but clear wobbles along the straight part remain if the second approach was used. The principle of estimation of the confidence limits of phases of a cycle is explained in Figure 6. The corridor of standard errors is built corresponding to the approximating curve. The difference between the value of the peak point estimation and the value which begins to be different from it at the appointed P-value is computed (shown as lateral vertical arrows). At this level, the horizontal cut-off line is plotted, crossing the approximating curve away from the peak. From the intersections, vertical lines lead to the abscissa axis (in Figure 6 drawn above).

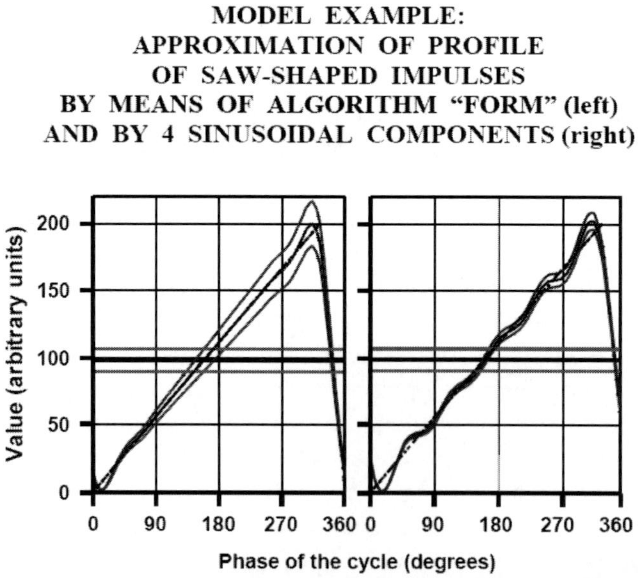

Figure 4. Distortions of a signal depend on the algorithms used. Algorithm "Form" produces them only close to the sharp extrema, and sinusoidal approximation does it also along the slopes. Dots: original data folded according to length of the cycle taken as 360 degrees. Lines: blue: 95% confidence corridor (CC) of data, black curve: approximation, horizontal black: mid-level, parallel brown: its 95% CC. Data: saw-shaped impulse.

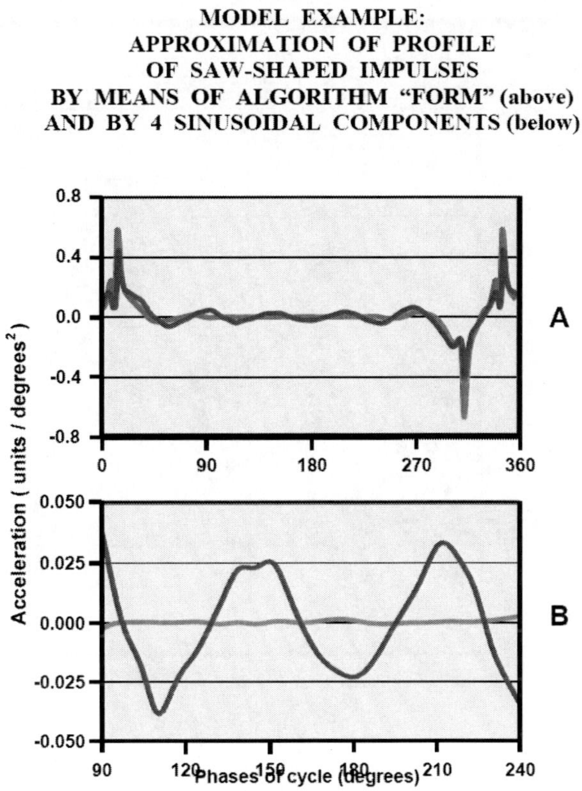

Figure 5. Accelerations during the saw-shaped impulse computed by means of algorithm "Form" (green) and by 4 sinusoidal components (red). "Form" reveals extrema of the signal better than trigonometric approximation (A) and does not produce artificial wobbles at most parts of the phases corresponding to straight slopes (B).

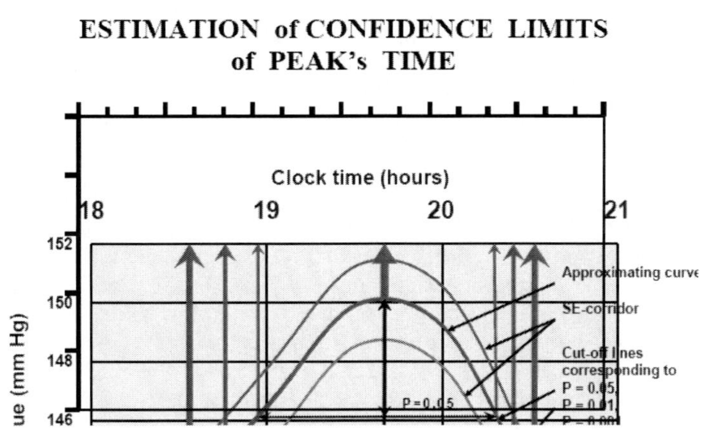

Figure 6. Width of confidence limits of peak's duration can be computed at selected probability level using principles of algorithm formulated by C. Bingham et al. [14]. Central red arrow corresponds to the top of the peak. Arrows next to it mark confidence limits at various P-values.

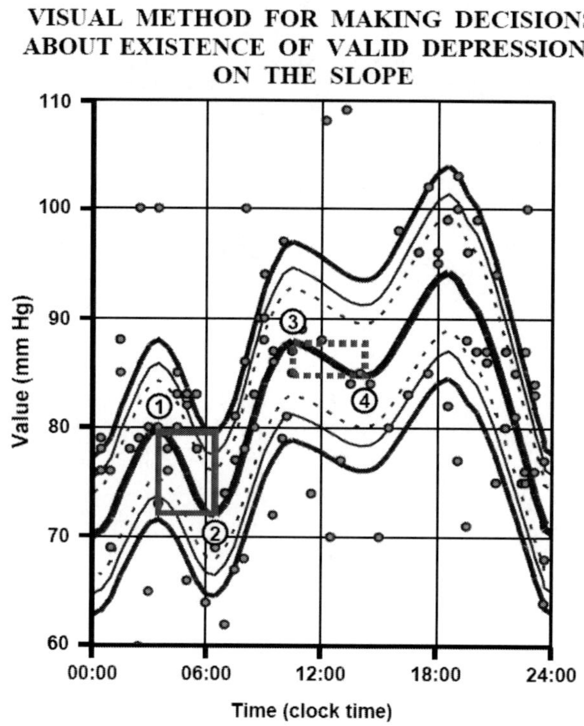

Figure 7. Visual method for making decisions about existence of validity of depressions on the slope. Blue dots: original data folded according to 24 hours. Middle thickest line: approximation of the process. Congruent lines above and below: cut-off lines. Any ordinate (e.g., y_1) of approximating line (having a slope) differs from the neighboring ones. Validity of this difference can be evaluated by taking into account the standard error (SE) of this dot (SE_{y1}) and appoint the P-value. SE of other ordinates ($SE_{y2} – SE_{yn}$) may be taken approximately equal to SE_{y1}. One can appreciate an ordinate $y_k$as differing from y_1 if (Δy / SQRT(2)*SE) > t, where t depends on P. This condition determines the ordinates of cut-off lines which are congruent to the approximating curve. In Figure 7, cut-off lines correspond to P = 0.05, 0.01 and 0.001 (shown dashed, thin and bold). Thus, if value of any neighboring ordinate of approximating curve hits inside the limits of the dashed corridor of ordinate y_1, Δy should be neglected. To decide whether the ordinates of two neighboring extrema differ with statistical significance, a rectangle can be used. Let us consider the upper left corner of the red rectangle with the extremum 1 and drag its lower right corner to the extremum 2. Position which becomes the lower left corner shows that p=value is close to 0.001, and thus the extremun 2 should be appreciated as a depression. The analogous comparison of extrema 3 and 4 shows that the lower left corner is situated between the approximating and dashed lines, thus P > 0.05 and extremum 4 should be neglected (to stress it, the contours of the testing rectangle are drawn dotted). Data of diastolic blood pressure during 12 – 14 Dec 2007, from ~10-year long series of automatic monitoring started on March 31, 1998 by GSK, a man 72 years old at start of half-hourly measurements for over 10 years (1997-2009).

Their crossings of the abscissa show the confidence limits of the peak's phase. The ideas of principles of this algorithm were described in relation to periodograms for computing confidence limits of the period length; C. Bingham et al. [14] used the F-criterion for this purpose; use of the t-criterion after Fisher's z-transformation was suggested by V. Karp and G. Katinas [11]. Ordinates of the cut-off lines are computed for every ordinate of the approximating curve. It allows appreciating validity of difference between any of its points (Figure 7).

EXCEL SHEET WITH RESULTS OF DIASTOLIC BLOOD PRESSURE ANALYSIS BY MEANS OF PROGRAM "FORM" AND MACROS "M-FORM"

	A	B	C	D	E	F	G	H	I	J
1		FILE			F.DAT	Time	Mid_Series			
2	Variable		GSK-DBP-12-14-Dec-2007			series	39429.52	13.12.2007		
3	Initial trial period				23.9933	Beginning	N TimePoints	113		
4	Final trial period				24.02	39428.00	Orig. data			
5	Initial period increase				0.0134	12.12.07	Maximum	113		
6	Interval (as part of period length)				0.4	End	Minimum	45		
7	Best fit period				23.9933	39431.04	Swing	68		
8	Process was computed as being cyclic					15.12.07	Approximation		Upper	Lower
9	General sum of squared deviations				20196.938		Determ_coe	0.368	limit	limit
10	Residual sum of squared deviations				12764.996	Repetitions	Mid-Level	80.7	78.2	83.2
11	Mid-Level				80.716814	per	Peak	94.2	89.0	99.5
12	Determination coefficient				0.367974	TimePoint	Pit	70.0	66.1	73.9
13	Number of measurements				113	*1*	Swing	24.3	22.9	25.6
14	Winter time				P(Corridor)	*0.050*	Peak (time)	18.5	16.0	20.6
15	Clock	Degrees			P(Cut-off)	*0.001*	Peak (degr.)	278	240	308
16	time	of cycle	Folded				Pit (time)	0.0	23.0	1.5
17	(hours)	360° ?	original	Approximated			Pit (degr.)	0	345	23
18		24_hours	data	data						
19	-1.98	-29.7		80.25866						
20	-1.64	-24.6		78.191445						
21	-1.493333	-22.4		77.32491		Mid-Level				
22	-1.48	-22.2		77.240634		80.716814				
23	-1.453333	-21.8		77.069977		80.716814				
24	-1.413333	-21.2		76.819372		80.716814				
25	-0.993333	-14.9		74.244902		80.716814				
26	-0.986667	-14.8		74.19237		80.716814				
27	-0.966667	-14.5		74.019572		80.716814				
28	-0.486667	-7.3		71.535883		80.716814	Deviations			
29	-0.48	-7.2		71.416353		80.716814	from			
30	-0.446667	-6.7		70.869707	Residuals	80.716814	Mid-Level			
31	0	0	45	70.363233	-25.36323	80.716814	-35.71681			
32	0.0133333	0.2	56	70.102774	-14.10277	80.716814	-24.71681			
33	0.02	0.3	60	69.985615	-9.985615	80.716814	-20.71681			
34	0.0266667	0.4	76	69.958688	6.041312	80.716814	-4.71681			
35	0.5	7.5	79	70.537708	8.462292	80.716814	-1.71681			

Figure 8. Results of analysis of diastolic blood pressure by means of macros "M-FORM" after folding and filtering by program "FORM". Season time definition shown in blue (cell A14) should be typed by hand. Cells F13-F15 should also be filled by hand because they are settings for computing confidence limits.

REALIZATION OF COMPUTATIONS

Steps 1 and 2 (stacking data and filtering them as well as iteration of trial periods) are performed by the program "FORM" developed by G.S. Katinas, G.I. Pupysheva and D.E. Korzhevsky [12]. For step 3, a special Excel macro is used. It concurrently computes parameters of 4 variables, e.g., for systolic, diastolic and pulse BP as well as for heart rate. Macros insure automatic computation of most characteristics mentioned above; only phases (and/or clock time) of confidence limits of peak, pit, elevations and depressions have to be estimated interactively. By default, confidence limits for the corridor of original values and approximating curve are usually computed at the 95% and the cut-off lines for both peak and pit at the 99.5% of statistical significance; these P-values also may be chosen voluntarily. Figure 8 demonstrates the beginning of the Excel sheet with the numerical results of the analysis. Such a sheet is created for every variable. For every variable, the graph is also plotted (Figure 9). The main results are included in the synoptic table (Figure 10).

24-HOUR PROFILE OF SYSTOLIC BLOOD PRESSURE PLOTTED BY MEANS OF "M-MACROS"

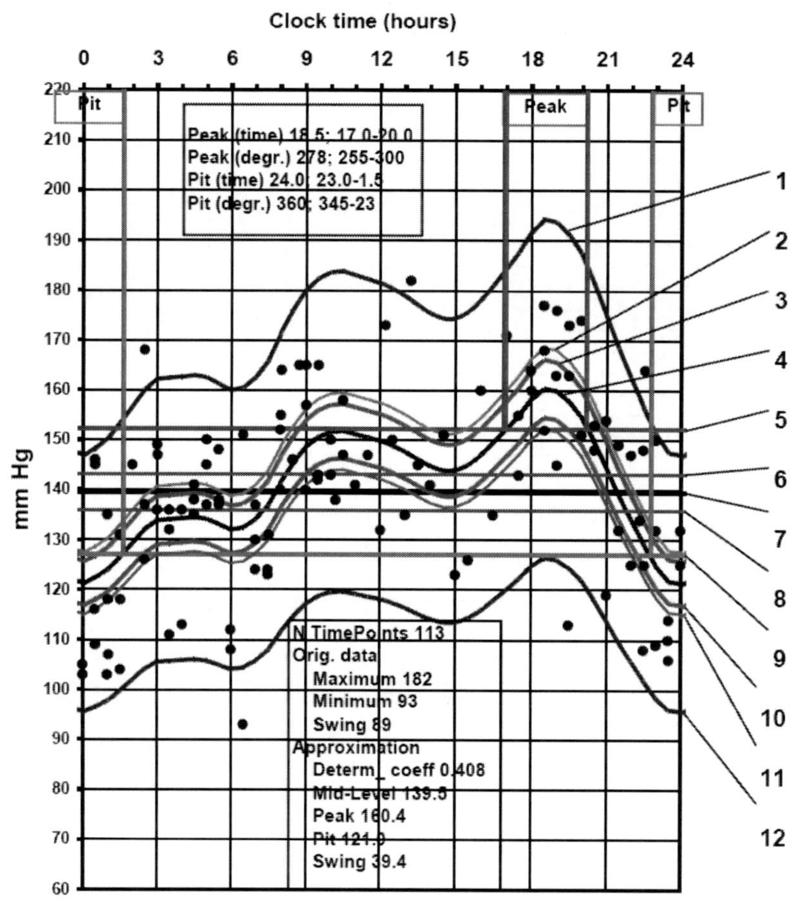

Figure 9. Visual information is created by "M-MACROS" for every cardiovascular variable analyzed: systolic, diastolic and pulse pressure as well as for heart rate. Boxes contain the quantitative information about the waveform. Vertical red lines delimit 99.9% confidence limits for the peak span, violet ones for the pit span (to show that it begins before and finishes after midnight marks "Pit" are placed partially out of graph's borders). Black dots: original data folded according to clock time. Curves: 1 and 12 (blue): 95% confidence limits of circadian variations (width of the corridor can be changed in cell F15 of Excel sheet, see Figure 8); 2 and 11: cut-off lines at P = 0.001 (lines are colored differently for easier visual comparison consecutive elevations and depressions); 3 and 10: confidence limits of approximation (width of the corridor can be changed in cell F16 of Excel sheet, see Figure 8); 4: approximation (9.6-hours and longer spectral components are kept); 5 and 9: upper and lower cut-off lines for the peak (red) and pit (violet); 6 and 8: 95% confidence limits for the mid-level (width of the corridor can be changed in cell F15 of Excel sheet, see Figure 8); 7: mid-level. Data: systolic blood pressure on Dec 12 – 14, 2007 from ~10-year long series of automatic BP monitoring started on March 31, 1998 by GSK a man 72 years old at start of half-hourly measurements for over 10 years (1997-2009). 9.6-hour and longer spectral components are kept.

SYNOPTIC TABLE WITH MAIN RESULTS OF ANALYSIS OF CARDIOVASCULAR VARIABLES BY MEANS OF PROGRAM "FORM" AND MACROS "M-FORM"

Patient	GSK						
Gender	M	Age (y)	83				
	PROFILE of CIRCADIAN OSCILLATIONS (Period is taken as 24 h[1])						
Variables:				SBP[2]	DBP[3]	HR[4]	PBP[5]
Observations started at:				12.12.07 0:00	12.12.07 0:00	12.12.07 0:00	12.12.07 0:00
Observations stopped at:				15.12.07 1:00	15.12.07 1:00	15.12.07 1:00	15.12.07 1:00
Series length (days):				3.04	3.04	3.04	3.04
Middle of series:				13.12.07 12:30	13.12.07 12:30	13.12.07 12:30	13.12.07 12:30
N Time_Points:				113	113	113	38
N Measurements:				113	113	113	113
Original data		Maximum		182	113	85	84
		Minimum		93	45	44	26
		Swing		89	68	41	58
Approximation							
	Mid-Level			**139.5**	**80.7**	**56.0**	**58.8**
		95%_CI[6]		135.9	78.2	54.6	56.4
		95%_CI		143.2	83.2	57.3	61.2
	Pit Time (h)			**24.0**	0.0	7.0	**23.3**
		99.9%_CI		1.5	23.0	5.0	0.3
		99.9%_CI		23.0	1.5	8.7	23.8
	Peak Time (h)			**18.5**	**18.5**	**19.0**	**18.8**
		99.9%_CI		17.0	16.0	18.5	0.3
		99.9%_CI		20.0	20.6	20.0	23.8
	Pit Value			**121.0**	**70.0**	**50.7**	**50.6**
		95%_CI		113.6	66.1	49.0	40.1
		95%_CI		128.3	73.9	52.5	61.2
	Peak Value			**160.4**	**94.2**	**64.9**	**66.7**
		95%_CI		150.6	89.0	62.7	52.8
		95%_CI		170.2	99.5	67.1	80.6
	Swing of oscillations			**39.4**	**24.3**	**14.1**	**16.1**
		99.9%_CI		37.0	22.9	13.6	12.7
		99.9%_CI		41.8	25.6	14.6	19.4
	Determ. Coeff.[7]			0.408	0.368	0.339	0.194
Comments	1) Period of oscillations is taken as 24 h because their real length						
		does not differ statistically significant from this estimation.					
	2) SBP - Systolic Blood Pressure (mm Hg).						
	3) DBP - Diastolic Blood Pressure (mm Hg).						
	4) HR - Heart Rate (beats/minute).						
	5) PBP - Pulse Blood Pressure (mm Hg).						
	6) 95%_CI и 99.9%_CI - Confidence Intervals (limits inbetween of which						
		the real values of parameter may vary at the shown probability);					
		If "0" is typed - the limirs were not computed.					
	7) Determ. Coeff. - Determination Coefficient characterizing the power of oscillation						
		in relation to the total power of the signal (including jams and noise).					
Additional characteristics							
	Pit	Phase (degr.)		**360.0**	**0.4**	**105.0**	**349.6**
		99.9%_CI		22.6	345.1	74.9	4.2
		99.9%_CI		22.6	22.9	131.2	357.1
	Peak	Phase (degr.)		**278**	**278**	**285**	**282**
		99.9%_CI		255	240	277	4
		99.9%_CI		300	308	300	357

Figure 10. Synoptic table combines the main information about quantitative estimation of profiles of cardiovascular variables. Name of the patient, gender and age (shown here in blue) should be typed by hand.

ACCELERATIONS OF CARDIOVASCULAR VARIABLES PLOTTED BY MEANS OF "M-MACROS"

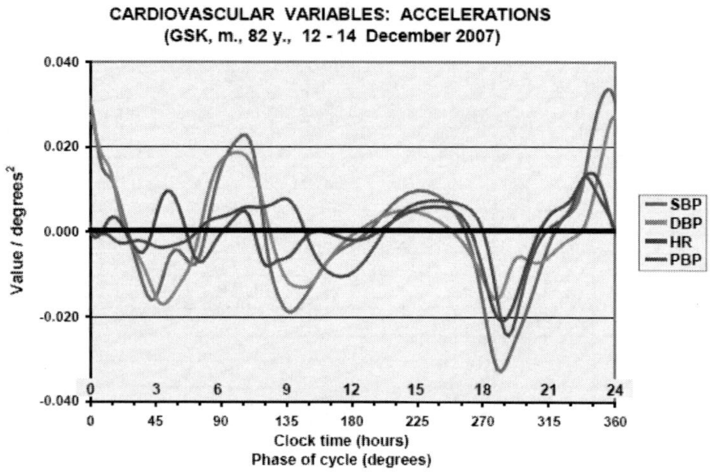

Figure 11. Acceleration graph plotted according to computing the second finite differences of ordinates of approximating curves. SBP: systolic blood pressure, DBP: diastolic blood pressure, HR: heart rate, PBP: pulse pressure. Data: cardiovascular variables on Dec 12 – 14, 2007 p from ~10-year long series of automatic BP monitoring started on March 31, 1998 by GSK a man 72 years old at start of half-hourly measurements for over 10 years (1997-2009). 9.6-hour and longer spectral components are kept.

PHASE PORTRAITS OF HEART RATE PLOTTED BY MEANS OF "M-MACROS"

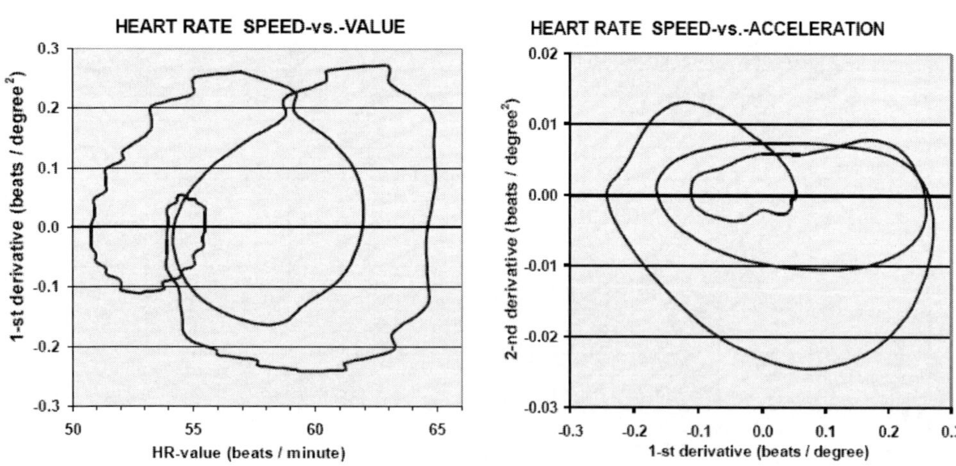

Figure 12. Phase portraits may help analyze the structure of oscillators involving feedsidewards. Heart rate records on Dec 12 – 14, 2007 from ~10-year long series of automatic BP monitoring started on March 31, 1998 by GSK a man 72 years old at start of half-hourly measurements for over 10 years (1997-2009). 9.6-hour and longer spectral components are kept.

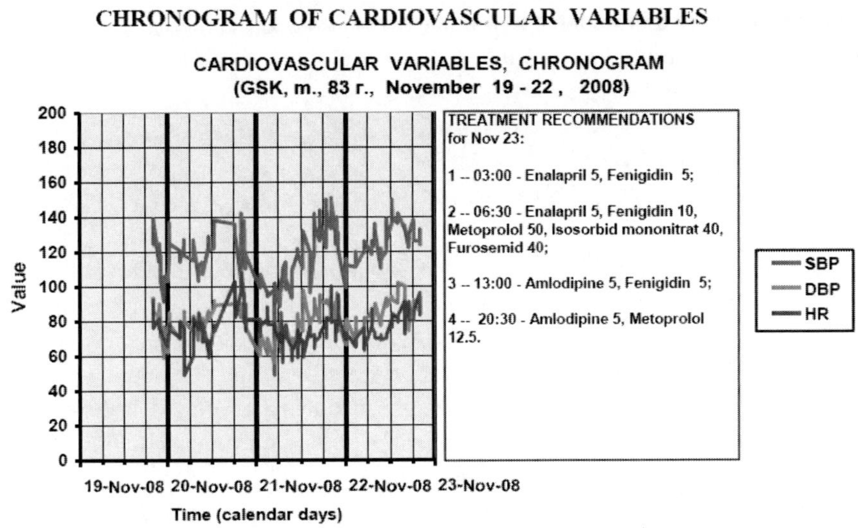

Figure 13.Chronogram (left) and the panel for typing recommendations (right). SBP: systolic blood pressure, DBP: diastolic blood pressure, HR: heart rate. Manual self-measurements by GSK a man 72 years old at start of half-hourly measurements for over 10 years (1997-2009).

Accelerations of all variables are plotted together (Figure 11); such a composition permits analyzing the mutual coordination of physiological mechanisms which are involved in the feedsidewards. Phase portraits (Figure 12) may help analyze oscillator structure. Results of most of these analyses are useful for physicians. For patients, some details (e.g., such as confidence intervals) may be excluded, but the chronogram with recommendations is of interest for them (Figure 13).

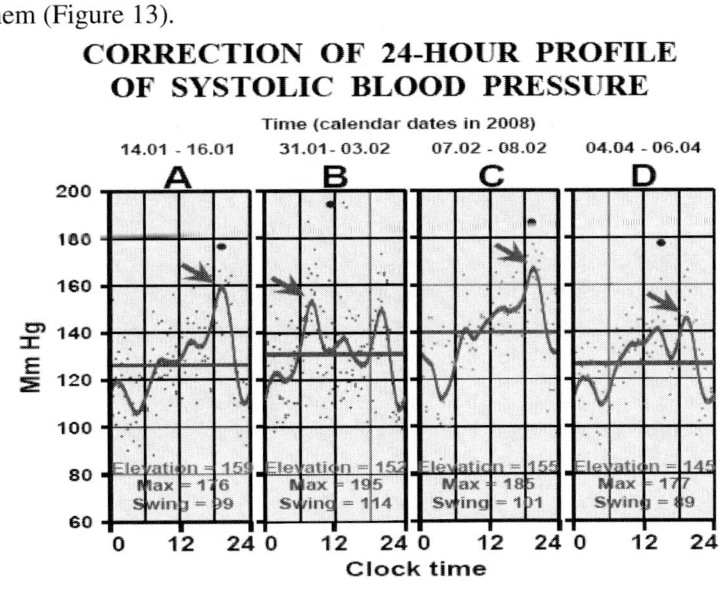

Figure 14. Adjusting profile of systolic blood pressure after changing treatment regimen. A: corrected only in timing. B: "Standard" taking medications in the morning, at daytime and in the evening. C: After a new timing correction for the form restoration. D: After following dosage correction. Data from ~10-year long series of automatic BP monitoring started on March 31, 1998 by GSK a man 72 years old at start of half-hourly measurements for over 10 years (1997-2009).

Within the limits of this paper, the medical problems of treatment and concrete methods for the optimization of timing (17) are not discussed. Not only automatic but also manual self-measurements may be used for computing.

SOME POSSIBLE FIELDS OF PRACTICAL APPLICATION

The correcting actions during diseases for obtaining the optimal results with the least dosages of medicines using sinusoidal description of the waveform were described and discussed earlier [15-25]. Now we suggest taking into account also the real shape of the curve and especially the time distribution of acceleration impulses [26, 27].

GSK was suffering from MESOR-hypertension during several decades (diagnosis was established in 1959). On 30 March 1998, he began BP and heart rate (HR) monitoring (mostly TM-2421 monitor from A&D, Tokyo, Japan).

Measurements were performed every 30 minutes with few gaps. In 2007-2008, regular corrections in timing of treatment (Nifedipine 10 mg several times during 24 hours) were applied. The SBP profile remained typical for so-called dippers, a criterion with flaws that, as compared to the cosinor, not only can fail [18-20], but can even mislead [21]: depression about 3-6 a.m., progressive (increase rather even) elevation, peak at about 6-8 p.m., and thereafter a return to minimal values. Results of computations on January 14-16 are shown in Table 2 and in Figure 14A.

At the end of January, GSK was consulted by a cardiologist who proposed to use retard medicines (Enalapril and Amlodipine) 3 times a day according to traditional mode: after awakening, in the middle of the day time and before going to bed. As a result (see Figure 14B) during January 21 – February 03, the profile became bimodal with the peak in the morning. The time of the evening elevation corresponded to the peak registered earlier. Its value was equal to the peak.

The correction of treatment timing was made taking into account both the profile of the curve and the acceleration time distribution, and the typical profile was restored on February 07 – 08 (Figure 14C), but the mid-level was too high. Correction of the dosage involved decreasing it (Figure 14D); only depression up to 12% as compared with the peak occurred during 2.7 hours after 15:00.

Successful correction of the treatment schedule lasted till May 27. On that day, according to consultation with the cardiologist, it was decided to explore how mono-therapy with the long-acting Mycardis (40 mg / day) taken at the standard morning time would impact on the BP dynamics. During the next 4 weeks, BP was gradually increasing but, to continue investigation, no corrections had been performed. Most parameters characterizing BP oscillation were drastically increased already within 20 days (Table 2). During the next 10 days, values of BP remained high, and the determination coefficient was low: the regular oscillations became relatively weak against the background of occasional fluctuations (Table 2). It can be shown also by comparing swings of original and approximated data (117 vs. 33 mm Hg). Subjectively such jumps of BP were very unfavorable. On June 26, the experiment was finished on behalf of patient's health. Subsequently, medicines were changed, and in several weeks, satisfactory results were obtained (Table 2).

Taking into account both the profile of the curve and the acceleration time distribution) was applied also in the clinic of the Almasov Federal Heart, Blood and Endocrinology Centre (St-Petersburg, Russia), and results were shortly published in 2011 [28].

COMMENT

The relative merits of different methods remain to be further tested in the clinic and by simulations.

ACKNOWLEDGMENT

The authors thank Emily Reimer, Public Services Assistant, Bio-Medical Library, University of Minnesota, Minneapolis.

REFERENCES

[1] Halberg F. Chronobiology. *Ann. Rev. Physiol.* (1969), 31, 675-725.

[2] Halberg F., Carandente F., Cornelissen G., Katinas G. Glossary of Chronobiology. *Chronobiologia*, 1977, 4 (Suppl. 1), 189 pp.

[3] Cornelissen G, Halberg F, Schwartzkopff O, Delmore P, Katinas G, Hunter D, Tarquini B, Tarquini R, Perfetto F, Watanabe Y, Otsuka K. Chronomes, time structures, for chronobioengineering for "a full life". *Biomed. Instrum. Technol*, 1999, 33, pp. 152-187.

[4] Halberg F, Cornelissen G, Chen C-H, Katinas GS, Otsuka K, Watanabe Y, Herold M, Loeckinger A, Kreze A, Kreze E, Perfetto F, Tarquini R, Maggioni C, Sothern RB, Schwartzkopff O. Chronobiology: time structures, chronomes, gauge aging, disease risk syndromes and the cosmos. *J. Anti-Aging Med.*, 2000, 3, -pp 67-90.

[5] Halberg F, Cornelissen G, Katinas G, Hillman D, Schwartzkopff O. Season's Appreciations 2000: Chronomics complement, among many other fields, genomics and proteomics. *Neuroendocrinol. Lett.*, 2001, 22, pp 53-73.

[6] Halberg F, Cornelissen G, Otsuka K, Katinas GS, Schwartzkopff O, Halpin C, Mikulecky M, Revilla M, Siegelova J, Homolka P, Dusek J, Fiser B, Singh RB. Chronomics* (*the study of time structures, chronomes) detects altered vascular variabilities constituting risks greater than hypertension: with an illustrative case report. In: Mitro P, Pella D, Rybar R, Valocik G, editors. Proceedings, 2nd Congress on Cardiovascular Diseases, Kosice, Slovakia, 25-27 April 2002. Bologna: *Monduzzi Editore*; 2002, pp 223-258.

[7] Cornelissen G, Masalov A, Halberg F, Richardson JD, Katinas GS, Sothern RB, Watanabe Y, Syutkina EV, Wendt HW, Bakken EE, Romanov Y. Multiple resonances among time structures, chronomes, around and in us. Is an about 1.3-year periodicity in solar wind built into the human cardiovascular chronome? *Human Physiology* 2004; 30 (2): 86-92.

[8] Halberg F., Cornélisen G., Katinas G., Dušek J., Homolka P., Karpíšek Z., P. Sonkowsky R., Schwartzkopf O. Chronomics and genetics. *Scripta Medica (Brno)*, 2007, 80 (4), pp 133 – 150.

[9] Chree C., Stagg JM. Recurrence Phenomena in Terrestrial Magnetism. *Philos Trans R. Soc. Lond Series, Ser. A*, 1928, 227: 21-62.

[10] Savitzky A, Golay JEM. Smoothing and differentiation of data by simplified least squares procedures. *Analytical Chemistry*, 1964, 36, pp 1627-1639.

[11] Karp V.P., Katinas GS. Mathematical methods for analysis of biorhythms. In: Komarov FI, ed. Chronobiology and Chronomedicine (Manual). Moscow, Medicine, 1987, pp 29-45 [In Russian].

[12] Karp V.P., Katinas GS. *Computative methods of analysis in chronobiology and chronomedicine*. St-Petersburg., Vostochnaja corona, 1997, 116 pp. [In Russian].

[13] Halberg F, Cornelissen G, Katinas GS, Watanabe Y, Otsuka K, Maggioni C, Perfetto F, Tarquini R, Schwartzkopff O, Bakken EE. Feedsidewards: intermodulation (strictly) among time structures, chronomes, in and around us, and cosmo-vasculo-neuroimmunity. About ten-yearly changes: what Galileo missed and Schwabe found. In: Conti A, Maestroni GJM, McCann SM, Sternberg EM, Lipton JM, Smith CC (eds.), Neuroimmunomodulation (Proc. 4th Int. Cong. International Society for Neuroimmunomodulation, Lugano, Switzerland, September 29-October 2, 1999). Ann NY *Acad. Sci.*, 2000, 917, pp 348-376.

[14] Bingham C., Cornelissen G., Halberg Erna, Halberg Franz. Testing period for single cosinor: extent of human 24-h cardiovascular "synchronization" on ordinary routine. (1984) *Chronobiologia*, 1984, 11, pp 263-274.

[15] Halberg F, Cornelissen G, Katinas G, Tvildiani L, Gigolashvili M, Janashia K, Toba T, Revilla M, Regal P, Sothern RB, Wendt HW, Wang ZR, Zeman M, Jozsa R, Singh RB, Mitsutake G, Chibisov SM, Lee J, Holley D, Holte JE, Sonkowsky RP, Schwartzkopff O, Delmore P, Otsuka K, Bakken EE, Czaplicki J, International BIOCOS Group. Chronobiology's progress: Part II, chronomics for an immediately applicable biomedicine. *J. Applied Biomedicine* 2006; 4: 73-86. http://www.zsf.jcu.cz/vyzkum/jab/4_2/halberg2.pdf.

[16] Otsuka K, Cornelissen G, Halberg F. Predictive value of blood pressure dipping and swinging with regard to vascular disease risk. *Clinical Drug Investigation* 1996; 11: 20-31.

[17] Cornelissen G, Halberg F, Otsuka K, Singh RB, Chen CH. Chronobiology predicts actual and proxy outcomes when dipping fails. *Hypertension* 2007; 49: 237-239. doi:10.1161/01.HYP.0000250392.51418.64.

[18] Gupta AK, Greenway FL, Cornelissen G, Pan W, Halberg F. Prediabetes is associated with abnormal circadian blood pressure variability. *J. Human Hypertension* 2008; 22: 627-633. doi:10:1038/jhh.2008.32.

[19] Cornelissen G, Halberg F, Beaty L, Kumagai Y, Halberg E, Halberg J, Lee J, Schwartzkopff O, Otsuka K. Cugini's syndrome in statunascendi: Oratio contra morem prevalentem et pro chronobiologica ratione ad pressione sanguinis curandam. *La Clinica Terapeutica* 2009; 160 (2): e13-e24.

[20] Elfeky MG, Aref WG, Elmagarmid AK. Using convolution to mine obscure patterns in one pass. In: Bertino E, Christodoulakis S, Plexousakis D, Christophides V, Koubarakis M, Böhm K, Ferrari E (eds). Advances in Database Technology -- EDBT 2004. 9th

International Conference on Extending Database Technology, Heraklion, Crete, Greece, March 2004: Proceedings. Berlin/Heidelberg: *Springer-Verlag*; 2004. p. 605-620.

[21] Elfeky MG, Aref WG, Elmagarmid AK. Periodicity detection in time series databases. IEEE *Trans Knowledge and Data Engineering* 2005; 17: 875-887.

[22] Glynn EF, Chen J, Mushegian AR. Detecting periodic patterns in unevenly spaced gene expression time series using Lomb-Scargleperiodograms. *Bioinformatics* 2006; 22: 310-318.

[23] Rasheed F, Alhajj R. Using suffix trees for periodicity detection in time series databases. 4th International IEEE Conference "Intelligent Systems" 2008; 2: 11-8-11-13. doi: 10.1109/IS.2008.4670501.

[24] Ploberger W, Reschenhofer E. Testing for cycles in multiple time series. *J Time Series Analysis* 2010; 31: 427-434. doi: 10.1111/j.1467-9892.2010.00675.x.

[25] Tominaga D. Periodicity detection method for small-sample time series datasets. *Bioinformatics and Biology Insights* 2010; 4: 127-136. doi: 10.4137/BBI.S5983.

[26] Katinas G.S. Revealing profiles of non-sinusoidal oscillations. In: Adapt. Physiol. And Life Quality: Problems of Traditional and Innovative Medicine. Moscow, Russian University of People Friendship, 2008, pp 141-143. [Russian, abstract in English].

[27] Katinas G.S. Possibilities of correction the 24-hour profile of arterial blood pressure according to results of monitoring. In: 1-st Russian Congress for Chronobiology and Chronomedicine. Vladikavkaz, *IPO SOIGSMI*, 2008, pp. 30-33 [Russian].

[28] Katinas GS, Mamontov OV, Konrady AO, Shlyahto EV, Brodskaya IS, Soboleva AB, Ertman AE, Lebedeva SV. Optimization of chronotherapy of hypertension according to ambulatory blood pressure monitoring (ABPM) after modeling circadian rhythms (CR) by means of Sawittzky-Goley filters. 14th Congress of the International Society for Holter and Noninvasive Electrocardiology (ISHNE2011) 26 – 28 April 2011. *Book of Abstracts*, Moscow, 2011, p. 8.

In: Cardiovascular Health and Chronomics
Editor: Ram B. Singh
ISBN: 978-1-62948-976-6

Chapter 8

THE GUT, BRAIN, AND HEART CONNECTION

R. B. Singh[1*], Jayeeta Chaudhury[2], Fabien De Meester[1], Agnieszka Wilczynska[1], Douglas W. Wilson[3], A. P. S. Hungin[3], Branislav Milovanovic[4] Toru Takahashi[5] S. Sarrafi Zadeh[6] and Shabnam Omidvor[6]

[1]The Tsim Tsoum Institute, Krakow, Poland
Center of Nutrition Research, Ahamadabad, India
[2]School of Medicine and Health, Durham University, UK
Department of Internal Medicine and Cardiology,
University Clinical Hospital, Belgrade, Serbia
[5]Graduate School of Human Environment Science, Fukuoka Women's University, Japan
[6]Faculty Member of Nursing and Midwifery,
Department of Babol Medical Sciences University, Iran

ABSTRACT

Background. Many experts have long accepted the brain-gut axis and its influence on obesity, insulin resistance and cardiovascular function. *Methods.* Internet search via Medline and Pub-Med and discussion with colleagues. *Results.* In experimental studies, it has been shown, how moderate intestinal inflammation affects complex behavior and how much treatment of such inflammation can influence factors from mind to matter, or brain to gut and the heart. It is likely that presence of long chain polyunsaturated fatty acids (PUFAs) in the duodenum, stimulates the release of cholecystokinin which in turn acts on cholecystokinin receptors, present on the abdominal vagus. After the food is ingested, it stimulates the secretion of incretins from the gut which increase the insulin secretion and initiate a gut-brain-liver axis in response to small amounts of triglycerides in the duodenum. Long chain PUFAs, particularly w-3 fatty acids, are cleaved from the dietary triglycerides by the gastrointestinal enzymes. Further studies indicated that a long chain fatty acid metabolite known as long chain fatty acid coenzyme A (LCFA CoA) is sensed by the gut. However, increased intake of trans fat, saturated fat and possibly w-6

* Correspondence: Dr Ram B Singh, The Tsim Tsoum Institute, Ul.Golebia 2, 31-007, Krakow, Poland Email: rbs@tsimtsoum.net, tel 0091 591 2417437.

fat and refined carbohydrates, particularly in presence of physical inactivity, may suppress gut-brain and liver axis by decreasing incretins, as well as cholecystokinin and leptin, resulting in insulin resistance and metabolic syndrome which is known to predispose cardiovascular diseases (CVDs). There is a rapid stimulation of leptin secretion from the gastric fundus on food ingestion, an effect which could be reproduced by cholecystokinin administration. Leptin gene expression and immune-reactivity have been reported in the gastric fundus. Leptin is known to enhance satiety–inducing effect of cholecystokinin indicating that cholecystokinin and leptin may function in concert with each other to induce satiety and regulate food consumption. Mediterranean diet and w-3 fatty acids have been reported to be protective against metabolic syndrome and CVDs but the exact mechanism of benefits is not known. It is possible that a part of the benefit may be due their protective effects on cell membrane phospholipids of the neuron, gut, liver and beta cells of pancreas apart from their direct effects on cardiovascular function.

Keywords: Metabolic syndrome, fatty acids, type 2 diabetes mellitus, coronary disease, hypertension, endothelial dysfunction, obesity, atherothrombosis, and inflammation, amino acids

INTRODUCTION

An experimental study in mice indicated that gastrointestinal inflammation can induce anxiety-like behaviour and altered biomarkers related to the brain dysfunction [1]. There is evidence from neuro-immunology that treating viral hepatitis with interferon, could trigger severe depression which indicated that circulating proinflammatory cytokines can play a significant role in mood and mood disorders [2]. This relationship may also contribute to the high prevalence of depression in patients with inflammatory bowel disease as well as among patients with cardiovascular diseases and type 2 diabetes mellitus [3]. In fact, the gut has been called the mirror of mind-brain function as well as dysfunction in most of the disciplines of internal medicine. In ancient times, Indians were aware of the adverse role of diet on mind-body interactions, which is evident from the following verse from an ancient scripture Bhagwatgeeta: "Foods which are bitter, acid, salted, burnt, fried and pungent, give rise to pain, mental stress and diseases" (3100 BCE). Charaka (600 BCE), a great physician of India, knew about the role of diet and lifestyle in the pathogenesis relating to brain-heart connection, which would be clear from the following verse: "Heart attack is born by the intake of fatty meals, overeating, excess of sleep, lack of exercise and anxiety" (Charaka Sutra, 600 BCE). Sushruta (Sushruta Samhita 600BCE) who was a surgeon, gave a more lucid description of atherosclerosis (Madaroga) " Excess eating of fatty foods and excess of sleep in day time, causes obesity, resulting in narrowing of the channels carrying blood to the heart. It is useful to give guggul, triphla and shilajeet in the treatment."

In experimental studies, animals did not display signs of general malaise, but still exhibited patterns that are a rodent correlates of anxiety and treatments ranging from probiotics to anti-tumor necrosis factor (TNF) antibodies, differentially altered inflammation, and behavior [1-4]. Blocking of TNF signaling blunted inflammation and decreased the level of circulating cytokines leading to normalized behavior [1,2]. The steroid budesonide was not quite as effective, but showed a similar direction of change(s). The probiotics decreased the markers of gut inflammation, but did not affect circulating cytokine levels. However, bifido bacteria was about as effective as the immunosuppressants in improving behavior which was associated with changed expression of an important neurotrophic factor in a brain region that is important in learning and memory [5]. In contrast, a murine strain of lactobacillus was ineffective. It seems that the brain-gut axis is bidirectional and includes more than just the sensory input that certainly influences mood and behavior [6]. It may be possible to target the neuroimmune connection by influencing intestinal inflammation, permeability, gut contents and/or the microbial flora [1-6]. Increased intestinal permeability has been associated with subtle changes in cytokine levels as well as chronic fatigue and fibromyalgia (4], which was selective for one of the two organisms tested, and this was similar to results obtained in humans [6, 7].

However, the benefit was independent of the changes in the circulating cytokines, indicating that other factors related to neural mechanisms may be important. Activation of N-methyl-D-aspartate (NMDA) receptors in the dorsal vagal complex lowers glucose production and neutralizes fast food induced hyperglycemia [9]. Long chain polyunsaturated fatty acids (PUFA) are known to inhibit neurodegeneration and proinflammatory cytokine induced inflammation [9, 10]

Leptin and cholecystokinin may also have interactions with incretins and may be inversely associated with long chain polyunsaturated fatty acid (PUFA) in the diet and in the tissues. Omega-3 fatty acid deficiency of the brain can independently have adverse effects on myocardium and vascular cells, in the form of a trigger that lies in the brain, and produce a chronic electrical instability in arterial cells causing coronary artery spasm and in the myocardium causing ventricular arrhythmias [11-13].

Lipoprotein associated phospholipase A2 [14] is a new proinflammatory marker under the influence of proinflammatory diets characterized with high w-6/w-3 ratio, trans fat and high saturated fat and refined carbohydrates [15-20]. Recent studies indicate that w-3 fatty acids in the tissues were inversely associated with CAD [21-23]. Further studies showed that Mediterranean diets rich in w-3 fatty acids were protective against cardiac events and metabolic syndrome [24-31].

However, no study has examined the role of brain related mechanisms in the protection provided by Mediterranean diets. It is quite possible that these diets act via gut-liver-brain and heart connection, at least partially in providing such benefits. It appears that the gut may act as a neuroendocrine organ that responds to food intake, and large meals rich in trans fat, w-6 fat and refined carbohydrates can cause inflammation and have adverse effects on the brain-gut and heart connection.

THE GUT-BRAIN-LIVER CONNECTION

The sight, taste and smell of foods are the first few characteristics which are conveyed to the brain as soon as we see the food and begin to eat. The first point of contact after ingestion of food, initiates a series of homeostatic mechanisms to regulate plasma glucose levels at near physiological concentration during fasting and postprandial periods. The foods ingested stimulate the secretion of incretins from the gut which increase the insulin secretion and initiate a gut-brain-liver axis in response to a small amount of triglycerides in the duodenum to rapidly increase insulin secretion, as reported by Wang and co-workers. [6] There is evidence that long chain polyunsaturated fatty acids (PUFA); such as oleic acid (w-9), linoleic acid (w-6), arachidonic acid (w-6), as well as w-3 fatty acids; alpha-linolenic acid (ALA), eicosapentaenoic acid (EPA), docosahexaenoic acid (DHA) are cleaved from the triglycerides content of foods by the gastrointestinal enzymes. When these fatty acids are administered in a dose dependent method in small amounts, there is marked and rapid increase in insulin sensitivity [6, 7]. These studies also indicated that a long chain fatty acid metabolite known as long chain fatty acid coenzyme (LCFA CoA) is sensed by the gut, possibly by specific receptors that are yet to be discovered. It seems that this lipid sensing in the gut is relayed to the liver in such a way that the homeostatic mechanisms in place are activated to maintain blood glucose levels within physiological range despite increased secretion of insulin by the beta cells of pancreas released by the incretins and to some extent by the inhibition of gluconeogenesis by the liver.

The human brain is quite rich in long chain PUFA such as arachidonic acid, EPA and DHA. These fatty acids are neuroprotective and constitute 30% to 50% of the total fatty acids in the brain, where they are predominantly associated with membrane phospholipids [11,12]. Therefore, if there is a deficiency of long chain PUFA, especially during the critical period of brain growth, from the third trimester to infant age (2 years), it may cause increased levels of tumor necrosis factor (TNF)-alpha which is known to damage the neurons [12]. TNF-alpha may also damage suprachiasmatic nucleus, pineal and pituitary glands, olfactory bulb and the hypothalamus, the last three are rich in insulin receptors, which are important in the metabolic syndrome (11-13]. Although, gut-brain connection could be independent of this association, PUFA deficiency could be a common mechanism for both the pathways.

In the gut-brain mechanism, the brain appears to play its role via parasympathetic pathways; mainly vagal network. The LCFA CoA, sensed by the gut, signals the brain through the vagus nerve via the hindbrain and then back down the vagal efferents pathways that terminate in the liver. The exact mechanisms, how the brain controls the gut are not known. It is possible that incretins have important roles which may be via their influence on neuronal cells or other gut hormone messengers such as cholecystokinin and/or leptin [32]. This concept is supported by the finding that intraduodenal perfusion of long chain PUFAs, but not of medium chain fatty acids, reduced the caloric intake that could be abolished by inhibition of fat hydrolysis. There was an increase in plasma cholecystokinin as well as decrease in energy intake and food consumption on perfusion of long chain PUFA. Administration of a potent cholecystokinin antagonist was associated with complete abolition of satiation effect of fatty acids. It is possible therefore that presence of long chain PUFAs in the duodenum stimulates the release of cholecystokinin which in turn acts on cholecystokinin receptors, present on the abdominal vagus. It is also possible that leptin may have a role in this mechanism, because leptin gene expression and immunoreactivity have been reported in the gastric fundus [33]. There is a rapid stimulation of leptin secretion on food ingestion, an effect which could be reproduced by cholecystokinin administration. Leptin is known to enhance satiety–inducing effect of cholecystokinin indicating that cholecystokinin and leptin may function in concert with each other to induce satiety and food consumption [34]. Leptin and cholecystokinin may also have interactions with incretins. It is clear that gut may be a neuroendocrine organ that responds to food intake, and both size of the meals and nutrient content can influence the production of satiety factors. These factors; leptin, cholecystokinin and incretins that enhance insulin secretion from pancreas and send messages to brain via the gut-vagus pathway appear to be important in this mechanism. The vagal mediator acetylecholine, in turn modulates the secretion and actions of various hypothalamic neurotransmitters and peptides [35-38]. The hypothalamus integrates all the messages received from the gut to maintain homeostasis of plasma glucose concentrations in the gut-brain connection.

In the gut-brain-liver axis, the vagus nerve plays a major role and it is important in the communication between liver and brain as well as between liver and beta cells of the pancreas. The LCFA CoA in the hypothalamus activates neural pathways which increase the insulin sensitivity in the liver, and reduce food intake [39-41]. Thus, long chain PUFAs and their metabolite LCFA-CoA function as a signal of nutrient intake and trigger counter-regulatory responses that originate in the brain-hypothalamus and the gut to regulate plasma glucose levels. But this regulatory system quickly fades in the face of continued ingestion of a diet rich in saturated fat, w-6 fat, trans fat and refined carbohydrates [6, 39, 40] which may worsen in presence of low physical activity. In other words, continued ingestion of proinflammatory foods blunts the feedback regulation from the hypothalamus to the gut, pancreas and liver. The result is that the inhibitory effect of proinflammatory foods, on further food intake is decreased, by inducing satiety or suppression of appetite. Thus, the individual continues to take food that ultimately results in obesity and the development of metabolic syndrome and CVDs.

Diabetes is characterized by hyperglycemia due partly to increased hepatic glucose production. The hypothalamus regulates hepatic glucose production in rodents. However, it is currently unknown whether other regions of the brain are sufficient in glucose production regulation. The N-methyl-D-aspartate (NMDA) receptor is composed of NR1 and NR2 subunits, which are activated by co-agonists glycine, glutamate or aspartate, respectively. There is direct administration of either co-agonist glycine or NMDA into the dorsal vagal complex [DVC), targeting the nucleus of the solitary tract, lowered glucose production in vivo [9]. Direct infusion of the NMDA receptor blocker MK-801 into the DVC negated the metabolic effect of glycine. To evaluate whether NR1 subunit of the NMDA receptor mediates the effect of glycine, NR1 in the DVC was inhibited by DVC NR1 antagonist 7-chlorokynurenic acid or DVC shRNA-NR1. Pharmacological and molecular inhibition of DVC NR1 negated the metabolic effect of glycine. To evaluate whether the NMDA receptors mediate the effects of NR2 agonist NMDA, DVC NMDA receptors were inhibited by antagonist D-2-amino-5-phosphonovaleric acid (D-APV). DVC D-APV fully negated the ability of DVC NMDA to lower glucose production [9]. Finally, hepatic vagotomy negated the DVC glycine ability to lower glucose production. These findings demonstrate that activation of NR1 and NR2 subunits of the NMDA receptors in the DVC is sufficient to trigger a brain-liver axis to lower glucose production, and suggest that DVC NMDA receptors serve as a therapeutic target for diabetes and obesity as well as vascular diseases.

Energy and glucose homeostasis are regulated by food intake and liver glucose production, respectively. The upper intestine has a critical role in nutrient digestion and absorption. However, studies indicate that upper intestinal lipids inhibit food intake as well in rodents and humans by the activation of an intestine-brain axis. In parallel, a brain-liver axis has recently been proposed to detect blood lipids to inhibit glucose production in rodents. Thus, this hypothesis led to the finding that upper intestinal lipids activate an intestine-brain-liver neural axis to regulate glucose homeostasis [6, 7]. Here it was demonstrated that direct administration of lipids into the upper intestine increased upper intestinal long-chain fatty acyl-coenzyme A (LCFA-CoA) levels and suppressed glucose production. Co-infusion of the acyl-CoA synthase inhibitor triacsin C or the anaesthetic tetracaine with duodenal lipids abolished the inhibition of glucose production, indicating that upper intestinal LCFA-CoAs regulate glucose production in the pre-absorptive state. Sub-diaphragmatic vagotomy or gut vagal deafferentation interrupts the neural connection between the gut and the brain, and blocks the ability of upper intestinal lipids to inhibit glucose production. Direct administration of the N-methyl-d-aspartate ion channel blocker MK-801 into the fourth ventricle or the nucleus of the solitary tract where gut sensory fibres terminate abolished the upper-intestinal-lipid-induced inhibition of glucose production. Finally, hepatic vagotomy negated the inhibitory effects of upper intestinal lipids on glucose production. These findings indicate that upper intestinal lipids activate an intestine-brain-liver neural axis to inhibit glucose production, and thereby reveal a previously unappreciated pathway that regulates glucose homeostasis.

Long Chain Polyunsaturated Fatty Acids and Insulin Resistance

Long chain PUFAs have anti-inflammatory action because they suppress the production of IL-6, TNF-alpha, HMGB1 (high-mobility group protein 1), and MIF (macrophage migration inhibitory factor), whereas Western diet characterized with excess of w-6 fat, saturated fat, trans fat and refined carbohydrates and oxidized cholesterol possess pro-inflammatory action due to their ability to block the metabolism of unsaturated fatty acids and triggering of macrophages and other immunocytes to produce pro-inflammatory cytokines [15-20].

It is possible to conclude that the modern Western diet which is dense in energy and low in micronutrients, promotes obesity, metabolic syndrome and CVDs, at least in part, by impairing nutrient-sensing systems in the gut-brain axis that are originally designed to limit food intake and enhance insulin sensitivity. It is also possible that a diet rich in polyunsaturated fatty acids, as used in the studies reported [24-31], is more important to trigger the gut-brain-liver circuit to modulate increases in plasma glucose concentration and restrict the development of obesity and type 2 diabetes. However, with modern diets, rich in trans fat, w-6 fat, saturated fats and refined starches and cola drinks, the gut-brain-liver circuit described above may not function efficiently to restrict food intake. This may also explain why long chain PUFAs are beneficial compared to modern fast foods rich in energy from proinflammatory foods [21-23].

Long chain w-3 fatty acids, when given orally, protect pancreatic beta cells from chemical-induced apoptosis and thus prevent the development of insulin resistance and also protect the gut-liver-brain axis [42]. These PUFAs also form precursors to various endocannabinoids that have been shown to play a role in the pathology of obesity, insulin resistance and type 2 diabetes mellitus [43, 44]. Long chain PUFAs are anti-inflammatory, whereas saturated fats, w-6 fats, trans fat and refined carbohydrates are pro-inflammatory. Diets rich in these nutrients can cause low-grade systemic inflammation and insulin resistance seen in obesity, type 2 diabetes mellitus, and the metabolic syndrome. The insulin resistance seen thus may override the acute-insulin sensitizing effects of intestinal long chain PUFAs. After bariatric surgery, the gut-brain-liver circuit described above may also play a significant role in the improvement in insulin sensitivity, decrease in blood glucose, and decrease in food intake and weight loss, because these beneficial effects are seen before the weight loss is seen. It is possible that these distinct changes in the hypothalamic neurotransmitters and peptides that could account for some, if not, all of the beneficial actions seen after bariatric surgery [45, 46]. It is clear that there is a close interaction between insulin and hypothalamic peptides and monoamines, in the pathophysiology of insulin resistance and metabolic syndrome. Lipoprotein-associated phospholipase A2 (Lp-PLA2) is a proinflammatory marker which is associated with the risk of cardiovascular (CV) events in population-based studies [14, 47]. The prognostic value of Lp-PLA2 in patients with acute coronary syndromes (ACS) has not been established. It is possible that Lp-PLA2 may also be regulated by the vagus nerve and hypothalamus as well as by gut-liver-brain and heart connection. Plasma levels of Lp-PLA2 activity were measured at baseline (n=3648) and 30 days (n=3265) in patients randomized to atorvastatin 80 mg/d or pravastatin 40 mg/d after ACS in the PROVE IT-TIMI 22 trial (Pravastatin or Atorvastatin Evaluation and Infection Therapy-Thrombolysis in Myocardial Infarction) . The primary end point was a composite of death from any cause, myocardial infarction, unstable angina, revascularization, or stroke (mean follow-up 24 months). At baseline after ACS, the risk of recurrent CV events was similar across all quintiles of Lp-PLA2 activity (P trend = 0.88). Overall, mean levels of Lp-PLA2 were lower at 30 days of follow-up than at baseline [35.7 versus 40.9 nmol.min(-1).mL(-1), P<0.001). In particular, treatment with atorvastatin 80 mg/d was associated with a 20% reduction in Lp-PLA2 activity (P<0.001), whereas Lp-PLA2 rose 3.6% with pravastatin 40 mg/d (P<0.001). Patients with 30-day Lp-PLA2 activity in the highest quintile were at significantly increased risk of recurrent CV events compared with those in the lowest quintile (26.4% versus 17.6%, P trend =0.002). After adjustment for cardiac risk factors, treatments, achieved low-density lipoprotein (LDL), and C-reactive protein, Lp-PLA2 activity in the highest quintile remained independently associated with a higher risk of recurrent CV events (adjusted hazard ratio 1.33, 95% confidence interval [CI] 1.01 to 1.74). Lp-PLA2 is not useful for risk stratification when measured early after ACS. At 30 days, Lp-PLA2 activity is significantly lowered with high-dose statin therapy and is associated with an increased risk of CV events independent of C-reactive protein and LDL cholesterol levels.

Acetylcholine, the principal neurotransmitter of the vagus is known to modulate the secretion and action of various monoamine peptides. It is possible that gut-pancreas-liver-brain axis is under the influence of vagal tone as well as under the influence of the incretins, BDNF, LCFA-CoA, cholecystokinin, leptin, endocannabinoids, and cytokines that may have their origins from the gut [6,7, 48-51]. Since both orexigenic and anorexigenic molecules are secreted by the gut, hypothalamus, and adipose tissue, the final responses in the form of satiety or hunger and food consumption depend on the balance between these regulatory and counter-regulatory stimuli. Leptin and cholecystokinin may also have interactions with incretins. It seems that gut should be considered, a neuroendocrine organ that responds to food intake and both size of the meals and nutrient content can influence the brain-gut and heart connection leading to development or prevention of obesity, insulin resistance, metabolic syndrome and cardiovascular diseases.

Brain Derived Neurotrophic Factor (BDNF) and Insulin Resistance

Brain-derived neurotrophic factor (BDNF) is present in brain neurons as well as in other tissues; salivary glands, stomach, duodenum, ileum, colon, lung, heart, liver, pancreas, kidney, bladder, uterus, oviduct, monocytes and eosinophils [52 ,53, 54]. BDNF is also involved in all the inflammatory diseases; myocardial injury, rheumatological conditions, inflammatory bowel disease and atopic dermatitis [55-58], apart from obesity, type 2 diabetes, metabolic syndrome and CAD. The deficiency of BDNF in association with these diseases indicates that it can inhibit inflammation or inflammation can decrease BDNF. Since the hypothalamus plays a critical role in energy homeostasis, it is important to identify factors that are relevant to glucose homeostasis. BDNF is one such factor produced by neuronal cells of the brain that regulate function of the gut and pancreatic beta islet activity in response to plasma levels of glucose, protein, fatty acids, insulin, and leptin and thus can modulate insulin resistance. BDNF, present in the hippocampus, cortex, basal forebrain, many nuclei in the brain stem, and catecholamine neurons, including dopamine neurons in the substantia nigra, regulates food intake and body weight in both experimental animals and humans. Treatment with BDNF decreased random blood glucose and body weight in obese, non-insulin-dependent diabetic C57BLKS-Lepr(db)/lepr(db) (db/db) mice. The effects of BDNF on non-fasted blood glucose levels are not caused by effects that persisted for weeks after cessation of BDNF treatment. It also reduced the hepatomegaly present in db/db mice, in association with reduced liver glycogen and reduced liver enzyme activity in serum, supporting the involvement of liver tissue in the mechanism of action for BDNF [48]. Treatment with BDNF once or twice per week (70mg/kg/week) to db/db mice for 3 weeks significantly reduced blood glucose and hemoglobinA1c, (HbA1c= glycosylated hemoglobin) as compared with control, indicating that BDNF reduced blood glucose as well as restored systemic glucose balance and hence it can prevent insulin resistance.

The hypothalamic nuclei is associated with satiety and locomotor function, consisting of BDNF and its tyrosine kinase receptor expresses, depending on food intake and sedentary behaviour. BDNF gene expression is reduced in experimental mice and they exhibit abnormalities in eating behaviors and locomotor activity, which may be reversed by BDNF or neurotrophin 4/5 infusion [51]. Gene transfer of BDNF can cause a marked reduction in weight and decrease in insulin resistance [44]. In one study, plasma levels of BDNF were inversely associated with type 2 diabetes mellitus independent of obesity [59]. BDNF activates the sympathetic nervous system via the central nervous system and regulates energy expenditure in obesity with diabetes [60]. It is possible that BDNF can be used as hypoglycemic agent by administering it once in a week [61]. The decrease in hyperglycemia continuously for 7 days by administration of BDNF can provide better control of vascular complications of diabetes by decreasing superoxide anion, dyslipidemia, insulin resistance, free fatty acids, proinflammatory cytokines and endothelial dysfunction resulting in a decline in CVDs. In a mouse model of obesity and type 2 diabetes [62], therapeutic efficacy of BDNF has been demonstrated. Gene transfer of BDNF may cause weight loss and alleviation of obesity associated insulin resistance. BDNF plays a significant role in the regulation of appetite, obesity, and in the pathogenesis of insulin resistance by its action on hypothalamic neurons as well as by modulating the release and activity of leptin, ghrelin, insulin, neuropeptide Y, melanocortin, serotonin, dopamine and other neuropeptides, neurotransmitters and gut hormones. It is possible that it would be a novel approach, in the management of metabolic syndrome, if BDNF can be delivered to the hypothalamus.

It is possible that a combination of BDNF with w-3 fatty acids` could be of additional benefit in the management of metabolic syndrome, because this complex of two agents would allow BDNF to pass the blood-brain barrier easily, to reach the hypothalamus for its activity against metabolic syndrome. W-3 fatty acids are present in large amount in the brain and are known to increase the production of BDNF as they are natural enhancers of BDNF expression [63].

There is a low-grade systemic inflammation occurring in obesity, metabolic syndrome, CVDs and type 2 diabetes mellitus. Since BDNF is involved in their pathobiology, it is tempting to speculate that BDNF may also have a role in the regulation of inflammation and immune reactivity. There is evidence that peripheral inflammation induced an increased expression of BDNF mRNA which was mediated by nerve growth factor in the dorsal root ganglion. In an experimental study by an intraplantar injection of Freund's adjuvant into rat paws, significant increases in the percentage of BDNF-immunoreactive neuron profiles in the L5 dorsal root ganglion and marked elevation in the expression of BDNF-IR terminals in the spinal dorsal horn were observed following the peripheral tissue inflammation produced. These findings suggest that peripheral tissue inflammation induces an increased BDNF synthesis in the dorsal root ganglion and elevated anterograde transport of BDNF to the spinal dorsal horn [64]. It is possible that BDNF might have a role in inflammation and hyperalgesia because nerve growth factor is a mediator of hyperalgesia. After 2 hours of induction of bladder inflammation by turpentine, there were significant increases in levels of nerve growth factor, BDNF, and neurotrophin-3 mRNAs, which also support this concept. It seems that neurotrophic factors might have a significant role in the inflammatory response because there is rapid elevation of nerve growth factor and BDNF and neurotrophin-3 corresponding to the sensory and reflex changes during bladder inflammation [65].

LEPTIN

Leptin is produced by the white adipose tissue, stomach, mammary gland, placenta, and skeletal muscle and is known as adiposity hormone. Leptin shows similar traits to that of insulin in action and reflects total fat mass, especially subcutaneous fat of the body.

It is important to note that a close interaction appears to exist between leptin and BDNF [66, 67]. Leptin prevents obesity by inhibiting appetite. Rodents and patients lacking leptin or functional leptin receptors developed hyperphagia and obesity [55]. Leptin acts on the hypothalamus and other areas in the brain through the neuronal circuits and also stimulates the enzymes involved in lipid metabolism. It also decreases feeding and increases energy expenditure by directly suppressing neuropeptide Y and increasing pro-opiomelanocortin. Leptin also acts centrally to increase insulin action in liver. Arcuate neurons expressing these peptides, project to the paraventricular nucleus and lateral hypothalamic area, resulting in increases in corticotrophin-releasing hormone (CRH) and thyrotropin-releasing hormone (TRH) and reduction in melanotropin-releasing hormone (MRH) and orexins [68]. Congenital leptin deficiency decreases brain weight, impairs myelination, and reduces several neuronal and glial proteins [69]. These deficits are partially reversible in adult Lepoblob mice by the administration of leptin [69].

Ingestion of high-fat, high-calorie diets is associated with hyperphagia, increased body fat, and obesity. The mechanisms responsible are currently unclear; however, altered leptin signaling may be an important factor. Vagal afferent neurons (VAN) integrate signals from the gut in response to ingestion of nutrients and express leptin receptors. Therefore, we tested the hypothesis that leptin resistance occurs in VAN in response to a high-fat diet. Sprague-Dawley rats, which exhibit a bimodal distribution of body weight gain, were used after ingestion of a high-fat diet for 8 wk. Body weight, food intake, and plasma leptin levels were measured. Leptin signaling was determined by immunohistochemical localization of phosphorylated STAT3 (pSTAT3) in cultured VAN and by quantification of pSTAT3 protein levels by Western blot analysis in nodose ganglia and arcuate nucleus in vivo. To determine the mechanism of leptin resistance in nodose ganglia, cultured VAN were stimulated with leptin alone or with lipopolysaccharide (LPS) and 'suppressor of cytokine signaling 3 protein' (SOCS-3) expression measured. We conclude that VAN of diet-induced obese rats become leptin resistant; LPS and SOCS-3 may play a role in the development of leptin resistance [87].

GHRELIN

Ghrelin is a novel 28 amino acid or exigenic peptide [88] gut hormone that increases food intake, and is produced in the epithelial cells lining the fundus of the stomach, with smaller amounts produced in the placenta, kidney, pituitary, and hypothalamus. . Circulating ghrelin levels are elevated by fasting and suppressed following a meal. [190] .The secretion and action of ghrelin are regulated by plasma glucose, insulin, acetylcholine levels in the brain, leptin, BDNF, and various others neurotransmitters and peptides [70,71]. Ghrelin administration acutely increases energy intake in lean and obese humans and chronically induces weight gain and adiposity in rodents. [190] . Ghrelin stimulates growth hormone secretion and regulates energy balance by acting on the arcuate nucleus of hypothalamus [72]. The main function of ghrelin is to increase hunger through its action on hypothalamus feeding centers, in both rodents and humans. Blood concentrations of ghrelin are lowest shortly after consumption of a meal, and then rise during the fast, just prior to the next meal. Intracerebroventricular injection of ghrelin increased the glucose utilization rate of the white and brown adipose tissue and strongly stimulated feeding in rats with increase in body weight [73]. Circulating concentrations of ghrelin are lower in obese people and are negatively correlated with body mass index. [190].

GHRELIN AND OREXINS

Orexins A (OXA) and B, 33 and 28 amino acid peptides, respectively, originate from 130 amino acid preproorexin produced by neurons located in the hypothalamus and in the spinal cord [89-91]. However, neurons and endocrine cells in the gut and pancreas also exhibit orexin-like immunoreactivity [92]. The projection of central orexin nerves in humans leads to the locus coeruleus, dorsal raphe nuclei, amygdala, suprachiasmatic nucleus, basal forebrain, cholinergic brainstem and spinal cord [93]. Orexins are involved in developing behavior patterns controlled by the hypothalamic region [94].

The Greek origin name of these peptides clearly reflects basic function of orexins (orexigenic peptides) - orexos means "to eat" - i.e., positive stimulation of energy balance. Similarly to ghrelin, a single intracerebroventricular injection of OXA increased caloric intake when administered during the light phase but not at the beginning of the dark phase, indicating that sensitivity to orexins might be controlled by circadian variations [95].

Orexin neurons interact with hypothalamic feeding pathways and monoaminergic/cholinergic centers, thus probably, linking energy balance with sleep/wakefulness phases and they motivate food seeking [96]. The studies revealed that orexins cooperate also in regulation of blood pressure and heart rate and release of some hypothalamic hormones [97]. Further, ghrelin and orexins with some other peptides constitute jointly an orexigenic part of gut-brain axis which integrates and regulates the homeostasis of energy balance and feeding [98].

CAN INFLAMMATION PREDISPOSE GUT-BRAIN DYSFUNCTION?

Inflammation appears to be a unifying hypothesis in the pathogenesis of obesity, hyperinsulinemia, insulin resistance, hypertension, coronary artery disease (CAD), and diabetes mellitus that are important components of metabolic syndrome. Gut hormones; leptin, incretins, cholecystokinin, PUFA CoA and BDNF appear to be crucial in the gut-brain-liver and beta cell interactions and in the development of inflammation.

This poses the possibility that inflammation in conjunction with some other unknown mechanisms related to diet and lifestyle factors and gut-brain dysfunction, could be underlying the development of metabolic syndrome leading to CVD. Obesity is a condition with low grade systemic inflammation which predisposes insulin resistance, increases perilipins, intramyocellular lipids and lipoprotein phospholipase A2 that are also proinflammatory markers. Mediterranean diet and physical activity are known to modulate these proinflammatory markers indicating that diet can also regulate this neuroendocrine gut-brain and heart connection.

In obesity, insulin resistance, essential hypertension, type 2 diabetes, and CAD, plasma levels of C-reactive protein (CRP), tumor necrosis factor (TNF)-alpha, and interleukin (IL)-6, 12, 18 and 1 markers of inflammation, are elevated both before and after the onset of these diseases [74-76]. IL-18 is responsible for rupture of atheroma resulting in thrombosis and acute coronary syndrome. Overweight children and adults showed an increase in CRP concentration compared with normal-weight children [51], and a direct correlation was observed between the degree of adiposity and plasma CRP levels. CRP concentrations were positively associated with an increased risk of CAD, ischemic stroke, peripheral arterial disease and CAD mortality in healthy men and woman.

A strong relation between elevated CRP levels and the cardiovascular risk factors fibrinogen and high-density lipoprotein cholesterol was also reported. Increased expression of IL-6 in adipose tissue and its release into the circulation is responsible for elevated CRP levels. There is evidence that IL-6 stimulates the production of CRP in the liver. Overweight and obese subjects have significantly higher serum levels of TNF-alpha compared to lean subjects. Weight reduction and/or exercise decreases serum concentration of TNF-alpha and other proinflammatory markers.

The negative correlations observed between plasma TNF-alpha and HDL cholesterol, glycosylated hemoglobin, and serum insulin concentrations explain why CAD is more frequent in obese compared to healthy or lean subjects [51]. High CRP levels were associated with a two-fold greater risk for developing diabetes after 3 to 4 years of the follow-up period [77] and significantly enhanced incidence of myocardial infarction, stroke, coronary revascularization, or cardiovascular death [78].

Dietary glycemic load is significantly and positively associated with plasma CRP in the healthy middle-aged woman [79], suggesting that hyperglycemia is proinflammatory in nature and induces inflammation. There is evidence that high CRP binds to ligands exposed in damaged tissue and activates complement [80] that could increase the size of myocardial and cerebral infarcts [81,82]. Human CRP activates complement; hence, neutralization or inhibitors of CRP could be of significant therapeutic value. Injection of human CRP in rats and 1, 6-bis (phosphocholine) hexane, a specific small-molecule inhibitor of CRP, was capable of abrogating the increase in infarct size and cardiac dysfunction [83,84].

In brief, It is possible that inhibition of inflammation might modulate gut-brain-liver dysfunction leading to inhibition of insulin resistance and metabolic syndrome. There is evidence that an acute rise in plasma glucose in normal subjects and with impaired glucose tolerance can cause increase in proinflammatory cytokines; IL-6, IL-18 and TNF-alpha levels which remain higher in subjects with impaired glucose tolerance compared to controls. The concentration of TNF-alpha was much lower among younger subjects compared to elderly in response to glucose challenge and hyperglycemia induced the production of the acute phase reactants from the adipose tissue in the these subjects [85, 86].

Role of Enteral Nutrition in Gut Protection

Multiple data clearly indicate that the function of the gut is by far not restricted to food processing and subsequent nutrient and fluid uptake [99]. Protein intake is critical for intestinal mucosal growth. Burrin *et. al.,* [122] have suggested that at least 40% of total calories need to be taken by the enteral route to demonstrate increased mucosal growth. However, smaller amounts have been found to have beneficial effects on the intestine, including a modulation of inflammation that can affect distal organs [123].

Animal experiments and some human data have shown that the gut communicates with bacteria that support digestion by their enzymatic capacity [100-101], that the gut regulates major epithelial and immune functions of importance for gut health and health in general [102,103] and that the gut reports to the brain via the N. vagus and hormones about energy uptake and other conditions that might affect mood and general well-being [104]. Any impairment of the GI barrier can increase the risk of developing infectious, inflammatory and functional GI diseases, as well as extra intestinal diseases such as immune-mediated and metabolic disorders.

However, interest in the importance of the gut after injury or operation has waxed and waned over the past century. Changes in the gut in sick patients include stress ulceration, bacterial overgrowth from stress ulceration prophylaxis, mucosal atrophy, loss of barrier function, increased permeability, and bacterial translocation. Alterations in the gut in such conditions of starvation and injury may increase mucosal permeability and allow bacteria to translocate from the lumen into mesenteric lymph nodes, resulting in systemic bacteremia, endotoxemia, and inflammatory mediator activation.

The normal gut contains enough bacteria and endotoxin to kill the host thousands of times over and that even small increases in gut permeability could have profound physiologic consequences. Since the mechanisms by which intestinal bacteria crossed the intestinal mucosal barrier and spread systemically is unknown, and so this process is termed "bacterial translocation. Several clinical studies showed a correlation between loss of gut barrier function and the development of systemic infection and multiple organ dysfunction syndrome (MODS). The concept of gut-induced sepsis was that during shock or stress states, the body's response was to decrease blood flow to the intestines, thereby ensuring sufficient blood flow to critical organs such as the heart and brain. This decrease in intestinal blood flow, if sufficient in magnitude or depth, would lead to gut injury and loss of normal gut barrier function, causing the gut to become "leaky". This gut barrier failure in turn would allow bacteria and their toxic products, such as endotoxin, to escape from the gut and enter the systemic circulation, thereby causing systemic sepsis and leading to MODS. Gut mucosal injury seen in septic rodents is not associated with concurrent mucosal perfusion deficits, suggesting that ischemia may not play a prominent role in normotensive septic gut dysfunction [159]. However, gut mucosal atrophy occurs during sepsis [160]. Studies suggest that one possible mechanism for sepsis-related gut dysfunction is increased apoptosis [160].

In i*nflammatory states*, tumor necrosis factor induces *c*ysteine-*asp*artic prote*ases* (caspase) activation within enterocytes, causing enterocyte detachment in the lumen and resulting in villus atrophy [160]. Several animal studies show that the endotoxin-induced inflammatory response in mice also induces increased apoptosis in the gut and this is associated with gut barrier dysfunction.[160,161]. Early gut wall integrity loss and local intestinal inflammation are associated with the development of inflammatory complications even in surgical and trauma patients. Prevention of these intestinal events is a potential target for therapies aimed to control systemic inflammation.

Thus, during these conditions, many organ systems such as the circulation and cardiovascular function, the ventilatory system, and kidneys fail quickly during sick conditions and require urgent support. The GI tract remains dormant for a longer period of time, and evidence of failure is not as dramatic or apparent as with other vital processes. Markers of gut failure, such as elevated bilirubin level, are late findings. In any case, methods of assessing, supporting, improving and maintaining gut health-related GI functions and prevention of failure can be important. Two functional entities are key to achieving and maintaining gut health [105,106,107,108]. These entities are the GI microbiome and the GI barrier . The GI microbiome consists of about 10^{14} bacteria that are mainly located in the large intestine [109,110]. The GI microbiome prevents colonisation by potentially pathogenic microorganisms, provides energy for the gut wall from undigested food (for example, carbohydrates and other nutrients) and it regulates the mucosal immune system, not only educating the naïve infant immune system but also serving as an important source of immune stimulators throughout life [111,112,113,114,115,116]. Thus, the GI microbiota contribute to energy homeostasis, prevents mucosal infections and likely mitigates immune system hypersensitivity. Most important, it contributes to the maintenance of an intact GI barrier, which seems to be closely related to infectious, inflammatory and allergic diseases [108,117]

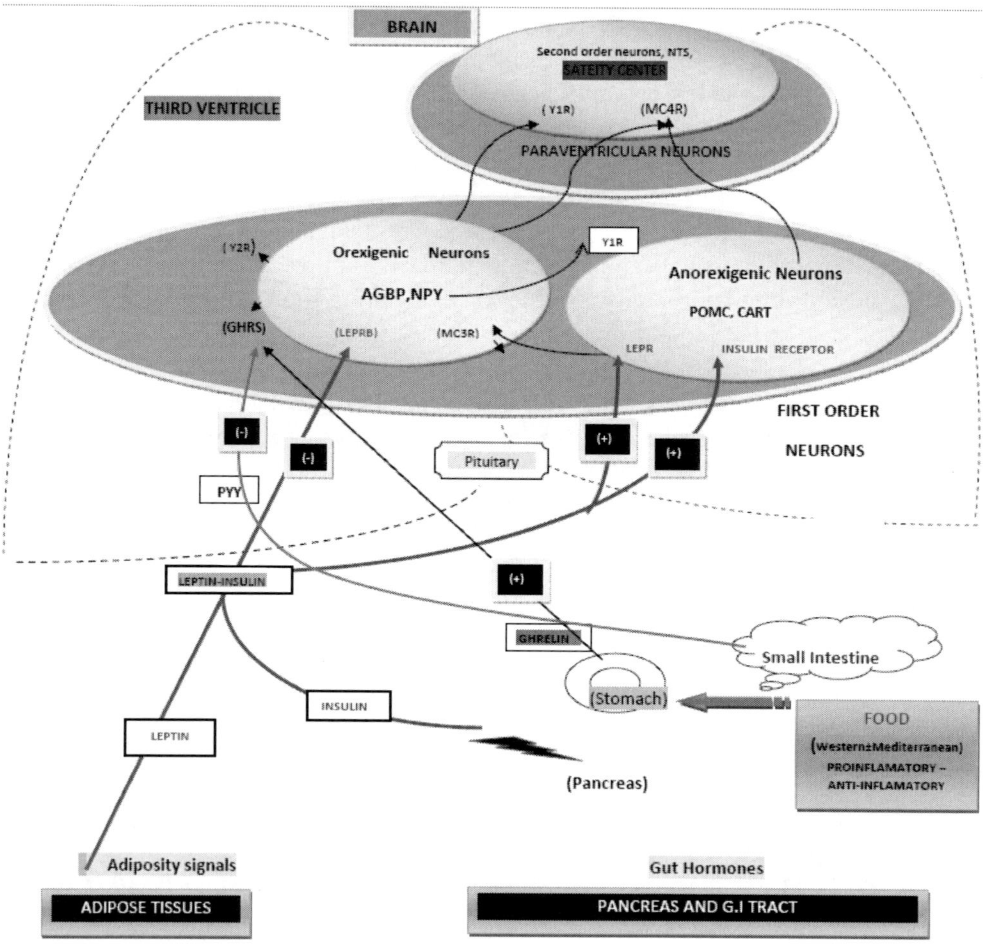

Figure 1. Gut-Brain Interactions in Controlling of Feeding Behavior in human. Hormonal circuits from the GUT involving stomach, small intestine, and pancreas and ADIPOSE TISSUE influence the sensations of hunger and satiety exerted via hypothalamic neuroendocrine pathways. GHSR(growth hormone secretagogue receptor to which ghrelin binds). MC3R and MC4R (melanocortin 3 receptor and melanocortin 4 receptor). Y1R (NPY receptors 1) and Y2R (NPY receptors 2). Ghrelin from the stomach, leptin from adipose tissue, insulin from the pancreas, and peptide tyrosine tyrosine (PYY) from the small intestine bind to receptors on orexigenic and/or anorexigenic neurons in the ARC of the hypothalamus. The effects of these peptide hormone-receptor interactions are release of either the orexigenic neuropeptides NPY and AgRP or the anorexigenic neuropeptides CART and the POMC-derived peptide α-MSH.

Maintenance of Gut Health

The conditions that might disturb the intestinal microbiome and the mucosal immune system should be avoided. Unbalanced diet and lack of exercise as well as extreme exercise and any type of chronic stress may have adverse effects. In recent publications, high-fat as well as high-fructose diets have been shown to disturb the GI barrier and, in this way, to induce fatty liver disease and subclinical inflammatory conditions associated with metabolic disturbances [169,171,172]. On the other hand, dietary changes have been shown to help prevent major diseases such as allergy, obesity and cancer [108,172,173]. Therefore, lifestyle characteristics, such as a balanced diet that includes high vegetable and fibre content and moderate consumption of red meat to prevent colon cancer [174,175], or an individualized elimination diet in selected individuals with food intolerances, food allergy or coeliac disease [176,177], moderate but regular exercise and avoidance of chronic stress, but also defined products such as select pre- and probiotics might all contribute to gut health [99]. Moreover, tobacco abstinence, moderate alcohol consumption, maintenance of normal body weight, avoidance of nonsteroidal anti-inflammatory drugs (NSAIDs) ingestion and control of stress can support gut health.

Systematic strategies to improve lifestyle and to avoid or reduce stress that have been validated in controlled trials are rare; however, meditative methods that often originate from traditional Chinese medicine and other Asiatic cultures (for example, ayurveda and tai-chi) are enjoying growing popularity and becoming increasingly accepted by health professionals as valuable tools to maintain gut health and general well-being [178,179].

Indeed, it has been shown in studies that chronic bowel diseases such as IBD are associated with adherence of commensal bacteria to the otherwise sterile intestinal epithelium [196] and that selected probiotics can prevent the adhesion of pathogenic bacteria to the intestinal mucosa [181] or restore leaky gut by improving the molecular composition of tight junctions[182,183].

Moreover, probiotic bacteria can support the normal development of the mucosal immune system, such as through the C-C chemokine receptor 6 (CCR6) gene expressed on lamina propria lymphocytes [111] and the production of protective IgA and antimicrobial defensins such as the CCR6 ligand human β-defensin 2 (or hBD2) [184,185] that are both lacking in many chronic IBDs [170]. These few selected examples indicate the strong rationale for using probiotics, possibly also in synergistic combinations with prebiotics, to maintain gut health [99].

Role of Enteral Nutrition in Gut Protection During Critical Care

The only medical research that considers this issue is the critical care arena. In critical care they are often tube-feeding patients directly into the stomach desperately trying to maintain their nutritional status following, surgery, trauma or other life threatening conditions.

The main focus of the nutritional support effort is to maintain adequate protein input to these patients; because if it cannot be provided outside sources of protein, then the body will start to break down. Specifically, the body will begin to break down muscle protein and other proteins within the body for essential needs. The cells of the gut wall are fed locally rather than systemically, meaning they get their nutritional support from nutrients as they are absorbed through the mucosal lining. In critical care they are concerned about bacterial translocation and maintaining the integrity of a gut wall.

Provision of adequate nutrition in critical illness is generally accepted as desirable and early feeding is considered superior to delayed feeding [197]. Appropriate enteral nutrition given at a time when the patient's GI tract can tolerate it, should best protect and preserve the gut. Enteral nutrition activates humoral and neural pathways to regulate food intake and sustain energy balance [125]. Enteral nutrition or luminal nutrition maintains intestinal health and enteral amino acids are more effective in supporting splanchnic protein synthesis. [99). Many preclinical animal studies performed in my laboratory and human clinical trials has established an important relationship between the route of nutrient administration, infection and gut barrier function. These studies established the importance of enteral feeding in high-risk patients, especially burn, trauma and ICU patients, and also the mechanisms by which lack of enteral feeding led to gut barrier dysfunction and bacterial translocation. Recent studies demonstrate that enteral nutrition and in particular lipid-rich nutrition modulates inflammation, preserves intestinal integrity and prevents organ damage. The data demonstrated that enteral lipid-rich nutrition modulates inflammation and preserves intestinal integrity via CCK release which activates cholecystokinin (CCK)-1 receptors located on afferent vagal fibers. Taken together, the current study reveals a novel gut-brain-immune axis and provides new insight into the applicability of enteral nutrition to treat inflammatory conditions [124].

Sepsis and the systemic inflammatory response_in both animals and humans are associated with gut mucosal damage and dysfunction [162,163]. Thus gut dysfunction during sepsis is a common problem, resulting in loss of gut mucosal barrier selectivity, increased permeability to various hydrophilic solutes [164], and translocation of bacterial products into the circulation [165,162], which may then further increase the inflammatory response in distant organs, resulting in multiple organ dysfunction and death [166]. As a result, the gut is viewed by many as an "engine" that drives sepsis. Controlling the inflammatory cascade during sepsis remains a major clinical challenge. One possible contributory mechanism to endotoxin-induced gut mucosal damage is increased apoptosis [160]. Enteral feeding significantly decreases gut macromolecular permeability during endotoxemia.[190]. Enteral feeding may decrease the distant organ inflammation by reducing gut apoptosis, thereby maintaining gut mucosal function during endotoxemia. [191]. Fed or fasted CD-1 mice were studied 6 h after intraperitoneal injection of either saline (sham) or endotoxin (4 mg/kg of 0111:B4 Escherichia coli lipopolysaccharide). Endotoxin increased gut permeability by 44% ($P < 0.05$), an effect reduced 36% by feeding ($P < 0.05$). Similarly, endotoxin increased pulmonary neutrophil infiltration (6.0 ± 1.0-fold, $P< 0.001$) and increased lung interleukin (IL)-6 (5.9 ± 0.1-fold, $P < 0.001$) and macrophage inflammatory protein (MIP)-2 expression (290 ± 40-fold, $P < 0.001$), whereas feeding decreased this effect by 43% for neutrophils, 40% for IL-6 ($P < 0.05$), and 35% for MIP-2 ($P < 0.05$) [191]. Thus, In experimental models of sepsis, enteral feeding has been shown to reduce mucosal atrophy and improve gut immunity[123], and clinically, enteral feeding reduces septic morbidity in postoperative and post trauma patients [187,188,189] . Recently, it has become more evident that the autonomic nervous system reduces inflammation through the vagus nerve. The current study investigates whether nutritional stimulation of the autonomic nervous system effectively attenuates the inflammatory response in murine Gram-negative sepsis. Male C57bl6 mice were intraperitoneally challenged with lipopolysaccharide derived from *Escherichia coli*. Before lipopolysaccharide administration, mice were fasted or enterally fed either lipid-rich nutrition or low-lipid nutrition. The anti-inflammatory effect of both nutritional compositions was mediated through cholecystokinin receptors ($p < .01$), activation of mesenteric vagal afferents ($p < .05$), and peripheral nicotinic receptors ($p < .05$). : The result demonstrates that lipid-rich nutrition attenuates intestinal damage and systemic as well as organ-specific inflammation in murine Gram-negative sepsis through the nutritional vagal anti-inflammatory pathway. These findings implicate enteral administration of lipid-enriched nutrition as a promising intervention to modulate the inflammatory response during septic conditions [125].

Thus today many of the new commercial enteral formulas were developed and the risks of intravenous nutrition were recognized as well as the benefits of "feeding the gut" in addition to the rest of the body.

Role of Immunonutrients in Gut Protection

Recently, it was demonstrated that macrophages in the gut wall are controlled by the vagus nerve, the so-called cholinergic anti-inflammatory pathway [168]. More recently, the concept of brain-gut dialogue has extended to the relationship between the nervous system and mucosal immune function.[167]. There is compelling evidence for a reciprocal or bi-directional communication between the immune system and the neuroendocrine system. This is mediated, in part, by shared ligands (chemical messengers) and receptors that are common to the immune and nervous systems. Although the concept of psychoneuroimmunology and neuroimmune cross-talk has been studied primarily in the context of the systemic immune system, it is likely to have special significance in the gut. The mucosal immune system is anatomically, functionally, and operationally distinct from the systemic immune system and is subject to independent regulatory signals. Furthermore, the intestinal mucosal immune system operates in a local milieu that depends on a dense innervation for its integrity, with juxtaposition of neuroendocrine cells and mucosal immune cells [167]. Moreover, The addition of enriching immunonutrients to early enteral feeding helps preserve an almost normal gut mucosa [124]. A number of adjuvants promote gut structure and function. Early enteral nutrition supplemented with glutamine, arginine, omega-3 fatty acids, and nucleotides [immune-enhancing nutrients (IENs)] has been shown to decrease infection complications in critically injured patients.[118]. These IENs, when added to standard enteral diets, enhance immune responsiveness, independent of their presumed nutritional effects [119].

During sepsis, activated macrophages are released from the gut into mesenteric lymph. However, an enteral diet with lipid is able to suppress the inflammatory cytokine release from mesenteric lymph leukocytes. [168].

Liquid diets containing omega-3 fatty acids used in hospitals for critically ill people have been shown to improve immune function and reduce infections[194]. Parenteral infusion of omega-3 PUFAs is advantageous, particularly in severely ill patients, because the fatty acids are rapidly incorporated by cells. In addition, when fatty acids are given parenterally, there are no losses from digestion and absorption as there are with enteral infusion.[194]. In a study (Omega-3s in Med Diet Protect Brain Health*)*, the scientists credited the better cognitive function of those eating the "Med" way to higher levels of DHA (an omega-3 fatty acid) and lower ratios of omega-6s to omega- 3s.[195].

IMMUNONUTRIENT GLUTAMINE, ARGININE – ROLE IN INTESTINAL BARRIER PROTECTION

However, among these immunonutrients glutamine is the most important. It is the most abundant amino acid in the body, and is the primary oxidizable fuel for intestinal mucosal cells, lymphocytes, and macrophages. The gastrointestinal tract is by far the greatest user of glutamine in the body, as enterocytes in the intestinal epithelium use glutamine as their principal metabolic fuel. Glutamine is metabolized by the gut and therefore, can contribute back energy in the form of ATP, whereas arginine is a non-metabolisable nutrient, using but not contributing energy. As arginine is one of the identified immune-enhancing nutrients and is also not metabolized by the gut, changes similar to those of alanine with compromised gut barrier function would be anticipated [118]. The metabolisable nutrient, glutamine, demonstrated protection by a significant reduction in mucosal injury, inflammation, and preservation of gut barrier function compared with the non-metabolisable, immune-enhancing nutrient, arginine. [121]. In animal studies, glutamine-enriched formulas enhanced nitrogen balance, decreased intestinal mucosal damage, improved gut function, decreased translocation and bacteremia, and improved survival in certain circumstances. Studies in human volunteers have shown that glutamine(Gln) decreases IL-6 and -8 in the intestine [148]. A recent review about intestinal permeability and systemic infections in critically ill patients concluded that glutamine (i.v. and enteral administration) reduced the frequency of systemic infections, presumably by maintaining intestinal barrier function [120] .

L-Glutamate is a multifunctional amino acid involved in taste perception, intermediary metabolism. L-glutamate acts as the dominant excitatory neurotransmitter in the brain [133,134] and appears also to function as a transmitter outside the brain (e.g., in the enteric neurons of the gut [135]). In addition, recent studies have uncovered new roles for Lglutamate in gut-brain axis activation and energy homeostasis. [132].

In particular, dietary L-glutamate is now known to stimulate L-glutamate receptors in the stomach and intestines, which produces local actions on gut function, and also, through the release of signaling molecules (nitric oxide and serotonin(5-HT)], activates the afferent vagus nerve and consequently a number of areas in the brain. Recent data suggest that one of the molecular mechanisms responsible for the gut-protective effects of enteral glutamine is the activation of peroxisome proliferator-activated receptor γ. [118]. L-Glutamate receptors and their cellular transduction molecules have recently been identified in gut epithelial cells. Stimulation of such L-glutamate receptors by luminal L-glutamate activates vagal afferent nerve fibers and then parts of the brain that are targeted directly or indirectly by these vagal inputs. Notably, 3 areas of the brain—the medial preoptic area, the hypothalamic dorsomedial nucleus, and the habenular nucleus—are activated by intragastric L-glutamate but not by glucose or sodium chloride [132]. In addition to these roles, recent studies have identified other functions for L-glutamate. L-Glutamate receptors and their cellular transduction molecules have now been identified in gut epithelial cells [136,137]. Stimulation of these receptors by luminal Lglutamate activates vagal afferent nerve fibers [138,139] and subsequently parts of the brain targeted directly or indirectly by these vagal inputs (e.g., the cerebral cortex, basal ganglia, limbic system, and hypothalamus) [140,141]. In addition, the chronic, elective ingestion of a palatable solution of monosodium L-glutamate (MSG) by rats has been found to reduce the development of obesity, fat deposition, and elevated plasma leptin concentrations through an enhancement of energy expenditure [142]. Such effects of L-glutamate may be initiated by gut L-glutamate receptors [136,137] that are linked to afferent branches of the vagus nerve [138,139, taste L-glutamate receptors in the oral cavity [143,144,145,146,147], or both. [132].

BRAIN AND GLUTAMATE RESPONSE

One current hypothesis is that this signaling cascade informs the brain of the amount of protein ingested [140]. It has repeatedly been suggested that the sensing of protein ingestion by animals might be linked to the L-glutamate content in foods [193]. It is unlikely that such protein "sensing" would be tied (at least exclusively) to L-glutamate taste in the mouth because taste is limited to free, but not to protein-bound, L-glutamate, and the free L-glutamate content of natural food items is not tied to their protein content (e.g., tomatoes are high in free L-glutamate, but not protein) [149,151]. But the total L-glutamate content of foods (free + protein bound) might provide a reasonable index of protein ingestion because L-glutamate is the most abundant amino acid in almost all dietary proteins [151]. Mediterranean soup is rich in tomatoes so is beneficial for gut – epithelial protection. We have recently observed that the brain activation induced by L-glutamate, but not by glucose, is abolished after sub-diaphragmatic total vagotomy [104]. This finding suggests that afferent vagus signaling is critical to gut L-glutamate "sensing" in the brain, but not to gut glucose sensing by the brain. Presumably because of its importance as an energy source in many cells in the body, and especially in the brain, gut epithelial cells metabolize relatively little glucose after the absorption of glucose [154,153]. In contrast, most luminal L-glutamate (and glutamine) is metabolized in the small intestine [154,152]. As a consequence, blood glucose concentrations increase after the ingestion of glucose, whereas blood L-glutamate concentrations generally do not rise after the ingestion of glutamate (unless a very large dose is administered) [154,142,155]. Before absorption, luminal glucose stimulates "sweet" receptors in the gut, which triggers the release of incretin hormones into the blood [156], whereas L-glutamate stimulates L-glutamate receptors to activate the vagus nerve (through the release of nitrous oxide and 5-HT) [158]. In concert with the incretins, the elevation of blood glucose by glucose ingestion enhances the release of insulin from pancreatic β cells [156,157]. Hence, the dominant pathway by which glucose ingestion signals the brain appears to be by the circulation (i.e., a humoral pathway involving the direct sensing of blood chemicals by the brain), whereas the pathway by which L-glutamate signals the brain is primarily neural, i.e., vagus mediated. If this difference holds, it is interesting to note that the mechanism by which the gut manages ingested glucose (with relatively little metabolism) and L-glutamate (with almost complete metabolism) harmonizes with the manner in which each signals the brain.

ENTERAL VS PARENTERAL NUTRITION AND GUT-BRAIN AXIS

Enteral nutrition activates humoral and neural pathways to regulate food intake and sustain energy balance [190]. Many clinical and experimental studies have demonstrated that total parenteral nutrition (TPN) can promote more cytokine production compared to enteral nutrition (EN) [126,127], including IL-6, IL-8, TNF, and CRP. Deprivation of enteral feeding by substituting it with a parenteral nutrition may negatively influence the gut-brain axis in new-born infants. [88]. Total parenteral nutrition has failed to show any clinical benefits for the patients, as it cannot protect the gut mucosa. Studies performed on animals and human beings show that TPN may worsen intestinal atrophy and intestinal and systemic immunity, and thereby it may also contribute to impair the intestinal mucosa permeability and facilitate bacterial translocation. [192]. On the contrary, enteral feeding repairs the mucosal damage of fasting. If enteral feeding is given very early, it may preserve epithelial integrity and bacterial ecology, thereby helping to maintain gut barrier function [128]. Doley et al. [130] reported that patients given TPN were more often infected by Gram-positive organisms and fungi as compared to those given EN who were more often infected by a Gram-negative organism. [129] Thus enteral nutrition prevents early gut barrier loss, enterocyte damage, and local intestinal inflammation. [131].

Conclusion

In brief, available evidence indicate that there is possible influence of gut hormones on brain function leading to increased risk of CVDs. The risk may be modulated by nutritional factors; fruits, vegetables, nuts, oils rich in w-3 fatty acids that are characteristics of a wild food hypothesis. While Western diet is known to have adverse effects on gut functions leading to increased risk of brain dysfunction and CVDs, Mediterranean diet may be protective. It may be possible to develop recopies for enteral nutrition based on protective diets rich in antioxidants, vitamins and amino acids. Further studies indicate that cocoa flavanol and CoQ10 might also benefit in brain-body interactions. [198,199] Nutrition in transition from Homo sapiens to Homo economicus indicate that environmental influences such as dietary patterns appear to be important in brain-body iteractions. [200]

Recent evidence indicates the role of food ingestion on release of various gut enzymes and hormones and their interaction with w-3 fatty acids as well as flavanols and gut-brain-liver axis causing insulin resistance, metabolic syndrome and heart disease.

Acknowledgments

Acknowledgements are due to International College of Nutrition and International College of cardiology and The Tsim Tsoum Institute for providing logestics to write this article.

References

[1] Bercik P, Verdu EF, Foster JA, Macri J, Potter M, Huang X, Malinowski P, Jackson W, Blennerhassett P, Neufeld KA, Lu J, Khan W, Corthesy-Theulaz I, Cherbut C, Bergonzelli GE, Collins SM. Chronic gastrointestinal inflammation induces anxiety-

like behavior and alters central nervous system biochemistry in mice. *Gastroenterology* 2010; doi:10.1053j.gastro.2010.06.063

[2] Loftis JM, Huckans M, Morasco BJ. Neuroimmune mechanisms of cytokine-induced depression: current theories and novel treatment strategies. *Neurobiol. Dis.* 2010; 37(3):519-33.

[3] Graff LA, Walker JR, Bernstein CN. Depression and anxiety in inflammatory bowel disease: a review of comorbidity and management. *Inflamm. Bowel. Dis.* 2009;15(7):1105-18.

[4] Maes M. Inflammatory and oxidative and nitrosative stress pathways underpinning chronic fatigue, somatization and psychosomatic symptoms. *Curr. Opin. Psychiatry.* 2009 ;22(1):75-83.

[5] O'Mahony L, McCarthy J, Kelly P, Hurley G, Luo F, Chen K, O'Sullivan GC, Kiely B, Collins JK, Shanahan F, Quigley EM. Lactobacillus and bifidobacterium in irritable bowel syndrome: symptom responses and relationship to cytokine profiles. *Gastroenterology*. 2005;128(3):541-51.

[6] Wang PY, Caspi L, Lam CK, Chari M, Li X, Light PE, Gutierrez-Juarez R, Ang M, Schwartz GJ, Lam TK. Upper intestinal lipids trigger a gut-brain-liver axis to regulate glucose production. *Nature* 2008;452:1012-16.

[7] Thaler JP, Cumming DE. Metabolism: food alert. *Nature* 2008; 452: 941-942.

[8] Venters HD, Tang Q, Liu Q, VanHoy RW, Dantzer R, Kelley KW. A new mechanism of neurodegeneration: proinflammatory cytokines inhibits receptor signaling by a survival peptide. *Proc. Natl. Acad. Sci. USA* 1999;96:9879-84.

[9] Lam CK, Chari M, Su BB, Cheung GW, Kokorovic A, Yang CS, Wang PY, Lai TY, Lam TK. Activation of N-methyl-D-aspartate (NMDA) receptors in the dorsal vagal complex lowers glucose production. *J. Biol. Chem.* 2010;285:21913-21.

[10] Inger Lauritzen I, Nicolas Blondeau N, Catherine Heurteaux C, Catherine Widmann C, Georges Romey G, and Michel Lazdunski M. Polyunsaturated fatty acids are potent neuroprotectors. *EMBO J.* 2000;19:1784-88.

[11] Singh RB, Pella D, Otsuka K. Can brain dysfunction may be a predisposing factor for metabolic syndrome. *Biomed. Pharmacother* 2004; 58(Supple) 52- 68.

[12] Crawford MA, Bazinet RP, Sinclair AJ. Fat intake and CNS functioning: ageing and disease. *Ann. Nutr. Meta.* 2009; 55: 202-228. DOI: 10.1159/000229003.

[13] Singh RB, Kartik C, Otsuka K, Pella D, Pella J. Brain-heart connection and the risk of heart attack. *Biomed. Pharmacother* 2004; 56:(Supple) 257-265.

[14] O'Donoghue M, Morrow DA, Sabatine MS, Murphy SA, McCabe CH, Cannon CP, Braunwald E. Lipoprotein-associated phospholipase A2 and its association with cardiovascular outcomes in patients with acute coronary syndromes in the PROVE IT-TIMI 22 (PRavastatin Or atorVastatin Evaluation and Infection Therapy-Thrombolysis In Myocardial Infarction) trial. *Circulation* 2006;113:1745-52.

[15] Esposito K, Glugliano D. Diet and inflammation:a link to metabolic and cardiovascular diseases. *Euro Heart J.*, 2006 ; 27:15-20.

[16] Vogel RA. Eating vascular biology and atherosclerosis: a lot to chew on. *Euro Heart J.* 2006,27:13-14.

[17] Gramenzi A, Gentile A, Fasoli M, Negri E, Parazzini F. La Vecchia C. Association between certain foods and risk of acute myocardial infarction in women. *BMJ* 1990; 300:771–3.

[18] Singh RB, Pella D, Sharma JP, Rastogi S, Kartikey K, Goel VK, Sharma R, Neki NS, Kumar A, Otsuka K. Increased concentrations of lipoprotein(a), circadian rhythms and metabolic reactions, evoked by acute myocardial infarctions, in relation to large breakfast. *Biomed. Pharmacother* 2004:58(Supple)116-122.

[19] Singh RB, Pella D, Neki NS, Chandel JP, Rastogi S, Mori H, Otsuka K. Mechanism of acute myocardial infarction study. *Biomed. Pharmacother* 2004:58(Supple)111-115.

[20] Fung TT, Malik V, Rexroad KM, Manson JE, Willett WC, Hu FB. Sweetened beverage consumption and risk of coronary heart in women. *Am. J. Clin. Nutr.* 2009; 89:1037–42.

[21] Smith PJ, Blumenthal JA, Babyak MA, Georgiades A, Sherwood A, Sketch MH Jr, Watkins LL. Association between n-3 fatty acid consumption and ventricular ectopy after myocardial infarction. *Am. J. Clin. Nutr*. 2009; 89:1315-20.

[22] Campos H, Baylin A, Willett WC. Alpha-Linolenic acid and risk of nonfatal acute myocardial infarction. *Circulation*. 2008; 118:339-345.

[23] Harris WS, Reid KJ, Sands SA, Spertus JA. Blood omega-3 and trans fatty acids in middle aged acute coronary syndrome patients. *Am. J. Cardiol*. 2007;99:154–158.

[24] Singh RB, De Meester F, Pella D, Basu TK, Watson R. Globalization of dietary wild foods protect against cardiovascular disease and all cause mortalities? A Scientific statement from the International College of Cardiology, Columbus Paradigm Institute and the International College of Nutrition. *The Open Nutra. Jour.*, 2009, 2, 42-45.

[25] de Lorgeril M, Renaud S, Mamelle N, Salen P, Martin JL, Monjaud I, Guidollet J, Touboul P, Delaye J. Mediterranean alpha-linolenic acid-rich diet in secondary prevention of coronary heart disease. *Lancet.* 1994, 343(8911):1454-9. Erratum in: *Lancet* 1995,345 (8951):738.

[26] Renaud S, de Lorgeril M, Delaye J, Guidollet J, Jacquard F, Mamelle N, Martin JL, Monjaud I, Salen P, Touboul P. Cretan Mediterranean diet for prevention of coronary heart disease. *Am. J. Clin. Nutr.* 1995, 61(6 Suppl):1360S-1367S.

[27] Singh RB, Dubnov G, Niaz MA, Ghosh S, Singh R, Rastogi SS, Manor O, Pella D, Berry EM. Effect of an Indo-Mediterranean diet on progression of coronary disease in high risk patients:a randomized single blind trial. *Lancet* 2002,360:1455-1461.

[28] Esposito K, Marfella R, Ciotola M, Di Palo C, Giugliano G, D'Armiento M, D'Andrea F, Giugliano D. Effect of a Mediterranean-style diet on endothelial dysfunction and markers of vascular inflammation in the metabolic syndrome: a randomized trial. *JAMA* 2004; 292:1440-1446.

[29] Singh RB, Niaz MA, Agarwal P, Beegum R, Rastogi SS. Effect of antioxitant rich foods on plasma ascorbic acid, cardiac enzyme and lipid peroxide levels in patients hospitalized with acute myocardial infarction. *J. Am. Diet. Assoc*. 1995; 95:775-80.

[30] Singh RB, Rastogi SS, Verma R, Bolaki L, Singh R,Ghosh S. An Indian experiment with nutritional modulation in acute myocardial infarction. *Am. J. Cardiol*. 1992; 69 : 879-85.

[31] Singh RB, Rastogi SS, Verma R, Laxmi B,Singh R, Ghosh S, Niaz MA. Randomized, controlled trial of cardioprotective diet in patients with acute myocardial infarction : results of one year follow up. *BMJ* 1992;304:1015-9.

[32] Metzinser D, Digen L, Drewe J, Meuli J, Dubendorpher R, Rukstuhl N, Amato MD, Rovati L, Beglinger C. The role of long chain fatty acids in regulating food intake and cholecystokinin release in humans. *Gut* 2000; 47: 688-693.

[33] Bado A, Levasseur S, Attoub S, Kermorgant S, Laigneau JP, Bortoluzzi MN, Moizo L, Lehy T, Guerre-Millo M, Le Marchand-Brustel Y, Lewin MJ. The stomach is a source of leptin. *Nature* 1998; 394: 790-793.

[34] Barrachina MD, Martinez V, Wang L,Wei Jyand Tache Y Synergistic interaction between leptin and cholecystokinin to reduc.e short-term food intake in mice. *Proc. Natl. Acad. Sci. USA* 1997;94: 10455-10460.

[35] Hirano H day J, Fibiger HC. Serotonergic regulation of acetylcholine release in rat frontal cortex. *J. Neurochem.* 1995; 65:1139-1145.

[36] Zhou FM, Liang Y, Dani JA. Endogenous nicotinic cholinergic activity regulates dopamine release in the striatum. *Nat. Neurosci.* 2001; 4:1224-29.

[37] Bartness TJ, Kay Song C, Shi H, Bowers RR, Foster MT. Brain-adipose tissue cross talk. *Proc. Nutr. Soc.* 2005; 64: 53-64.

[38] Huang LZ, Winzer-Serhan UH. Nicotine regulates mRNA expression of feeding peptides in the arcuate nucleus in neonatal rat pups. *Dev. Neurobiol* 2007; 67: 363-377.

[39] Obici S, Feng Z, Morgan M, Karkanias G, Stein D, Rossetti L. Central administration of oleic acid inhibits glucose production and food intake. *Diabetes* 2002; 51: 271-275.

[40] Obici S, Feng Z, Arduini A, Conti R. and Rossetti, L. Inhibition of hypothalamic carnitine palmitoyltransferase-l decreases food intake and glucose production. *Nat .Med.* 2003; 9: 756-761.

[41] Lam TK, Pocai A, Gutierrez-Juarez R, Obici S, Bryan J, Aguilar-Bryan L, Schwartz Jand Rossetti L .Hypothalamic sensing of circulating fatty acids is required for glucose homeostasis. *Nat. Med.* 2005; 11:320-327.

[42] Suresh Y, Das UN. Differential effect of saturated, monounsaturated, and polyunsaturated fatty acids on alloxan-induced diabetes mellitus. Prostaglandins Leukot Essen Fatty Acids 2006; 74: 199-213.

[43] Richard D, Guesdon B, Timofeeva E. The brain endocannabinoid system in the regulation of energy balance. *Best Pract. Res. Clin. Endocrinol. Metab.* 2009; 23: 17-32.

[44] Di Marzo V. The endocannabinoid system in obesity and type diabetes. *Daibetologia* 2008; 51: 1356-1367.

[45] Romanova IV, Ramos EJB, Xu Y, Quinn R, Chen C, George ZM, Inui A, Das U, Meguid MM. Neurobiologic changes in the hypothalamus associated with weight loss after gastric bypass. *J. Am. Coll. Surg.* 2004; 199: 887-895.

[46] Xu Y, Ramos EJB, Middleton F, Romanova I, Quinn R, Chen C, Das U, Inui A, Meguid MM. Gene expression profiles post Roux-en-Y gastric bypass. *Surgery* 2004; 136: 246-252.

[47] Madjid M, Ali M, Willersion JT. Lipotrotein associated phospholipase A2, a novel risk marker for cardiovascular disease. *Texas Heart J.* 2010;37:25-39.

[48] Tonra JR,_Ono M., Liu X, Garcia K, Jackson C. Brain-derived neurotrophic factor improves blood glucose control and alleviates fasting hyperglycemia in C57BLKS-Lepr (db)/lepr (db) mice. *Daibetes* 1999; 48: 588-594.

[49] Kernie SG, Liebl DJ, Parada LF. BDNF regulates eating behavior and locomotor activity in mice. *EMBO J.* 2000;19:1290-1300.

[50] Tsuchida A, Nonomura T, Ono-Kishino M, Nakagawa T, Taiji M, Noguchi H. Acute effects of brain-derived neurotrophic factor on energy expenditure in obese diabetic mice. *Int. J. Obes. Relat Metab. Disord.* 2001;25:1293.

[51] Das UN. Pathophysiology of type 2 diabetes mellitus with particular reference to hypothalamus. In editor UN Das, *Metabolic Syndrome, Pathophysiology, The Role of Essential Fatty acids.* Wiley-Blackwell, Singapore 2010, p 130-35.

[52] Lommatzsch M, Braun A, Mannsfeldt A, Botchkarev VA, Botchkareva NV, Paus R, Fischer A, Lewin GR, Renz H. Abundant production of brain-derived neurotrophic factor by adult visceral epithelia. *Am. J. Pathol.* 1999; 155: 1183-1193.

[53] Rost B, Hanf G, Ohnemus U. Monocytes of allergics and non-allergics produce, store and release the neurotrophins NGF, BDNF and NT-3. *Regul. Pept.* 2005; 124: 19-25.

[54] Noga O, Englmann C, Hanf G, Grützkau A, Seybold J, Kunkel G. The production, storage and release of the neurotrophins nerve growth factor, brain-derived neurotrophic factor and neurotrophins-3 by human peripheral eosinophils in allergics and non-allergics. *Clin. Exp. Allergy* 2003; 33: 649-654.

[55] Cai D, Holm JM, Duignan IJ, Zheng J, Xaymardan M, Chin A.BDNF-mrdiated enhancement of inflammation and injury in the aging heart. *Physiol. Genomics* 2006; 24: 191-197.

[56] Johansson M, Norrgard O, Frogmen S. Study of expression patterns and levels of neurotrophins and neurotrophin receptors in ulcerative colitis. *Inflamm. Bowel. Dis.* 2007; 13: 398-409.

[57] di Mola F, Friess H, Zhu ZW, Koliopanos A, Bley T, Di Sebastiano P, Innocenti P, Zimmermann A, Büchler MW. Nerve growth factor and Trk high affinity receptors (TrkA) gene expression in inflammatory bowel disease. *Gut* 2000; 46: 670-679.

[58] Raap U, Werfel T, Goltz C, Deneka N, Langer K, Bruder M, Kapp K, Schmid-Ott G, Wedi B. Circulating levels of brain-derived neurotrophic factor correlate with disease severity in the intrinsic type of atopic dermatitis. *Allergy* 2006; 61: 1416-1418.

[59] Krabbe KS, Nielsen AR, Krogh-Madsen R, Plomgaard P, Rasmussen P, Erikstrup C, Fischer CP, Lindegaard B, Petersen AM, Taudorf S, Secher NH, Pilegaard H, Bruunsgaard H, Pedersen BK.Brain derived neurotrophic factor and type 2 diabetes. *Diabetologia* 2007; 50:431-38.

[60] Nonomura T, Tsuchida A, Ono-kishino M, nakagawa T, Taiji M and Noguchi H. Brain derived neurotrophic factor regulates energy expenditure through the central nervous system in obese diabetic mice. *Int. J. Exp. Diab. Res.* 2001; 2:201-209.

[61] Ono M, Itakura Y, Nonomura T. Intermittent administration of brain derived neurotrophic factor ameliorates glucose metabolism in the obese diabetic mice. *Metabolism* 2000; 49:129-133.

[62] CaoL, D Lin EJ, Cahill MC, Wang C, Liu X and During MJ. Molecular therapy of obesity and diabetes by a physiological autoregulatory approach . *Nat. Med.* 2009; 15: 447-454.

[63] Rao JS, Ertley RN, J Lee H, DeMar JC, Arnold JT, Rapoport SI, Bazinet RP. n-3 polyunsaturated fatty acid deprivation in rats decreases frontal cortex BDNF via a p38 MAPK-dependent mechanism . *Mol. Psychiatry* 2007; 12: 36-46

[64] Cho HJ, Kim JK, Zhou XF, Rush RA. Increased brain derived neurotrophic factor immunoreactivity in rat dorsal root ganglia and spinal cord following peripheral inflammation . Brain Res 1997;764:269-272.

[65] Oddiah D, Praveen A, MahonMc, Stephen B , MarcusR. Rapid increase of NGF, BDNF, and NT-3 mRNAs in inflamed bladder. *Neuroreport* 1998; 9:1455-58.

[66] Komori, T, Morikawa, Y, Nanjo, K, and Senba, E. Induction of brain-derived neurotrophic factor by leptin in the ventromrdial hypothalamus. *Neuroscience* 2006; 139: 1107-1115.

[67] Nakagawa T, OgawaY, Ebihara K, Yamanaka M, Tsuchida A, Taiji M, Noguchi H and Nakao K. Antiobesity and antidiabetic effects of brain derived neurotrophic factor in rodent models of leptin resistance. *Int. J. obesity Rel. Meta. Disor.* 2003; 27:557-565.

[68] Zhang Y, Proenca R, Maffei M, Barone M, Leopold L, Friedman JM. Positional cloning of the mouse obese gene and its human homologue. *Nature* 1994; 372: 425-432. Zhang Y et. al., Positional cloning of the mouse obese gene and its human homologue. *Nature* 1994; 372: 425-432.

[69] Huang Q, Viale A, Picard F, Nahon JL, Richard D. Effects of leptin on melanin-concentrating hormone expression in the brain of lean and obese Lep(ob)/Lep(ob) mice. *Neuroendocrinology* 1999; 69: 145-153.

[70] Saad MF, Bernaba B, Hwu C, Jinagouda S, Fahmi S, Kogosov E, Boyadjian. Insulin regulates plasma ghrelin concentration. *J. Clin. Endocrinol. metab.* 2002; 87: 3997-4000.

[71] Broglio F, Gottero C, Koetsveld PV, Prodam F, Destefanis S, Benso A, Gauna C, Hofland L, Arvat E, van der Lely AJ and Ghigo E. Acetylcholine regulates ghrelin secretion in humans. *J. Clin. Endocrinol. Metab.* 2004; 89: 2429-2433.

[72] Tamura H, Kamegai J, Shimizu T,Ishii S,Sugihara H, and Oikawa S Ghrelin stimulates GH but not food intake in arcuate nucleus ablated rats. *Endocrinology* 2002; 143: 3268-3275.

[73] Kamegai J, Tamura H, Shimizu T, IshiiS, Sugihara H and Wakabayashi I. Chronic central infusion of ghrelin increases hypothalamic neuropeptide Y and agouti-related protein mRNA levels and body weight in rats. *Diabetes* 2001; 50: 2438-2443.

[74] Albert MA, Glynn RJ, Ridker PM. Plasma concentration of C-reactive protein and the calculated Framingham coronary heart disease risk score. *Circulation* 2003; 108: 161-165.

[75] Luc G, Bard J, Juhan-Vague I. C-reactive protein, interleukins-6, and fibrinogen as predictors of coronary heart disease. The PRIME study. *Arterioscler. Thromb. Vase. Biol.* 2003; 23: 1255-1261.

[76] Barzilay JI, , Abraham L, Heckbert SR, Cushman M, Kuller LH, Resnick HE and Tracy RP. The relation of markers of inflammation to the development of glucose disorders in the elderly. *Diabetes* 2001; 50: 2384-2389.

[77] Kim MJ,Yoo KH, Park HS, Chung SM, Jin CJ, Lee Y, Shin YG, Chung CH. Plasma adiponectin and insulin resistance in Korean type 2 diabetes mellitus. *Yonsei Med. J.* 2005; 46: 42-50.

[78] Mora S, Glynn RJ, Hsia J, MacFadyen JD, Genest J, Ridker PM. Statins for the primary prevention of cardiovascular events in women with elevated high-sensitive C-reactive protein or dyslipidemia 2010, DOI:10.1161/CIRCULATIONAHA.109.906479.

[79] Liu S, MansonJE, BuringJE, Stampfer MJ, Willett WC. Relation between a diet with a high glycemic load and plasma concentrations of high-sensitivity C-reactive protein in middle-aged women. *Am. J. Clin. Nutr.* 2002; 75: 492-498.

[80] Pepys MB, Hirschfield GM. C-reactive protein: a critical update. *J. Clin .Invest.* 2003; 111: 1805-1812.

[81] Griselli M, Herbert J, Hutchinson WL, Taylor KM, Sohail M, Krausz T. C-reactive protein and complement are important mediators of tissue damage in acute myocardial infarction. *J. Exp. Med.* 1999: 190: 1733-1739.

[82] Gill R, Kemp JA, Sabin C. Human C-reactive protein increases cerebral infract size after middle cerebral artery occlusion in adult rats. *J Cereb Blood Flow Metab* 2004; 24: 1214-1218.

[83] Pepys MB Mark B. Pepys, Hirschfield GM, TennentGA, Gallimore JR, KahanMC, Bellotti V, HawkinsPN, Myers RM, Smith MD, Polara A, Alexander J. A., CobbA, Ley SV, Aquilina JA, Robinson CV, Sharif I, Gray GA, Sabin CA, Jenvey MC, Kolstoe SE, Thompson D and Wood SP. Targeting C-reactive protein for the treatment of cardiovascular disease. *Nature* 2006; 440: 1217-1221.

[84] Esposito K ,Nappo F, Marfella R, Giugliano G, Giugliano F; Ciotola M, Quagliaro L, A, Giugliano D. Inflammatory cytokine concentration are acutely increased by hyperglycemia in humans. Role of oxidative stress. *Circulation* 2002; 106: 2067-2072.

[85] Kirwan J, Krishnan RK, Weaver JA, Del Aguila LF, Evans WJ.P. Human aging is associated with altered TNF-a production during hyperglycemia and hyperinsulinemia. *Am. J. Physiol* .2001; 281: E1137-E1143.

[86] Lin Y, RajalaMW, Berger JP, Moller DE, Barzilai N, and Scherer PE. Hyperglycemia-induced production of acute phase reactants in adipose tissue. *J. Biol. Chem.* 2001; 276: 42077-42083.

[87] de Lartigue G, Barbier de la Serre C, Espero E, Lee J, Raybould HE Diet Induced obesity leads to the development of leptin resistance in vagal afferent neurons. *Am. J. Physiol. Endolcrinol. metab.* 2011, 301,11: E187-95

[88] Tomasik PJ and Satefko K. The effect of enteral and parenteral feeding on secretion of orexigenic peptides in infants .BMC *Gastroenterology* 2009,9:92 doi 10.1186/1471-230x-9-92

[89] Sakurai T, Amemiya A, Ishii M, Matsuzaki I, Chemelli RM, Tanaka H, Williams SC, Richardson JA, Kozlowski GP, Wilson S, Arch JR, Buckingham RE, Haynes AC, Carr SA, Annan RS, McNulty DE, Liu WS, Terrett JA, Elshourbagy NA, Bergsma DJ, Yanagisawa M: Orexins and orexin receptors: a family of hypothalamic neuropeptides and G protein-coupled receptors that regulate feeding behavior. Cell 1998, 92:573-585.

[90] De Lecea L, Kilduff TS, Peyron C, Gao X, Foye PE, Danielson PE, Fukuhara C, Battenberg EL, Gautvik VT, Bartlett FS, Frankel WN, Pol AN, Bloom FE, Gautvik KM, Sutcliffe JG: The hypocretins: hypothalamus-specific peptides with neuroexcitatory activity. Proc. Natl. Acad. Sci. USA. 1998, 95:322-327.

[91] Date Y, Ueta Y, Yamashita H, Yamaguchi H, Matsukura S, Kangawa K, Sakurai T, Yanagisawa M, Nakazato M: Orexins, orexigenic hypothalamic peptides, interact with autonomic, neuroendocrine and neuroregulatory systems. Proc. Natl. Acad. Sci. USA 1999, 96:748-753.

[92] Kirchgessner AL, Liu M-T: Orexins synthesis and response in the gut. Neuron 1999, 24:941-951.

[93] Peyron C, Tighe DK, Pol AN, de Lecea L, Heller HC, Sutcliffe JG, Kilduff TS: Neurons containing hypocretin (orexin) project to multiple neuronal systems. J. Neurosci. 1998, 18:9996-10015.

[94] Ohno K, Sakurai T: Orexin neuronal circuitry: Role in the regulation of sleep and wakefulness. Front Neuroendocrinol 2008, 29:70-87.

[95] Rodgers RJ, Halford JC, Nunes de Souza RL, Canto de Souza AL, Piper DC, Arch JR, Blundell JE: Dose-response effects of orexin-A on food intake and the behavioural satiety sequence in rats. Regul Pept 2000, 96:71-84.
[96] Sakurai T. Roles of orexin/hypocretin in regulation of sleep/wakefulness and energy homeostasis. Sleep Med. Rev. 2005, 9:231-241.
[97] Sutcliffe JG, De Lecea L: The hypocretins: excitatory neuromedulatory peptides for multiple homeostatic systems, including sleep and feeding. J. Neurosci. Res 2000, 62:161-168.
[98] Willie JT, Chemelli TM, Sinton CM, Yanagisawa M: To eat or sleep? Orexin in the regulation of feeding and wakefulness. Annu. Rev. Neurosci. 2001, 24:429-458.
[99] Bischoff SC Gut Health: a new objective in medicine? *BMC Med* 2011;9:24 Doi 10.1186/1741-7015-9-24
[100] Blaut M, Clavel T. Metabolic diversity of the intestinal microbiota: implications for health and disease. *J. Nutr.* 2007;137:751S–755S.
[101] Vrieze A, Holleman F, Zoetendal E.G., de Vos WM, Hoekstra JB, Nieuwdorp M. The environment within: how gut microbiota may influence metabolism and body composition. *Diabetologia.* 2010;53:606–613. doi: 10.1007/s00125-010-1662-7.
[102] Sharma R, Young C, Neu J. Molecular modulation of intestinal epithelial barrier: contribution of microbiota. *J. Biomed. Biotechnol.* 2010;2010:305879.
[103] Chung H, Kasper DL. Microbiota-stimulated immune mechanisms to maintain gut homeostasis. *Curr Opin Immunol.* 2010; 22:455–460. doi: 10.1016/j.coi.2010.06.008.
[104] Tsurugizawa T, Uematsu A, Nakamura E, Hasumura M, Hirota M, Kondoh T, Uneyama H, Torii K. Mechanisms of neural response to gastrointestinal nutritive stimuli: the gut-brain axis. *Gastroenterology.* 2009;137:262–273. doi: 10.1053/j.gastro.2009.02.057.
[105] Fasano A, Shea-Donohue T. Mechanisms of disease: the role of intestinal barrier function in the pathogenesis of gastrointestinal autoimmune diseases. *Nat Clin Pract Gastroenterol Hepatol.* 2005;2:416–422. doi: 10.1038/ncpgasthep0259.
[106] Nieuwenhuis EE, Blumberg RS. The role of the epithelial barrier in inflammatory bowel disease. *Adv. Exp. Med. Biol.* 2006; 579:108–116. full_text.
[107] Meddings J. The significance of the gut barrier in disease. *Gut.* 2008;57:438–440. doi: 10.1136/gut.2007.143172.
[108] Groschwitz KR, Hogan SP. Intestinal barrier function: molecular regulation and disease pathogenesis. *J. Allergy Clin. Immunol.* 2009;124:3–20. doi: 10.1016/j.jaci.2009.05.038.
[109] Mai V, Morris JG Jr. Colonic bacterial flora: changing understandings in the molecular age. *J. Nutr.* 2004;134:459–464.
[110] Macfarlane GT, Macfarlane S. Human colonic microbiota: ecology, physiology and metabolic potential of intestinal bacteria. *Scand. J. Gastroenterol Suppl.* 1997;222:3–9.
[111] Bouskra D, Brézillon C, Bérard M, Werts C, Varona R, Boneca IG, Eberl G. Lymphoid tissue genesis induced by commensals through NOD1 regulates intestinal homeostasis. *Nature.* 2008;456:507–510. doi: 10.1038/nature07450.
[112] Round JL, Mazmanian SK. The gut microbiota shapes intestinal immune responses during health and disease. *Nat. Rev. Immunol.* 2009;9:313–323. doi: 10.1038/nri2515.

[113] Mai V, Draganov PV. Recent advances and remaining gaps in our knowledge of associations between gut microbiota and human health. *World J. Gastroenterol.* 2009;15:81–85. doi: 10.3748/wjg.15.81.

[114] Kinross JM, von Roon AC, Holmes E, Darzi A, Nicholson JK. The human gut microbiome: implications for future health care. *Curr. Gastroenterol Rep.* 2008;10:396–403. doi: 10.1007/s11894-008-0075-y.

[115] Hsiao WW, Metz C, Singh DP, Roth J. The microbes of the intestine: an introduction to their metabolic and signaling capabilities. *Endocrinol. Metab. Clin. North Am.* 2008;37:857–871. doi: 10.1016/j.ecl.2008.08.006.

[116] Possemiers S, Grootaert C, Vermeiren J, Gross G, Marzorati M, Verstraete W, Van de Wiele T. The intestinal environment in health and disease: recent insights on the potential of intestinal bacteria to influence human health. *Curr. Pharm.Des.* 2009;15:2051–2065. doi: 10.2174/138161209788489159.

[117] Preidis GA, Versalovic J. Targeting the human microbiome with antibiotics, probiotics, and prebiotics: gastroenterology enters the metagenomics era. *Gastroenterology.* 2009;136:2015–2031. doi: 10.1053/j.gastro.2009.01.072.

[118] Ben K and Kozar RA Enteral Glutamine : a novel mediator of PPARy in the postischemic gut; *J Leukoc Biol* 200884(3) : 595 -599 doi: 10.1189/jlb

[119] Daly J M, Lieberman M D, Goldfine J, Shou J, Weintraub F, Rosato E F, Lavin P. Enteral nutrition with supplemental arginine, RNA, and omega-3 fatty acids in patients after operation: immunologic, metabolic, and clinical outcome. *Surgery.* 1992;112:56–67.

[120] De-Souza D A, Greene L J. Intestinal permeability and systemic infections in critically ill patients: effect of glutamine. *Crit. Care. Med.* 2005;33:1125–1135.

[121] Kozar R A, Schultz S G, Hassoun H T, DeSoignie R, Weisbrodt N W, Haber M M, Moore F A. The type of sodium-coupled solute modulates small bowel mucosal injury, transport function, and ATP after ischemia/reperfusion injury in rats. *Gastroenterology.* 2002;123:810–816.

[122] Burrin DG, Stoll B, Jiang R, Chang X, Hartmann B, Holst JJ, Greeley GHJ, and Reeds PJ. Minimal enteral nutrient requirements for intestinal growth in neonatal piglets: how much is enough? Am. J. Clin. Nutr. 71: 1603–1610, 2000.

[123] Kudsk KA. Current aspects of mucosal immunology and its influence by nutrition. Am. J. Surg 183: 390–398, 2002.

[124] Lubbers T, De Haan JJ, Luyer MD, de Haan JJ, Hadfoune M, Buurman WA, Greve JW cholecystokinin/ Cholecystokinin-1 Receptor- Mediated peripheral activation of the afferent vagus by enteral nutrients attenuates inflammation in rats. *Ann Surg* 2010 ;252,2 : 376-82

[125] Lubbers T, De Haan JJ, Hadfoune M, Zhang Y, Luyer MD, Grundy D, Buurman WA, Greve JW. Lipid enriched Enteral nutrition contrils the inflmmatory response in murine gram- negative sepsis, *Crit Care Med* 2010, 38(10): 1996 – 2002.

[126] D. G. Kelly and C. R. Fleming, "Nutritional considerations in inflammatory bowel diseases," *Gastroenterology Clinics of North America,* vol. 24, no. 3, pp. 597–611, 1995.

[127] D. B. A. Silk, "Medical management of severe inflammatory disease of the rectum: nutritional aspects," *Bailliere's Clinical Gastroenterology*, vol. 6, no. 1, pp. 27–41, 1992.

[128] C. Dervenis, "Enteral nutrition in severe acute pancreatitis: future development," *Journal of the Pancreas,* vol. 5, no. 2, pp. 60–63, 2004.
[129] P. E. Marik and G. P. Zaloga, "Meta-analysis of parenteral-nutrition versus enteral nutrition in patients with acute pancreatitis," *British Medical Journal*, vol. 328, no. 7453, pp. 1407–1410, 2004.
[130] R. P. Doley, T. D. Yadav, J. D. Wig_JD, Kochhar. R, Singh G, Bharathy KG, Kudari A, et al., "Enteral nutrition in severe acute pancreatitis," *Journal of the Pancreas,* vol. 10, no. 2, pp. 157–162, 2009.
[131] Quan H, wang X and Guo C A Meta-analysis of enteral nutrition and total parenteral nutrition in patients with acute pancreatitis . *Gastroenterology Research practice* 2011 doi: 10.1155/2011/698248
[132] Kondoh T, mallick N and Toriik K .Activation of gut- brain axis by dietary glutamate and physiologic significance in energy homeostasis. *Am. J Clin Nutr* 2009 ,90,3,8325-8375.doi 10.3945/ajcn.2009.27462V
[133] Meldrum BS Glutamate as a neurotransmitter in the brain: review of physiology and pathology. J. Nutr. 2000;130:1007S–15S.
[134] Tsumoto T, Hagihara K, Sato H, Hata Y NMDA receptors in the visual cortex of young kittens are more effective than those of adult cats. Nature 1987;327:513–4.
[135] Liu MT, Rothstein JD, Gershon MD, Kirchgessner AL Glutamatergic enteric neurons. J. Neurosci. 1997;17:4764–84.
[136] San Gabriel AM, Maekawa T, Uneyama H, Yoshie S, Torii K mGluR1 in the fundic glands of rat stomach. FEBS Lett 2007;581:1119–23.
[137] Bezençon C, le Coutre J, Damak S Taste-signaling proteins are coexpressed in solitary intestinal epithelial cells. Chem Senses 2007;32:41–9.
[138] Niijima A Reflex effects of oral, gastrointestinal and hepatoportal glutamate sensors on vagal nerve activity. J Nutr 2000;130:971S–3S.Medline.
[139] Uneyama H, Niijima A, San Gabriel A, Torii K Luminal amino acid sensing in the rat gastric mucosa. Am J. Physiol. Gastrointest Liver Physiol. 2006;291:G1163–70.
[140] Kondoh T, Torii K Brain activation by umami substances via gustatory and visceral signaling pathways, and physiological significance. Biol Pharm Bull 2008;31:1827–32.
[141] Tsurugizawa T, Kondoh T, Torii K Forebrain activation induced by postoral nutritive substances in rats. Neuroreport 2008;19:1111–5
[142] Kondoh T, Torii K. *MSG* intake suppresses weight gain, fat deposition, and plasma leptin levels in male Sprague-Dawley rats. Physiol. Behav 2008;95:135–44.
[143] Chaudhari N, Landin AM, Roper SD. A metabotropic glutamate receptor variant functions as a taste receptor. Nat. Neurosci. 2000;3:113–9.
[144] Li X, Staszewski L, Xu H, Durick K, Zoller M, Adler E. Human receptors for sweet and umami taste. Proc Natl. Acad. Sci. USA 2002;99:4692–6.
[145] Nelson G, Chandrashekar J, Hoon MA,_Feng L, Zhao G, Ryba NJ, Zuker CS. An amino-acid taste receptor. Nature 2002;416:199–202.
[146] San Gabriel A, Uneyama H, Yoshie S, Torii K. Cloning and characterization of a novel mGluR1 variant from vallate papillae that functions as a receptor for L-glutamate stimuli. Chem. Senses 2005;*30* Suppl 1:i25–6.
[147] Toyono T, Seta Y, Kataoka S, Kawano S, Shigemoto R, Toyoshima K. Expression of metabotropic glutamate receptor group I in rat gustatory papillae. Cell Tissue Res 2003;313:29–35.

[148] Li N, Liboni K, Fang MZ, Samuelson D, Lewis P, Patel R, Neu. J. Glutamine decreases lipopolysaccharide-induced intestinal inflammation in infant rats *AJP- GI,* 2004 ; 286,6,G914- G921. Doi 10.1152.

[149] Yamaguchi S, Ninomiya K. Umami and food palatability. J Nutr 2000*;*130*:*921S–6S.

[150] Kawamura Y, Kare MR ,Torii K, Mimura T, Yugari Y. Biochemical mechanism of umami taste perception and effect of dietary protein on the taste preference for amino acids and sodium chloride in rats. In*:* Kawamura Y, Kare MR, eds. *Umami: a basic taste*. New York, NY*:* Marcel Dekker, 1987*:*513–63.

[151] Giacometti T. Free and bound glutamate in natural products. In*:* Filer LJ Jr., Garattini MR, Kare MR, Reynolds WA, Wurtman RJ, eds. Glutamic acid: *advances in biochemistry and physiology.* New York, NY*: Raven Press*, 1979*:*25–34.

[152] Reeds PJ, Burrin DG, Stoll B, Jahoor F. Intestinal glutamate metabolism. J. Nutr. 2000*;*130*:*978S–82S.

[153] Young VR, Ajami AM. Glutamate: an amino acid of particular distribution. J. Nutr. 2000*;*130*:*892S–900S.

[154] Reeds PJ, Burrin DG, Jahoor F, Wykes L, Henry J, Frazer EM. Enteral glutamate is almost completely metabolized in first pass by the gastrointestinal tract of infant pigs. Am. J. Physiol. 1996*;*270*:*E413–8.

[155] Stegink LD, Ann Reynolds W, Filer LJ Jr., Baker GL, Daabees TT, Pitkin RM. Comparative metabolism of glutamate in the mouse, monkey, and man. In*:* Filer LJ Jr., Garattini S, Kare MR, Reynolds WA, Wurtoman RJ, eds. *Glutamic acid: advances in biochemistry and physiology.* New York, NY*:* Raven Press, 1979*:*85–102.

[156] Jang HJ, Kokrashvili Z, Theodorakis MJ, Carlson OD, Kim BJ, Zhou J., Kim, HH, Xu, X., Chan. Gut-expressed gustducin and taste receptors regulate secretion of glucagon-like peptide-1. Proc. Natl. Acad. Sci. USA 2007*;*104*:*15069–74.

[157] Gautier JF, Choukem SP, Girard J. Physiology of incretins (GIP and GLP-1) and abnormalities in type 2 diabetes. Diabetes Metab. 2008*;*34*:*S65–72.

[158] Uneyama H, Niijima A, San Gabriel A, Torii K. Luminal amino acid sensing in the rat gastric mucosa. Am J Physiol Gastrointest Liver Physiol 2006*;*291*:*G1163–70.

[159] Neviere RR, Pitt-Hyde ML, Piper RD, Sibbald WJ, Potter RF. Microvascular perfusion deficits are not a prerequisite for mucosal injury in septic rats. Am. J. Physiol. Gastrointest Liver Physiol. 1999, 276*:*G933–G940.

[160] Piguet PF, Vesin C, Donati Y, Barazzone C. TNF-induced enterocyte apoptosis and detachment in mice: induction of caspases and prevention by a caspase inhibitor, ZVAD-fmk. Lab. Invest *.1999;*79*:*495–500.

[161] Sun Z, Wang X, Wallen R, Deng X, Du X, Hallberg E, Andersson R. The influence of apoptosis on intestinal barrier integrity in rats. Scand. J. Gastroenterol.*1998* 33*:*415–422.

[162] Deitch EA, Ma L,Ma WJ, Grisham MB, Granger DN, Specian RD, Berg RD Inhibition of endotoxin-induced bacterial translocation in mice. J. Clin. Invest.*1989;* 84*:*36–42.

[163] O'Dwyer ST, Michie HR, Ziegler TR, Revhaug A, Smith RJ, Wilmore DW. A single dose of endotoxin increases intestinal permeability in healthy humans. Arch. Surg *1988;* 123*:*1459–1464.

[164] Ciancio MJ, Vitiritti L, Dhar A, Chang EB. Endotoxin-induced alterations in rat colonic water and electrolyte transport. Gastroenterology.*1992;* 103*:*1437–1443.

[165] Deitch EA, Berg R, Specian R. Endotoxin promotes the translocation of bacteria from the gut. *Arch. Surg*. 1987*;* 122*:*185–190 .

[166] Nieuwenhuijzen GA, Deitch EA, Goris RJ. The relationship between gut-derived bacteria and the development of the multiple organ dysfunction syndrome. *J. Anat.* 1996*;* 189*:*537–548.

[167] ShanahnF. Brain-gut axis and mucosal immunity: prospective on mucosal psychoneuroimmunology. *Semin Gastrointestinal Dis* 1999; 10,1 :8-13.10065767. cit:17.

[168] Eisner F, Jacob P, Frick JS, Feilitzsch M, Geisel J, Mueller MH, Küper MA, Raybould HE, Königsrainer I, Glatzle J. Immunonutrition with long-chain fatty acids prevents activation of macrophages in the gut wall. *J. Gastrointest Surg*. 2011 May;15(5):853-9. Epub 2011 Mar 8.

[169] Spruss A, Bergheim I Dietary fructose and intestinal barrier: potential risk factor in the pathogenesis of nonalcoholic fatty liver disease. J. Nutr. Biochem. 2009, 20:657-662.

[170] Wehkamp J, Fellermann K, Herrlinger KR, Bevins CL, Stange EF. Mechanisms of disease: defensins in gastrointestinal diseases. *Nat. Clin. Pract. Gastroenterol Hepatol.* 2005;2:406–415. doi: 10.1038/ncpgasthep0265.

[171] Sauter NS, Schulthess FT, Galasso R, Castellani LW, Maedler K. The antiinflammatory cytokine interleukin-1 receptor antagonist protects from high-fat diet-induced hyperglycemia. *Endocrinology*. 2008;149:2208–2218. doi: 10.1210/en.2007-1059.

[172] Lê KA, Ith M, Kreis R, Faeh D, Bortolotti M, Tran C, Boesch C, Tappy L. Fructose overconsumption causes dyslipidemia and ectopic lipid deposition in healthy subjects with and without a family history of type 2 diabetes. *Am. J. Clin. Nutr.* 2009;89:1760–1765.

[173] World Cancer Research Fund/American Institute for Cancer Research. Food, Nutrition, Physical Activity, and the Prevention of Cancer: A Global Perspective. http://www.dietandcancerreport.org/.

[174] Miller PE, Lesko SM, Muscat JE, Lazarus P, Hartman TJ. Dietary patterns and colorectal adenoma and cancer risk: a review of the epidemiological evidence. *Nutr Cancer.* 2010;62:413–424. doi: 10.1080/01635580903407114.

[175] Randi G, Edefonti V, Ferraroni M, La Vecchia C, Decarli A. Dietary patterns and the risk of colorectal cancer and adenomas. *Nutr. Rev.* 2010;68:389–408. doi: 10.1111/j.1753-4887.2010.00299.x.

[176] Chafen JJ, Newberry SJ, Riedl MA, Bravata DM, Maglione M, Suttorp MJ, Sundaram V, Paige NM, Towfigh A, Hulley BJ, Shekelle PG. Diagnosing and managing common food allergies: a systematic review. *JAMA.* 2010;303:1848–1856. doi: 10.1001/jama. 2010.582.

[177] Tack GJ, Verbeek WH, Schreurs MW, Mulder CJ. The spectrum of celiac disease: epidemiology, clinical aspects and treatment. *Nat Rev Gastroenterol Hepatol.* 2010;7:204–213. doi: 10.1038/nrgastro.2010.23.

[178] Tan S, Tillisch K, Bolus SR, Olivas TI, Spiegel BM, Naliboff B, Chang L, Mayer EA. Traditional Chinese medicine based subgrouping of irritable bowel syndrome patients. *Am. J. Chin. Med.* 2005;33:365–379. doi: 10.1142/S0192415X05002989.

[179] Chopra A. Ayurvedic medicine and arthritis. *Rheum. Dis. Clin. North Am.* 2000;26:133–144. doi: 10.1016/S0889-857X(05)70127-7.

[180] Swidsinski A, Ladhoff A, Pernthaler A, Swidsinski S, Loening-Baucke V, Ortner M, Weber J, Hoffmann U, Schreiber S, Dietel M, Lochs H. Mucosal flora in inflammatory bowel disease. *Gastroenterology.* 2002;122:44–54. doi: 10.1053/gast.2002.30294.

[181] Kim Y, Kim SH, Whang KY, Kim YJ, Oh S. Inhibition of Escherichia coli O157:H7 attachment by interactions between lactic acid bacteria and intestinal epithelial cells. *J. Microbiol. Biotechnol.* 2008;18:1278–1285.

[182] Ukena SN, Singh A, Dringenberg U, Engelhardt R, Seidler U, Hansen W, Bleich A, Bruder D, Franzke A, Rogler G, Suerbaum S, Buer J, Gunzer F, Westendorf AM. Probiotic Escherichia coli Nissle 1917 inhibits leaky gut by enhancing mucosal integrity. *PLoS One.* 2007;2:e1308. doi: 10.1371/journal.pone.0001308.

[183] Qin H, Zhang Z, Hang X, Jiang Y. L. plantarum prevents enteroinvasive Escherichia coli-induced tight junction proteins changes in intestinal epithelial cells. *BMC Microbiol.* 2009;9:63. doi: 10.1186/1471-2180-9-63.

[184] Schlee M, Wehkamp J, Altenhoefer A, Oelschlaeger TA, Stange EF, Fellermann K. Induction of human β-defensin 2 by the probiotic Escherichia coli Nissle 1917 is mediated through flagellin. *Infect Immun.* 2007;75:2399–2407. doi: 10.1128/IAI.01563-06.

[185] Möndel M, Schroeder BO, Zimmermann K, Huber H, Nuding S, Beisner J, Fellermann K, Stange EF, Wehkamp J. Probiotic E. coli treatment mediates antimicrobial human β-defensin synthesis and fecal excretion in humans. *Mucosal Immunol.* 2009;2:166–172.

[186] Kudsk KA, Li J, Renegar KB.*(*1996*)* Loss of upper respiratory tract immunity with parenteral feeding. Ann. Surg 223*:*629–635.

[187] Kudsk KA, Croce MA, Fabian TC, Minard G, Tolley EA, Poret HA, Kuhl MR, Brown RO.Enteral versus parenteral feeding. Effects on septic morbidity after blunt and penetrating abdominal trauma. Ann. Surg. *1992;* 215*:*503–511.

[188] Moore FA, Feliciano DV, Andrassy RJ, McArdle AH, Booth FV, Morgenstein-Wagner TB, Kellum JM Jr., Welling RE, Moore EE. Early enteral feeding, compared with parenteral, reduces postoperative septic complications. The results of a meta-analysis. Ann. Surg.*1992; 216:172–183.*Medline.

[189] Singh G, Ram RP, Khanna SK. Early postoperative enteral feeding in patients with nontraumatic intestinal perforation and peritonitis. J. Am. Coll Surg. *1998; 187:142–146.*

[190] Patterson M , Bloom SR Gardiner J. V. Grelin and appetite control in human- potential application in the treament of obesity 2011 Doi 10.1016/ jpeptides 2011.07.021.

[191] Alscher KT, Phang PT, Mcdonald et.al., Enteral feeding decreases gut apoptosis permeability and lung inflammation during murine endotoxemia *AJP-G*I ,2001.81:2,G569 –G576.

[192] Modena JT , Cavasco LB, Basto CA Vicuña AO, Ramírez MP. Total enteral nutrition as prophylactic therapy for pancreatic necrosis infection in severe acute pancreatitis. *Pancreatology.* 2006 6:1-2.58-64.

[193] Torii K, Mimura T, Yugari Y. *Biochemical mechanism of umami taste perception and effect of dietary protein on the taste preference for amino acids and sodium chloride in rats.* In*:* Kawamura Y, Kare MR, eds. Umami: a basic taste. New York, NY*:* Marcel Dekker, 1987*:*513–63.

[194] Waitzberg DL, Torrin RS . Fish Oil lipid emulsion and immune response : what Clinicians need to know. *Nutr.Clin.. Pract.* 2009 .24:4.487-99.

[195] Féart C, Torrès MJ, Samieri C, Jutand MA, Peuchant E, Simopoulos AP, Barberger-Gateau P. Adherance to a Meditterranean diet and plasma fatty acids : data from Bordeux sample of the three- city study. *Br. J. Nut.* 2011.106,1:149-58.Epub 2011, Feb 8.

[196] Swidsinski A, Ladhoff A, Pernthaler A, Swidsinski S, Loening-Baucke V, Ortner M, Weber J, Hoffmann U, Schreiber S, Dietel M, Lochs H: Mucosal flora in inflammatory bowel disease. Gastroenterology 2002, 122:44-54.

[197] Deane A, Chapman MJ, Fraser RJ Bryant LK, Burgstad C, Nguyen NQ. Mechanisms underlying feed intolerance in critically ill: Implications for treatment. *World. J. Gastroenterology*.2007;13.29: 3909-3917.

[198] Tokunaga M,Takahashi T, Singh RB, Niaz MA, De Meester F, Wilson DW. New developments in coenzyme Q10 research contributed by a single group.Editorial. The Open Nutra J 2013;6:113-116.

[199] Hristova K, Nakaoka T, Otsuka K, Fedacko J, Singh R, Singh RB et al. Perspectives on chocolate consumption and risk of cardiovascular disease ad cognitive function. The Open Nutr J 2012;5:207-212.

[200] Singh RB, Takahashi T, Nakaoka T, Otsuka K, Toda E, Shin HH, Kyu Lee M, Beeharry V, Hristova K, Fedacko J, Pella D, De Meester F, Wilson DW, Juneja LR. Nutrition in transition from Homo sapiens to *Homo economicus*. The Open Nutra J 2013;6:6-17.

In: Cardiovascular Health and Chronomics
Editor: Ram B. Singh
ISBN: 978-1-62948-976-6

Chapter 9

EVOLUTION: A MODIFIED SYNTHESIS IN RELATION TO INHERITANCE OF DISEASES

R. B. Singh[1*], Fabien De Meester[1], Agnieszka Wilczynska[1], D. W. Wilson[2] and A. P. S. Hungin[2]
[1]The Tsim Tsoum Institute, Krakow, Poland,
[2]School of Medicine and Health, Durham University, UK

INTRODUCTION

These authors have discussed, for the last two decades, those environmental factors in general and nutrition in particular could be important in the pathobiology of epigenetic inheritance [1-4]. This knowledge was unknown to Julian Huxley who published his landmark Monograph; 'Evolution: The Modern Synthesis in 1942'. This monograph brought Darwin's ideas into the 20th century and incorporated a knowledge of genes that was emerging in this century in the light of Gregor Mendel's experiments on inheritance (' Monograph: *Experiments with Plant Hybrids)* at Hynčice (Vražné) in the now Czech Republic. In the mid-century, Barbara McClintock discovered transposable elements where parts of the genome can jump around and cause mutations or alter the gene expression, skewing Mendelian ratios and inheritance patterns. This report disrupted the predictable Mendelian system that went into building the Modern Synthesis.

In The Origin of Species, Darwin wrote: "Natural selection can act only by taking advantage of slight successive variations; it can never take a leap, but must advance by the shortest and slowest steps". It is possible, he believed that organisms gradually adapt to their environments *via* minute physical adjustments. Biologists have since found evidence that dramatic adaptations can also occur, for instance in the form of major morphological changes. These adaptations appear to be the cause of changes in structure from apes to man and various human races. It is not clear whether such adaptations may be responsible for the development of diseases or health or for development of man to humans.

*Corresponding author: Dr Ram B Singh, The TsimTsoum Institute, Ul.Golebia 2, 31-007, Krakow, Poland, Email: rbs@tsimtsoum.net, tel 0091 591 2417437.

The Modern Synthesis emphasizes that "populations containing some level of genetic variations evolve *via* changes in gene frequency, induced mostly by natural selection and phenotypic changes are gradual with speciation and diversification into higher taxonomic levels come about over long periods of change. It seems that these ideas remained largely unchallenged for more than 50 years from now, although phenotypic changes have occurred rapidly in the last 100 years [1-6]. Furthermore, evolutionary biology, biochemistry, genomics, developmental biology, systems biology and the impact of the environment on genes concerning mechanism of evolution have grown significantly beyond the Modern Synthesis. It is possible that some lineages have greater 'evolvability' than others, independent of how much baseline genetic variation is present. In this connection, heritable phenotype variations may depend on the biology and biochemistry of the genes. Since some populations have more genetic variation than others and are expected to generate phenotypic variation more rapidly, the Modern Synthesis addresses 'evolvability' in a population which may not be a distinct trait, independent of genetic variation. The Modern synthesis also does not place emphasis on the influence of photosynthesis, flight, multicellularity which stands out against a backdrop of slow and steady evolution. It is therefore important that Modern Synthesis should be modified by the experts because Life Sciences are dynamic, and which continue to change.

The ability to evolve at a different speed than other species needs some particular characteristics. A prion that results from the mis-folding of the Sup35 protein in the yeast; *Saccharomyces cerevisiae*, may serve as a conduit for evolution of novel traits and as molecular vehicle for evolvability. The functional domain of Sup35 is highly conserved in a variety of organismal groups which serves as a translation termination factor. It may also help ribosomes to recognize stop codons on mRNA and thus mediates the normal translation of proteins. The yeast cells, which have aggregates of prions, read through about 5 to 10 % of stop codons in a given cell, so the mis-folded prions are not able to function correctly. The mRNA may stick around longer in cells enabling the expression of more proteins, so the cells with prions can express normally silent sequences beyond the termini of genes or express different levels of normal proteins. These cells are capable of expressing a wide variety of phenotypes [6]. This study [6] found that nearly half of the conditions having the prion (PSI) led to significant phenotypic effects in some of the strains, which uncovered previously-hidden phenotypic variation in the yeast cells. This variation was advantageous in some of the conditions to which yeast cells were subjected. It is possible that prion (PSI) may act as capacitor and potentiator of 'evolvability', because switching into the PSI state sensitizes the yeast more efficiently to produce phenotypic diversity, with change in nutrition and other environmental conditions [6].

The low ω-6/ω-3 fatty acid ratio and coenzyme Q10 present in the cell membrane to which humans adapted during evolution, may have beneficial effects on this phenotypic diversity. The prion (PSI) can pass from mother to daughter yeast cells when they divide either mitotically or meiotically despite lineage where they revert back to non-prion state selection, resulting into beneficial adaptations [6]. Depending on the strain, this occurs naturally at every 100,000-1000,000 cell divisions or so depending on the strain. There are more extreme phenotypes, if the cell is stressed-out of an organism, compared to a benign environment which may possibly be independent of natural selection but this is not yet established. It is also possible that "natural selection" may not actually be "natural" but it is our ignorance about the environmental factors that are causing phenotypic variations, which would be clearer from the evidence on the facilitation of genetic variation.

FACILITATED GENETIC VARIATION

A complex set of physical and chemical factors, coming into play during development , which can influence structures and functions that are beyond simple genome to phenotype translation, may be called facilitated variation. Several mechanisms have been proposed to explain facilitated variations including those related to oscillations of certain regulatory elements affecting segmentation in embryo and chemicals acting during development which can cause patterns of stripes or spots in the organism. In this connection, the role of nutrients causing birth defects and other biochemical phenotypes appear to be important. It seems that nutritional status of mothers and the psychological environment during the perinatal period could be important influencers in the facilitated variations. These variations may spark faster evolutionary alterations compared to random mutations, because developmental changes can create additional phenotypes upon which selection can act. However, many of these random mutations may be due to less known environmental factors and astrochronobiological factors like solar activity, geomagnetic forces and nutrients excess or deficiency, developing according to time structures.

The Modern Synthesis emphasizes that genotypes translate directly into phenotypes and evolutionary changes stemmed from the slow, gradual accumulation of the random genetic mutations. It seems that this simple explanation is quite complex in view of the developmental biology and biochemistry and astrochronobiology in the light of nutritional content of the cell. The Modern Synthesis ignores the complex gene interactions and sudden morphological reorganizations that occur during development, because in experiments and natural setups, there is discordance between genotype and phenotype. In one study, house finches which were natives to deserts began spreading throughout the United States in the 1940s through the pet trade [7]. The birds were tracked for 19 generations over a span of 15 years at a study site in Montana which showed that the population of birds was developing unique beak morphologies as adaptations to the new environment related to availability of foods, at a surprisingly rapid rate [7].This finding is also against the view expressed in the Modern Synthesis that it may be a random mutation. This evidence clearly shows the capability of the flinch to adapt rapidly to survive, against the environmental factors related to housing, the new habitats. The interacting embryonic processes resulted in an initial level of phenotypic variation greater than that predicted from underlying genotypic variation [7]. The selection was essentially blind, during development of the egg, to the creation of the initial pool of phenotypic variation because it occurred only later when young birds began feeding on the foods available in their new habitat. The selection was able to determine which beaks were more or less suited to the environment which was related to the availability of foods, without looking into developmental process, by which this beak was produced, due to opportunity for diversification. It is not clear as to what extent the quality of the food or method of eating or nutrient content was responsible for phenotypic variation.

From Environmental Cycles to the Non-Origin of the Universe (Fabien De Meester 2011 Personal Communication)

If one agrees that the East could not be wrong with the yin-yang conceptual representation of the Universe and everything it contains, and that the West must be right with the Theory of Evolution and everything it selects, then the notion of origin is irrelevant as "Everything" is about revolving "Nothing". Human can make the difference.

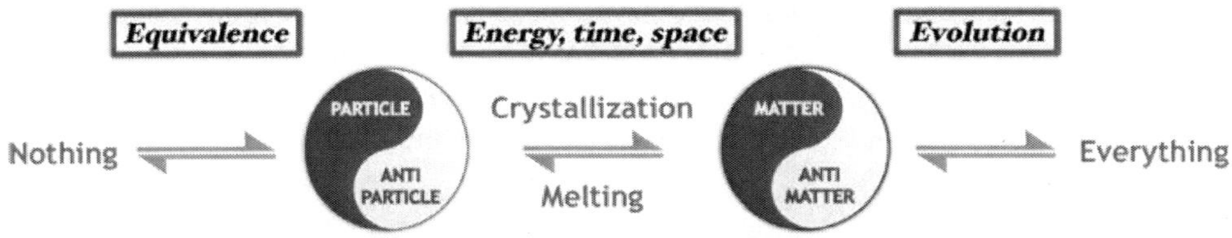

The representation of nothing as yin-yang interaction between an elementary particle and its anti-particle counterpart. In a pure state of nothing, crystallization is spontaneous and exothermic. It leads to matter with release of energy, space and time. Natural selection does the rest, with matter and anti-matter submitted to a single rule, the fitter the better. Gravity is a manifestation of anti-matter and remains proportional to matter at all times. The system is entirely reversible with no time/space constraints.

Starting from that advanced though basic evidence-based scientific, philosophical and theological merged principle, one can analyze the world we live in with wisdom and rejuvenated curiosity as Human; body and mind are naturally selected advanced interacting components reminiscent of earlier more basic assemblages of matter and anti-matter, which evolved independently according to the same rule, i.e., Natural Selection. As Man shares a common ancestor with Stone that fills in the Universe, one perceives the true nature of matter, which naturally evolved to the complex human genome. Anti-matter is no more complicated to perceive; it started with gravity as its most basic anti-space manifestation in Stone and naturally evolved to mind in its most sophisticated anti-genome manifestation in Human.

The non-origin hypothesis of the Universe does require additional validation studies that keep on-going at all levels, i.e., scientific, philosophical and theological. So far, it is an extrapolation of common knowledge. It helps though at taking today the right decision for the future of Humanity; (i) at scientific level, it appears that dark holes are of infinite low – not high – density, (ii) at philosophical level, it sounds like the aim of a human lifetime is to create Value, and (iii) at theological level, Human appears as a promising alternative concept to God.

In fact, Nature always keeps trace of its past and current man affairs follow the same cyclic, repeatable pattern of development as all phenomena in the entire Universe. By creating Value, Humans have the unique ability to counteract Nature and the today predictable future collapse of the Universe. The size of the challenge is to the height of Humanity as a Talent.

Multilevel Epigenetic Inheritance

Passing of phenotypic change to subsequent generations in ways that are outside the genetic code of DNA are described as multilevel inheritance. It is known that chromatin structure, remodeled histone proteins or methylated DNA, often mediated by environmental factors, can pass from parent to offspring without changing the actual sequence of the inherited genome. The Modern Synthesis proposed that genes were the primary units of inheritance and that evolution may be a change in the genetic composition of the population, although molecular mechanisms were unknown in 1942. If epigenetics play any role in evolutionary change, we must demonstrate that the changes last and are stable and cause heritable effects through several generations [8]. Two groups of genetically identical Arabodopsis plants (*Brassicaceae*) were exposed to either hot or cold conditions for two; P and F1 generations in an experiment [9]. The next generation; F2 from both groups was grown at normal temperatures, but the offspring (F3) from both groups were grown in either hot or cold conditions. The F3 plants grown in hot conditions and descended from P and F1 plants also grown in hot conditions produced five-fold more seeds compared to F3 plants grown in hot conditions but descended from cold treated ancestors ([9]. It seems that epigenetic factors affecting flower productions and early stage seed survival, due to molecular adaptations, in these plants may have caused these mutations.

Can Nutrient Influence Inheritance?

The poster child for epigenetic changes in mammals is the yellow agouti mouse, an epigenetic biosensor for nutritional and environmental changes. These fat and yellow mice owe their appearance to epigenetic modification that removes methyl groups from the normally methylated agouti gene [10]. In a developing mouse fetus, if the above modification occurs shortly after fertilization, the baby mouse may exhibit the yellow fur and obese phenotype with greater risk of developing cancer and diabetes [10]. However, the genetic code remains unchanged from normal mice. In one study, Waterland and Jirtle [10] altered the nutrient intake to serve as methyl group donors in mouse mothers, to cause methylation or demethylation of the agouti gene. Increased supplementation of choline, betaine, folic acid and vitamin B12 in the diet of pregnant yellow agouti mice was able to decrease the incidence of deleterious phenotypes in offspring, by allowing for the remethylation of the agouti gene. If these mice be born with the agouti phenotype, they can pass that deleterious epigenetic trait in their offspring, regardless of their diet during pregnancy. This landmark study indicate that nutrients can cause phenotypic changes which can pass on through cell division and mating to the offspring due to their possible influence on (natural) selection [10].

It is possible therefore to say; that we are what we eat and what our parents ate, and potentially what our grand parents ate which would be modification of the old Sanskrit saying '*Aham Annam'* from the ancient Vedas (5000BCE). It may be also important to emphasize advances in Astrochronobiology particularly; time of eating, accordingly, therefore we are not only 'what we eat' but also 'we are when we eat' and when our father and grandfather were eating [11]. There is a need to study the effects of low ω-6/ω-3 ratio, arginine, taurine, cysteine, coenzyme Q10 on the remethylation of the agouti gene and their effects on phenotypic variations. However this mode of inheritance needs to penetrate more than a few generations before it earns a place in evolutionary concept. Epigenetic inheritance appears to be widespread but it does not mean that it lasts and causes 'evolutionarily' important effects. It is not clear until now, that if epigenetic changes are not stable for 20 to 30 generations, it would be relevant to evolution.

The biologists should be bold enough to revise the Modern Synthesis published in 1942, because several experiments related to epigenetics are underway to establish the role of environmental factors in inheritance. There is need for modifications of this landmark monograph which may be written by multiple authors including physicians.

Observing inheritance in 3 to 4 or more generations among patients of diabetes, hypertension and coronary artery disease. The evolutionary theory appears to be quite flexible to incorporate well-substantiated new concepts, which continue to arise during the last few decades. In 1975, we (RBS) treated a patient of type 2 diabetes mellitus whose father also had a history of diabetes mellitus and died a sudden death. This 48 years old patient was, while under treatment (RBS) developed tightness in the chest and his electrocardiogram showed changes indicating myocardial ischaemia. He was treated and recovered but after a few months died a sudden cardiac death in 1977. His son who was about 40 years, presented with type 2 diabetes mellitus, while under treatment, also died a sudden cardiac death next year. Diet and lifestyle history showed that they have /had a wholesale business of selling clarified butter (anhydrous milk fat) during the time of the 1st generation, clarified butter and hydrogenated fat, during the 2nd generation which continued during the 3rd generation. Thus, albeit a single family, all the three generations were sedentary and were consuming more than 40% of calories from clarified butter which may have caused inheritance of deleterious phenotypes in the offspring, resulting in to type 2 diabetes and sudden death.

The adverse effects of trans fat (unsaturated fats with trans-isomer fatty acids). and oxidized cholesterol present in the clarified butter consumed by the last three 'degenerations', on proinflammatory cytokines are well known and inflammation has been demonstrated to cause genetic damage. Sedentary behavior also enhances inflammation by its adverse effects on adiponectin, leptin and brain derived neurotrophic factor which are anti-inflammatory. Inflammation could be important in the pathobiology of epigenetic inheritance. Now RBS is tracking the son (4th generation) who is about 30 years, to look for type 2 diabetes mellitus and coronary artery disease who may be a future candidate for sudden cardiac death in the 4th degeneration because he and his baby are consuming clarified butter and high ω-6/ω-3 ratio diet by using sun flower oil and they continue to be sedentary. We need to study genetic markers in this 4th generation to support this view. However, we need many more such observations and research because sufficient examples of transgenerational epigenetics are lacking. Most of the time, epigenetic characters are not inherited past one or two generations.

The Evolution of Diet and the Tsim Tsoum Concept

A fundamental view of evolution in relation to human development from apes to man has been that the genetic makeup of contemporary humans which shows minor difference from that of the modern humans who appeared in Africa between 100,000 and 50,000 years ago [12, 13]. The human evolution has been comparatively rapid during the past 50,000 years, as revealed by the molecular geneticists. There have been marked changes in the food supply with the development of agriculture about 10,000 years ago from now. However, only non-significant change in our genes occurred, during the past 10 centuries, due to presence of ω-3 fatty acids, amino acids, vitamins and minerals in the diet and non-significant changes in the environment [12-15]. There have been some structural modifications in individual DNA sequences, and altered gene regulation has been the dominant mechanism involved. Increased evolutionary rapidity in humans compared with rates for other primates, has resulted from unprecedented demographic expansion, which has provided a far larger pool of mutations, upon which natural selection can operate. The human Diaspora which has exposed humans to environmental changes appears to be quite different from those of their ancestral land in Africa and this rapid change may predispose deleterious epigenetic inheritance.

The spontaneous mutation rate for nuclear DNA is estimated at 0.5% per million years. Hence, over the past 10,000 years there has been time for very little change in our genes, possibly 0.005%. Our genes appear to be similar to the genes of our ancestors during the Paleolithic period 40,000 years ago, the time when our genetic profile was established. Man appears to live in a nutritional environment which completely differs from that for which our genetic constitution was selected. However, it was only during the last 100-160 years that dietary intakes and environment have changed significantly, causing increased intake of saturated fatty acids (SFA) and linoleic acid, and decrease in ω-3 fatty acids, from grain fed cattle, tamed at farmhouses, rather than meat from running animals. There is marked decrease in the intake of vitamins and antioxidants. The food and nutrient intake among hunter-gatherers and during the Paleolithic period showed no remarkable differences. However, during the last 160 years, there is marked reduction in consumption of ω-3 fatty acids, vitamins and minerals and proteins and significant increase in the intakes of carbohydrates, (mainly refined,), fat (saturated, trans fat, linoleic acid) and salt compared to Paleolithic period [1-4,12-15]. There are also marked changes in the environmental factors which may also have deleterious effects on epigenetic inheritance.

The Columbus concept of diet means that humans evolved on a diet that was low in saturated fat and the amount of ω-3 and ω-6 fatty acids was quite equal.[14] Nature recommends to ingest fatty acids in a balanced ratio (polyunsaturated:saturated: ω-6:ω-3=1:1) as part of dietary lipid pattern in which monounsaturated fatty acids is the major fat(P:M:S=1:6:1).These ratios represent the overall distribution of fats in a natural untamed environment. (www.columbus-concept.com). The Columbus foods include egg, milk, meat, oil , and bread, all rich in ω-3 fatty acids, similar to wild foods, consumed about 160 years ago from now. Blood lipid composition does reflect one's health status: (a) circulating serum lipoproteins and their ratio provide information on their atherogenicity to blood vessels and (b) circulating plasma fatty acids, such as ω-6/ω-3 fatty acid ratio, gives an indication of proinflammatory status of blood vessels. (a) and (b) are phenotype-related and depend on genetic, environmental and developmental factors. As such, they appear as universal markers for holistic health. Blood cholesterol is central to this approach. Its 3D-representation shows how circulating lipoproteins affect blood vessel integrity arising from their circulating throughout the body. Of major importance appear to be the essential dietary nutrients (essential amino acids, fatty acids, antioxidant vitamins and minerals) and the functional component of the regimen(diet, sport, spiritualism, etc.) because these factors may be important in the epigenetic inheritance . An example is given of an essential dietary nutrient and of a functional component of man's regimen that affect health in a predictive way derived from the 3D representation of blood cholesterol. Caption The Columbus Concept and its 3D representation of blood lipoprotein behaviours. "Bad" LDL-C, "good" HDLC, and "healthy" LDL-CC:HDL-CC ratios. CC= Columbus Concept. The TsimTsoum Concept is an extension of the Columbus concept. The word TsimTsoum is derived from Hebrew and it is similar to 'ying yang' in Chinese. (http://www.tsimtsoum.net/ editorials/tsimtsoum_editorial_2009-Kosice-14th-WCCN-and-5th-ICCD.pdf). It includes the simultaneous approach of controlling of Mind-brain- body connection and interactions of leptin, cholecystokinin, polyunsaturated fatty acid CoA, brain derived neurotrophic factor and neuropeptide Y secreted in the hypothalamus [16-18]. The TsimTsoum Concept also presumes that if, and only if, diet and environmental factors are designed holistically, for individuals and populations or for any species, for several generations, they may be responsible for the alteration in biological functions causing epigenetic inheritance, resulting in the development of man to human and thence to superhuman. This is possible due to nutritional transition from Homo sapiens to Homo *economicus(19).(Figure 1).*

Figure 1. Effects of diet and lifestyle on body composition from *Homo sapiens* to *Homo erectus* and modern man in indicating body mind index.

In brief, it is possible that diet and other environmental changes, can modulate biological functions including genetic variations. If these environmental changes are subjected to species for several generations, they might predispose epigenetic inheritance. However, more evidence is necessary to demonstrate the role of environment in the epigenetic inheritance. The Modern Synthesis needs modification in the light of above Advances.

Acknowledgments

Due to TsimTsoum Institute, Krakow, Poland and International College of Nutrition for the support to produce this work.

References

[1] Singh, RB, Mori, H. Risk factors for coronary heart disease: synthesis of a new hypothesis through adaptation. *Med. Hypoth*, 1992, 39, 334-341.

[2] Mishra, S, Singh, RB, Dwivedi, SP, De Meester, F, Rybar, R, Pella, D, Fedacko, J, Juneja, LR. Effects of nutraceuticals on gene expression. *The Open Nutra J.*, 2009, 2, 70-80.

[3] Singh, RB. Darwin, evolution and origin of species. *The Open Nutr J.*, 2009, 2, 86-87.

[4] Singh, RB;Moshiri, M; De Meester, F, Juneja, L;Muthusamy, V, Manoharan, S. The evolution of low w-6/w-3 ratio dietary pattern and risk of cardiovascular diseases and diabetes. *JAMR*,2011, 3:45-70.

[5] Verhoeven, KJF, Jansen, JJ, van Dijk, PJ, Biere, A. Stress induced DNA methylation, changes and their heritability in asexual dendilions. *New Phyto*, 2010, 1108-1118.

[6] True, HL, Lindquist, SL. A yeast prion provides a mechanism for genetic variation and Phenotypic diversity. *Nature*, 2000, 407, 477-483.

[7] Badyaey, A. Homeostatic Hitchhiking: A Mechanism for the Evolutionary Retention of Complex Adaptations. *Integr. Comp. Biol.* 2013; 53 (6): 913-922. doi:10.1093/icb/ict084.

[8] Jablonka, F, Raz, G. Transgenerational epigenetic inheritance: prevalence, mechanisms and implications for the study of heredity and evolution. *Q Rev Biol*, 2009, 84, 131-176.

[9] Whittle, CA, Otto, SP, Johnston, MO, Krochko, JE. Adaptive epigenetic memory of ancestral temperature regime in Arabidopsis thaliana. *Botany*, 2009, 87, 650-657.

[10] Waterland, RA, Jirtle, RL. Transposable elements: targets for early nutritional effects on epigenetic gene regulation. *Mol. Cell Biol.*, 2003, 23, 5293-300.

[11] Halberg, F, Cornélissen, G, Otsuka, K, Siegelova, J, Fišer, B, Dušek, J, Homolka, J, Salvador, S, Peña, DL, Singh, RB and the BIOCOS project. Extended consensus on need and means to detect vascular variability disorders (vvds) and vascular variability syndromes (vvss). *Intl. J. of Geronto-Geriatrics*, 2008, 11, 119-146.

[12] Eaton, B. Evolution and cholesterol. *World Rev. Nutr. Diet*, 2009, 100, 46-54.

[13] De Meester, F. Progress in lipid nutrition: the Columbus concept addressing chronic diseases. *World Rev. Nutr. Diet,* 2009, 100, 110-21.

[14] Simopolous, AP. Genetic variation and dietary response: nutrigenetics/ nutrigenomics. *Asia Pac. J. Clin. Nutr.*,2002,11, S117-S128.

[15] Eaton, SB, Eaton, SB III, Sinclair, AJ, Cordain, I, Mann, NJ. Dietary intake of long chain polyunsaturated fatty acids during the Paleolithic period In Simopoulos AP edition. The return of w-3 fatty acids in the food supply. Land based Animal Food Products and their Health Effects. *World Rev. Nutr. Dietetics*, 1998, 83, 12-23.

[16] Kwiatek, AW, Singh, RB, De Meester, F. Nutrition and behaviour: the role of w-3 fatty acids. *The Open Nutra J.*, 2010, 3,119-128.

[17] Thaler, JP; Cummings, DF. Metabolism: food alert. *Nature*, 2008, 452, 941-942.

[18] Wang, PY, Caspi, L, Lam, CK, Chari, M, Li, X, Light, PE, Gutierrez-Juarez, R;Ang, M; Schwartz, GJ; Lam, TK. Upper intestinal lipids trigger a gut-brain-liver axis to regulate glucose production. *Nature,* 2008, 452, 1012-16.

[19] Singh RB, Takahashi T, Nakaoka T, Otsuka K, Toda E, Shin HH, Kyu Lee M, Beeharry V, Hristova K, Fedacko J, Pella D, De Meester F, Wilson DW, Juneja LR. Nutrition in transition from Homo sapiens to *Homo economicus*. The Open Nutra J 2013;6:6-17.

In: Cardiovascular Health and Chronomics
Editor: Ram B. Singh
ISBN: 978-1-62948-976-6

Chapter 10

CUSUM-ASSESSED PERSONALIZED HYPOTENSIVE TREATMENT SURVEILLANCE

D. W. Wilson[1], F. Halberg[2], G. Cornélissen[2], Y. Watanabe[3] and A. P. S. Hungin[1]

[1]Durham University, School of Medicine and Health, Stockton-on-Tees, UK
[2]University of Minnesota, Halberg Chronobiology Center, Minneapolis, MN, US
[3]Tokyo Women's Medical University, Department of Medicine, Tokyo, Japan

ABSTRACT

An optimization approach for essential circadian timing of treatment of high blood pressure, diagnosed as one of the described vascular variability disorders, VVDs, is presented that specifically lowers the MESOR (M=Midline Estimating Statistic Of Rhythm) without inducing any other VVDs, such as CHAT (Circadian Hyper-Amplitude [A] Tension). This has been achieved using Cumulative sums and parameter tests applied to half-hourly automatic around-the-clock measurements for weeks at a time or longer. Hypotensive treatment (Rx) with a popular drug, at the wrong time was associated with CHAT, while the same dose when given at another time to the same patient was beneficial by lowering the M while leaving the 2A unaffected. In conclusion it is best to guard against an iatrogenic, excessive blood pressure overswing, systolic and/or diastolic, and against inducing other VVDs. "First do no harm" by not flying blind, i.e., by adequate surveillance of ambulatory blood pressure and heart rate measurements and concomitant sequential process control combined with statistical parameter testing.

Keywords: Human BP, antihypertensives, rhythmometry, process-control

INTRODUCTION

Process control procedures have been used by these authors to monitor, evaluate, and to exercise control, whencesoever possible, of laboratory and biological assays/measurements using, inter alia, cumulative sum (CUSUM) techniques. These procedures [1-3] have been applied to several concomitant parameters such as a mean response of a measured variate and a co-parameter each measured at the same time point sequentially generally over a frequency e.g., daily, weekly, monthly, that facilitates interpretation notably significant change points. In the case of laboratory measurement it is often the mean, imprecision and drift across the assay batch that is monitored over time. In this paper we demonstrate the importance of monitoring time series parameters for circadian ambulatory blood pressure that are in the armamentarium of the chronobiologist, specifically those developed at the University of Minnesota [3-5]. It is important to optimize the circadian timing of the treatment of high blood pressure, diagnosed as one of the Vascular Variability Disorders (VVDs) such as those shown in Figure 1.

These time series parameters include the MESOR (M=Midline Estimating Statistic Of Rhythm); the amplitude (A) (or the double amplitude, 2A, a measure of the total swing predictable by the fitted cosine curve); the timing of the peak of the rhythm, called the acrophase; the period; and other parameters such as pulse pressure and the standard deviation of heart rate variability.

Surveillance of these parameters over time can be used to assess whether subjects become normotensive following treatment or indeed whether the timing, dose and type of anti-hypertensive drug are adequate.

As often happens, for example, with laboratory tests, intervention may change more than one parameter. We focus on the premise that control of blood pressure does indeed contribute towards beneficial outcome, but the drug itself may also contribute problems in their own right.

This is evident, for example, from the work of Otsuka et al. [6] who followed-up treated and untreated patients for 6 years and recorded the incidence of adverse outcomes.

Other potential factors [7] are also important. A recent publication by Rothwell et al. [8] has focused on visit-to-visit variability and a much broader spectrum of rhythms and to that extent predictable variation has been recorded at the Halberg Center for a number of years.

The implementation of blood pressure data capture strategies, and the use of on-going statistical analyses using automated procedures, will probably guard against potential iatrogenic effects. Otherwise non-surveillance will fail to detect an excessive blood pressure overswing in systolic and/or diastolic pressure and other VVDs. Paraphrased, this means "First do no harm" by not flying blind [11].

Objectives

- To optimize the circadian timing of the treatment of high blood pressure, diagnosed as one of the Vascular Variability Disorders (VVDs), namely as MESOR-Hypertension (M=Midline Estimating Statistic Of Rhythm);
- To achieve this optimization of M-lowering without inducing any other VVDs, such as CHAT (Circadian Hyper-Amplitude-Tension), an odd timing (circadian ecphasia), an odd period or frequency (circadian ecfrequentia), etc.

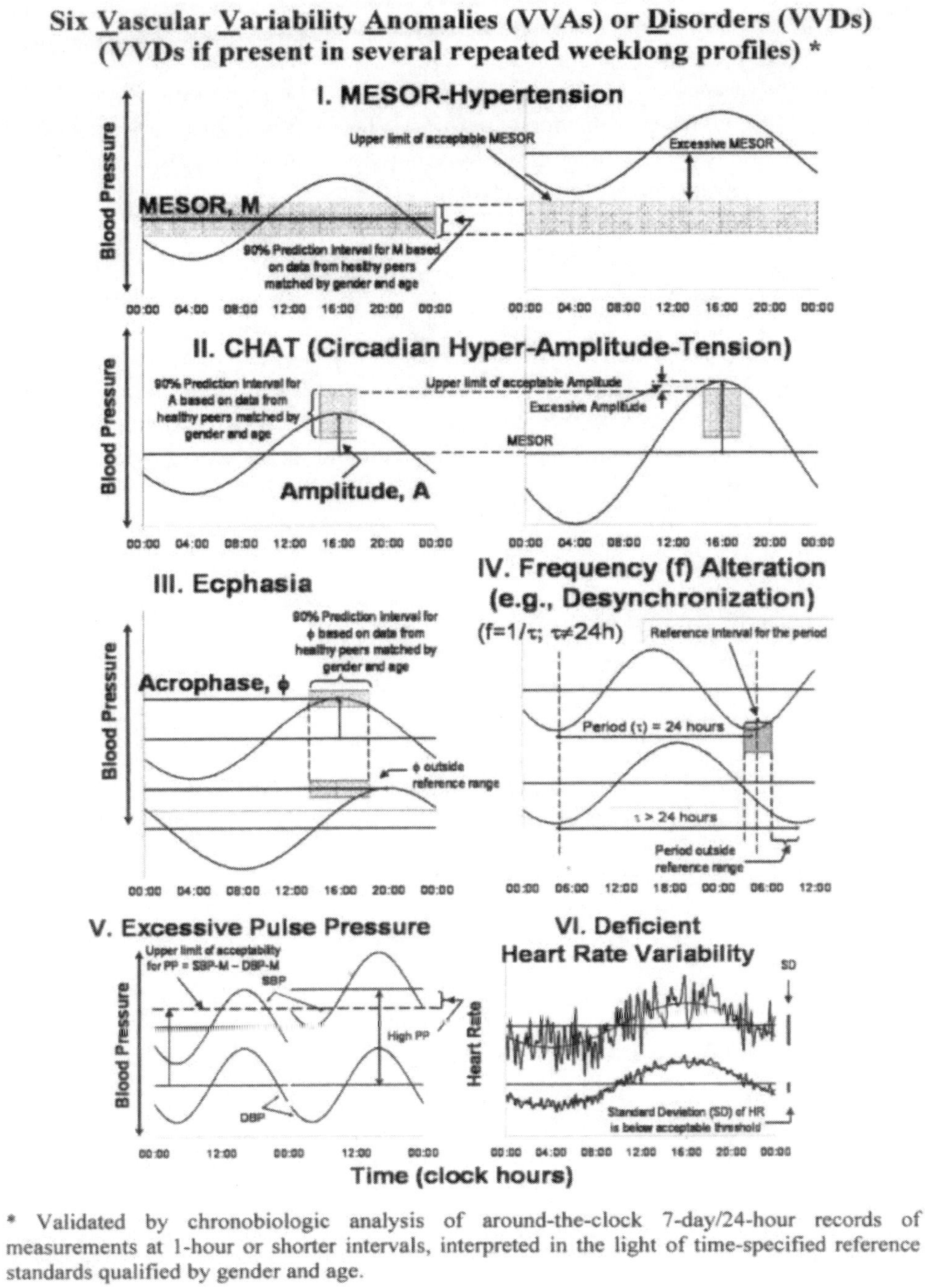

Figure 1. Abstract limits for acceptable I. MESOR, II. amplitude, III. acrophase, IV. frequency (period), V. pulse pressure and VI. heart rate variability associated with blood pressure and heart rate surveillance and their corresponding VVD descriptions. © Halberg.

METHODS

Data from the monitoring of blood pressure and heart rate have been analyzed by model fitting and stacking, with particular focus on the VVDs (MESOR-hypertension and CHAT) that have been controlled using CUSUM techniques. Particularly relevant is that the avoidance of hypotension does not occur at the expense of increased risk caused by an adversely co-varying VVD such as an excessive circadian amplitude.

Data Collection

Figures 2 and 3 show the time course of the circadian systolic blood pressure MESOR and amplitude in the light of CUSUMs (blood pressure and heart rate data that were collected using automatic ambulatory equipment half-hourly around-the-clock for weeks at a time or longer for each timespan). Importantly, treatment was systematically changed in its circadian timing.

Data Analysis and Interpretation

Blood pressure data and heart rate variability were analysed and interpreted using the sphygmochron [12]. Based on cosinor methodology [4-5] and age- and gender-qualified reference ranges, the sphygmochron is a chronobiologically interpreted blood pressure and heart rate monitoring profile over time, with a computer comparison of model- and stacking-derived with individualized and/or group limits. Cumulative sums [3] and parameter tests [5] were applied to the data.

RESULTS

Figure 1a-f shows the circadian MESOR, amplitude, acrophase, period, pulse pressure and heart rate variability, most with their statistical confidence limits, and their corresponding VVDs.

Such diagnoses are based on the sphygmochron.

Figures 2 and 3 demonstrate the CUSUM profile when applied to the circadian MESOR and amplitude in a hypertensive patient given no treatment until 00.00 on 10/10/2004. Figure 2 demonstrates a lowering of the MESOR for therapy at different times, but Figure 3 shows that with a certain timing there is an adverse effect of an increased amplitude. Thus, examination of the MESOR's response to hypotensive treatment (Rx) with a popular drug, at the wrong time can be associated with CHAT, while the same dose when given at another time to the same patient was beneficial by lowering the M while leaving the 2A unaffected. This clearly demonstrates the need for ambulatory blood pressure surveillance to optimize circadian timing of hypotensive therapy. These data were not isolated cases. Figure 4 also shows that treatment given at one time of day is harmful, whereas at another time the same dose of the same drug is beneficial.

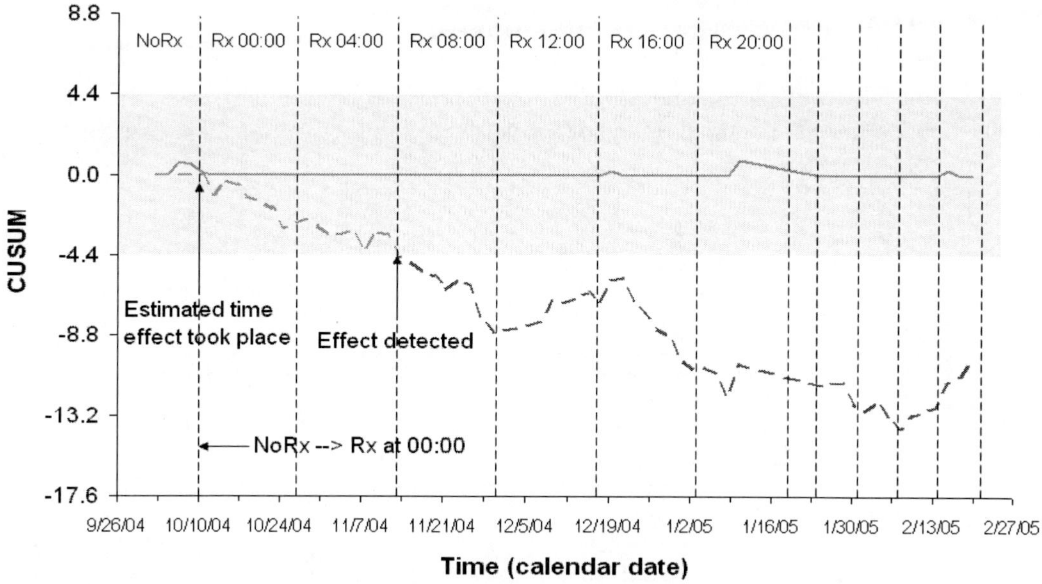

Figure 2. Application of CUSUM procedure to monitor the circadian MESOR of systolic blood pressure following antihypertensive therapy with a popular drug and to determine whether a lowering of the MESOR has occurred by a cumulative sum which reaches values outside a decision interval. © Halberg.

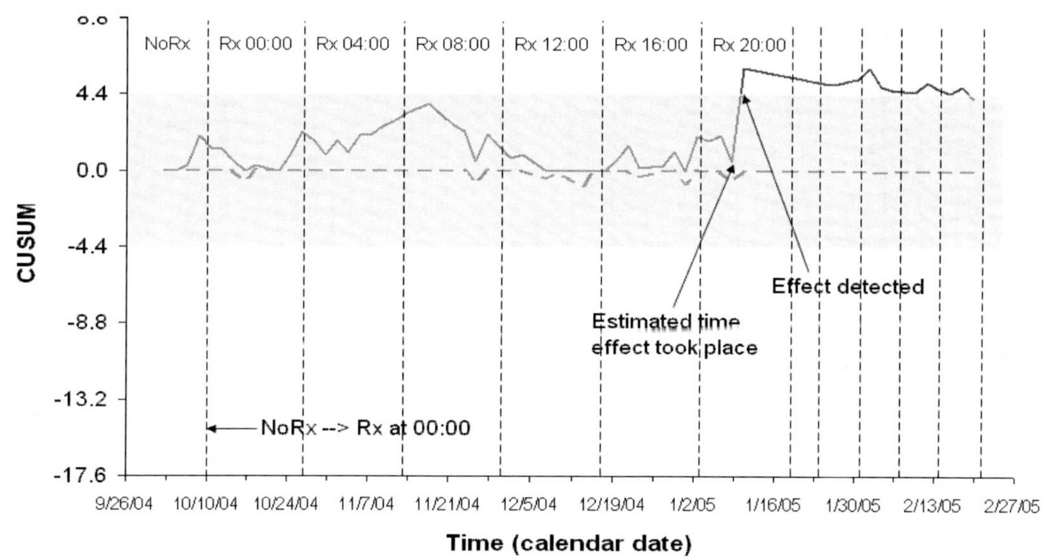

Figure 3. Application of CUSM procedure to monitor the circadian amplitude for systolic blood pressure following antihypertensive therapy with a popular drug and to determine whether an adverse increase in amplitude has occurred as adjudged from the CUSUM which reaches values outside the decision interval. © Halberg.

Discussion

Evidence is available to indicate that blood pressure monitoring surveillance must be individualized since the optimal timing of a given drug can vary from patient to patient.

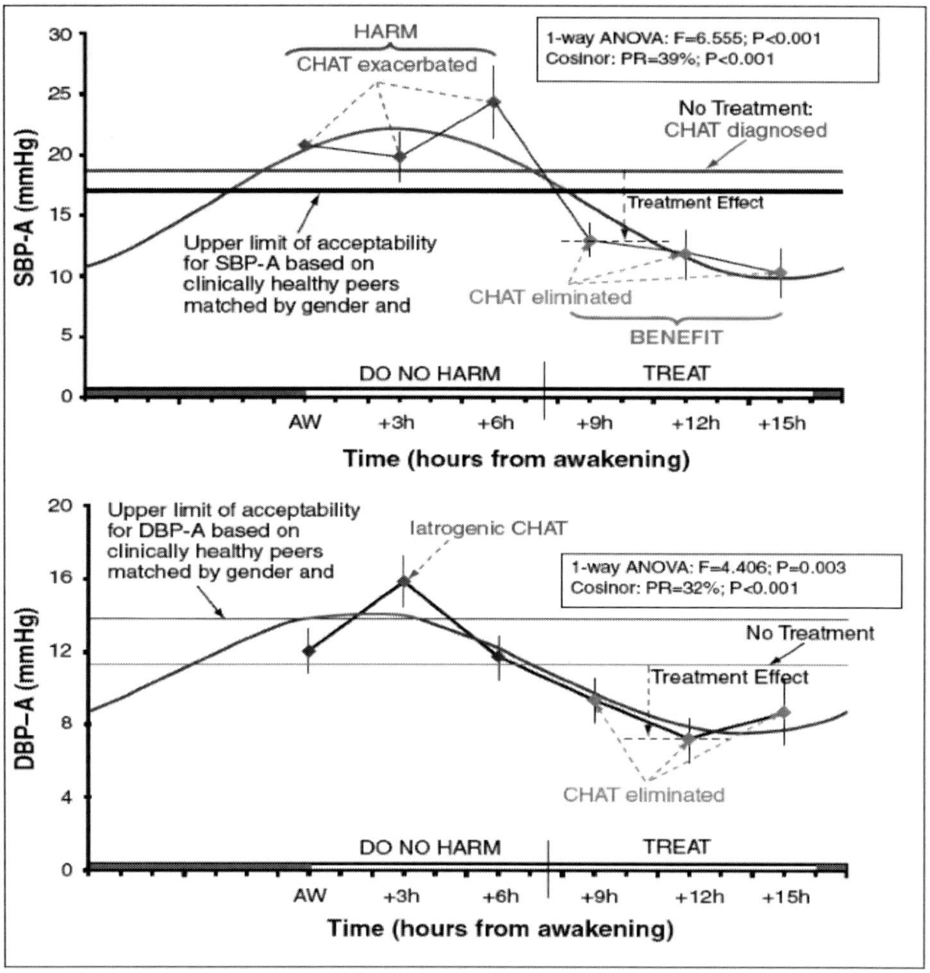

Figure 4. At one administration time (before noon), Hyzaar induces CHAT in diastolic BP and exacerbates a preexisting CHAT in systolic BP (red). At another time of administration, Hyzaar eliminates a pre-existing VVD (green). These opposite effects were found in tests at six medication times, each administered for about a month, with half-hourly surveillance of BP during the last week of each span. These differences occur as a function of the timing of the drug's use along the scale of 24 hours. Original study by Dr Yoshihiko Watanabe. © Halberg.

As this paper has demonstrated circadian timing can be crucial as reduction in MESOR may lead, subject to circadian timing, to an increased amplitude and therefore increased risk of a cardiovascular or cerebrovascular event. It is essential that patients should have their timing and treatments optimized according to their chronodiagnosis and subsequent chronotheranostics.

Individualized monitoring strategies using CUSUM procedures can now be implemented for chronodiagnosis and chronotheranostics for hypertensive patients. The design and evaluation of clinical trials, even for dietary studies, can now be placed on a firm chronobiological surveillance footing.

CONCLUSION

In hypotensive therapy it is important to guard against an iatrogenic, excessive blood pressure overswing in systolic and/or diastolic BP; "First do no harm" by not flying blind [11].

REFERENCES

[1] Rowlands RJ, Wilson DW, Nix ABJ, Kemp KW, Griffiths K. Advantages of CUSUM techniques for quality control in clinical chemistry. *Clinica Chimica Acta* 1980; 108: 393-397.

[2] Wilson DW, Griffiths K, Halberg F, Simpson HW, Griffiths R, Kemp KW, Nix ABJ, Rowlands RJ. Breast skin temperature rhythms in relation to ovulation. *Chronobiologia* 1983; 10: 231-243.

[3] Cornélissen G, Halberg F, Hawkins D, Otsuka K, Henke W. Individual assessment of antihypertensive response by self-starting cumulative sums. *Journal of Medical Engineering and Technology* 1997; 21: 111-120.

[4] Nelson W, Tong YL, Lee JK., Halberg F. Methods for cosinor-rhythmometry. *Chronobiologia* 1979; 6: 305-323.

[5] Bingham C, Arbogast B, Cornélissen Guillaume G, Lee JK, Halberg F. Inferential statistical methods for estimating and comparing cosinor parameters. *Chronobiologia* 1982; 9: 397-439.

[6] Otsuka K, Suzuki M, Omori K, Yamanaka T, Toyoshima T, Shinagawa M, Nakajima S, Kubo Y, Ihbachi N, Watanabe Y, Watanabe H, Kuwajima I, Suzuki Y, Kuramoto K, Kawamura H, Cornélissen G, Halberg F. Chronocardiological assessment of hypertension: a new dimension in diagnosis and treatment. In: Proc. Workshop on Computer Methods on Chronobiology and Chronomedicine, Tokyo, Sept. 13, 1990, Halberg F., Watanabe H. eds., Medical Review, Tokyo, 1992, pp. 55-67.

[7] Williams B. Drug Treatment of Hypertension. Most patients will need a treatment cocktail-including a thiazide diuretic. *BMJ* 2003; 326: 61-62.

[8] Rothwell PM, Howard SC, Dolan E, O'Brien E, Dobson JE, Dahlöf B, Sever PS, Poulter NR. (2010). Prognostic significance of visit-to-visit variability, maximum systolic blood pressure, and episodic hypertension. *The Lancet.* 2010; 375 (9718): 895-905.

[9] Halberg F, Cornélissen G, Otsuka K, Siegelova J, Fiser B, Dusek J, Homolka P, Sanchez de la Peña S, Singh RB, BIOCOS project. Extended consensus on need and means to detect vascular variability disorders (VVDs) and vascular variability

syndromes (VVSs). Geronto-Geriatrics: *International Journal of Gerontology-ChronomeGeriatrics* 2008; 11 (14): 119-146.

[10] Halberg F, Cornélissen G, Otsuka K, Siegelova J, Fiser B, Dusek J, Homolka P, Sanchez de la Peña S, Singh RB, BIOCOS project. Extended consensus on means and need to detect vascular variability disorders (VVDs) and vascular variability syndromes (VVSs). Leibniz-Online 2009; Nr. 5 (http://www2.hu-berlin.de/leibniz-sozietaet /journal/archiv_5_09.html). 35 pp.

[11] Halberg F, Cornélissen G, Wilson D, Singh RB, De Meester F, Watanabe Y, Otsuka K, Khalilov E. Chronobiology and chronomics: detecting and applying the cycles of nature. *Biologist* 2009; 56 (4): 209-214.

[12] Halberg F, Cornelissen G, Halberg J, Bakken E, Delmore P, Wu J, Sanchez de la Pena S, Halberg E. The sphygmochron for blood pressure and heart rate assessment: a chronobiologic approach. In: Medical Monitoring in the Home and Work Environment, Miles L.E., Broughton R.J. (eds.), Raven Press, New York, 1990, pp. 85-98.

In: Cardiovascular Health and Chronomics
Editor: Ram B. Singh

ISBN: 978-1-62948-976-6

Chapter 11

Association of Coronary Risk Factors with Blood Glucose Levels in Type-2 Diabetes Mellitus

K. S. N. Reddy[1,*], T. M. Reddy[2], K. K. Reddy[1], J. Lalu Naik[1], A. Papa Rao[1] and R. B. Singh[3]

[1]Department of Anthropology, Sri Venkateswara University, Tirupati, India
[2]Department of Medicine, S.V.R.R. Government General Hospital and Medical College, Tirupati, India

Abstract

Background and Aim: The aim of the present study is to assess the association of coronary risk factors with glucose levels among patients with type-2 diabetes mellitus.

Subjects and Methods: This study included 270 subjects with type 2 diabetes mellitus. We obtained primary risk factors, disease history, anthropometry, blood pressure, blood glucose and lipids by questionnaires following WHO guidelines and laboratory data by standard methods.

Results: Of 270 patients with type 2 diabetes, 43 percent of males and 50 percent of females are having family history of diabetes mellitus. None of the subjects fall in malnutrition category whereas, 30 percent were normal weight, 50 percent were overweight and 18 to 30 percent were obese. The risk factor prevalence rates are: systolic hypertension; 40%, diastolic hypertension; 60%, hypertriglyceridemia; 55-60% and hypercholesterolemia; 32-37%. Both fasting blood glucose (FBS) and post prandial blood glucose (PPBS) levels increased from<39 yrs age group to 40-54 yrs age group and then decreased in >55 yrs age group ($p<0.05$). Blood glucose levels were unaltered within the categories of BMI, while WHR elevated the FBS levels ($p<0.05$). Both FBS and PPBS elevated from latent diabetics to diabetics with 5-9 yrs duration, there after a decrease was observed in FBS alone. Sixty percent of the variance in FBS levels and 58 percent of the variance in PPBS was explained by independent variables. The ODD's of FBS are: 40-54 yrs age group (1.023, 95% CI: 1.011-1.034), >55 yrs (1.020, 95% CI: 1.009-

* Corresponding Author: Dr. K.K. Reddy, Ph.D, FICN, Division of Human Genetics, Department of Anthropology, Sri Venkateswara University. Tirupati- 517 502, India. Ph: 91 94 904 91 276. E-mail: kanalakr@yahoo.com.

1.032), duration of diabetes for 5-9 yrs (1.024, 95% CI:1.010-1.039), family history (1.084, 95% CI: 1.063-1.106), WHR>1 (1.028, 95% CI: 1.006-1.050), systolic hypertension (1.023, 95% CI:1.016-1.030), diastolic hypertension (1.022, 95% CI: 1.015-1.030). Similar ODD's were obtained for PPBS levels also.

Conclusion: In conclusion the results indicate that family history, duration of the disease and blood pressure are significant risk factors in type-2 diabetes mellitus. Hence preventive strategies warranted towards the management of hyperglycemia in type 2 diabetes to prevent vascular disease.

Keywords: Diabetes, blood glucose, lipids, blood pressure, family history

INTRODUCTION

Diabetes mellitus is a metabolic disorder and its prevalence is an important global health problem [1-4]. In 1985 an estimated 30 million people world-wide had diabetes, and in 2000, there were 171 million cases. This number is projected to increase to about 366 million by 2030 [2]. The prevalence of diabetes is higher in developed countries than in developing countries, but the incidence of diabetes in developing countries is likely to increase at a higher rate 3-5. International Diabetes Federation estimated that about 69.9 million people in India will be suffering from diabetes by 2025 [4]. The prevalence of diabetes is 12% in urban and 3% in rural Indians, and this trend is likely to continue (3-6). Rapid increase in urbanization, economic development and job related sedentary habits in developing countries are likely the contributory factors for the expected rise in diabetes prevalence3-5. With its present rate of increase, diabetes is poised to become one of the world's major diseases which are a major concern of public health [5, 6]. Diabetes is a chronic disorder of carbohydrate, fat and protein metabolism characterized by the increased level of blood sugar and the potential to develop into heart disease, stroke, kidney disease and loss of nerve function [7]. The prevalence of type-2 diabetes has shown to vary across the population groups [8]. Some studies have indicated that obesity is a significant contributor for the development of type-2 diabetes routed through reduction in physical activity and change in diet [9]. Subjects with family history of diabetes develop diabetes earlier compared to subjects without family history [10]. The macrovascular complications of diabetes account for more than 70% of all deaths in individuals with diabetes [11]. Lipid abnormalities in type 2 diabetes are characterized by high triglyceride concentrations and low high density lipoprotein-cholesterol concentrations [12]. Since type-2 diabetes is a multi-factorial problem, and vascular disease is the most important complication, there is a need to assess the magnitude of the risk exerted by hyperglycemia, hyperlipidemia and oxidative stress in the progression of the disease. In the present study, we assess the association of coronary risk factors with blood glucose levels among patients with type 2 diabetes mellitus.

MATERIALS AND METHODS

We selected patients suspected of type 2 diabetes mellitus for this study. 0f 300 subjects screened 270 volunteered to participate in this study, 14 had no diabetes, 10 with complications were excluded and rest 6 refused to participate so were excluded. All the subjects (n=270) were attending the out patient department at SVRR Medical College Hospital and four other prominent clinics in Tirupati town of Andhra Pradesh during the period of January 2011 to July 2011. Fasting and post prandial blood glucose levels were obtained after written informed consent. Latent diabetics with symptoms of polyphasia and polydipsia and polyurea were also taken into consideration. The exclusion criteria are pregnant women, chronic bedridden patients, smokers, alcoholics and chewers of tobacco related products. The protocol, case report forms and consent forms were approved by the Sri Venkateswara University Ethics Committee on human subject Research.

A total of 270 type-2 diabetics (Males=139 and Females= 131) in the age range of 28 to 79 yrs were enrolled in the study. After obtaining the written informed consent, subject's demographics (age, sex, education, occupation and income), clinical history and family history were recorded. For the purpose of comparison, the subjects were divided into three age groups like <39 yrs, 40-54 yrs and >55 yrs. Socioeconomic status was assessed by taking the individuals attributes; education, occupation, housing, family income and consumer durables as per the procedure specified by Singh et al. [13], and classified them into SES 1 to V (SES-1 is the highest and SES-5 is lowest class). Selection of one Government Clinic and 4 private clinics allowed us to secure the sample representing all sections of the people. Physical activity is assessed based on subjects occupational and leisure time activities [13].

Subject's physical assessment included height, weight, waist and hip circumference. Weight was recorded with the subjects wearing light clothing without shoes. Height was measured and rounded off to the nearest centimetre, with the subjects standing in erect posture. Body Mass Index (BMI) was calculated as weight in kg divided by the height in meter square. Based on the BMI, subjects were categorized as normal weight (17.9-24.99), over weight (25.0-29.9) and obese (>30.0). Waist girth was measured at the level of umbilicus with person breathing silently and hip girth measured at inter-trochanteric girth while standing according to the method specified by Weiner and Lourie [14].

Waist Hip Ratio (WHR) was calculated as waist circumference/hip circumference and categorized the subjects into 0.7-0.79, 0.8-0.89, 0.9-099 and >1. Blood pressure was recorded with sphygmomanometer as per the procedure of Rose et al. [15]. Systolic hypertension (>140 mmHg) and Diastolic hypertension (>90 mmHg) was calculated as per JNC VII committee recommendations [16].

All subjects were requested to come for Out Patient Unit with 11 hr fasted condition. 5 ml of intravenous blood sample was drawn from each subject and serum was separated upon blood collection and processed for various biochemical parameters like blood glucose, cholesterol, HDL cholesterol and triglycerides. Subjects second blood sample was taken after 2 hr and the serum is used for the estimation of Post prandial blood sugar. Blood glucose was determined by the glucose-oxidase method with an intra- and inter-assay coefficient of variation of <1%.

Serum lipid levels (total cholesterol, HDL cholesterol and triglycerides) were measured by enzymatic methods with an intra- and inter assay coefficient of variation of <3%. The HDL cholesterol was estimated after selective precipitation of non-HDL fractions.

Statistical analysis was carried out by SPSS 16.0 version and p values were set at 0.05 as significant. Percentages were calculated for discontinuous variables. Descriptive statistics with 95%CI in mean difference were provided for continuous variables and test of significant was seen with t-test. Analysis of variance (ANOVA) was carried out to see the variance within and between the groups.

To assess the variation explained by the independent variables in FBS and PPBS, a linear regression model was fitted with an entry at 0.05 and removal at 0.10. Further logistic regression model (binary and multinominal) was fitted with FBS sand PPBS as dependent variables and other factors served as independent variables.

RESULTS

This case control study included 270 patients with type-2 diabetis mellitus. We assessed the prevalence of risk factors and their association with blood glucose levels among these patients. Information on demographics and risk factor prevalence rates were presented in table 1. Around 12 to 14 percent of the sample falls in >39 yrs age group, 40 percent fall 40-54 yrs age group and the remaining 45 percent fall in >55 yrs age group. 14 to 20 percent of the sample comes under SES-V category, 20 to 30 percent in SES-IV category, 40 to 48 percent in SES-III category and 8 to 19 percent in SES-II category.

Based on individuals diabetes history, subjects were classified into groups like latent, <4 years, 5-9 yrs, 10-14 yrs and >15 yrs. Eight percent of the sample are latent diabetics, 40 to 46 are <4 yrs of duration, 31 to 35 percent are 5-9 yrs duration, 9 to 13 percent are 10-14 yrs duration and 1 to 8 percent are in above >15 yrs duration. In the present study 43 percent of males and 50 percent of females do possess the family history of diabetes.

Physical activity pattern indicates that 15 to 17 percent are practicing sedentary life style, 24 to 70 are medium, 15 to 56 percent are moderate and only 4 percent of males involved in heavy physical activity.

In the present sample 21 to 32 percent of diabetics are normal weight, 46 to 50 are overweight and 18 to 33 percent are obese.

Regarding WHR, one to 8 percent are in 0.7-0.79 category, 40 to 57 percent in 0.8-0.89 category, 35 to 58 percent in 0.9-0.99 and only 1 percent of the sample comes under >1 category. Systolic hypertension is 38 percent in males and 42 percent in females. Diastolic hypertension is 59 percent in males and 63 percent in females. Hypertriglyceridemia is 55 percent in males and 60 percent in females. Hypercholesterolemia is 37 percent in males and 32 percent in females.

Descriptive statistics for anthropometry, blood pressure and biochemistries were shown in table 2. Males were found to be taller than females ($p<0.05$), while no differences in weight and body mass index. Males possess higher waist circumference (90.14 ± 9.85 vs 88.70 ± 10.13; $p<0.05$) and WHR (0.91 ± 0.05 vs. 0.88 ± 0.06; $p<0.05$) than females. No sex differences were observed in blood pressure, FBS, PPBS and triglycerides. Males possess higher cholesterol levels and females possess higher HDL cholesterol levels ($p<0.05$).

One way analysis of variance show the changes in FBS and PPBS levels across the age groups, duration of the diabetes, categories of BMI, WHR, physical activity and socio-economic status and the results were presented in graphical form in Figures 1 to 6. Both FBS and PPBS levels initially increased from <39 yrs age group to 40-54 years age group and then FBS decreased and PPBS increased in >55 yrs age group (F-value: FBS, fasting: 8.251; PPBS, 11.859; $p<0.05$). When the duration of disease were taken into consideration, both

FBS and PPBS increased from latent diabetics to diabetics with a duration of 5-9 yrs and then FBS decreased and PPBS maintained with minor fluctuations in the subsequent periods of exposure to diabetics (F-value: FBS, 3.757; PPBS, 5.437; $p<0.05$).

Table 1. Demographics and risk factor profile of the study subjects

SNo	Variable	Male n=139		Female n=131	
		N	%	N	%
1	Age groups (yrs)				
	<39	17	12.2	19	14.5
	40-54	59	42.4	53	40.5
	>55	63	45.3	59	45.0
2	SES				
	SES-5	20	14.4	26	19.80
	SES-4	27	19.4	42	32.10
	SES-3	66	47.5	52	39.70
	SES-2	26	18.7	11	8.40
3	Duration of diabetes				
	Latent	11	7.90	11	8.40
	<4 yrs	56	40.30	61	46.60
	5-9 yrs	43	30.90	46	35.10
	10-14 yrs	18	12.90	12	9.20
	>15 yrs	11	7.90	1	0.80
4	Family history of the disease				
	Yes	60	43.2	66	50.4
	No	79	56.8	65	49.6
5	Physical activity				
	Sedentary	24	17.30	20	15.30
	Medium	33	23.70	92	70.20
	Moderate	78	56.10	19	14.50
	Heavy	4	2.90	0	0.00
6	Body mass index				
	Normal weight (18.5-24.9)	45	32.40	28	21.40
	Over weight (25.0-29.9)	69	49.60	60	45.80
	Obese (>30.0)	25	18.00	43	32.80
7	Waist hip ratio				
	0.7-0.79	1	0.70	10	7.60
	0.8-0.89	55	39.60	74	56.50
	0.9-0.99	81	58.30	46	35.10
	>1	2	1.40	1	0.80
8	Systolic hypertension				
	<140	86	61.90	76	58.00
	>140	53	38.10	55	42.00
9	Diastolic hypertension				
	<90	57	41.00	49	37.40
	>90	82	59.00	82	62.60
10	Hypertriglyceridemia (>120mg%)	77	55.39	79	60.30
11	Hypercholesterolemia(>200mg%)	52	37.41	42	32.06

Table 2. Descriptive statistics for anthropometry, blood pressure and lipids in type-2 diabetes

Variable	Males (n= 139)		Females(n=131)		95%CI		t-value
	Mean	SD	Mean	SD	Lower	Upper	
Age in yrs	53.24	10.86	51.01	10.08	-0.29	4.74	4.92 *
Height in cm	162.18	7.91	150.27	6.69	10.15	13.67	-3.31 *
Weight in Kgs	70.08	10.55	63.88	10.14	3.72	8.68	1.19
Body mass index kgm^{-2}	26.68	3.84	28.34	4.37	-2.64	-0.67	-1.47
Waist circumference in cm	90.14	9.85	88.70	10.13	-0.95	3.84	-3.18 *
Hip circumference in cm	99.24	8.14	100.86	9.93	-3.79	0.55	-1.42
Waist hip ratio	0.91	0.05	0.88	0.06	0.02	0.04	4.37 *
Systolic blood pressure mmHg	134.60	11.36	136.56	11.25	-4.67	0.75	-1.58
Diastolic blood pressure mmHg	90.40	8.61	91.85	7.35	-3.37	0.48	-0.81
Fasting blood sugar mg%	162.14	43.46	171.28	51.61	-20.58	2.28	0.55
Post prandial blood sugar mg%	234.42	65.01	240.93	66.56	-22.30	9.29	0.17
Triglycerides mg%	156.97	90.02	149.12	64.76	-20.60	36.29	-1.24
Cholesterol mg%	187.27	43.49	185.93	42.97	-14.33	17.01	4.92 *
HDL-cholesterol mg%	46.27	3.64	47.02	2.86	-1.96	0.45	-3.31 *

* $p<0.05$.

Fasting blood glucose slightly elevated from normal weight diabetics to obese diabetics and PPBS unaltered within the categories of BMI, and these changes were insignificant. On the other hand FBS significantly elevated from 0.7-0.79 to >1 category of WHR (F-value: 4.606; $p<0.05$), but the elevation in PPBS within the categories of WHR is insignificant. Though the levels of FBS and PPBS are decreasing from lower SES to higher SES, the decrease is significant only in FBS (F-value: fasting: 2.649; $p<0.05$). Physical activity pattern shows a significant decrease in PPBS (F-value: PPBS, 2.270; $p<0.05$).

In order to assess the effect of variation exerted by the independent variables on dependent variables like FBS and PPBS, a linear regression analysis was carried out with a probability of entry of variables at 0.05 and removal at 0.10, and the results were shown in tables 3 and 4. Out of the total variables entered into the model independent variables like age, sex, duration of disease, family history, physical activity, BMI, WHR, blood pressure and lipid levels fitted in the model. 60 percent of the variance in FBS is explained by the independent variables. Family history and systolic blood pressure were positively associated towards the elevation of FBS levels. Similarly 58 percent of the variance in PPBS is explained by the independent variables. Duration of the disease, family history and diastolic blood pressure were shown positive impact on PPBS.

Logistic regression analysis provided the odds ratios to explain the relative risk of independent variables in the elevation FBS and PPBS (tables 5 and 6). Subjects in the age of 40-54 are 1.023 times risk (95%CI: 1.011-1.034) in the elevation of FBS when compared to >39 yrs age. The risk is 1.020 times when the age increased (>55 yrs). Similarly 40-54 yrs age group subjects are 1.016 times risk in the elevation of PPBS (95%CI: 1.009-1.024) when compared to <39 yrs age group, and the risk elevated to 1.017 times in >55 yrs age group.

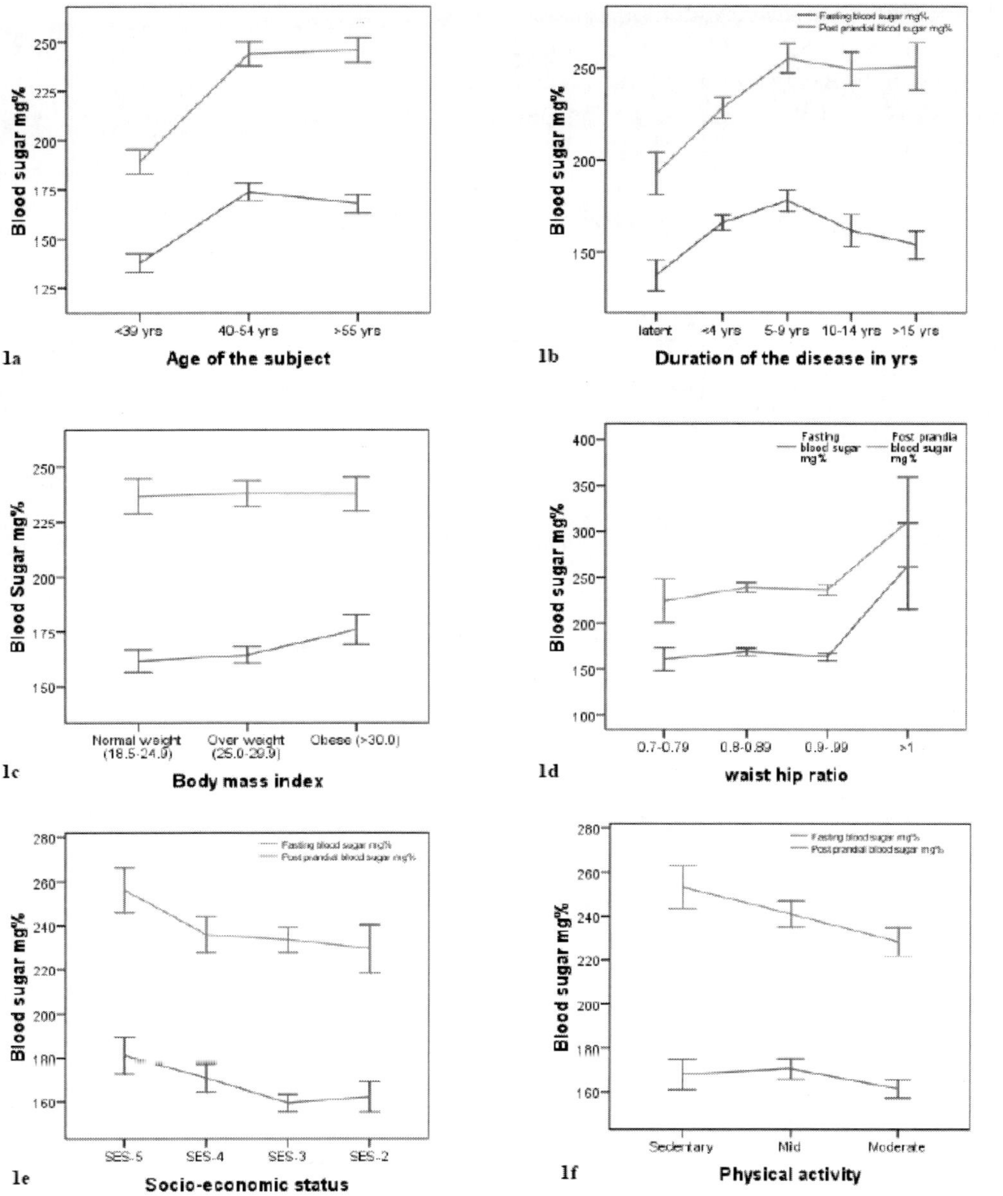

Figure 1. Changes in blood sugar levels. a) Age vs. Blood sugar: F-value, FBS: 8.251*, PPBS, 11.859*. b) Duration of disease vs. Blood sugar: F-value: FBS, 3.757*, PPBS: 5.437*. c) Body mass index vs. Blood sugar: F value: FBS: 1.806, PPBS: 0.009. d) Waist hip ratio vs. Blood sugar: F-value: FBS: 4.606*, PPBS: 1.418. e) Socio-economics status vs. Blood sugar: F-value: FBS: 2.649*, PPBS:1.598. f) Physical activity vs. Blood sugar: F-value: FBS: 1.069, PPBS: 2.270*. * $p<0.05$.

When compared to latent diabetics, the diabetic with a disease duration of <4 yrs are 1.019 times risk in the elevation of FBS, and the risk reached to 1.024 times in 5-9 yrs duration. Further exposure to the disease resulted a decrease in the odds ratios like 1.017 times in 10-14 yrs and 1.013 times in >15 yrs duration. Similar findings were noticed in case of FPBS also. Subject's with a family history of diabetes are 1.084 times risk (95%CI: 1.063-

1.106, $p<0.05$) in the elevation of FBS and 1.033 times risk (95% CI: 1.025-1.041, $p<0.05$) in the elevation of PPBS. WHR >1, simultaneously show an odds ratio of 1.028 with FBS and 1.020 with PPBS. Diabetics with systolic hypertensives were 1.023 times risk in the elevation of FBS and 1.012 times risk in the elevation of PPBS when compared to normotensive diabetics.

Table 3. Linear regression analysis for fasting blood glucose levels in type-2 diabetes subjects

Variable	Regression coefficients		t-value	Significance	95% CI	
	Beta	S.E			Lower	Upper
(Constant)	167.77	84.21	1.99	0.05	0.78	334.75
age	-0.39	0.38	-1.03	0.30	-1.14	0.36
sex	-4.39	6.23	-0.71	0.48	-16.74	7.95
Duration of the disease	0.40	0.74	0.54	0.59	-1.07	1.86
Family history	58.83	5.78	10.19	0.00	47.38	70.28
Physical activity	-8.00	5.05	-1.59	0.12	-18.01	2.01
body mass index	0.24	0.69	0.34	0.73	-1.14	1.62
waist hip ratio	5.29	51.70	0.10	0.92	-97.22	107.79
Systolic blood pressure	0.52	0.30	1.98	0.05	-0.08	1.12
Diastolic blood pressure	0.63	0.40	1.59	0.11	-0.16	1.42
Triglycerides	0.04	0.04	0.87	0.39	-0.05	0.12
Cholesterol	0.02	0.06	0.29	0.77	-0.11	0.14
HDL-cholesterol	-0.35	0.95	-0.37	0.72	-2.22	1.53

Adjusted r^2=60.0.

Table 4. Linear regression analysis for Post prandial blood glucose in type-2 diabetes subjects

Variable	Regression coefficients		t-value	Significance	95% CI	
	Beta	S.E			Lower	Upper
(Constant)	99.60	128.68	0.77	0.44	-155.54	354.74
age	-0.52	0.58	-0.90	0.37	-1.66	0.63
sex	-13.19	9.51	-1.39	0.17	-32.05	5.68
Duration of the disease	2.11	1.13	1.87	0.06	-0.13	4.35
Family history	75.47	8.82	8.55	0.00	57.98	92.97
Physical activity	-10.01	7.71	-1.30	0.20	-25.30	5.29
body mass index	-1.70	1.06	-1.60	0.11	-3.80	0.41
waist hip ratio	34.78	78.99	0.44	0.66	-121.84	191.40
Systolic blood pressure	0.17	0.46	0.38	0.71	-0.74	1.09
Diastolic blood pressure	2.45	0.61	4.03	0.00	1.24	3.65
Triglycerides	0.03	0.06	0.49	0.63	-0.09	0.16
Cholesterol	0.11	0.10	1.13	0.26	-0.08	0.30
HDL-cholesterol	1.15	1.45	0.79	0.43	-1.72	4.01

Adjusted r^2=57.8.

Similarly diastolic hypertensives are 1.022 times risk in the elevation of FBS and 1.021 times risk in the elevation of PPBS when compared to normotensive diabetics.

Table 5. Logistic regression analysis for fasting blood glucose in type-2 diabetes subjects

Variable	Exp (B)	S.E	Significance	95% CI	
				Lower	Upper
Age					
<39 yrs (reference)	1.000				
40-54 yrs	1.023	0.006	0.000	1.011	1.034
>55 yrs	1.020	0.006	0.000	1.009	1.032
Duration of the disease					
Latent (reference)	1.000				
<4 yrs	1.019	0.007	0.006	1.005	1.033
5-9 yrs	1.024	0.007	0.001	1.010	1.039
10-14 yrs	1.017	0.008	0.031	1.002	1.033
>15 yrs	1.013	0.010	0.197	0.994	1.032
Family history	1.084	0.010	0.000	1.063	1.106
Waist hip ratio					
0.7-0.79 (reference)	1.000				
>1	1.028	0.011	0.013	1.006	1.050
Systolic hypertension	1.023	0.004	0.000	1.016	1.030
Diastolic hypertension	1.022	0.004	0.000	1.015	1.030

Table 6. Logistic regression analysis for Post prandial blood glucose levels in type-2 diabetes subjects

Variable	Exp (B)	S.E	Significance	95% CI	
				Lower	Upper
Age					
<39 yrs (reference)	1.000				
40-54 yrs	1.016	0.004	0.000	1.009	1.024
>55 yrs	1.017	0.004	0.000	1.009	1.025
Duration of the disease					
Latent (reference)	1.000				
<4 yrs	1.011	0.004	0.015	1.002	1.020
5-9 yrs	1.018	0.005	0.000	1.008	1.027
10-14 yrs	1.016	0.005	0.002	1.006	1.026
>15 yrs	1.016	0.006	0.008	1.004	1.029
Family history	1.033	0.004	0.000	1.025	1.041
Waist hip ratio					
0.7-0.79 (reference)	1.000				
>1	1.020	0.010	0.050	0.999	1.041
Systolic hypertension	1.012	0.002	0.000	1.007	1.016
Diastolic hypertension	1.021	0.003	0.000	1.015	1.027

Table 7. Prevalence of type-2 diabetes in India

Year	Author	Place	Urban %	Rural%
1971	Tripathi et al.	Cuttack(Central)	1.2	-
1972	Ahuja et al.,	New Delhi(North)	2.3	-
1979	Gupta et al.	Multicentre	3.0	1.9
1984	Murthy et al.	Tenali(South)	4.7	-
1986	Patel Bhadran	(West)	3.8	-
1988	Ramchandran et al.	Kudremukh(South)	5.0	-
1989	Kodall et al.	Gangavathi(South)	-	2.2
1989	Rao et al.	Eluru(South)	-	1.6
1991	Ahuja et al.	New Delhi(North)	6.7	-
1992	Ramchandran et al.	Madras(South)	8.2	2.4
1995	Singh et al.	Moradabad	7.9	2.5
1997	Ramchandran et al.,	Madras(South)	11.6	-
1998	Singh et al.	Moradabad	6.0	2.9
2000	Rammurthy et al.	Kerala(South)	12.4	2.5
2001	Ramchandran et al.	National(DESI)	12.1	-
2001	Iyer et al.	Dombivill	6.2	-
2004	Sadokot et al.	National	5.9(4.3)	2.7
2006	Mohan et al.	Chennai	14.3	-
2006	Menon et al.	Chenai	19.5	-
2008	Kumar et al.	Kolkatta	11.5	-

Modified from Ramchandran 2007.

DISCUSSION

This controlled case study shows that duration of diabetes, family history and blood pressure are the significant risk factors associated with glucose levels in type-2 diabetes. Stronger the family history higher the chance of elevation in blood glucose levels leading to diabetes related micro and macro vascular complications. Elevation in blood pressure or uncontrolled hypertension which may occur due to vascular disease in type 2 diabetes, may also lead to accumulation of multiple complications. Interestingly in the present study, exposure to the hyperglycemia for a period of 9 years was associated with greater frequency of elevated blood glucose levels and further exposure to the condition have shown negative trend. This may be due to modification of lifestyle which may reduce percentage (10%) of overweight/obesity in diabetics with a disease exposure of >10 years. The higher risk of hyperglycemia may be due to decrease in insulin secretion by the beta cells of pancreas and loss of weight may occur due to ketabolism in type 2 daibetes [3-6].

The risk factor profile for type-2 diabetes differs from population to population [3-5, 17]. Our results are in best agreement with the above findings. The duration of the disease was more than 10 years among 25% of the patients. Whereas, 93 percent of the people fall in <9 years of exposure to toxicity of hyperglycemia, 36 percent in <4 yrs of duration in the age range of <54 years. When age and onset of diabetes are taken into consideration, it is inferred that subjects are more likely to expose to diabetes in <40yrs in Indian situation. In western communities the onset of diabetes is observed at >50 years [8]. Thus the present findings indicate that duration of diabetes is an important determinant of complications in type 2 diabetes.

Population studies indicated that diabetes is more prevalent in Western communities compared to developing countries which may be due to sedentary life styles and increased consumption energy dense foods [3-6, 18]. The distribution of socio-economic grading in the present study emphasizes that diabetes is no longer confined to affluent communities in Indian settings, and low SES people are also likely to be affected in the changing nutritional and life style transition. However, this finding may be limitation of a case control study, because in population surveys, lower social classes 4 and 5 have a much lower prevalence of type 2 diabetes compared to social classes 1-3. [3-6]

Hyperglycemia in diabetes is a serious concern because it enhances complications due to glucose toxicity to smooth muscle cells and endothelial cells as well as cardiomyocytes. Hyperglycemia may be associated with hypertriglyceridemia, rise in free fatty acids,low HDL cholesterol, increase in oxidative stress. In the present study, nearly half of the sample had a positive history of diabetes mellitus, irrespective of duration of the disease. When age and onset of diabetes is taken into consideration, subjects with a family history of disease develop diabetes earlier than those subjects without family history of diabetes. Similar findings were observed in other studies [10]. Poor glycemic control is noticed in the present study, despite treatment with hypoglycemic drugs. This may be due to lack of awareness and health education about complications of hyperglycemia. Personal interviews with all the subjects indicated that the sense of restriction in food habits is meager. The number, content and size of meals have important role in the management of hyperglycemia. Physical activity patterns positively reduce the post prandial blood glucose levels [3-6].

Overall and abdominal adiposity is a risk factor in the development of diabetes [19]. Many studies have emphasized about the positive role of BMI and WHR in type-2 diabetes. Although generalized obesity appeared to ensure type-2 diabetes, the results of the present study failed to show such an association which may be due ketabolic phase in type 2 diabetes. Majority of the patients (70%) in the present study had overweight and obesity which is risk factor of type 2 diabetes. Overweight/obese subjects with WHR >0.90 for men and >0.85 for women may have central obesity. WHR showed positive relationship with diabetes than BMI Similar results were published else where [20, 21].

The association of BMI with type-2 diabetes appears to differ in different ethnic groups. Epidemiological data from Asian Indians, Mexican Americans and non-Hispanic Whites from San Antonio heart study showed that Mexican Americans had the highest rate of obesity and highest prevalence of diabetes (men 19.6%; women 11.8%). But Non Hispanic Whites had high rates of obesity but a low prevalence of diabetes (men 4.4%; women 5.7%). Although Asian Indians have lower BMI than Mexican Americans, the risk conferred by BMI was similarly high in Asian Indians and Mexican Americans than Non Hispanic whites [22-32]. WHR ratio appeared to be significantly associated with the occurrence of diabetes in men but not in women. Extensive review by McKeigue [23] showed that average waist/hip circumference ratios are higher in South Asians than in Europeans of similar BMI. A study in Bangladesh also showed positive relation of diabetes and WHR [17].

Hypertension had positive correlation with hyperglycemia. According to one study from Orissa, India, most common disease associated with diabetes was hypertension [24]. Evidence suggest that the increased rate of hypertension in diabetics is due to possible mechanism of insulin resistance leading to abnormalities in renal tubular ion exchange, rennin angiotension system, prostaglandin kallikrein/kinin system., diabetic nephropathy [25-32]. Systolic hypertension (>140 mmHg) and diastolic hypertension (>90 mmHg) was associated with diabetes in our subjects as demonstrated by logistic regression analysis. Sayeed et al. [26] observed the rate of systolic hypertension and diastolic hypertension as 23.2 and 13.6 percent in newly diagnosed type-2 diabetes subjects. Similar results were published in population groups of southern Taiwan [27], Nigeria [28], China [29] India, and Australia [30-33]

In brief, this study shows that coronary risk factors are significantly associated with hyperglycemia in diabetes subjects. Family history, duration of disease and blood pressure are the most commonly observed risk factors. It is possible that this would sensitize health care practitioners and diabetes subjects towards the management of hyperglycemia in type 2 diabetes. Further studies in the population would be necessary to confirm our findings.

Acknowledgments

The authors greatly acknowledge the financial support rendered by the University Grants Commission, New Delhi to the Department of Anthropology, Sri Venkateswara University, Tirupati under UGC-SAP-DRS Phase-III. We are grateful to International College of Nutrition for logistic support.

References

[1] King H, Aubert RE, Herman WH. Global burden of diabetes, 1995–2025: prevalence, numerical estimates, and projections. *Diabetes Care* 1998; 21: 1414–1431.

[2] Wild S, Roglic G, Green A, King H. Global prevalence of Diabetes: Estimates for the year 2000 and projection for 2030. *Diabetes Care* 2004; 27: 1047- 1053.

[3] Singh RB, Bazaz S, Niaz MA, Rastogi SS, Moshiri M. Prevalence of type 2 diabetes mellitus and risk of hypertension and coronary artery disease in rural and urban population with low rates of obesity. *Int. J. Cardiol.* 1998; 66:65-72.

[4] Rastogi SS, Singh RB, Singh NK, Alams SM, Singh RG, Tripathi KK, Srivastav SK, Vajpeyee SK, Hirdesh Gupta, V. V. Muthusamy, R. K. Goyal, M. E. Yeolekar, Adarsh Kumar, N. S. Neki, G. S. Sainani, V. Shantaram, Daniel Pella, Jan Fedacko, Fabien De Meester, T. K. Basu1, Surya Acharya and Lech Ozimek. Prevention of cardiovascular disease and diabetes mellitus In low and middle income countries. *The Open Nutra.* 2011,4:97-106.

[5] Singh RB, Rastogi SS, Niaz MA, Postiglion A. Association of central obesity and insulin resistance with high prevalence of diabetes and cardiovascular disease in an elderly population with low fat intake and lower than normal prevalence of obesity. The Indian paradox. *Coro. Art Dis.* 1998; 9: 559-65.

[6] World Health Organization: Global health risks: mortality and burden of disease attributable to selected major risks., *WHO*, Geneva, 2009, ISBN 978-924-156387-1.

[7] Kaur J, Singh P, Sowers JR. Diabetes and cardiovascular diseases. *Am. J. Ther.* 2002; 9: 510–515.

[8] Diamond J. Diabetes in India. *Nature* 2011; 469: 478-479.

[9] Mahanta, BN, Mahanta, TG. Clinical Profile of Persons with Family History of Diabetes Mellitus with Special Reference to Body Fat Percentage. *JAPI* 2009; 57: 641-843.

[10] Mohan V, Shanthirani CS, Deepa R. Diabetes and IGT in a selected South Indian population with special reference to family history, obesity and lifestyle factors–The Chennai Urban population study (CUPS 14). *JAPI* 2003; 51: 771-777.

[11] Ken G, Catherine CC, Maureen IH. Mortality in adults with and without diabetes in a national cohort of the U.S. population 1971–1993. *Diabetes Care* 1998; 21:1138-1145.

[12] Mooradian AD, Dyslipidemia in type 2 diabetes mellitus. *Nature Clinical Practice Endocrinologyand Metabolism* 2009; 5: 150-159.

[13] Singh RB, Ghosh S, Niaz MA, Rastogi V. Validation of physical activity and socioeconomic status questionnaire in relation to food intakes for the five city study and proposed classifications for Indians. *J. Assoc. Physicians India* 1997; 45: 603–607.

[14] Weiner JS and Lourie JA. *General medical examinations, Techniques. Practical Human Biology.* Academic press, London. 1981.

[15] Rose GA, Blackburn HG, Gillum RF, Prineas RJ. *Cardiovascular survey methods.* WHO Publication, 2nd Edn, No. 56. Geneva: World Health Organization. 1982.

[16] *The Seventh Report of the Joint National Committee on Prevention, Detection, Evaluation, and Treatment of High Blood Pressure*, 2004.

[17] Hussain A, Vaaler S, Sayeed MA, Mahtab H, Keramat Ali SM, Azad Khan AK. Type 2 diabetes and impaired fasting blood glucose in rural Bangladesh: a population based study. *European Journal of Public Health* 2006; 17: 291–296.

[18] Singh RB, Beegom R, Mehta AS, Niaz M A, De AK, Mitra RK. Social class, coronary risk factors and under nutrition, a double burden of disease, in women during transition, in five Indian cities. *Int. J. Cardiol* 1999; 69: 139-147.

[19] Shils ME, Olson JA, Shike M, editors. *Modern nutrition in health and disease.* Vol 1, 8th edition. Philadelphia: Lea and Febiger, 1994.

[20] Hadaegh F, Shafiee G, Azizi F. Anthropometric predictors of incident type-2 diabetes mellitus in Iranian women. *Ann. Saudi Med.* 2009: 29:194-200.

[21] Nyamdorj R. BMI Compared With Central Obesity Indicators in Relation to Diabetes and Hypertension in Asians: The Decoda Study *Group. Obesity* 2008; 16: 1622–1635.

[22] Ramachandran A, Snehalatha C, Viswanathan V. Risk of noninsulin dependent diabetes mellitus conferred by obesity and central adiposity in different ethnic groups: a comparative analysis between Asian Indians, Mexican Americans and Whites. *Diabetes Res. Clin. Pract.* 1997; 36:121–125.

[23] McKeigue PM. Metabolic consequences of obesity and body fat pattern: lessons from migrant studies. *Ciba. Found. Symp.* 1996; 201:54–64

[24] Sahoahoo S, Dash KK, Sahoo S, SahuD, Prescription analysis in diabetes mellitus, *International Journal of Pharma and Bio Sciences* 2010;1:1-8.

[25] Das S, MohantyR, Patnaik UK, Management of Hypertension in Diabetes Mellitus, *Indian Academy of Clinical Medicine* 2001; 2:30-33.

[26] 26. Sayeed MA, Khan AR, Banu A, Hussain MZ. Prevalence of diabetes and hypertension in a rural population of Bangladesh. *Diabetes Care* 1995; 18:555–558.

[27] Lu FH, Yang YC, Wu JS. A population-based study of the prevalence and associated factors of diabetes mellitus in southern Taiwan. *Diabet Med.* 1998; 15:564–72.

[28] Olatunbosun ST, Ojo PO, Fineberg NS, Bella AF. Prevalence of diabetes mellitus and impaired glucose tolerance in a group of urban adults in Nigeria. *J. Natl. Med. Assoc.* 1998;90:293–301.

[29] Pan XR, Yang WY, Li GW, Liu J. Prevalence of diabetes and its risk factors in China, 1994. National Diabetes Prevention and Control Cooperative Group. *Diabetes Care* 1997; 20:1664–1669.

[30] Donnelly R, Molyneaux L, McGill M, Yue DK. Detection and treatment of hypertension with non-insulin-dependent diabetes mellitus: does the "rule of halves" apply to a diabetic population? *Diabetes Res. Clin. Pract* .1997;37:35–40.

[31] Singh RB, Rastogi SS, Rao PV, Das S, Madhu SV, Das AK, Sahay BK, Fuss SM,Begum R, Sainani GS, Shah NA. Diet and lifestyle guidelines and desirable levels of risk factors for the prevention of diabetes and its risk factors in Indians: A scientific statement of the International College of Nutrition. *J. Cardiovasc. Risk* 1997; 4:201-08.

[32] Singh RB, Rastogi SS, Rastogi V, Niaz MA, Madhu SV, Chen Min, Shoumin Z. Blood pressure trends, plasma insulin levels, and risk factors in rural and urban elderly populations of north India. *Coro. Art Dis.* 1997; 8:463-468.

[33] Singh RB, Beegum R, Rastogi V, RastogiSS,Madhu V. Clinical characteristics of hypertension among patients of noninsulin dependent diabetes mellitus in North and Sounth India. *J. Diab. Asso. India* 1996; 36:45-50.

In: Cardiovascular Health and Chronomics
Editor: Ram B. Singh
ISBN: 978-1-62948-976-6

Chapter 12

PLASMA FIBRINOGEN LEVELS IN ACUTE MYOCARDIAL INFARCTION

***Asna Urooj*[*], *K. Kusuma, P. P. Preetham*[1], *B. Mohan*[1] *and P. Upadya*[1]**

Department of Studies in Food Science and Nutrition,
University of Mysore, Mysore, India
[1]CSI Holdsworth Memorial Hospital, Mysore, India

ABSTRACT

Research evidence suggests that high plasma fibrinogen represents a major risk factor for cardiovascular diseases. The increased occurrence of CVD necessitates the need to analyze fibrinogen levels in ischemic heart disease(IHD) patients and study its relationship with physiological and behavioral risk factors. The present study analyses plasma fibrinogen levels in relation to known cardiac risk factors, in 72 patients (63 men and 9 women) afflicted with myocardial infarction. Parameters studied were anthropometry, plasma fibrinogen and lipid profile. Anthropometric data indicated a lower muscle mass status in majority of the subjects. Although the body mass index (BMI) was normal in majority of subjects, waist-to-hip ratio (WHR) was higher than normal range. Clinical data revealed atherogenic lipid profile in 58% of the subjects. Mean plasma fibrinogen values were 3.8g/L, high normal (>3.0g/L) in 28% and above normal (>4.0g/L) in 30% of the patients. Fibrinogen levels were higher ($p<0.01$) in patients with diabetes, while they were lower in alcohol users ($p<0.05$). Also, mean fibrinogen, cholesterol, LDL, triglycerides, WHR and BMI were higher in smokers than in non-smokers. The results show that fibrinogen level may be the risk factor of IHD in presence of known CVD risk factors. Since most of the risk factors are amenable to changes in lifestyle, it is worthwhile to modulate the levels of these parameters to achieve reductions in plasma fibrinogen.

[*] Corresponding author: Email: asnaurooj@foodsci.uni-mysore.ac.in. Tele fax : 0821 2419632(o).

INTRODUCTION

Chronic diseases like cardiovascular disease (CVD), cancer and diabetes are well recognized as major health problems in the world today. The involvement of lipid and thrombotic component in the pathogenesis of ischemic heart disease (IHD) is well known [1,2] . Clinical and epidemiological evidence indicate that the pre-thrombotic state characterized by hypercoagulability, is associated with increased risk of myocardial infarction [3-5]. Components of haemostatic system which could influence IHD due to their thrombotic characteristics have been identified viz., plasma fibrinogen, factor VII, coagulant activity, increased platelet aggregation as well as impaired fibrinolysis [3-5].

Table 1. Prevalence of Classic CVD Risk Factors

Risk factor	%
High Cholesterol (>200mg/dl)	38
Elevated Triglycerides (>160mg/dl)	50
Hypertension	42
Obesity	16
WHR (>0.8)	40
Smoking	56
Alcohol users	41
Diabetes	51

Table 2. Anthropometric measures of the Subjects

Parameters	Male (n= 63)	Female (n=9)
Body Mass Index	24 ± 2.9	24 ± 3.5
MUAC (cm)	28.2 ± 2.7	27.1 ± 3.1
MUAMC (cm)	15.9 ± 4.6	34. 1 ± 3.7
TSF (cm)	12.1 ± 6.0	17. 6 ± 8.7
Body fat (%)	27 ± 6.0	38 ± 3.0

Values are mean ± SD

<u>Normal ranges</u>

BMI : > 18.5-25

WHR : ≤ 0.8

TSF : M = 12.5, F = 16.5 (Triceps skinfold)

MUAC : M = 29.3, F = 28.5

(Mid upper arm circumference)

MUAMC : M = 25.3, F = 23.5

(Mid upper arm muscle circumference)

Prospective epidemiologic investigations indicate that plasma fibrinogen is a powerful predictor for myocardial infarction and stroke [6]. Clinical findings suggest that hyperfibrinogenemia favors atherothrombotic events of IHD by predisposing to thrombosis and by enhancing atherogenesis [7]. The present study is planned with the objective of analyzing plasma fibrinogen levels in MI patients

METHODOLOGY

The study included a total of 72 patients (63M, 9W) of acute myocardial infarction(AMI) admitted at the CSI Holdsworth Memorial Hospital, Mysore, over a period of 10 months. The research protocol included general examination and personal history, anthropometric measurements and biochemical tests. The study was approved by the hospital ethical committee.

Personal history – detailed enquiries about past and present complaints, personal and other habits (alcohol use and smoking) and about family history were made using a formulated questionnaire.

Anthropometric measures – Body weight and height were measured in the subjects in minimum clothing and without shoes on 2 occasions. BMI was calculated as body weight (kg) divided by height squared (m^2). The waist and hip circumference were also measured and waist-to-hip ratio (WHR) was calculated. Skinfold measurements were taken using Lange skinfold caliper at the biceps, triceps, suprailiac and subscapular levels and body fat percentage was calculated according to the method of Durnin and Womersly [8].

Biochemical tests – Plasma fibrinogen was analyzed using Nycotest ™ test (Nycomed Pharma, Oslo) based on the method of Clausse [9]. Other relevant and routine tests including Hb, cardiac enzymes, lipid profile and blood sugars were carried out in each patient using standard diagnostic kits. Serum LDL-cholesterol was calculated by the formula LDL-C = serum cholesterol – HDL-C (serum triglycerides/5). Blood samples for fibrinogen analysis were collected immediately on admission to the ICU, drawn in tubes containing trisodium citrate.

Statistical analysis – The anthropometric and biochemical data is expressed as mean and SD. Linear correlation coefficients between plasma fibrinogen and each of anthropometric and biochemical parameters were calculated.

RESULTS

The age of the subjects ranged from 25-69y. Women comprised a small number in the study group. For majority of subjects the disease had occurred after 40y, the percentage of men being affected was greater in the age-group of 41-50 and 51-60y. The disease history of the patients revealed the occurrence of classical risk factors like hypertension, diabetes in > 40% of the subjects. Also, cigarette smoking and alcohol consumption was quite common among male subjects.

The BMI ranged from 21-29 (mean 24$\pm$ 3.9). Majority of subjects (70%) had BMI within the normal range (18.5-25), only 27% were found to be overweight indicating grade I obesity.

WHR was more than 0.8, in both men and women. Female subjects tended to have higher body fat when compared to males.

The mean plasma fibrinogen and lipid profile are given in Table 3. The mean plasma fibrinogen was 3.8 $\pm$ 1.9 g/L, ranging from 1.9 – 6.8 g/L. Although the overall mean value for plasma fibrinogen was well within the normal range (2-4g/L), a majority of subjects had a high normal value i.e., > 3.0g/L, while 30% of them had values higher than 4g/L. Data pertaining to the lipid profile, classified according to NCEP Guidelines[10], revealed a distinctive atherogenic profile in majority of subjects. It was observed that mean total cholesterol and triglycerides were higher than the desirable range in > 50% of the subjects, while LDL cholesterol was within the desirable range in only 29% of the subjects. Mean body weight, BMI, %body fat, fibrinogen and lipid profile were higher in subjects with WHR >0.8 (mean 0.95).

The influence of smoking and alcohol consumption in men on plasma fibrinogen is shown in table 4. In this study, IHD developed in 70% of current smokers, compared to ex-smokers and non-smokers. Mean plasma fibrinogen in smokers was significantly higher than in non-smokers. As expected, plasma fibrinogen in alcohol users was significantly lower ($p<0.05$) compared to non-alcohol users. Alcohol consumption appears to lower fibrinogen.

Comparison of mean fibrinogen levels between different groups is shown in Figure 1. It was observed that subjects with diabetes and hypertension had significantly ($p<0.01$) higher levels compared to non-diabetic and normotensive IHD subjects. Significant correlation between fibrinogen and age of the subjects, body fat % and triglyceride levels were observed (table 5).

Discussion

The results demonstrate that fibrinogen was a risk factor in subjects with at least one of the known CVD risk factors. Fibrinogen was higher in these subjects compared with subjects having no previous history of diabetes, hypertension and lipid abnormalities. The presence of well known CVD risk factors might contribute to increased risk of IHD through increases in plasma fibrinogen. Fibrinogen has been reported to be positively associated with the incidence of IHD in epidemiological studies [3,5,7].

Several studies have reported increases in fibrinogen with age [11-13]. These studies, however, were based on >200 patients. Although the total number was small (n=72), the results indicate an age effect. It is reported that moderate alcohol consumption is protective against heart disease [2,13]. In this study, maximal beneficial effect on fibrinogen were observed in subjects with moderate levels of alcohol consumption.

Obesity is associated with elevated fibrinogen levels [2, 12-14]. Although, the obese subjects had higher values than non-obese group, the correlation did not reach statistical significance. Elevated fibrinogen was more closely related to body fat than to the degree of obesity expressed as BMI. This observation is in agreement with reported data [12].

Table 3. Plasma fibrinogen and lipid profile data of the subjects

Biochemical parameter overall mean Population within ranges (n=72) and their mean values

1. Plasma fibrinogen 3.8 ± 1.9 (g/l)

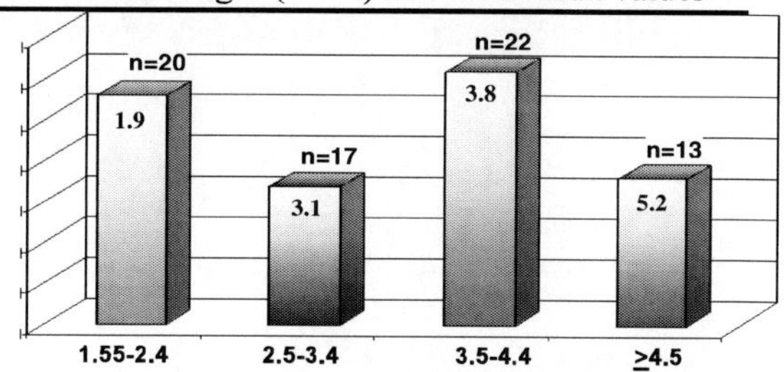

2. Total cholesterol 211.6 ± 30.4 (mg %)

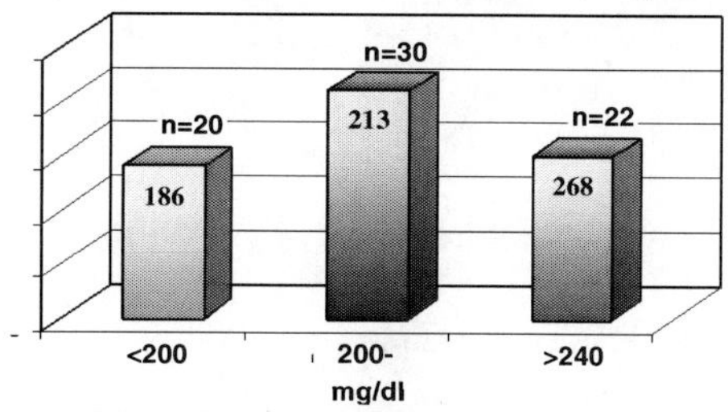

3. LDL cholesterol 110. ± 30.3 (mg %)

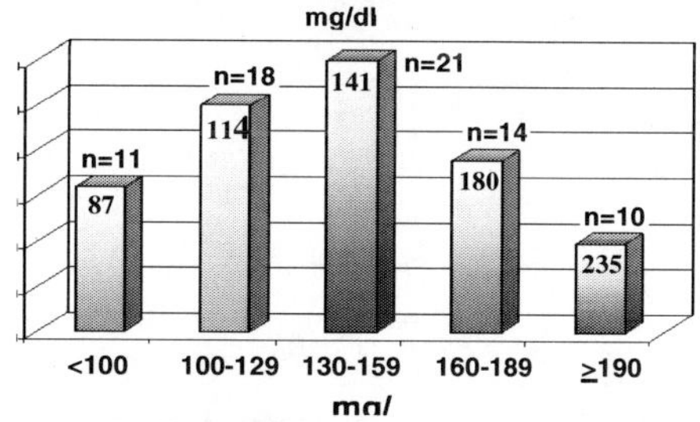

4.HDL cholesterol 39.3 ± 7.4 (mg %)

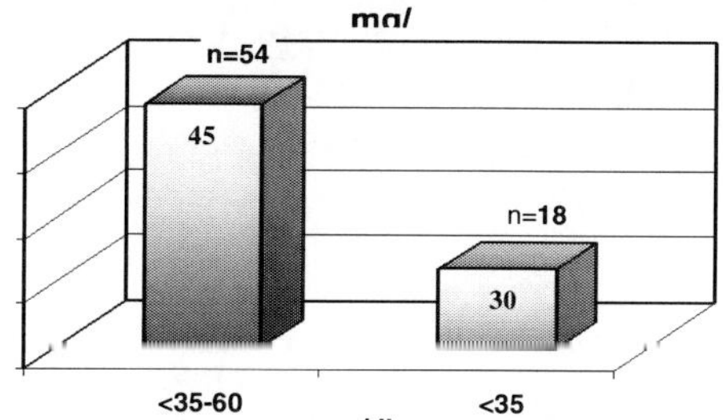

5. Triglycerides 210 ± 53.1 (mg %)

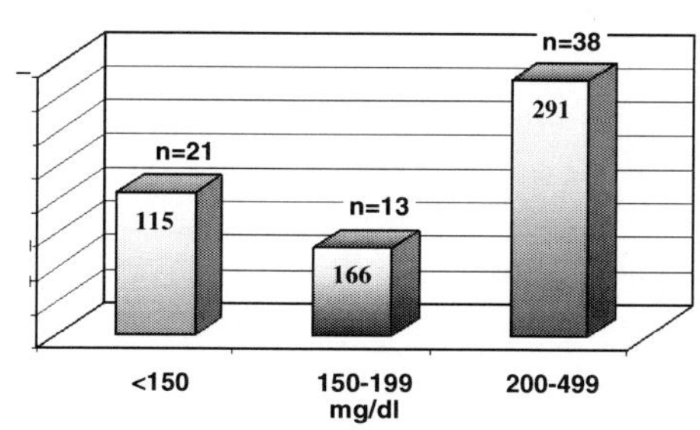

Table 4. Influence of smoking habit and alcohol consumption on plasma fibrinogen

Smoking Habit	n	Fibrinogen g/L	Alcohol Consumption	n	Fibrinogen g /L
Non-smokers	22	2.86	Non-drinkers	33	3.0
Ex-smokers	7	2.92	Very light	5	4.2
Light smokers	18	3.41	Light	8	2.6
Heavy smokers	16	3.82	Moderate	9	2.0
			Heavy	8	2.2

Values are mean ± SD, $p < 0.05$.

Also, it emphasizes the importance of body fat in the development of metabolic abnormalities. Similarly, significant correlation between fibrinogen and triglycerides (TG) reiterates the involvement of hypertryglyceridemia in atherosclerosis. Increase in TG may be diet induced or due to genetic and metabolic defects. Also, it is well known that carbohydrate rich diets increase production of triglycerides. Elevations in TG levels may have an effect on fibrinogen synthesis through increases in free fatty acids. Lipolysis of TG liberates fatty acids and it is known that free fatty acids (FFA) stimulate hepatic fibrinogen synthesis [2,15]. This biosynthetic control manifested by FFA appears to provide a connecting metabolic link between TG metabolism and thrombosis in the genesis of atherosclerosis. Indians are reported to have high levels of TG [16], which may be due to habitual high carbohydrate diets and / or genetic susceptibility [14]). In the present study, the carbohydrate intake of the subjects was high, contributing >75% of total calories. The clinical presentation of diseases, environmental conditions and nutritional status of Indians are quite different compared to Western population [17-19]. Indians may also have higher lipoprotein(a) which enhances the atherogenecity of LDL cholesterol and more thrombogenic [19]. Increased intake of Western diet is known to cause rise in triglycerides and free fatty acids which are precursors of fibrinogen. Increased concentration of fibrinogen can occur in between 6.00 AM to 12.00 noon as circadian rhythm which may become worst in patients with ACS [19]. Thus, the excess risk of Indians to IHD also results from diet and lifestyle related conventional risk factors. Furthermore, few population based studies have mainly focused on the relation of plasma fibrinogen with single metabolic abnormalities [1-3].

In conclusion, the results of this observational study show that fibrinogen increased the risk of IHD in presence of known CVD risk factors. Since most of the risk factors are amenable to changes in lifestyle, it is worthwhile to reduce the levels of these parameters to achieve reductions in plasma fibrinogen.

Table 5. Correlations between fibrinogen, somatic and biochemical parameters

Parameters	Fibrinogen r	 P level
1. Somatic		
BMI	0.20	NS
WHR	0.03	NS
MUAC	0.07	NS
TSF	1.75	NS
Body fat	0.39	<0.05
2. Biochemical		
Cholesterol	0.03	NS
HDL-C	0.17	NS
LDL-C	0.05	NS
TG	0.37	<0.05
3. Others		
Age	0.31	<0.05
Blood pressure (systolic)	0.02	NS

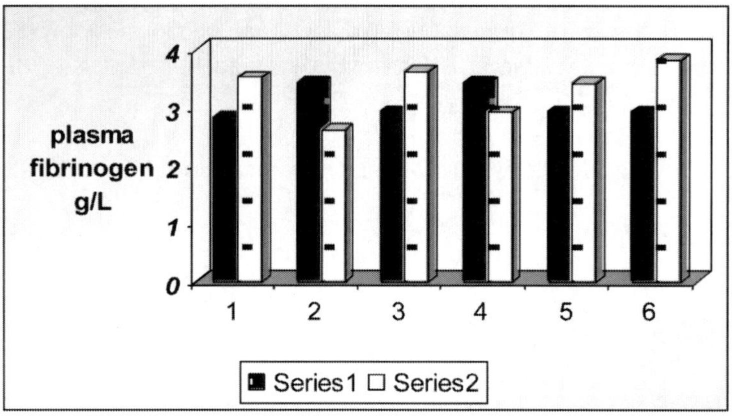

	Series 1	Series 2
1-	Non-smoker	Smoker
2-	Non-alcohol user	Alcohol user
3-	Non-obese	Obese
4-	Non-vegetarian	Vegetarian
5-	Normotensive	Hypertensive
6-	Non-diabetic	Diabetic

Figure 1. Comparison of fibrinogen in different subject groups.

REFERENCES

[1] Khader V and Sunitha M. Relationship between diet and plasma lipids in ischemic heart disease and angina patients. *The Cardiothoracic J.* 1997; 3 (6): 52-57.

[2] Vorster HH, Cummings JH and Veldman FJ. Diet and haemostasis: time for nutrition science to get more involved. *Brit. J. Nutr.* 1997; 77: 671-684.

[3] Meade TW, North WRS, Chakrabarti R, Stirling Y, Haines AP, Thompson SG. Haemostatic function and cardiovascular death: Early results of a prospective study. *Lancet* 1980, i: 1050-1054.

[4] Meade TW, Brozovic M, Chakrabarti R, Haines AP, Imeson JD Mellows S, Miller G.J, North W.R.S, Sterling Y and Thompson S.G. Haemostatic function and ischemic heart disease: principal results of Northwick Park Heart study. *Lancet* 1986; 2: 533-537.

[5] Fehily AM, Milbank JE, Yarnell JWG, Hayes TM, Kubiki AJ and Eastham RD. Dietary requirements of lipoprotein, total cholesterol, viscosity, fibrinogen and blood pressure. *Amer J. Clin. Nutr.* 1982 ; 36: 890 - 896.

[6] Wilhelmsen L, Svardsudd K, Korsan-Bengsten K, Larson B, Welin L and Tibblin G. Fibrinogen as risk factor for stroke and MI. *New Engl. J. Med.* 1984; 311 (8): 501-505.

[7] Ernst E and Resch KL. Fibrinogen as a cardiovascular risk factor; A meta-analysis and review of literature. *Annals of Internal Med.* 1993; 118: 956-963.

[8] Durnin JVGA and Womersly J. *Brit J. Nutr* .1974 ;, 32: 77-97.

[9] Clauss A. Gerinnungsphysiologische Schnell -methods zur Bestimmung des Fibrinogens. *Acta Hemat* 1957; 17: 237-246.

[10] Cholesterol guidelines – A call for more aggressive lipid lowering. *Current Medical Science – A Cipla update on therapeutics* 2001; 16(3): 1-8.

[11] Doolittle RF. The transformation of fibrinogen into fibrin, In: *Haemostasis Biochemistry Physiology and Pathology,* Ed., Ogston D, Bennett B, John Wiley and sons, London, 1977; pp 169-185.

[12] Landin K, Stigendal L, Eriksson E, Krotkiewski M, Risberg Bo, Tengborn L Smith L. Abdominal obesity is associated with an impaired fibrinolytic activity and elevated plasma activator inhibitor-1. *Metabolism* 1990; 39 (10) : 1044 -1048.

[13] Meade TW, Chakrabarti R, Haines A P, North W R S and Stirling Y. Characteristics affecting fibrinolytic activity and plasma fibrinogen concentrations. *Brit. Med. J.* 1979, 1: 153-156.

[14] Vague P, Juhan-Vague I, Aillaud MF, Badier C, Viard R Alessi MCand Collen. Correlation between blood fibrinolytic activity, plasminogen activator inhibitor level, plasma insulin level and relative body weight in normal and obese subjects. *Metabolism* 1986; 35(3): 250-253.

[15] Howard B V, Bekker N A, Flack J D, Jacobs J M, Lewis D R, Macrovina C E and Perkins L L. Concentrations of Lp(a) in black and white young adults: relations to risk factors for cardiovascular disease. *Ann. Epidem.* 1994; 4: 341- 350.

[16] Latheef SAA, Subramanyam G, Reddy K N and Vijaya J.. Adiposity in relation to blood pressure and lipids in elderly. *The Cardiothoracic. J.* 1998; 3 (11): 39- 45.

[17] Singh RB, De Meester F, Pella D, Basu TK, Watson R. Globalization of dietary wild foods protect against cardiovascular disease and all cause mortalities? A Scientific statement from the International College of Cardiology, Columbus Paradigm Institute and the International College of Nutrition. *The Open Nutraceuticals Journal*, 2009, 2, 42-45.

[18] Singh RB, Fadecko, J, Pellad, D, De Meester F, Moshiri M, Aroussy WE. Superfood dietary approaches for acute myocardial infarction. *World Heart J.* 2010;2: 13-23.

[19] Singh RB, Fedacko J, Sharma JP, Vargova V, Sharma, Moshiri M, De Meester F, Otsuka K. Association of inflammation, heavy meals, magnesium, nitrite, and coenzyme Q10 deficiency and circadian rhythms with risk of acute coronary syndromes. *World Heart J.* 2010;2:219-228.

In: Cardiovascular Health and Chronomics
Editor: Ram B. Singh

ISBN: 978-1-62948-976-6

Chapter 13

SYNERGISTIC ANTILIPIDEMIC AND ANTIHEPATOTOXIC ACTION OF *GONGRONEMA LATIFOLIUM* AND *NAUCLEA LATIFOLIA* IN STREPTOZOTOCIN DIABETIC RAT MODELS

P. E. Ebong*[1], G. S. Effiong[2], I. J. Atangwho[1] and E. U. Eyong[1]
[1]Department of Biochemistry, College of Medical Sciences,
University of Calabar, Calabar, Nigeria
[2]Department of Clinical Pharmacy and Biopharmacy, Faculty of Pharmacy,
University of Uyo, Uyo, Nigeria

ABSTRACT

Purpose: The combination effect of *Gongronema latifolium*(GL) and *Nauclea latifolia* (NL), promising vegetables used in traditional management of metabolic disorders on serum lipid components and some liver marker enzymes was investigated in diabetic rat models. *Method:* The study design consisted of thirty six (36) rats divided into 6 groups of 6 rats each. Whereas groups 1 and 2, non diabetic and diabetic controls received placebo treatment, groups 3-5 respectively received 200mg/kg b.w. of GL, NL, and 100mg/kg b.w. each of GL and NL per day, for 21 days. The 6th group received subcutaneous insulin, 5IU/kg b.w. per day. Thereafter, the animals were sacrificed, and serum prepared from blood collected was used to assay lipid components and liver function enzymes, using standard analytical kits. *Result:* Measured blood glucose in diabetic animals decreased significantly from initial by 66.34%, 18.12% and 67.73% respectively upon treatment with GL, NI and GL plus NI. Whereas diabetes induction caused significant increases (p<0.05) in total cholesterol (TC), 54.42% and low density lipoprotein (LDL), 55.0% compared to the normal control (NC), treatment with extracts of GL or NI significantly decreased (p<0.05) these by (58.70% and 24.79%, TC) and (71.70% and 33.38%, LDL) respectively. However, the combined extracts produced more steady reduction (59.70% and 70.26, TC and LDL), similar to the NC. Also the

* Correspondence: Prof Dr PE Ebong, Department of Biochemistry, College of Medical Sciences, University of Calabar, P.M.B. 1115, Calabar, Nigeria, patebong@yahoo.com.

amino transferases (ALT and AST) activities which increased by 66.83% and 72.87% in the DC rats indicating hepatotoxicity, secondary to hyperglycemia, became reduced upon treatment with GL, NI and GL plus NI. Again, the extent of reduction was better with combined extracts (85.58% and 53.77%), and compared favorably with NC. *Conclusion:* Lower dosages of the two extracts when combined provides higher efficacy in protection against atherosclerosis and hepatotoxicity in diabetes. Hence the need to adopt this strategy in bio-prospecting for antidiabetic natural products.

Keywords: Synergism, diabetes mellitus, lipid profile, aminotranaminases, *Gongronema latifolium*, *Nauclea latifolia*

INTRODUCTION

Diabetes mellitus is a multifactorial disease, which is characterized by hyperglycemia and glucosuria (Atangwho, 2008) among others. Diabetes mellitus is also a non-communicable disease, which is considered as one of the five leading cause of death in the world. About 100 million people around the world have been diagnosed with diabetes and by the year 2010, it was projected that 215 million people will have the disease (Zimmet, 1999).

Recently, the search for appropriate hypoglycemic agents has been focused on plants used in traditional medicine partly because of leads provided by traditional medicine to natural products that may be better treatments than currently used drugs (Rates, 2001). Many minor components of foods such as secondary plant metabolites have been shown to alter biological processes, which may reduce the risk of chronic disease in human (Nwanjo et al. ,2006). Polyherbal therapy, sometimes called polyherbalism a combination of herbs or phytochemicals from more than one source, a concept originally peculiar to Ayurveda (Singh, 2005) has today been accepted as an effective therapeutic approach in sourcing medicament for degenerative ailments. The herbs are selected according to the disease and have several advantages over monotherapies namely; reduced toxicity and side effects, and maximum/synergistic therapeutic efficacy (Singh, 2005 and Tiwari and Rao, 2002). Singh et al. have succesfully used commiphora mukul, fruits and vegetables, fenugreek and stargooseberry in the management of patients with hyperlipidemia in 1990s.

Gongronema latifolium (Asclepiadaceae) is a tropical rainforest plant primarily used as spice and vegetable in traditional folk medicine (Ugochukwu and Babady, 2002 and Ugochukwu et al., 2003). Reports from various authors show that it contains essential oils, saponins and pregnans among others (Schneider et al., 1993; Morebise and Fafunwa, 1998 and Morebise et al., 2002). The leaves of *Gongronema latifolium* have protective role against diabetes, hypertension, stomach upsets and pains, typhoid fever (Etim et al., 2008).

Nauclea latifolia (Rubiaceae) commonly known as Pin cushion tree is a strangling shrub or small tree native to tropical Africa and Asia. It grows in Akwa Ibom and Cross River states of Nigeria and is called "Mbom-mbong" whilst in the Northern Nigeria; it is called "Tabasiya. Parts of the plant are commonly prescribed traditionally as a remedy for diabetes mellitus and hypertension (Akpanabiatu et al. 2005; Nworgu et al., 2008 and Okwori et al., 2008).

However, there are no empirical data or scientific reports to support the antidiabetic effect of the combined administration of the two plants. This present study was designed to test the hypoglycemic effect of polyherbal therapy of ethanolic extracts of *Nauclea latifolia* and *Gongronema latifolium* in normoglycemic and STZ-induced diabetic rats.

MATERIALS AND METHODS

Plant material: Fresh *Gongronema latifolium* leaves were collected from a cultivated land at Ibiaku Itam in Itu Local Government Area of Akwa Ibom State, Nigeria while *Nauclea latifolia* was collected by the fence of Industrial Training Fund office in Calabar, Cross River State of Nigeria. The sample was authenticated by Dr. E. G. Amanke, a Botanist in the Department of Botany, University of Calabar, Nigeria and voucher specimen deposited at the Department of Botany herbarium, University of Calabar.

Preparation of ethanol extract of plants: The ethanol extract was prepared using the wet method of extraction. One kilogramme of the fresh leaves of the plants were cut into pieces separately, blended in1.5litres of ethanol (96%) with an electric blender and transferred into amber coloured bottle and kept in cool (4°C) dark compartment for 72hours. The mixture/blend was filtered using a cheese material and thereafter with Whatman No 1 filter paper. The extract was concentrated *in vacuo* using a rotary evaporator at 37-40°C and dried completely in a desiccator containing a self-indicating silica gel.

ANIMALS

Albino Wistar rats of (150-250g) of both sexes obtained from the animal house of the Department of Pharmacology and Toxicology, Faculty of Pharmacy of the University of Uyo, Nigeria were used. They were kept in clean cages (wooden bottom and wire mesh top), maintained under standard laboratory conditions (Temperature 25± 5oC, Relative humidity 50-60%, and a 12/12h light/dark cycle) and were allowed free access to standard diet (Vital Feed from Grand Cereals and Oil Mills Limited, Jos, Plateau State of Nigeria) and water ad libitum. Animals were acclimatized for 14 days in the animal house of the Department of Biochemistry, University of Calabar, Nigeria before each of the experiments.

Experimental Induction of Diabetes

The diabetes was induced in the overnight fasted animals by a single intraperitoneal injection of freshly prepared solution of streptozotocin (Sigma, USA) 50mgkg^{-1}b.w. in 0.1M cold sodium citrate buffer pH 4.5 (Ghoraishian, 2006 and Rao and Naidu, 2010). The animals were considered as being diabetic if the blood glucose value were >200mg dl^{-1} on the third day after streptozotocin injection and were used in the experiment. This was estimated using One Touch Glucometer (Lifescan, inc 1995 Milpas, Califonia, USA) with blood obtained from the tail vein of the overnight fasted rats.

Experimental Protocol

Albino rats weighing 150-250g were used in this study. All the rats were kept at an average room temperature 0f 30°C in the animal room of Biochemistry Department, University of Calabar, Nigeria. They were allowed free access to water and feed (product of Pfhizer Nigeria, ltd, Jos) throughout the experimental period.

Thirty (30) rats included in the study were divided into six (6) groups of six (6) animals each (five diabetic and one non diabetic). Diabetes induction was done by intraperitoneal injection of 50mg/kg.b.w of Streptozotocin (STZ) (Sigma, St. Loius, MO, U.S.A) dissolved in sodium citrate buffer (0.01M, pH 4.5).

Diabetes was confirmed by the determination of fasting blood glucose concentration on the third day of post administration of STZ, showing glucose levels above 200mg/dl.

The experimental groupings were as follows: G.l = Diabetic rats treated with 200mg/kg.b.w. of *Gongronema latifolium* twice daily, N.l = Diabetic rats treated with 200mg/kg.b.w of *Nauclea latifolia* twice daily, G.l/N.l = Diabetic rats treated with 200mg/kg.b.w of Nauclea latifolia and Gongronema latifolia, morning and evening respectively.

Insulin = diabetic rats treated with 5 units/kg.b.w of insulin, DC = diabetic control; diabetic rats given placebo, NC = non-diabetic rats given placebo treatment.

At the end of twenty one days, the body weights, and fasting blood glucose of the animals were again determined.

Blood was collected on every 3 days through the rat's tail vein for glucose estimation. At the end of the experimental period, food was withdrawn from the rats and they were fasted overnight but had free access to water. They were then euthanized under chloroform vapor and sacrificed.

Immediately, overnight fasting blood samples were collected for sera preparation by cardiac puncture into sterile plain tubes. Serum samples were separated from the clot by centrifugation at 3,000rpm for 10 minutes using bench top centrifuge (MSE Minor, England) and stored frozen until needed for analysis. All analysis was completed within 24 hours of sample collection.

Biochemical Assays

Assay kits used for the biochemical assays were obtained from Randox Laboratories Ltd., Admore Diamond Road, Crumlin, Co., Antrim, United Kingdom, Qt 94QY: Lipid profile-Triglycerides (TG), Total cholesterol (TC), High density lipoprotein (HDL), Low density lipoprotein (LDL) and Very low density lipoprotein (VLDL), Aspartate aminotransferase (AST), and Alanine aminotransferase (ALT and were determined using Reitman and Frankel method, 1956., Glucose concentration were determined by the use of One Touch Glucometer (Lifescan, Inc., 1995 Milpitas, Califonia 95035, USA).

Statistical Analysis

All data were expressed as mean ± SD of the number of experiments. The results were analysed for statistical significant by one way ANOVA using the SPSS statistical program of Post Hoc Test (LSD) between groups using MS excel program. P values < 0.05 were considered significant.

RESULTS

The effects of *Gongronema latifolium* (G.l) and *Nauclea latifolia* (N.l) treatment on blood glucose, serum lipid profile and some enzymes of diabetic and non-diabetic rats are shown on tables 1-3 respectively.

Table 1. Effect of Plant Extracts on Blood Glucose levels ($mgkg^{-1}$) of Streptozotozin-induced Diabetic Rats

	Initial	Final	% change
GL	$504.00±30.54^{*, a}$	$169.67±16.62^{a}$	66.34
NL	$406.67±27.68^{*, a, c}$	$333.00±23.08^{*, a, c}$	18.12
GL + NL	$502.67±2.76^{*, a}$	$164.33±32.76^{a}$	67.73
Insulin (SD)	578.00±6.34*	77.33±10.36*	86.62
DC	292.00±21.05	214.67±14.42	26.43

*p<0.05 vs DC; a = p<0.05 vs SD; c = p<0.05 vs GL+NL.

Table 2. Effect of Plant Extracts on Serum Lipid Profile of Streptozotozin-induced Diabetic Rats

	TG(mg/dL)	TC(mg/dL)	HDL(mg/dL)	LDL(mg/dL)	VLDL(mg/dL)
GL	$66.20±7.57^{d}$	$105.07±5.74^{a, b}$	$60.20±2.94^{*,a,b,d}$	$58.10±5.36^{a, b}$	$13.24±1.51^{d}$
NL	$61.07±4.60^{d}$	$191.34±21.70^{*,b, d}$	$66.83±0.14^{*, d}$	$136.72±20.79^{b}$	$12.21±0.92^{d}$
GL +NL	$25.57±0.78^{*,a, b}$	$102.54±2.18^{a, b}$	$46.61±1.00^{*,b}$	$61.04±2.95^{a, b}$	$5.11±0.16^{*, a, b}$
Insulin	82.86±9.91	448.59±36.04*	66.36±0.15*	$398.80±38.17^{*, a}$	16.57±1.98
DC	75.23±14.56	254.40±29.84*	64.22±2.33*	205.23±31.01*	15.05±2.91
NC	76.41±9.65	115.95±14.59	8.26±1.04	92.40±15.06	15.28±1.93

*P<0.05 vs NC; a = P<0.05 vs DC; b = P<0.05 vs Insulin; d = P<0.05 vsGL+NL.

Table 3. Effect of Plant Extracts on some Serum Enzymes in Streptozotozin-induced Diabetic Rats

	AST (U/L)	ALT(U/L)	AMYLASE(U/L)
GL	$41.33±2.01^{*, a, b}$	$18.27±5.01^{*, a, b}$	$261.19±36.52^{b, d}$
NL	$50.33±12.23^{*, a, b}$	$14.33±1.38^{*, a, b}$	$216.37±5.74^{a, b}$
GL +NL	$32.67±1.17^{*, a, b}$	$10.00±0.73^{*, a, b}$	$171.371±9.97^{a, b}$
DC	70.67*±0.42*	69.33±18.09	254.14±45.41
NC	19.17±2.52	83.00±4.02	166.59±26.65

*P<0.05 vs NC; a = P<0.05 vs DC; b = P<0.05 vs Insulin; d = P<0.05 vsGL+NL.

Measured blood glucose in diabetic animals decreased significantly from initial by 66.34%, 18.12% and 67.73% respectively upon treatment with GL, NI and GL plus NI. Whereas diabetes induction caused significant increases ($p<0.05$) in total cholesterol (TC), 54.42% and low density lipoprotein (LDL), 55.0% compared to the normal control (NC), treatment with extracts of GL or NI significantly decreased ($p<0.05$) these by (58.70% and 24.79%, TC) and (71.70% and 33.38%, LDL) respectively. However, the combined extracts produced more steady reduction (59.70% and 70.26, TC and LDL), similar to the NC. Also the amino transferases (ALT and AST) activities which increased by 66.83% and 72.87% in the DC rats indicating hepatotoxicity, secondary to hyperglycemia, became reduced upon treatment with GL, NI and GL plus NI (Figure 1).

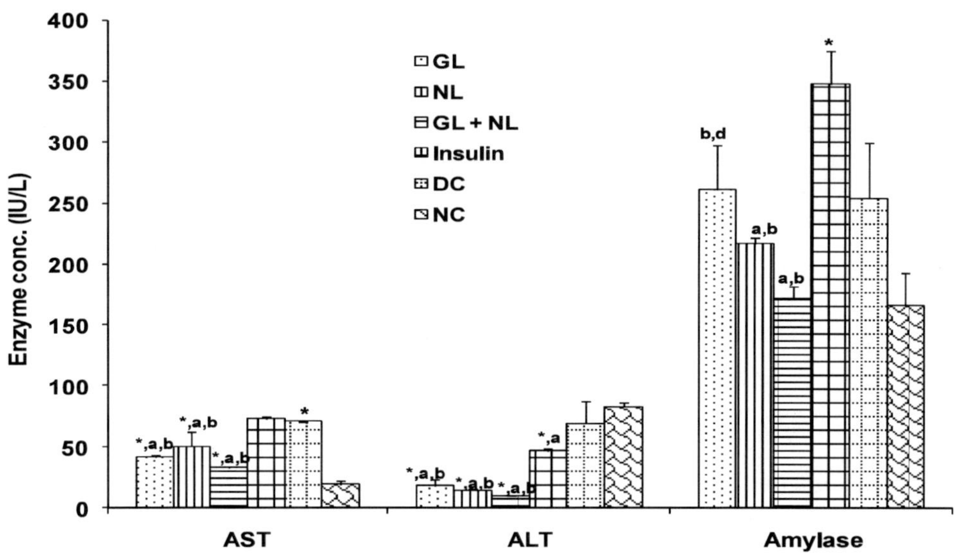

Figure 1. Effect of treatment on some serum enzymes (U/L) in diabetic rats. Values are mean ±SEM, n=6. *$p<0.05$ vs NC; a=$P<0.05$ vs DC; b=$P<0.05$ vs Insulin; d=$P<0.05$ vs GL+NL.

Again, the extent of reduction was better with combined extracts (85.58% and 53.77%), and compared favorably with NC.

The result shows that induction of diabetes with STZ caused a significant increase in TC and LDL as seen in the DC rats. Intervention with individual extract of G.l and N.l however reduced these indices but was better impacted when the extract were combined, indicating a positive synergy. This observation was also seen in blood glucose levels, lipid profile and the enzyme activities.

DISCUSSION

The presence of high amount of cholesterol in the diet has been shown to increase plasma cholesterol and may elevate aortic atherosclerosis (Lipid Research Clinical Program (LRCP), 1984). Many investigations have shown that diet treatment or drug therapy to regulate cholesterol can decrease subsequent cardiovascular disease (CVD) associated mortality and morbidity (Kwiterovich, 1997). On the basis of this, great effort have been made to reduce the risk of CVD through the regulation of cholesterol, thus the therapeutic benefits of plant foods have been focused of many extensive dietary studies (Yoko-Zawa et al., 2006 and Zhang et al., 2007). Traditional plant remedies have been used for centuries in the treatment of diseases (Akhan and Ali, 1984), but only a few have been scientifically evaluated. Therefore, the effects of G.l and N.l leaves extracts on the lipid profile of rats were studied. The two plants reduced the serum TC and LDL-C levels but a combined administration of the two plants further reduced it better than in individual plant treatment.

Nwanjo, 2005 has shown that the administration of aqueous leaf extract of *Vernonia amygdalina* produced hypoglycaemic, hypolipidaemic and antioxidant effects in rats which is similar to the observation of this finding. The result is also in line with the findings of Ugochukwu et al., 2003) and Adaramoye et al., (2000). The cholesterol lowering effects of these plant extracts could be beneficial in preventing lipid abnormalities which may arise in certain metabolic disorders (Cho et al., 2002). Ezekwe and Obidoa (2001) have reported that flavonoids, tannins and saponins may play some roles in the hypolipidaemic effect of some plants. The mechanism of hypocholesterolaemic action of these plant leaves may be due to inhibition of the absorption of dietary cholesterol in the intestine or its production by the liver (Ahmed-Raus et al., 1991) or stimulation of the biliary secretion of cholesterol and cholesterol excretion in faeces (Anderson et al., 1991).

Serum enzymes in this case viz. alanine aminotransferase, (ALT), Aspartate aminotransferase (AST), are present in the hepatic and biliary cells (Jensen, et al., 2004). These enzymes are usually released from the hepatocytes and leak into circulation causing increase in their serum levels under hepatocellular injury or inflammation of the biliary tract cells. Serum levels of these enzymes are particularly high in acute hepatocellular damage caused by drug toxicity and xenobiotics (Norman, 1998). The extent of the enzymes changes is related to the nature, closes to toxic agent and duration of toxicity (Shi et al., 2003; Song et al., 2003; Brukner et al., 1984).

In an inflammatory condition, there is a leakage of cytoplasmic enzymes into circulation, hence ALT levels increased above that of AST. Thus, when there is gross cellular necrosis, as in STZ-induced diabetes-damaged to the pancreatic cells by STZ, the level of AST may rise higher than that of ALT (Recknagel, 1987). This is because ALT levels is increased in the serum due to conditions where cells of the liver have been inflamed or undergo cell death, and is specific for the liver cells (Jeasen et al., 2004) but the AST levels can be triggered in other conditions such as myocardial infarction apart from hepatocellular damage (Jensen et al., 2004).

In the present study, there was conformity in the levels of the enzymes with available literature (Kim et al., 2006 and Atangwho et al., 2007). The higher levels of AST and ALT observed in DC may be attributed to the fact that the STZ may have caused some injury to other organs, in addition to the liver and pancreatic duct (gland). This now potentiates the release of AST into circulation over the ALT. However, the increase in aminotransferase is generally produced by cellular necrosis (Buckner et al., 1986).

Induction of diabetes by STZ is observed to be associated with increased serum enzyme activities (Atangwo et al., 2008). Hence the elevated levels of the enzymes (AST and ALT), generally was attributed to injury caused to the hepatocytes by STZ, which now affects the normal function of the liver, since these enzymes are indisputably markers of liver injury as stated earlier; they are localized in the cytoplasm under normal conditions and are released into the circulation under abnormal conditions (e.g., cellular damage) (Mourelle et al., 1987 and Romeo et al., 1999). The increased level of AST and ALT in rats induced diabetes with STZ was a clear indication of a kind of injury or the other caused by STZ toxicity.

The beneficial or synergistic effect of combination therapy over monotherapy has also been indicated. The seemingly untoward action of G.l extract in both diabetic and non-diabetic rats-decreased AST and ALT activity has effectively been modulated by N.l extract when co-administrated, since results of G.l + N.l group compared well with normal control group. Combined extracts of G.l and N.l may therefore be safer for use as a diabetic management compared to the individual extracts. Considering the result of the lipid profile, antiatherogenic effect is better with combined extracts relative to individual extracts and insulin administration. Evidence from the present study confirms that G.L and N.l were found to be highly effective in reducing TC and LDL thereby exhibiting hypocholesterolaemic effects.

It is evident from the result of this investigation that lower dosages of the two extracts when combined provide higher efficacy in protection against atherosclerosis and hepatotoxicity in diabetes. Hence the need to adopt this strategy in bio-prospecting for antidiabetic natural products.

REFERENCES

Adaramoye, O.A., Achem, J., Akintayo, O.O. and Fafunso, M.A.(2007). Hypolipidaemic effect of *Telfairia occidentalis* (Fluted pumpkin) in rats fed a cholesterol rich diet. *Journal of Medicinal Food*. 10: 330-336.

Ahmed-Rawus, R. R., Abdul-Latif, E. A., Mohamed, J. I. (2001). Lowering of lipid composition in aorta of Guinea pigs by *Curcuma domestica*. *BMC Complementary and Alternative Medicine*, 1: 6 (Abstract).

Akhtar, M.S and Ali, M.R. (1984). Study of antidiabetic effect of a compound medicinal plant prescription in normal and diabetic rabbits. *Journal of Pakistan Medicinal Association*. 34: 239-244.

Anderson, K.M., Odelt, P.M., Wilson, P.W. and Kannel, W.B. (1991). Cardiovascular disease risk profile. *American Heart Journal*, 21: 293-298.

Atangwho I. J, Ebong P.E, Egbung G.E, Eteng M.U and Eyong E.U (2007). Effect of *Vernonia amygdalina* Del on liver function in alloxan-induced hyperglycemic rats. *J. Pharm. Bioresour*, 4(1): 1-7.

Bruckner, J.V, Mackenzie, W,F and Muralidhara, S. (1986). Oral toxicity of carbontetrachloride. *Acute, Fundamental and Applied Toxicology*, 6: 16-34.

Bruckner, J.V., Luthra, R.and Kyle, G.M. (1984). Influence of time of exposure to carbontetrachloride on toxic liver injury. *Ann. Rev. Chronopharmacol*. 1: 373-376.

Cho, S. U., Park, J.Y., Park, E.M., Cho, M.S., Lee, M.Y., Jeon, S.M., Jung, M.K., Kim, M.J and Parl, Y.B. (2002). Aiteration of hepatic and antioxidant enzyme activity and lipid profile in STZ-induced diabetic rats by supplementation of dandelion water extract. *Clin. Cim. Acta*. 317: 109-117.

Etim, O.E; Akpan, E. J and Usoh, I. F. (2008). Hepatotoxicity of carbon tetrachloride: Protective effect of *Gongronema latifolium Pakistan Journal of Pharmacological Sciences.* 21:268-274.

Jensen, J.E; Stainberg, S.E; Freese, P. and Marino, E. (2004). Liver function tests. *Journal of Digestive Disorder* 6: 1-3.

Kwiterovich, P.OJr. (1997).The effect of dietary food antioxidants,and proxidants of blood lipids. Lipoprotein and atherosclerosis. *Journal of American Diet Association* 97:531-541.

Lipid Research Clinical Program (LRCP)(1984). The lipid research coronary primary prevention trial results II. *Journal of American Medicinal Association* 251: 306-374.

Morebise, O., Fafuniso, M.A., Makinde, J. M., Olayide, O. A. and Awe, E. (2002). Anti-inflammatory property of *Gongronema latifolium. Phytotherapy Rsearch*, 16: 575-577.

Morebise,O. and Fafunso, M.A. (1998). Antimicrobial and Phytotoxic activities of sapanis extracts from two Nigerian edible medicinal plants. *Biokemistril* 8:69-77.

Mourelle, M., Ameeziua, J.L and Hong, E (1987). Effect of riprostol and colchines on carbontetrachloride acute liver damage in rats. Relationship with Plasma membrane Lipids. *Prostaglandins*, 33: 869-877.

Norman, J.J. (1998). *Common Laboratory Test in Liver Diseases.* Columbia Education. Columbia.

Nwanjo, H.U. (2005). Efficacy of aqueous leaf extract of *Vernonia amygdalina* on plasma lipoprotein and oxidative status in diabetic rats models. *Nigerian Journal of Physiological Sciences* 20: 39-42.

Nwanjo, H.U; Okafor, M.C. and Oze, G.O (2006). Anti-lipid Peroxidative Activity of *Gongronema latifolium* in Streptozotocin-induced Diabetic Rats. *Nigerian Journal of Physiological Sciences* 21 (1-2):61-65.

Rates, S. M. (2001). Plants as source of drugs. *Toxicon,* 39(5): 603-613.

Recknagel, R.O. (1987). Carbon tetrachloride hepatotoxicity. *Pharmacol. Rev..*, 19: 145-195.

Romeo, S., Lasher, S.B., and Genarruzzbeitia, E. (1999). Protective effects of calcium channel blockers in carbontetrachloride-induced liver toxicity. *Life Science*, 55: 981-990.

Schneider, C. R., Sheidt, K., Breitmaier, E. (2003). Four new pregnant glycosides from *Gongronema latifolium* extract on serum lipid profile and diabetic rabbit stress in hepatocyes of diabetic rats. *Journal of Biosciences*, 28(1): 1-5.

Shi, J. Asiski, K., Ikawa, Y. and Wakke, K. (2003). Evidence of hepatocyte apoptosis in rat liver after the administration of carbontetrachloride. *Journal of Medicinal Research* 4: 1-8.

Singh RB, Niaz MA, Ghosh S. Effect on central obesity and associated disturbances of low energy, fruit and vegetable enriched prudent diet in north Indians. *Postgrad. Med. J.* 1994; 70:895-900.

Singh RB, Niaz MA, Ghosh S. Hypolipidemic and antioxidant effects of commiphora mukul as adjunct to diet therapy in patients with hypercholesterolemia. *Cardiovas. Drug Ther.* 1994; 8:85-90.

Singh RB, Niaz MA, Rastogi V, Singh N, Postiglion A, Rastogi SS. Hypolipidemic and antioxidant effects of fenugreek seeds and triphala as adjunct to dietary therapy in patients with mild to moderate hypercholesterolemia. *Perfusion* 1998; 11: 124-130.

Song, I. Lee, Y., Chung, S. and Shim, C.(2003). Multiple alteration of canalicular membrane transport activities in rats with CCl_4 –induced hepatic injury. *Journal of Pharmacology and Toxicology* 8: 506-571.

Ugochukwu, N. H., Babady, N. E., Cobourne, M. I. and Gasset, S. R. (2003). The effect of *Gongronema latifolium* extract on serum lipid profile and oxidative stress in hepatocytes of diabetic rats. *Journal of Biosciences,* 28 (1): 1-5.

Ugochukwu, N.H and Babady, N.E. (2002). Antioxidant effects of *Gongronema latifolium* in hepatocytes of rat models of non-insulin dependent diabetes mellitus. Fitoterapia 73: 612-618.

Yokokawa,T., Ishida, A., Cho, E.J and Nakagawa, T. (2006). The effect of Coptidis rhizome extract on a hypercholesterolemic animal model. *Phytomedicine* 10: 17-22.

Zhang, H.W., Zhang, Y. H., Lu, M.J and Tongwei-Jun, C.A.O. (2007). Comparison of hypertension, dyslipidaemia and hyperglycaemia between buckwheat seed-consuming and non-consuming Mongolian-Chinese population. *Clinical Experimental Pharmacology and Physiology*. 34: 838-844.

Zimmet, P.Z. (1999). Diabetes epidemiology as a tool to trigger diabetes research and care. *Diabetologia* 42(5):99-518.

In: Cardiovascular Health and Chronomics
Editor: Ram B. Singh
ISBN: 978-1-62948-976-6

Chapter 14

24- HOUR CHRONOMICS OF AMBULATORY BLOOD PRESSURE MONITORING IN ROTATING NIGHT SHIFT WORKERS AND CONTROLS

Baby Anjum[1], Nar Singh Verma[2*], Sandeep Tiwari[3], Vinod Jain[3], Ranjana Singh[1], Shipra Bhardwaj[2], Qulsoom Naz[2], Abbas A. Mahdi[1], Ram B. Singh[4] and Raj K. Singh[5]

[1]Department of Biochemistry, C S M Medical University, Lucknow, India
[2]Department of Physiology, C S M Medical University, Lucknow, India
[3]Department of Surgery, C S M Medical University, Lucknow, India
[4]Halberg Hospital and Research Institute, Moradabad, India
[5]Department of Biochemistry, SGRRIM & HS, Dehradun, India

ABSTRACT

Background. Concerns about the health of night shift workers has long been debated. Shift work is associated with the disruption of circadian rhythms, where a person's internal body clock is in conflict with the rotating night shift schedule which causes internal desynchronization, sleep disturbances resulting in to increased risk of cardiovascular diseases and diabetes and other physiological disorders.

Objectives: We investigated the circadian pattern of blood pressure and heart rate in night shift workers and day shift controls, and whether the changes in circadian pattern produced by rotating night shift are reversible in due course of time.

Method: 14 healthy nursing professionals, aged 20-40 year, performing day and night shifts and 14 subjects as controls, performing day duty were recruited in the study. Circadian patterns of blood pressure and heart rate were evaluated in night shift workers during the work shift (Night & Day shift) and in controls.

Result: Night shift workers showed a very interesting altered circadian pattern of double amplitude when subject went back to the day shift. An extremely significant

*Address correspondence to: Dr N S Verma, Professor. Department of Physiology, C S M Medical University (Formerly King George's Medical University), Lucknow-226003 (Uttar Pradesh) India. E-mail: drnsvermakgmu@gmail.com; Phone: +91 9839064560 Fax No: 0522-2257539.

different pattern of SBP and DBP double amplitude was found between day shift and controls ($p<0.01$).

Conclusion: Effect of rotating night shift develops later due to rotating shift and desynchronization. Acrophase pattern was clinically significant when studied in individual subjects in different shifts with controls. Alterations in Acrophase were persistent during night as well as day shift due to incomplete recovery and ecphasia was very common among night shift workers.

Keywords: Rotating night shift, ambulatory blood pressure and heart rate monitoring, ecphasia

INTRODUCTION

Concerns about the health of night shift workers have been long debated. It was a matter of concern that these subjects, when deprived of night sleep, may have biological parameters that are affected by such activity and, whether this deprivation can be and associated parameter change can be compensated for by day time sleep. Researchers have found that there is an increase risk of accidents & human error, an increase in sleepiness and of fatigue due to deprivation of sleep, increased health problems; and disruption to family and social life. Night shift working is associated with a disruption of circadian rhythms, where a person's internal body clock is in conflict with the rotating shift schedule. The circadian rhythm of the human body is characterized with an alternating cycle of sleep and awakening [1]. Among healthy subjects, sleep tends to occur during a particular phase of circadian cycle [2]. Those who work during the night shift may attempt to sleep when their body clock is adjusted for the awakening phase [3]. This attempt disturbs the body clock resulting in a contradictory relationship between sleep time and circadian schedule. There is evidence that shift work affects both sleep and awakening by disrupting the circadian regulation which has adverse effects on family and societal life [4]. The night shift work alters both length and quality of sleep. The average sleep cycle for night shift workers is sleeping during the day which may be 2-4 hrs shorter than that of the day worker sleeping at night. Day sleep is light, fragmented, and more likely to be disrupted and hence, the insomnia can be severe in night shift workers [5]. It is possible that the circadian sleep propensity rhythm and hormonal rhythm are under influence of circadian pacemaker as well as sleep habit [6].

Most rhythms are driven by an internal biological clock located in the hypothalamic suprachiasmatic nucleus and can be synchronized by external signals such as light-dark cycles [7]. The rapidly rotating shift system including two consecutive night shifts, do not significantly alter the normal circadian rhythm of the body, particularly performance level, body temperature and hormone release [8].

Majority of the circadian rhythms in our body have both an endogenous component regulated by an internal clock, viz. the suprachiasmatic nuclei (SCN), and an exogenous component composed of a light-dark cycle [1, 5]. The disruption in natural time pattern, under influence of night dark cycle, acts upon the circadian system to bring it into synchronization with the new time pattern.

Normal circadian pattern of blood pressure- Blood pressure is higher during the day time (between 10 a.m. to 6 p.m. and lower at night .There is a characteristic dip in blood pressure between midnight and 3 a.m, however, between 3a.m. and 6 a.m. there is a slow and steady rise in blood pressure but after arousal there is sudden steep elevation of blood pressure. This rapid increase in blood pressure is continuous for 4-6 hour i.e., 6 a.m. to 12 p.m. An interaction between biorhythms, the biological clock and triggers, which may be important in the pathogenesis of altered heart rate variability (HRV) and blood pressure variability (BPV). Circadian rhythms are under the influence of, and physiological variables are mediated by the activation of the adrenals, sympathetic/parasympathetic, hypothalamic and pituitary activity [9, 10]. Circadian rhythms, coordinated in part by the parietal hypothalamic-pituitary and adrenal mechanisms, have been reported in almost all variables examined thus far, including the circulation. It is possible that all metabolic functions undergo circadian rhythms [11].

Previous reports indicate that diurnal blood pressure (BP) variation, in addition to high BP per sec, is related to target organ damage and the incidence of cardiovascular events. Physical activity is one of the determinants of ambulatory BP and diurnal variation. [12]. Day shift workers show typical circadian rhythms with a drop in both systolic and diastolic blood at night. This pattern is reversed in night–shift workers [13].

In the present case-control study, we investigated the circadian pattern of blood pressure and heart rate in night-shift workers and in controls, to determine whether these changes are reversible after change to day-duty schedules.

Methodology

Subjects

Out of 20 volunteers (night shift workers), 6 were excluded due to non-fulfillment of study protocol. The duration and pattern of shift work were the same among all the subjects and 14 healthy nursing professionals, aged 20-40 year, performing day and night shift duties (continuous 9 days night shift with alternate day shifts) {from last 6 years} and were randomly selected and recruited from the Trauma Center, GM and Associated Hospitals, Chhatrapati Shahuji Maharaj Medical University, Lucknow, UP, India. 14 subjects acting as controls, performing day duty (from 9:00 a.m. to 5:00 p.m.) alone were also recruited. The study was conducted during February to July, 2009 when the average temperature of the city ranged between 34°C and 38°C. At 26.50 N°, Lucknow is located just north of the tropic of cancer. The study was approved by the institutional ethic committee (Ref. code: XXXIV ECM/B-P3) and written, informed consent was obtained from all the subjects to participate in the study. Inclusion criteria (for night shift workers) were healthy nursing professional of both genders, aged between 20-40 years who performed night and day duty and inclusion criteria (for controls) was healthy working professional of both genders, aged between 20-40 years who performed day duty. Exclusion criteria weresubjects with any acute/chronic illness, known patients with diabetes mellitus, other endocrinological disorders, hypertension, coronary artery disease, and chronic renal.

24-Hour Ambulatory Blood Pressure and Heart Rate Monitoring in Night Shift Workers and in Controls

Blood pressure and heart rate were recorded by ambulatory blood pressure monitor TM-2430 (A&D, Tokyo, Japan) that measured repeated oscillatory blood pressure and heart rate at selected intervals. Taking serial measurements a few times each day is important to reduce the error associated with a single measurement. The chronobiologic characterization of the circadian amplitude and acrophase in addition to the midline estimating statistics of rhythm (MESOR) further reduces the errors in interpretation. Taking only one or two measurements a day, always at awakening and/or at bed time may fail to reveal abnormalities seen only at other times of the day, or abnormalities that apply only to the variability in blood pressure or heart rate [14].

In this study, the subjects wore an ambulatory blood pressure monitor TM-2430 programmed to automatically measure blood pressure and heart rate at 30 min intervals while awake and sleeping hours during night shift and again when they were shifted to day duties and similarly in controls. The data were downloaded after every monitoring span to a local PC *via* an interface (TM-2421, A&D). Each blood pressure and heart rate profile was analyzed by a sphygmochron, utilizing both a parametric and non-parametric approach. Ambulatory blood pressure monitoring records were sent to the Halberg Chronobiology Center, University of Minnesota, Minneapolis, MN, USA for further interpretations. Original oscillometric data from each blood pressure series was first synchronized according to the activity-rest cycle of each individual by recomputing all the records in hours, from bed time to avoid differences among subjects in actual time of daily activity and to express results in circadian time rather than in less meaningful clock hours. After synchronization, blood pressure and heart rate values were edited according to commonly used criteria for the removal of outliers and measurement errors. The remaining data were analyzed chronobiologically.

The study of human chronomes can serve the derivation of refined reference values to better define health and to identify pre-disease, so that prophylactic interventions can be instituted as early as possible, preferably before the onset of disease [15]. In the current implementation of the chronobiological recommendations, reference values have been specified for clinically healthy peers of a given gender and ethnicity in different age groups [16]. Ambulatory blood pressure monitoring was done during their day and night shifts and in controls. Some essential parameters which are directly under influence of night shift workers such as body temperature, time of arousal, time of going to bed, duration of nocturnal and diurnal sleep, mode of waking up, sleep latency, quality of sleep, feeding habits, menstrual history (for females), and family history were also recorded. Acrophase (hr:min), time of overall high/peak values, and hyperbaric index) were calculated for systolic blood pressure (SBP), diastolic blood pressure (DBP), and heart rate. The circadian amplitude, a measure of the extent of reproducible variability within a day, was obtained by linear curve fitting, which yielded added parameters: in midline-estimating statistics of rhythm, the MESOR (a time structure or chronome-adjusted mean), acrophase (the timing of overall high values), recurring in each cycles, and the amplitude and acrophase of the 12 hour (and higher order) harmonics of the circadian variation that with the characteristic of the fundamental 24 hours component, describe the circadian wave form. The MESOR is a more precise and more accurate estimate of location than the arithmetic mean [16].

RESULTS

We measured the effect of rotating night shift by comparisons of night shift workers with actual control group. Night shift workers worked on a different shift (night and day shift) therefore we compared the circadian pattern of blood pressure and heart rate in shift workers during their night shift with controls, again when these night shift workers came back to day shift circadian patterns were compared with controls and we also compared within the shift the circadian pattern of blood pressure and heart rate of night shift with day shift. Data were presented as means and standard deviations (SDs) and statistical significance between the appropriate groups by the *p*-value.

Table 1. Baseline characteristics of night shift workers and controls: Means (SDs)

Baseline Characteristics	Night shift workers (n=14)	Controls (n=14)
Age	25.71 (5.28)	30.14 (7.00)
Weight (kg)	51.42 (7.78)	57.00 (5.60)
Height (cm)	159.78 (10.26)	162.00 (6.43)
Body mass index (BMI)	20.08 (1.78)	21.76 (2.20)

n= number of subjects
Data are presented as Means(SDs).

Table 2. Midline statistics of rhythms (MESOR) of Night shift workers and Controls

MESOR of SBP,DBP and HR of Night shift worker (During night shift) and Controls			
	Night Shift (Mean (SD))	Controls (Mean (SD))	*p* Values
SBP	120.16 (8.02)	120.91 (11.74)	0.84
DBP	72.64 (5.22)	76.10 (7.05)	0.15
HR	75.49 (5.47)	77.59 (22.30)	0.733
MESOR of SBP,DBP and HR of Night shift worker (During day shift) and Controls			
	Day Shift (Mean (SD))	Controls (Mean (SD))	*p* Values
SBP	120.66 (6.43)	120.91 (11,74)	0,94
DBP	73.44 (4.72)	76.10 (7.05)	0.25
HR	75.45 (5.68)	77.59 (22.30)	0.73
MESOR of SBP,DBP and HR of Night shift worker (During night shift) and Night shift worker (During day shift)			
	Night Shift (Mean (SD)	Day Shift (Mean (SD))	*p* Values
SBP	120.16 (8.02)	120.66 (6.43)	0.85
DBP	72.64 (5.22)	73.44 (4.72)	0.67
HR	75.49 (5.47)	75.45 (5.68)	0.98

The circadian pattern of blood pressure and heart rate as adjudged from the MESOR was not statistically significant in night shift workers during their night as well as day shift and in controls. (p>0.05) (Table 2). Therefore changes were not found in MESOR values when we compared night shift to controls, day shift to controls and night shift with day shift (Figure 1). Similarly for hyperbaric index.

Significant difference was found in double amplitude or predictable change during day shift.

But these differences were not found between night shift and controls. However, altered circadian patterns of systolic blood pressure, diastolic blood pressure and heart were found between day shift and controls.

Controls showed a normal range of systolic and diastolic blood pressure for double amplitude.

Table 3. Double Amplitude (2A)/Predictable change of Night shift workers and Controls

Double Amplitude of SBP,DBP and HR of Night shift worker (During night shift) and Controls			
	Night Shift (Mean (SD))	Controls (Mean (SD))	p Values
SBP	27.36 (18.83)	22.83 (10.99)	0.55
DBP	19.50 (12.78)	17.10 (6.83)	0.54
HR	14.63 (9.28)	18.70 (10.04)	0.27
Double Amplitude of SBP,DBP and HR of Night shift worker (During day shift) and Controls			
	Day Shift (Mean (SD))	Controls (Mean (SD))	p Values
SBP	42.98 (20.75)	22.83 (10.99)	0.005**
DBP	29.19 (13.48)	17.10 (6.83)	0.006**
HR	18.87 (8.70)	18.70 (10.04)	0.96
Double Amplitude of SBP,DBP and HR of Night shift worker (During night shift) and Night shift worker (During day shift)			
	Night Shift (Mean (SD))	Day Shift (Mean (SD))	p Values
SBP	27.36 (18.83)	42.98 (20.75)	0.04*
DBP	19.50 (12.78)	29.19 (13.48)	0.02*
HR	14.63 (9.28)	18.87 (8.70)	0.22

Abbreviations: MESOR, midline estimating statistics of rhythm; 2A,Double amplitude; SBP, systolic blood pressure; DBP, diastolic blood pressure.

*significant; **extremely significant.

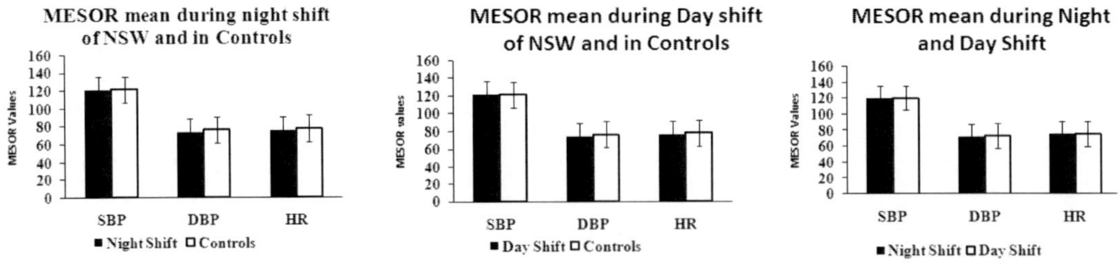

A B C

Figure 1. MESOR of SBP, DBP, and Heart rate during night shift, day shift and in controls. (A) MESOR of night shift workers during night shift and in controls. (B) MESOR of night shift workers during day shift and in controls. (C) MESOR of night shift workers during night shift and day shift. Abbreviations: MESOR, Midline statistics of rhythm; SBP, systolic blood pressure; DBP, diastolic blood pressure; HR, heart rate.

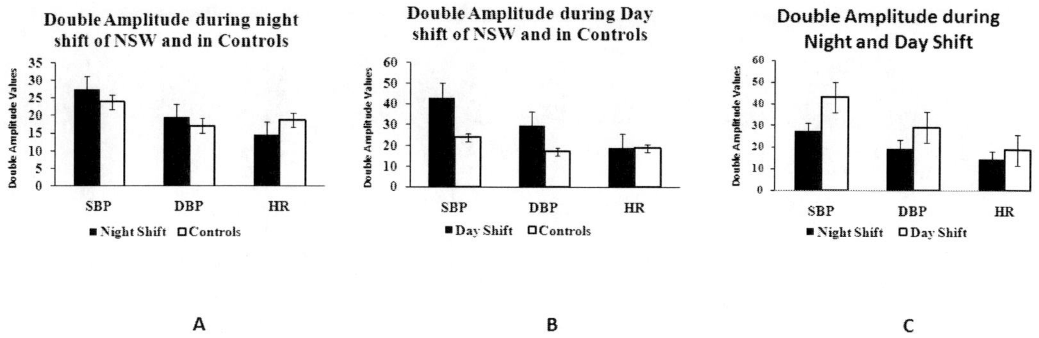

Figure 2. Double Amplitude of SBP, DBP, and Heart rate during night shift, day shift and in controls. (A) Double Amplitude of night shift workers during night shift and in controls. (B) Double Amplitude of night shift workers during day shift and in controls. (C) Double Amplitude of night shift workers during night shift and day shift. Abbreviations: MESOR, Midline statistics of rhythm; SBP, systolic blood pressure; DBP, diastolic blood pressure; HR, heart rate.

Night shift workers showed a very interesting altered circadian pattern when subject went back to the day shift (Figure 2) that means the effect of rotating night shift develops later due to frequently rotating shifts and associated desynchronization.

Extremely significant patterns of SBP and DBP of double amplitude were found between day shift and controls ($p<0.01$) (Table 3). Changes in heart rate during day shift were not found, therefore a significant difference was not found in heart rate between day shift and controls or between night shift with day shift. When we compared night shift with day shift within the group, significant changes were found in the circadian pattern of systolic and diastolic blood pressure due to effects of the rotating night shift developing later during day shift but not in night shift ($p<0.05$).(Table 3).

The acrophase was significant clinically. Chronobiological study needs to evaluate the data individually, not statistically, and this will be interesting when we estimate time patterns i.e., time of overall peak values. The acrophase pattern was clinically significant when we studied individual subjects in different shifts with controls. Very interesting patterns of systolic blood pressure, diastolic blood pressure and heart rate acrophase were found during night shift, however during day shift incomplete recovery was found in a few subjects (Figure 3). Normal patterns of SBP, DBP and HR acrophase (within the range) was found in controls. Hyperbaric index (three hours fractionated time interval/upper limit of tolerance interval) is a parameter by which we can measure the changes above the tolerance limits. Alteration in hyperbaric index of SBP and DBP (ecphasia) was found during night and day shift therefore recovery was not found during day shift. (Figure 4). A change in heart rate of hyperbaric index was not found in different shifts and in controls. Ecphasia (odd timing of circadian pattern of blood pressure not heart) was clearly found during night and day shift.

Significant differences in MESOR were not found for night and day shift. Therefore hypertension was not found in night shift workers. A significant difference was found in double amplitude or predictable change during day shift. We concluded that the effect of rotating night shift was not apparent during the night shift however it develops later when subjects came back to day shift. Very interesting altered circadian patterns was found in night shift workers when subjects came back to the day shift. It indicated that the effect of rotating night shift develops later due to frequently rotating shift and subsequent desynchronization.

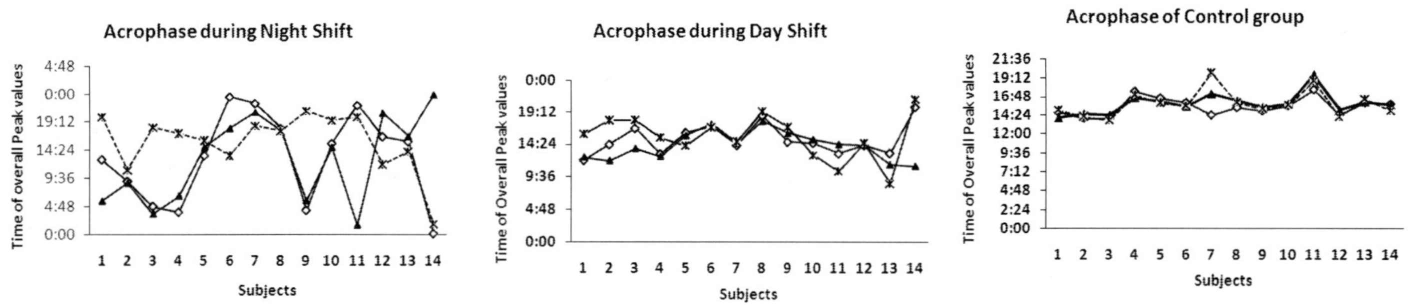

A B C

Figure 3. Acrophase (time of overall peak values) of SBP, DBP, and Heart rate. (A) Night shift. (B) Day shift. (**C**) In Controls. , Systolic blood pressure; Diastolic blood pressure; Heart rate.

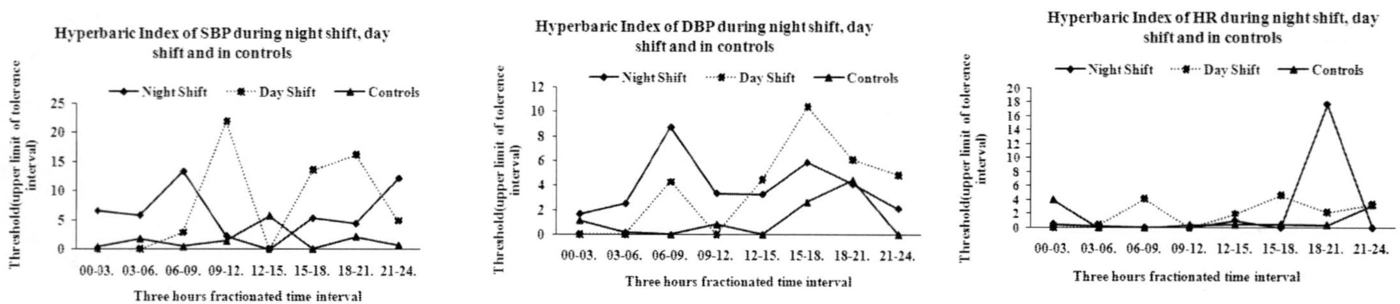

A B C

Figure 4. Hyperbaric index of night shift workers during night shift, day shift and in controls. (A) Systolic blood pressure. (B) Diastolic blood pressure. (C) Heart rate. Night Shift, Day Shift, Controls.

The present study concluded that alterations in acrophase were persistent during night as well as the day shift due to incomplete recovery and ecphasia was very common among night shift workers.

DISCUSSION

Night shift work has been reported to be associated with altered circadian patterns of blood pressure and desynchronization. The present study determines the effect of night shift on circadian patterns of blood pressure and heart rate and these effects were evaluated by comparison with actual controls who performed day duty. The study determines whether the changes in circadian pattern produced by night shift are reversible in due course of time.

Our results show the changes in the systolic and diastolic blood pressure double amplitude when subjects completed 8-9 days night shift and came back to day shift. The present observation confirmed the previous report of altered and reversed pattern of blood pressure and heart rate (acrophase) in night shift workers. [13, 17] Identical blood pressure and heart rate have been observed among night shift workers. [18] MESOR values were similar in shift workers and controls. On the basis of the findings of the altered circadian pattern of double amplitude we decided that effect of rotating night shift was not present during night shift however it develops later when the subjects came back to day shift. It indicated that the effect of rotating night shift develops later due to frequently rotating shift and desynchronization.

Night shift workers are awake when they are supposed to sleep and they attempt to sleep in day time when they are normally supposed to be awake. Previous findings were similar to our present study which showed that the odd timing of blood pressure (ecphasia) clearly indicates the non dipping pattern of blood pressure at night during night shift.

In view of these clinical manifestations c by our subjects it is possible that night shift work may be harmful for the health, family and social life. Other studies have also reported higher incidence of poorer sleep and its complications in night shift workers. (19-21)In our study, subjects complained of several sleep related problems viz, difficulty in sleep, sleepiness and other problems related to cognitive function like mental fatigue, difficulty in skilled work, problems remembering, decreased alertness, irritation and depression during night shift which are similar to previous study. [17] The most important physiological mechanisms regarding the shift work, particularly night shift work, is the problem of entrainment (resynchronization) of physiological functions after a phase shift of working and sleeping times. Other similar studies have shown that the shift work is associated with increased cardiovascular morbidity and mortality [22-24].

Singh et al. have reported for the first time from India in the years 1988-2001, the results of ambulatory blood pressure monitoring for 4 to 7 days, in several studies among healthy subjects [25-27]. In one study [25], normal systolic and diastolic blood pressures with morning rise and dipping in the night have been reported in 2002. In other studies, subjects doing active prayer had lower heart rate and lower blood pressures compared to those who were not doing regular prayer [26]. It is possible that shift workers may benefit by doing active prayer, although such sub-study could not be done in this subset of shift workers. Singh et al. also reported for the first time in the literature, a circadian decline in antioxidant vitamins, nitrite, coenzyme Q10 and magnesium and increase in oxidative stress during second quarter of the day compared to values in the evening [27,28]. The adverse effects of shift work have been reviewed but no study has reported if shift work would be associated with greater decline in vitamins and greater increase in oxidative stress, because such findings may explain the pathogenesis of increased blood pressure variability, type 2 diabetes mellitus, hypertension insulin resistance and mortality among shift workers [22-24,29].

An Pan et al. contributed substantially to this field of research by examining the association between rotating shift work (≥3 nights/month plus days and evenings) and type 2 diabetes mellitus among 177,000 female nurses aged 25–67 at baseline followed up for up to two decades (the Nurses' Health Study)(30). This large-scale study revealed a graded association between the duration of working life the nurses had been engaged in shift work and risk of developing type 2 diabetes. Compared with women who reported no shift work, participants with 1–2 years of shift work had a 5% excess risk of type 2 diabetes, rising to 20% after 3–9 years, 40% after 10–19 years, and almost 60% for ≥20 years. Further studies showed that shift work can predispose obesity, diabetes and insulin resistance as well as CAD [31-36]. Further studies also reported similar results indicating changes in SBP and DBP seemed to be peak during waking time at the same time on the day off, as they did on the working day.

Whereas HRV profiles usually returned to baseline level after each shift. The SBP and DBP of night shift workers did not completely returned to the baseline levels the following off duty day. 12-h night shift work may elevate BP and HR and decrease HRV which may be associated with delayed blood pressure recovery [37-39].

In brief, it is possible that rotating night shift appears to have adverse effects on workers health, quality of family and social life as well as to patient care. Further studies in a larger number of subjects and long term follow up are necessary to confirm our findings. However, it may be advised that rotating night shift work should be avoided due to the development of the stage of desynchronization. However, we are increasingly residing in a society which wants to work 7 days in a week and 24 hours in a day. Therefore, option to eradicate shift working is not realistic. If the observed association between rotating shift work and noncommunicable diseases is causal, as observed in our study, additional efforts to prevent CVDs and type 2 diabetes among shift workers through promotion of healthy lifestyles, weight control, and early identification and treatment of prediabetic and diabetic employees are needed. Rotating shift work comprises a range of alternative schedule patterns, such as backward- and forward-rotating shift systems, and the proportion of night and early morning shifts varies. Our study, addressed some of these variations and identified a pattern that may minimize the risk. However, large-scale randomized trials are necessary to provide insights into causality.

Acknowledgments

The financial support provided by Council of Science & Technology, UP Government is highly appreciated. The authors are thankful to Prof. G. Cornelissen, Director, Halberg Chronobiology Center, and University of Minnesota; USA, for advanced cosinor analysis and chronobiological interpretations. The authors wish to acknowledge with gratitude the nursing administration of the institution and each of the nurses whose participation made this study possible.

CONFLICTS OF INTEREST

The authors declare there is no conflict of interest.

REFERENCES

[1] Shneerson J M, Ohayon M M, Carscadon M A. Circadian rhythm, Rapid eye movement (REM) sleep; 2007.Armenian medical network.

[2] Aschoff J. Exogenous and endogenous components in circadian rhythms. *Cold Spring Harbor Symp. Quant Biol.*1960; 25:11-28.

[3] Lamond N, Dorrian J, Roach G D, McCulloch K, Holmes A L, Burgess, Fletcher A, Dawson D. The impact of a week of simulated night work on sleep, circadian phase and performance. *Occup. Environ. Med.* 2003; 60:el 13.

[4] Abdalkader R S. Effects of night shift on nurses working in intensive care units at Jorden University Hospital. *European journal of scientific research.* 2008; ISSN 1450-216 X vol.23 No.1 pp. 70-86.

[5] Santhi N, Jeanne F Duffy, Todd S Horowitz, Charles A, Czeiser. Scheduling of sleep/darkness affects the circadian phase of night shift workers. *Neuroscience letters.* 2005;384: 316-320.

[6] Kudo Y, Uchivama M, Okawa M, Shibui K,Kamei Y,Hayakawa T,Kim K, Ishibashi K. Correlation between the circadian sleep propensity rhythm and hormonal rhythm under ultra short sleep wake cycle. *Psychiatry Clin. Neurosci.* 1999; 53:253-255.

[7] Nagai K, Nagai N, Sugahara K, Niijima A, Nakagawa H. Circadian rhythms and energy metabolism and special reference to the suprachiasmatic nucleus. *Neurosci. Biobehav Rev.* 1994; 18:579-584.

[8] Costa G, Ghirlanda G, Tarondi G, Minors D, Waterhouse J. Evaluation of a rapidly rotating shift system of tolerance of nurses to night work. *Int. Arch. Occup. Environ Health.* 1994; 65:305-311.

[9] Singh R B, Weydahl A, Otsuka K, Watanabe Y, Yano S, Mori H, Ichimaru Y, Mitsutake G, Satu Y, Fanghong L, Zhao ZY, Kartic C, Gyozdiakoya A. Can nutrition influence circadian rhythm and heart rate variability. *Biomed. Pharmacother.* 2001:55; 1:115s 124s.

[10] Singh R B, Cornelissen G, Weydahl A, Schwartzkopff O, Katina G, Otsuka k, Watanabe Y, Yano S, Mori H, Ichimaru Y, Mitsutake G, Pella D, Fanghong L, Zhao Z, Rao RS, Gyozgiakoya A, Halberg F. Circadian heart rate and blood pressure variability considered for heart rate and patient care. *Int. J. Cardiol* 203, Jan: 87 (1):9-28.

[11] Singh R B, Pella D, Otsuka K, Halberg F, Cornelissen G. New Insights into circadian aspects of health and disease. *J. Assoc Physicians India*, 2002, Nov:50:1416-25

[12] Kario K, Schwartz JE, Pickering TG. Ambulatory physical activity as a determinant of diurnal blood pressure variation. *Hypertension.* 1999; 34:685-691.

[13] Sternberg H, Rosenthal T,Shamiss A, Green M. Altered circadian rhythm of blood pressure in shift workers. *J. Hum. Hypertens*. 1995; 9:349-353.

[14] Cornelissen G, Halberg F, Bakken EE, Wang Z, Tarquni R, Perfetto F, Laffi G. Maggioni C, Kumagai Y, Homolka P, Havelkova A, Dusek J, Svacinova H, Siegelova J, Fisher B. Chronobiology of high blood pressure. *Scripta Medica (Brno).* 2007; 80:157-166.

[15] Watanebe Y, Cornellisen G, Halberg F, Otsuka K, Ohkawa S, Kikuchi T, Siegelova J. Need for chronobiologic reference values (chronodesm) smotthed over age: A problem awaiting a BIOCOS solution. *Scripta Medica* (Brno). 2000; 73:105-110.

[16] Halberg F, Cornellisen G, Wall D, et al. Engineering and governmental challenge:7 day/24 hour chronobiological blood pressure and heart rate screening: *Part I. Biomed. Instrum Technol*. 2002; 36:89-122.

[17] Anjum B, Verma NS, Tiwari S, Singh R, Mahdi AA, Singh RB, Singh RK. Association of salivary cortisol with chronomics of 24 hours ambulatory blood pressure/heart rate among night shift workers. *Bioscience Trends. 2011; 5(4):182-188.*

[18] Goto T, Yokoyama K, Araki T, Saitoh H, Saitoh M, Satoh S. Identical blood pressure levels and slower heart rate among nurses during night work and day work. *J. Hum. Hyperten*. 1994; 8:11-14.

[19] Hennig J, Kieferdarf P, Moritz C, Huwe S, Netter P. Changes in cortisol secretion during shift work implication for tolerance to shift work. *Ergonomics*. 1998; 41:610-621.

[20] Akerstedt T. Sleepiness as a consequence of shift work. *Sleep*. 1988; 11:17-34.

[21] Rutenfranz J,Colquhoun WP, Knauth P, Ghata JN. Biomedical and physiological aspects of shift work. *Scand J. Work Environ Health*. 1977; 3:165-182.

[22] .Motohashi Y. Alteration of circadian rhythm in shift-working ambulance personnel monitoring of salivary cortisol rhythm. *Ergonomics*. 1992; 35:1331-1340.

[23] Knutsson A, Akerstedt T, Jonsson BG, Orth-Gomer K. Increased risk of ischemic heart disease in shift workers. *Lancet*. 1986; 2:89-92.

[24] Kawachi L,Colditz GA,Stampfer MJ, Willett WC, Manson JE,Speizer FE,Hennekens CH. Prospective study of shift work and risk of coronary heart disease in women. *Circulation*. 1995; 92:3178-3182.25.

[25] Singh RB, Cornelissen G, Siegelova J, Homolka P, Halberg F. About half weekly pattern of blood pressure and heart rate in men and women of India. *Scripta Medica (BRNO),* 2002,75:125-128.

[26] Singh RB, Cornelissen G, Kumar A, Bathina S, Halberg F. Larger Circadian Amplitude of Heart Rate Associated with Active Prayer in Hindu Indians in Asia. *World Heart J.* 2009;1: pp. 219-222

[27] Singh RB, Niaz MA, Cornelissen G, Otsuka K, Siegelova J, Fiser B, Halberg F. Circadian rhythmicity of circulating vitamin concentrations. *Scripta Medica (BRNO),* 2001,74:93-96

[28] Singh RB, Fedacko J,Sharma JP, Vargova V et al. Association of inflammation, heavy meals, magnesium, nitrite and coenzyme Q10 deficiency and circadian rhythms with risk of acute coronary syndromes. World Heart J 2010;2:219-228.29. Singh RB, Gupta A, Pella D, Kumar A. Circadian Cardiology. *World Heart J.* 2008;1:101-126.

[29] Pan A, Schernhammer ES, Sun Q, Hu FB. Rotating night shift work and risk of type 2diabetes: two prospective cohort studies in women. PLoS Med 2011; 9: e1001141. doi:10.1371/journal.pmed.1001141.

[30] Kivimäki M, Batty GD, Hublin C. Shift Work as a Risk Factor for Future Type 2 Diabetes: Evidence, Mechanisms, Implications, and Future Research Directions. PLoS Med 2011; 8(12): e1001138. doi:10.1371/journal.pmed.1001138.

[31] van Drongelen A, Boot CR, Merkus SL, Smid T, van der Beek AJ (2011) The effects of shift work on body weight change - a systematic review of longitudinal studies. *Scand. J. WorkEnviron. Health* 2011;37: 263–275.

[32] Froy O. Metabolism and circadian rhythms--implications for obesity. *Endocr. Rev.* 2010; 31: 1–24.

[33] Gale JE, Cox HI, Qian J, Block GD, Colwell CS, et al. Disruption of circadian rhythms accelerates development of diabetes through pancreatic beta-cell loss and dysfunction. *J. Biol. Rhythms* 2011;26: 423–433.

[34] Van Cauter E. Sleep disturbances and insulin resistance. *Diabet Med.*2011; doi:10.1111/j.1464-5491.2011.03459.x.

[35] Kivima¨ki M, Virtanen M, Elovainio M,Kouvonen A, Va¨a¨na¨nen A, et al. Work stress in the etiology of coronary heart disease—a meta-analysis. *Scand J. Work Environ Health* 2006;32:431–442.

[36] Lo SH, Lin LY, Hwang J S, Chang YY, Liau CS, Wang JD. Working the night shift causes increased vascular stress and delayed recovery in young women. *Chronobiol Int.* 2010 Aug, 27 (27): 1454-68.

[37] Lo SH, Liau CS, Hwang JS, wang JD. Dynamic blood pressure changes and recovery under different work shifts in young women. *Am. J.Hypertens* 2008 Jul 21(7):759-64.

[38] Su TC, Lin LY, Baker D, Schnalt PL, Chen NF, Hwang WC, Chen CF, wang JD. Elevated blood pressure, decreased heart rate variability and incomplete blood pressure recovery after a 12-hour night shift work. *J. Occup. Health*, 2008: 50(5): 380-86.

In: Cardiovascular Health and Chronomics
Editor: Ram B. Singh

ISBN: 978-1-62948-976-6

Chapter 15

ASSOCIATION OF OBESITY AND ABDOMINAL ADIPOSITY WITH BLOOD PRESSURE IN ADULTS

***K. S. N. Reddy*[*1], *K. K. Reddy*[1] and *T. P. K. Reddy*[2]**

[1]Department of Anthropology, College of Sciences, Sri Venkateswara University, Tirupati, Andhra Pradesh, India

[2]Dr. Parvathi Kumara Reddy Thavanati, Profesor y Investigador, Instituto de Genetica, Departamento de Biologia Molecular y Genomica, Centro Universitario de Ciencias de la Salud, Universidad de Guadalajara, Edificio P, Nivel II, Col., Independencia, Guadalajara, Mexico

ABSTRACT

Objective: Obesity is a well known risk factor for cardiovascular diseases (CVD).The present study is aimed to assess the relationship of obesity and abdominal obesity on blood pressure in an adult Indian population.

Subjects and Methods: Cross-sectional survey was conducted among students and staff members of the university. Six hundred and sixty adult subjects (303 males + 357 females) untreated for hypertension in the age range of 21-60 years. Physical assessment included height, weight, circumferences of waist and hip, systolic blood pressure (BP) and diastolic BP besides the information on demographic variables. Body mass index (BMI) was calculated as weight in kg/height in meter^2 (kgm^{-2}), waist-to-hip ratio (WHR) as waist circumference/hip circumference. Logistic regression analysis was carried out apart from one way anova and correlations. Categorical variables are shown in percentages and continuous variables as mean±S.D.

Results: Men are found to have higher abdominal obesity (0.90±0.07) than women (0.84±0.08) ($p<0.05$) while no difference in BMI. The prevalence of hypertension increased with age and BMI quartiles. The indicators of adiposity (BMI, WHR) were positively associated with blood pressure in males, while in females only BMI shown a positive association. Men with higher WHR are 2.988 times, and women with higher WHR are 1.177 times at risk to develop hypertension. The odds of hypertension were more than six fold among the elderly in male sex (OR=6.213: 95%CI 1.815, 21.273), but

* Address for correspondence: Dr. K. K. Reddy, Ph.D. Department of Anthropology. Sri Venkateswara University.Tiruapti- 517 502. Andhra Pradesh, India. E-mail: kanalakr@yahoo.com.

in females the odds of hypertension in elderly is only two fold (OR= 2.423: 95%CI 0.801, 7.334). The odds of hypertension rose steadily with increase in BMI reaching 7.579 (95CI; 1.510, 38.046) in males and 15.56 (95%CI; 1.883, 128.526) in females with BMI >25 kgm^{-2}. Adjustment for age decreased the odds of hypertension in males and increased in females in the BMI category of >25 kgm^{-2}, while no change in the remaining quartiles.

Conclusion: These findings suggest a linear relation of adiposity with Blood pressure.

Keywords: Body Mass Index, waist hip ratio, abdominal adiposity, blood pressure

INTRODUCTION

The prevalence of obesity and its associated coronary heart disease (CHD) risk factors are increasing in developing countries [1,2]. Worldwide about 58% of diabetes mellitus and 21% of ischemic heart diseases are attributable to BMI above 21 kg/m^2 [3-5]. Developing countries are increasingly faced with a double burden of hypertension and other cardiovascular diseases, along with infection and malnutrition [1-4]. An increasing number of developing nations are acquiring atherogenic lifestyles which include the adoption of atherogenic dietary habits similar to those seen in industrialized societies. This appears to be consistent with economic development [5]. Major coronary risk factors are smoking, hypertension, dislipidemia, diabetes and obesity.

Obesity and hypertension have been shown to increase in parallel across populations along with their degree of development and Western acculturation [5,6]. Clinical and epidemiological studies indicate that developing countries have a higher susceptibility for developing obesity, central obesity, hypertension and coronary disease.[1-7]. The increased risk of obesity and hypertension may be partly explained by rapid changes in diet and lifestyle factors. The present study is aimed examined the relationship between adiposity (BMI and WHR) and blood pressure in a population where the prevalence of obesity is growing rapidly. It is possible that this information would be relevant for prevention and control of hypertension in developing countries.

MATERIALS AND METHODS

The present study was conducted at Sri Venkateswara University, Tirupati, India. As a part of our University Silver Jubilee Celebrations, an Obesity Awareness Camp (OAC) was conducted by the Department of Anthropology, with an aim to create awareness about the adiposity among the staff and students of the University.

Formal written requests were sent to the individual faculty and circulars to student community to participate in the OAC. Of total 700 subjects who responded to publicity, 680 subjects volunteered and of these 20 subjects were excluded due to various reasons. Remaining 660 (303 males and 357 females) subjects, aged 21-60 years, volunteered given written informed consent to participate in the programme. Pregnancy (in women), usage of antihypertensive medication and any gross physical abnormality were the exclusion criteria against the physical measurements (n=20). The study protocol was approved by our Institutional Ethics Committee. The present study, focusing on adiposity and blood pressure (BP), has a cross-sectional descriptive design, allowing internal comparisons among the major socio-demographic groups, such as sex, education, occupation etc. The subjects were divided into 10 years age class intervals for internal comparison.

Data were collected using questionnaire and through physical measurements of weight, height, circumferences of waist and hip and Blood pressure. The survey instrument, mainly the questionnaire, has been validated in a small pilot study among our Departmental Staff and Students. The physical assessment included height, weight, circumferences of waist and hip were measured as specified by Reddy et al. [8]. Weight and height were measured participants standing without footwear and wearing light clothing. Participants stood upright with the head in Frankfort plane for height measurement. Height was recorded to the nearest 0.5 cm and weight was recorded to the nearest 100g. Body mass index (BMI) was calculated as weight in kg/height in metre^2 (kgm^{-2}). Over weight was defined as BMI >23 and obesity was defined as BMI > 25 [8].

Waist girth was measured at the level of umbilicus with person breathing silently and hip measured as standing inter-trochanteric girth according to the WHO guidelines. Waist hip ratio (WHR) was calculated from the circumferences of waist and hip. Abdominal adiposity was defined as WHR>0.90 for males and >0.85 for females (9). Blood pressure was measured with a random zero muddler sphygmomanometer at the study site in a sitting position after the participant rested for at least 5 min. Three consecutive measurements were taken with an interval of 3 min in between.

Average systolic blood pressure and diastolic blood pressure were determined from the second and third measurements. Hypertension was diagnosed when the systolic blood pressure was > 140 mmHg or the diastolic blood pressure was > 90 mmHg, as per the guidelines prescribed by the Joint National Committee on detection, evaluation and treatment of high blood pressure [10].

Statistical analysis was carried out via SPSS – 16.0 and alpha levels were set at $p< 0.05$. The prevalence rates were reported as percent. Continuous variables were reported as mean±1SD and differences between genders were tested by students 't' test. Further, logistic regression analysis was carried out to determine the odds of hypertension across the age groups, BMI and WHR categories, education and occupation, while controlling for possible confounding.

RESULTS

Descriptions of the demographic and coronary risk factors are presented in table.1. In both the sexes, around 20 percent of sample is overweight, 15 percent are obese and 13 percent of the males and 18 percent of the females categorized as undernutrition. Forty percent of the males and 36 percent of the females had abdominal adiposity. The prevalence of undiagnosed hypertension in the present sample is 15% in males and 10% in females.

Physical measurements were available for all the participants, for whom the BMI and WHR could be calculated (table 2).

Table 1. Description of the demographic and coronary risk factor characteristics

Variable	Males (n=303)	Females (n=357)
Age in years		
21-30	98 (32.34)	104 (29.13)
31-40	75 (24.75)	106 (29.69)
41-50	73 (24.09)	75 (21.01)
51-60	57 (18.81)	72 (20.17)
BMI Category		
<18.5	39 (12.87)	64 (17.93)
18.5 to 23.0	162(53.46)	162(45.38)
23.0-25	56 (18.48)	75 (21.00)
>25.0	46 (15.18)	56 (15.68)
WHR category		
Men <0.90; Women<0.85	180(59.41)	227(63.58)
Men >0.90; Women>0.85	123(40.59)	130(36.41)
Blood pressure		
Normotensive	258(85.15)	319(89.36)
Hypertensive	45 (14.85)	38 (10.64)

( ) = percentages.

Table 2. Descriptive statistics for the anthropometry and blood pressure in the study population

Variable	Males (Mean±SD)	Females (Mean±SD)	t-value
Height in cm	167.23±6.43 (151-186)	155.23±6.24 (145-176)	19.82*
Weight in Kg	62.18±9.95 (42-86)	52.21±8.50 (35-79)	11.33*
Body mass Index	22.19±3.09 (15.43-29.69)	21.65±3.19 (15.34-28.95)	1.80
Waist circumference in cm	82.11±11.43 (59-116)	75.59±10.39 (51-105)	6.26*
Hip circumference in cm	92.45±8.08 (66-121)	90.56±9.97 (54-121)	2.16*
WHR	0.89±0.07 (0.63-1.09)	0.84±0.08 (0.65-1.01)	6.81*
Systolic BP in mmHg	125.64±12.45 (91-152)	121.58±11.28 (86-153)	3.58*
Diastolic BP in mmHg	82.31±10.21 (50-108)	80.57±8.58 (54-104)	1.94

* $p<0.05$.

The mean weight and height varied between male and female subjects. Men were taller and heavier than female counterparts. However, the resulting mean BMI in men (22.19±3.09) didn't vary with women (21.65±3.19). Male gender are found to have higher waist circumference than female gender ($p<0.05$), while no significant difference in hip circumference. Men are found to have higher abdominal adiposity (0.90±0.07) than women (0.84±0.08). Although males possess higher systolic and diastolic BP than females, significant difference observed only with systolic blood pressure ($p<0.05$).

Correlation coefficients for age, anthropometry and blood pressure is shown in table 3. The unadjusted pair wise correlations were higher in males than females.

Age has shown a positive association with anthropometry and blood pressure except in height. The correlation coefficients of SBP and DBP for age varied from 0.153- 0.275 ($p<0.05$) in men, from 0.219- 0.171 in women. The indicators of adiposity (BMI, WHR) were positively associated with blood pressure in males, while in females only BMI shown a positive association with blood pressure.

Conventional BMI cutoff points were applied to classify the study populations into underweight (BMI<18.5 kg/m^2), normal (BMI:18.5-22.99 kg/m^2), overweight (BMI:23.0-24.99 kg/m^2) and obese (BMI>25 kg/m^2); abdominal adiposity (Men >0.90; Women>0.85), and the distribution of hypertensives were shown in table 3.

The prevalence of hypertension increased with age and BMI quartiles. However, the extent of this association varied between different age groups and BMI quintiles. The rise in the prevalence of hypertension was more drastic at age group 41-50 years in male sex and 51-60 in female sex.

There is a steep increase in hypertension in second BMI quartile and with no difference to third quartile between sexes and sudden elevation in fourth quartile is noticed in male sex. The prevalence of hypertension is elevated in males and decreased in females with increase in abdominal adiposity.

Selected socio-demographic and adiposity characteristics that are considered as possible determinants of hypertension were subjected to a logistic regression analysis. The analysis was conducted separately for each sex and the resulting OR and 95% CI are presented in table 4. The BMI along with WHR and age were found to be significant determinants of hypertension in the study population. The odds of hypertension were more than six fold among the elderly in male sex (OR=6.213: 95%CI 1.815, 21.273), but in females the same is only two fold (OR= 2.423: 95% CI 0.801, 7.334).

The odds of hypertension rose steadily with increase in BMI reaching 7.579 (95CI; 1.510, 38.046) in males and 15.56 (95%CI; 1.883, 128.526) in females with BMI >25 kgm^{-2}. When age is adjusted in the model, the odds of hypertension sharply declined in male sex (OR=4.339, 95%CI; .698, 26.966) and an increase in female sex (OR=22.019, 95%CI; 2.343, 206.939) in the BMI category of >25 kgm^{-2}, while no change in the remaining categories.

On the other hand when WHR and age independently and together adjusted in the model, no significant deviations taken place in the odds of hypertension among BMI categories of males, while in female sex the odds of hypertension increased in BMI >25 kgm-2. Men with higher WHR are 2.988 times at risk to develop hypertension than with lower WHR. Similarly women with higher WHR are 1.177 times at risk to develop hypertension when compared to lower WHR. When age and BMI independently and together adjusted to see the changes in odds of hypertension in WHR categories, a decline in the odds ratio is evident in both males and females. The odds of hypertension among the male employees (OR= 5.025, 95%CI 2.218, 11.383) is greater than female employees (OR= 1.262, 95%CI 0.550, 2.894) in developing hypertension.

Table 3. Distribution of hypertension across the age groups and BMI categories

Variable	Males (n=303)	Females (n=357)
Age in years		
21-30	6 (1.98)	9 (2.52)
31-40	6 (1.98)	12(3.36)
41-50	16(5.28)	3 (0.84)
51-60	17(5.61)	14(3.92)
Total	45 (14.85)	38(10.64)
BMI Category		
<18.5	3 (0.99)	2 (0.56)
18.5 to 23.0	12(3.96)	9 (2.52)
23.0-25	12(3.96)	12(3.36)
>25.0	18(5.94)	16 (4.48)
WHR Category		
Men <0.90; Women<0.85	16 (5.28)	22 (6.16)
Men >0.90; Women>0.85	29 (9.57)	16 (4.48)

( ) = percentages.

Table 4. Determinants of high blood pressure in the study population (logistic regression)

	Males		Females	
	OR	95%CI	OR	95%CI
Age in years				
21-30	1.00		1.00	
31-40	1.326	0.315, 5.585	1.333	0.437, 4.065
41-50	4.414	1.313, 14.862	0.437	0.085, 2.264
51-60	6.213	1.815, 21.273	2.423	0.801, 7.334
BMI Category				
<18.5	1.00		1.00	
18.5 to 23.0	0.960	0.191, 4.814	2.471	0.289, 21.152
23.0-25	3.310	0.641, 17.085	8.000	0.968, 66.089
>25.0	7.579	1.510, 38.046	15.56	1.883, 28.526
WHR category				
Men <0.90; Women<0.85	1.00		1.00	
Men >0.90; Women>0.85	2.988	1.336, 6.683	1.177	0.504, 2.748
Education				
Graduation	1.00		1.00	
Post graduation	0.833	0.345, 2.015	0.967	0.422, 2.217
Occupation				
Student	1.00		1.00	
Employee	5.025	2.218, 11.383	1.262	0.550, 2.894

DISCUSSION

In this study, we examined the relationship of hypertension with obesity and abdominal adiposity in a suburban population. The results indicate a unpretentious, but significant linear association of BMI and WHR with blood pressure in males, independent of age. No such association WHR was observed in females. However, BMI greater than 25 kg/m^{-2} was associated with increased risk of hypertension in presence of higher abdominal obesity. These findings are in agreement with other studies, supporting a consistent relationship between body mass and abdominal adiposity with Blood Pressure [11]. Logistic regression analysis revealed that obesity, abdominal obesity and age were significant determinants of hypertension in males and females. In presence of insignificant differences in the relationship of BP with adiposity in female gender provides substantial argument against a lower hypertensive effect of obesity than male gender. A few studies from India have also reported the adverse effects of obesity on coronary risk factors and mortality [1-4] A recent cross-sectional survey [1,3], screened 6940 subjects, (3507 men (M), 3433 women (W): 1993-96) aged 235 years and above, from cities located in five corners of India (Kolkata, n=900; Nagpur, n=894; Mumbai, n=1542; Thiruanantpuram, n=1602; Moradabad, n=2002). The overall prevalence of obesity was 6.8% (7.8 vs. 6.2%, P<0.05) and overweight 33.5% (35.0 vs. 32.0%, P 0.05) among women and men, respectively. The highest prevalence of obesity (7.8%) and overweight (36.9%) was found among subjects aged 35 to 44years in both sexes. The prevalence of obesity was significantly (P <0.05) greater in Trivandrum (8.5%), Calcutta (7.1%) and Bombay (8.3%) compared to Moradabad (6.2%) among women and in Trivandrum (7.4%) and Bombay (7.2%), compared to Nagpur (5.0%) among men. There was a significant decreasing trend in obesity (P <0.05) and overweight (P<0.05) with increasing age above 35-44years in both sexes. The overall prevalence of subjects>23kg/m2 was 50.8% and central obesity 52.6%.The overall prevalence of sedentary behaviour was 59.3% among women and 58.5% among men. Both sedentary behavior and mild activity showed a significant increasing trend in women after the age of 35-44years. In men, such a trend was observed above the age of 45years. Sedentary behaviour was significantly (P<0.05) greater in Trivandrum, Calcutta, and Bombay compared to Nagpur. Sedentary behaviour was significantly (P <0.001) associated with obesity in both sexes, compared to non-obese men and women. The overall prevalence of under nutrition was 5.5% (n=380). Diagnosis for prehypertension (BP 130-139/85-89 mm Hg) and hypertension (BP≥140/90 mm Hg) were based on European Society of Cardiology criteria [3]. Prevalence of pre-hypertension and hypertension, respectively, was significantly greater in South India (Trivandrum: W 31.5;31.9%; M 35.1;35.5%) and West India (Mumbai: W 30.0;29.1%; M 34.7;35.6%) compared to North India (Moradabad: W 24.6;24.5%; M 26.7;27.0%) and East India (Kolkata: W 20.9;22.4%; M 23.5;24.0%). Subjects with pre-hypertension and hypertension were older, with higher BMI, central obesity and of sedentary behavior. They had higher salt and alcohol intake, with greater oral contraceptive usage (W). Multivariable logistic regression analysis, revealed strong positive associations of hypertension with age, central obesity, BMI, sedentary lifestyle, salt and alcohol intake and oral contraceptive usage (W). Fruit, vegetable and legume intake showed inverse associations, tobacco intake showed none. In one survey [4] of death records among 2222 subjects, aged 25 to 64 years, majority of the decendents (n=792,35.6%) (men 31.1%,n=431; and women 43.1%,n=361) had normal BMI of 18.5-22.9Kg/m2. The prevalence of underweight victims was 14.2% (n=315), overweight 29.4%(n=654) and obese 20.8% (n=461). There was an overall increase in risk factors; diabetes mellitus, hypertension, and CAD among overweight and obese victims based on BMI criteria, and the trend was significant. However, tobacco intake showed nonsignificant trend, highest in the underweight victims, without significant differences in the other categories of BMI. BMI was positively associated with significant rising trend in the prevalence of circulatory causes of death, both among men and women [4].

The evidence presented above supports a common general physiopathological mechanism linking the excessive fat deposition to elevated BP independently of genetic and environmental background. The mechanism of obesity-associated hypertension appears to be an inadequate vasodilatation in the face of the increased blood volume and cardiac output, which are the natural consequences of an increased body mass. This defect in control of vascular resistance has been attributed to increased activity of the sympathetic nervous system, abnormal renin-angiotensin-aldosterone relations, and insulin resistance [12]. Obesity seems to accentuate the development of a cluster of metabolic disorders (including hypertension and dyslipidemia) in subjects presenting the syndrome X, referred to as the insulin resistance syndrome [13].

The relationship of BP to cardiovascular mortality has been found to be similar among different countries, continuous and linear, even at the lower range of BP, i.e, below the cut-off points (140/90 or 160/95 mmHg) generally used to define hypertension [14]. Therefore, changes in BP corresponding to defined gains in adiposity can directly be converted into their effect on the relative risk of death from cardiovascular diseases. Measuring the relation of the adiposity parameters (BMI, WC and WHR) to BP by only considering the prevalence of hypertension would underestimate their real impact on mortality. The prevalence of hypertension at different BMI quartiles revealed a steep rise at fourth quartile in both males and females. Significant associations between BMI and BP have also been documented in various populations [15].

Similar findings have been reported in other studies [16]. The relationship between adiposity and BP in this study might be potentially confounded by dietary salt intake and physical activity levels, both of which are not available for the present sample. The study demonstrated that adiposity is closely associated with BP in countries at different stages of socioeconomic and epidemiologic transition. Mean BP levels are increasing with categories of BMI and WHR. The risk of hypertension is higher with overall and abdominal obesity. Together with data from other studies [17], there is an overall convergence of evidence towards a steeper rise in BP with the advance of age in developing countries when compared to developed countries. In the present study, this mechanism proved to be independent of body mass and abdominal adiposity while some experts think that body mass is the dominant causal factor [18]. Further research is needed to determine the etiopathology of this mechanism which have been emphasized in other studies [19-25]. Oxidative stress, insulin resistance in association with obesity have been reported from India [20, 21]. In south Asia, only modest increase in BMI above 23Kg/M2 has been associated with insulin resistance and hypertension [21-24]. In view of these findings, it has been proposed that south Asians appear to need modified guidelines regarding diet and lifestyle changes on prevention of CVDs [24, 25].

In brief, the findings indicate age-independent linear association between BP and adiposity. A modest increase in BMI appear to be associated with central obesity and hypertension. The present study shares the views about the recommendations of WHO [19] for developing countries by promoting physical activity and healthy dietary habits, including the reduction of alcohol drinking and salt intake and increase in the intake of fruits, vegetables and legumes which are probably important risk factors of high BP with the advance of age.

ACKNOWLEDGMENTS

The authors greatly acknowledge the financial support extended by Sri Venkateswara University, Tirupati, India for conducting the Obesity Awareness Camp.

The authors KSN Reddy and KK Reddy greatly acknowledge the financial support rendered by the University Grants Commission, New Delhi to the Department of Anthropology, Sri Venkateswara University, Tirupati under UGC-SAP-DRS Phase-III.

REFERENCES

[1] Singh RB, Pella D, Kartikey K, DeMeester F, et al., and the Five City Study Group. Prevalence of obesity, physical inactivity and undernutrition, a triple burden of diseases, during transition in a middle income country. *Acta. Cardiol.* 2007, 62:119-127.

[2] Reddy KK, Rao AP, Reddy TPK. Socioeconomic status and the prevalence of coronary heart disease risk factors. *Asia Pacific J. Clin Nutr* 2002; 11: 98-103.

[3] Singh RB, Fedacko J, Pella D, Macejova Z et al. Prevalence and risk factors of pre-hypertension and hypertension in five Indian cities. *Acta Cardiol* 2011; 66:29-37.

[4] Vargova V, Singh RB, Singh S, Singh V, Kulshrestha SK, Fedacko J, Pella D. Association of increased mortality with overweight and obesity among urban descendents in North India dying due to various causes. *World Heart J.* 2010;2:133-140.

[5] World Health Organization. Diet, Nutrition and the Prevention of Chronic Diseases, Report of a Joint WHO/FAO Expert Consultation. WHO Technical Report Series No. 916 *World Health Organization*: Geneva, 2003.

[6] Cooper R, Rotimi C, Ataman S, McGee D, Osotimehin B, Kadiri S, Muna W, Kingue S, Fraser H, Forrester T, Bennett F, Wilks R. The prevalence of hypertension in seven populations of West African origin. *Am. J. Publ. Health* 1997; 87: 160-168.

[7] Forrester T, Cooper RS, Wetherall D. Emergence of Western diseases in the tropical world: the experience with chronic cardiovascular diseases. *Br. Med. Bull* 1998; 52: 463-473.

[8] Reddy KK, Papa Rao A and Reddy TPK: Effects of age, sex and life styles of CHD risk factors: Influence of obesity and body fat distribution. *J. Hum. Ecol.* 1998; 9:593-601.

[9] WHO Expert Committee. Physical Status: the use and interpretation of anthropometry. *WHO Technical Report Series* 1995;854:424-38.

[10] Chobanian AV, Bakris GL, Black HR, Cushman WC, Green LA, IzzoJr JL et al. The seventh report of the Joint National Committee on Prevention, Detection, Evaluation,

and Treatment of High Blood Pressure: The JNC 7 Report. *JAMA* 2003; 289: 2560–2572.

[11] Doll S, Paccaud F, Bovet P, Burnier M and Wietlisbach V. Body mass index, abdominal adiposity and blood pressure: consistency of their association across developing and developed countries. *Int. J. Obesity* 2002; 26: 48-57.

[12] Dustan HP. Obesity and hypertension in blacks. *Cardiovasc. Drugs Ther* 1990; 4:395-402.

[13] Srinivas K, Vijaya Bhaskar M, Aruna Kumari R, Nagaraj K and Reddy KK: Antioxidants, lipid peroxidation and lipoproteins in primary hypertension. *Ind. Heart J.* 2000; 52: 285-288.

[14] Van den Hoogen PGW, Feskens EJM, Nagelkerke NJD, Menotti A, Nissinen A, Kromhout D, for the Seven Countries Study Research Group. The relation between blood pressure and mortality due to coronary heart disease among men in different parts of the World. *New Engl. J. Med.* 2000; 342: 1-8.

[15] Hu FB, Wang B, Chen C, Jin Y, Yang J, Stampfer MJ et al. Body mass index and cardiovascular risk factors in a rural Chinese population. *Am. J. Epidemiol* 2000; 151: 88–97.

[16] Singh RB, Rastogi SS, Rastogi V, Niaz MA, Madhu SV, Chen M *et al.* Blood pressure trends, plasma insulin levels and risk factors in rural and urban elderly populations of north India. *Coron. Artery Dis.* 1997; 8: 463–468.

[17] Cooper R, Rotimi C, Ataman S, McGee D, Osotimehin B, Kadiri S, Muna W, Kingue S, Fraser H, Forrester T, Bennett F, Wilks R. The prevalence of hypertension in seven populations of West African origin. *Am. J. Publ. Health* 1997; 87: 160-168.

[18] Stamler R, Stamler J, Riedlinger WF, Algera G, Roberts RH. Weight and blood pressure: findings in hypertension screening of 1 million Americans. *JAMA*1978; 240: 1607–1610.

[19] Gyarfas I. Control of hypertension in the population: strategies in affluent and developing countries. *Clin. Exp. Hypertens* 1996; 18: 387-397.

[20] Reddy KK, Ramamurthy R, Somasekharaiah BV, Reddy TPK and Papa Rao A: Free radicals and antioxidants: Influence of obesity, body fat and life styles. *Asia Pacific J. Cin. Nutr. 1997;*6:296-301.

[21] Singh RB, Rastogi SS, Niaz MA, Postiglion A. Association of central obesity and insulin resistance with high prevalence of diabetes and cardiovascular disease in an elderly population with low fat intake and lower than normal prevalence of obesity. The Indian paradox. *Coro Art. Dis.* 1998; 9: 559-65.

[22] Singh RB, Bazaz S, Niaz MA, Rastogi SS, Moshiri M. Prevalence of type 2 diabetes mellitus and risk of hypertension and coronary artery disease in rural and urban population with low rates of obesity. *Int. J.Cardiol.* 1998; 66:65-72.

[23] Singh RB, Niaz MA, Beegum R, Wander GS, Thakur AS, Rissam HS. Body fat percent by bioelectrical impedance analysis and risk of coronary artery disease among urban men with low rates of obesity. The Indian paradox. *J. Am. Coll Nutr.* 1999; 18:268-273.

[24] Singh RB, Verma SP, Niaz MA. Social class and coronary artery disease in India. *Lancet* 1999; 353: 154.

[25] Singh RB. Coronary artery disease risk factors in South Asians and Americans. *Am. J. Clin. Nutr.* 1999; 70:112-113.

In: Cardiovascular Health and Chronomics
Editor: Ram B. Singh
ISBN: 978-1-62948-976-6

Chapter 16

CHRONOBIOLOGICALLY INTERPRETED CLINICAL AND PHYSIOLOGICAL MONITORING AND DIETARY VISTAS IN THE CARE OF THE ELDERLY

R. B. Singh[1,2], F. De Meester[2], A. Wilcynszka[2,3], G. Cornélissen[4], D. W. Wilson[5], A. P. S. Hungin[5], R. Singh[6], S. Sarrafi Zadeh[7], Shabnam Omidvor[7], Toru Takahashi[8] and F. Halberg[4]

[1]Halberg Hospital and Research Institute, IFTM University, Civil Lines, Moradabad, India
[2]TsimTsoum Institute, Kraków, Poland
[3]Institute of Psychology, University of Silesia, Katowice, Poland
[4]Halberg Chronobiology Center, University of Minnesota, Mayo Hospital, Minneapolis Campus, Minneapolis, MN United States
[5]School of Medicine and Health, Queen's Campus, Durham University, United Kingdom
[6]Department of Medicine, Darlington Memorial Hospital, Darlington
[7]Faculty Member of Nursing and Midwifery, Department of Babol Medical Sciences University, Iran
[8]Graduate School of Human Environment Science, Fukuoka Women's University, Japan

1. CHRONOBIOLOGY AND CHRONOMICS

The Studies made at the Halberg Chronobiology Center, Department of Laboratory Medicine and Pathology, University of Minnesota, have demonstrated that chronobiology is an essential tool for quantifying and understanding variability in diastolic (D) and systolic (S) blood pressure (BP), heart rate (HR) and pulse pressure and thus for identifying subjects at risk of myocardial infarction or stroke, and for choosing the appropriate treatment and timing thereof (chronotherapeutics).

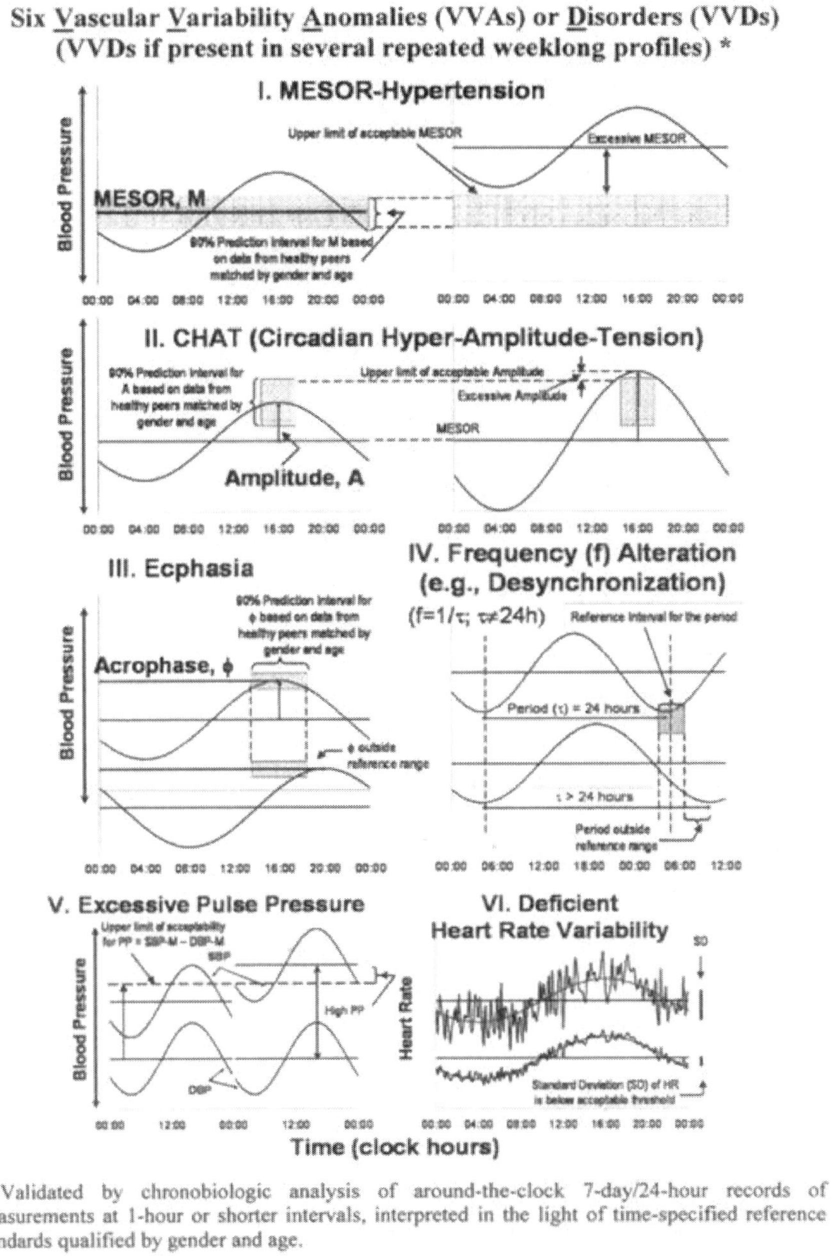

Figure 1. Vascular Variability Anomalies, VVAs, that are dubbed Vascular Variability Disorders, VVDs, when they are replicated in several consecutive 7-day around-the-clock spans after a known stress subsides (see Figure 5); if 2 or more VVDs coexist, they become Vascular Variability Syndromes, VVSs, associated with much higher risks of hard events than the risk of a high blood pressure in itself and sometimes produced by unsurveilled antihypertensive treatment that can generate VVDs. © Halberg.

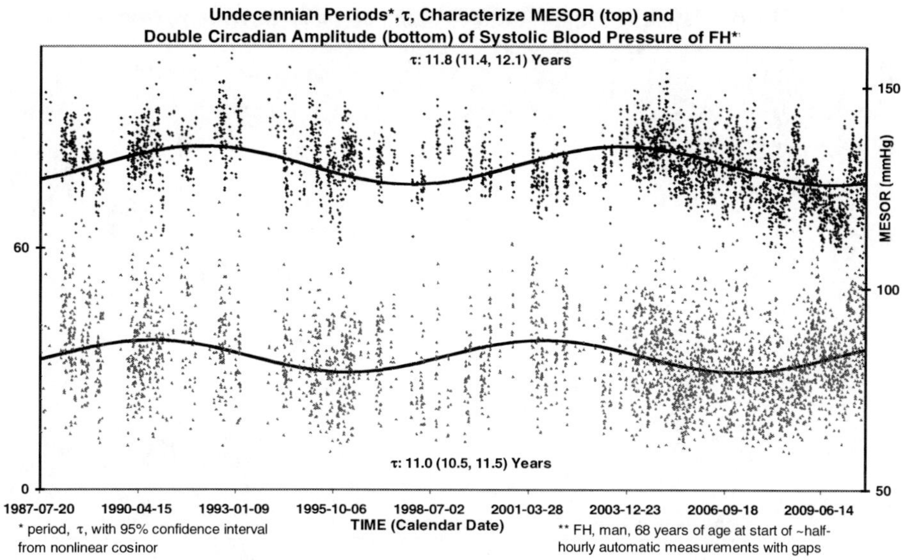

Figure 2. Sections of original half-hourly around-the-clock systolic blood pressure were fitted first by cosinor with a 24-hour cosine curve to obtain intermediate statistics, imputations, namely the **m**idline-**e**stimating **s**tatistic **o**f **r**hythm, MESOR, M, and the double circadian amplitude, 2A, each constituted into a separate time series and analyzed by the extended cosinor, which reveals congruent periods of about 11 years roughly corresponding in length to the Horrebow-Schwabe sunspot cycle for the MESOR curve on top (scale on the right) and for the circadian double amplitude curve below (scale on the left). © Halberg.

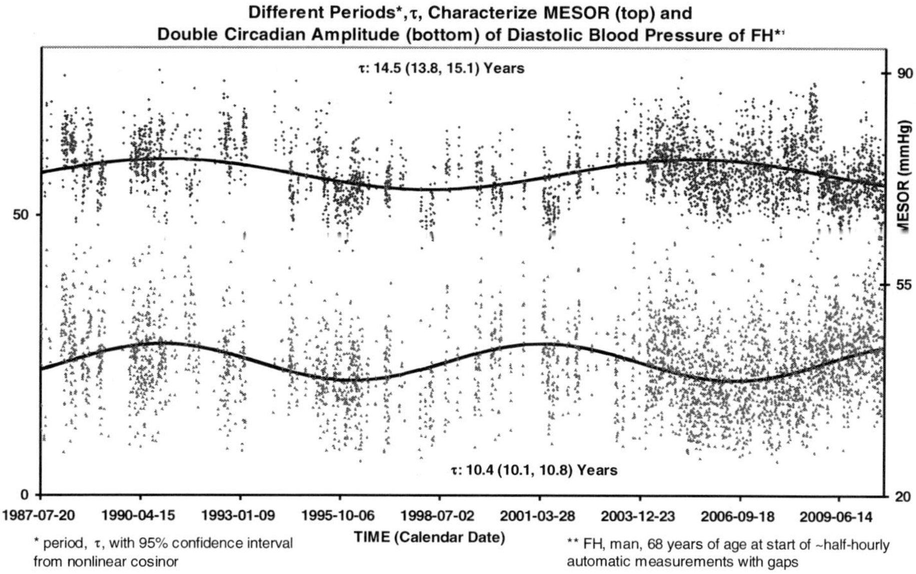

Figure 3. For diastolic blood pressure, the extended cosinor applied to imputations of M and 2A reveals different periods, one perhaps corresponding to the Horrebow-Schwabe sunspot cycle (curve below), the other clearly longer, perhaps a global cycle (curve above), corresponding perhaps to a global Makarov cycle. © Halberg.

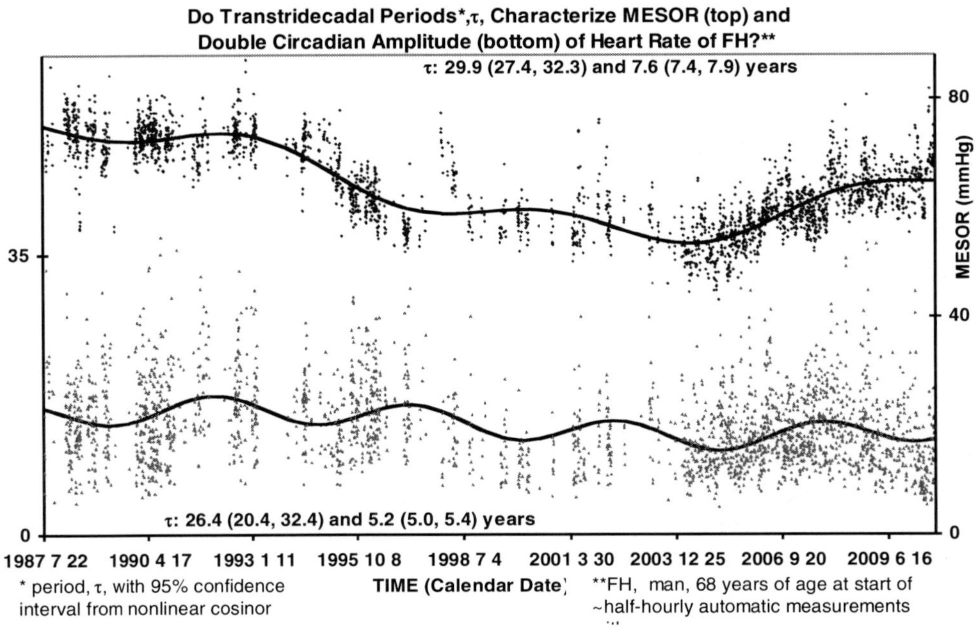

Figure 4. For heart rate, the extended cosinor imputation of M and 2A shows, a.o., a period longer than the record's length; with this qualification, the 95% confidence interval of the longest period overlaps 30 years, in keeping with a paratridecadal period along with an ~7.6-year period for curve on top that could be a harmonic of the ~21-year Hale cycle for M and a period of ~5.2 years for curve below, which could be a harmonic of the decadal Horrebow-Schwabe cycle. © Halberg.

Monitoring subjects, using automatic half-hourly round-the clock ambulatory BP and HR measurements, over a minimal timespan of 7 days at selected time intervals determined by infradian rhythms, is particularly relevant in the care of the elderly.

The identification and characterization of vascular variability disorders, Figure 1, such as MESOR-hypertension, MH, excessive circadian amplitude, A (CHAT), BP ecphasia, an odd BP phase but an acceptable HR phase, excessive pulse pressure, deficient heart rate variability and ecfrequentia -- one or several odd circadian periods, τ, between 20 and 28 hours (h), with a 95% confidence interval of τ not covering 24 h -- can be monitored with variables of the mind such as mood (positive and negative affect) and vigor. Decades-long research shows infradianτs, many of them reflecting τs of solar and/or geomagnetic activity. Aging can be gauged by circadian ($20 \leq \tau \leq 28$ h) rhythm characteristics that are modulated by infradians ($\tau > 28$ h); in the case of M and a circadian double amplitude, 2A, the time structures in and around us are best surveilled for diagnosis and therapy. Figures 2-4 (2), rather than assuming an imaginary linear increase or decrease in BP or HR with age.

The role of gaps in the data has been documented to create artifacts with the very data of Figure 2 [3], another reason to urge continuing monitoring with affordable, unobtrusive instrumentation for automatic BP and HR recording and eventually automatic analysis [1]. By contrast, ignoring rhythms and taking single, even clock-hour-specified observations can result in opposite inferences from two different fixed clock-hours of measurement, as an artifact from a change with age, e.g., of circadian As or ϕs [4]. It is then simpler to flip a coin and save the expenditure of resources on almost certainly controversial and hence wasteful results collected in aging studies, clinical or in research, at some arbitrary single, even if specified and fixed to be the same clock-hour.

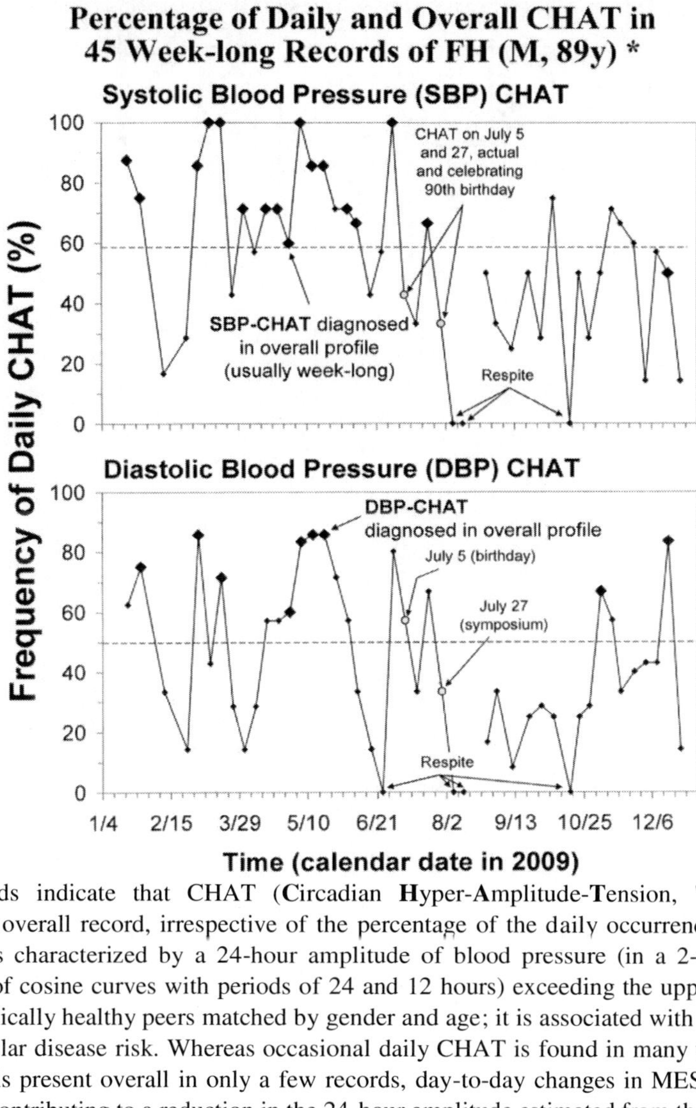

*Large diamonds indicate that CHAT (**C**ircadian **H**yper-**A**mplitude-**T**ension, "overswing") was detected in overall record, irrespective of the percentage of the daily occurrence of CHAT. This condition is characterized by a 24-hour amplitude of blood pressure (in a 2-component model consisting of cosine curves with periods of 24 and 12 hours) exceeding the upper 95% prediction limit of clinically healthy peers matched by gender and age; it is associated with a large increase in cardiovascular disease risk. Whereas occasional daily CHAT is found in many usually week-long profiles, it is present overall in only a few records, day-to-day changes in MESOR and circadian acrophase contributing to a reduction in the 24-hour amplitude estimated from the entire records.

Figure 5. A start in gauging strain by the recording of Vascular Variability Anomalies that subside after a mental load, the preparation of a conference with uncertainties is removed (by the symposium held on July 27). © Halberg.

Table 1. Results from chronomics pilot study*

		MESOR				Double amplitude			
Month of monitoring		SBP	DBP	HR	PP	SBP	DBP	HR	HR-SD
% of value on month 1									
2	Mean	101.92	101.16	101.24	102.65	115.44	105.67	111.77	103.43
	SE	1.14	1.13	1.59	1.73	13.77	10.50	14.39	8.89
	Paired t	1.69	1.02	0.78	1.53	1.12	0.54	0.82	0.39
	P (H_0: mean=100%)	0.13	0.34	0.46	0.16	0.29	0.60	0.44	0.71
3	Mean	102.35	100.51	102.72	104.51	98.89	90.70	106.42	105.70
	SE	2.07	1.55	1.99	3.24	16.25	10.32	10.37	6.18
	Paired t	1.13	0.33	1.37	1.40	0.07	0.90	0.62	0.92
	P (H_0: mean=100%)	0.29	0.75	0.21	0.20	0.95	0.39	0.55	0.38
Regression with time (months 1-3)									
	r	0.095	-0.041	0.168	0.166	-0.135	-0.211	-0.029	0.070
	P	0.699	0.867	0.492	0.497	0.580	0.387	0.907	0.775

*Endpoints during first month weeklong monitoring equated to 100% (used as single value in regression analyses).

MESOR: midline-estimating statistic of rhythm, a rhythm-adjusted mean; double amplitude: extent of predictable change from fit of 24-hour cosine curve to the data.

SBP: systolic blood pressure; DBP: diastolic blood pressure; HR: heart rate; PP: pulse pressure; SD: standard deviation.

2. Mind and Body Interactions As Gauges of Loads and in Particular Harbingers of Cardiovascular Disease (CVD)

The study of loads (stress) also benefits from a chronobiologic approach, which we dub stress-strain, in keeping with the stress-strain approach in physics, Hooke's law [5-8], to distance ourselves from the majority of stressor-stress studies that rely on one or at best several spotchecks along the 24-hour scale of a given variable such as BP or a hormone instead of assessing rhythm characteristics as endpoints in their own right to detect their alteration as a measure of loads. Figure 5 [8] shows that once the stress of uncertainties in preparing a symposium ended (with the completion of the conference), so did the long uninterrupted strings of VVAs.

Table 1 shows results from a Chronomics Project, an investigation now ongoing, and reveals the consistency of the kinds of endpoints investigated. As part of the 3x3 chronomic pilot study, each of 9 subjects (3 in each of 3 age groups: adolescents, adults and elderly), provided a weeklong or longer profile of around-the-clock BP and HR measurements at 30-minute intervals with the TM-2430 ambulatory monitor from A and D (Tokyo, Japan) on 3 occasions, about one month apart.

The data were analyzed by sphygmochron, yielding, a.o., estimates of the MESOR (M) and double circadian amplitude (2A) of SBP, DBP and HR, as well as the overall standard deviation (SD) of HR. The pulse pressure (PP) was obtained by subtracting the MESOR of DBP from that of SBP. With the brevity of the series, circadian τs were not yet studied, although circaseptan (~7-day) and ~3.5-day variabilities were mapped [9]. During the 3-month study, subjects were supplemented with daily 2.5 mg long-chain (C20-22) fatty acids and a mix of vitamins and minerals.

The omega-3 supplementation altered the blood omega 6:3 ratio as anticipated, but here we examine whether this had an effect on BP and HR endpoints. The latter's time course was first charted in original units, revealing large inter-individual variability, expected in part from the gender and age differences.

During the 3-month study, subjects were supplemented with daily 2.5 mg long-chain (C20-22) fatty acids and a mix of vitamins and minerals. The omega-3 supplementation altered the blood omega 6:3 ratio as anticipated, but here we examine whether this had an effect on BP and HR endpoints. The latter's time course was first charted in original units, revealing large inter-individual variability, expected in part from the gender and age differences. Each endpoint of each subject was then equated to 100% during the first monitoring session, and corresponding endpoints during the second and third month of study expressed as a percentage of the first month value. Using only a single 100% value for month 1 and all relative values for months 2 and 3, linear regression analyses versus time found no statistically significant change in the endpoints examined (P>0.30), Table 1. Deviations from start were assessed at months 2 and 3 and tested by paired t-test, Table 1. Whereas results did not reach statistical significance, there was an increase, rather than the anticipated decrease in the MESOR of both BP and HR. The PP was 4.5% higher in month 3 versus month 1. Any merit of a numerical decrease in the DBP-2A of nearly 10% in month 3 would depend on whether CHAT was initially present or absent. The numerical increase of 5.7% in month 3 for HR-SD represents the only potentially tangible benefit apparent thus far of the omega-3 supplementation. Larger numbers of individuals in investigations now planned for longer omega-3 supplementation -- at different circadian stages -- on clinically healthy subjects and eventually in those with an elevated left ventricular mass index (LVMI) may serve for treatment of VVDs, eventually found, to prevent heart failure (HF), notably HF with a preserved ejection fraction (HFPEF). The LVMI could be an important target for the prevention of HF [10, 11]. The aggressive timely treatment not only of MH [9] but of all VVDs is indicated for the prevention of HF, notably of HFPEF.

CONCLUSION

Chronomics, the study of interactions among cycles in and around us implies, beyond circadian VVAs, focus on infradian rhythms with periods longer than 28 hours. Some of the amplitudes of infradians are relatively large, as in the case of paratridecadals (components with a period of about 35 years). Other infradian cycles' amplitudes are small but possible clues to mechanisms of sudden cardiac death, as in the case of aeolian about 5-month rhythms corresponding to cycles in solar flares. Clinical and research scholars of aging must realize that at two consistently sampled times of day one can have opposite effects and hence any other endpoints than (at least the circadian) rhythm characteristics can lead to much wasted effort. Using circadian characteristics while mapping infradians, opens a new perspective of aging aimed at prevention rather than only treatment of CVD and in the latter case also at timed nutrition, a problem yet to be solved for omega-3 supplementation.

REFERENCES

[1] Halberg F, Cornelissen G, Otsuka K, Siegelova J, Fiser B, Dusek J, Homolka P, Sanchez de la Pena S, Singh RB, BIOCOS project. Extended consensus on need and means to detect vascular variability disorders (VVDs) and vascular variability syndromes (VVSs). Geronto-Geriatrics: *Int. J. Gerontology-ChronomeGeriatrics* 2008; 11 (14): 119-146; Leibniz-Online Nr. 5, 2009 (http://www2.hu-berlin.de/leibniz-sozietaet/journal/archiv_5_09.html). 35 pp.

[2] Halberg F, Hillman D, Schwartzkopff O, Cornelissen G. Selectively assorted decadal and multidecadal solar signatures investigated longitudinally in individuals and transversely in populations. In preparation.

[3] Halberg F, Cornélissen G, Katinas G, Tvildiani L, Gigolashvili M, Janashia K, Toba T, Revilla M, Regal P, Sothern RB, Wendt HW, Wang ZR, Zeman M, Jozsa R, Singh RB, Mitsutake G, Chibisov SM, Lee J, Holley D, Holte JE, Sonkowsky RP, Schwartzkopff O, Delmore P, Otsuka K, Bakken EE, Czaplicki J, International BIOCOS Group. Chronobiology's progress: season's appreciations 2004-2005. Time-, frequency-, phase-, variable-, individual-, age- and site-specific chronomics. *J. Applied Biomedicine* 2006; 4: 1-38. http://www.zsf.jcu.cz/vyzkum/jab/4_1/halberg.pdf

[4] Halberg F, Sothern RB, Cornélissen G, Czaplicki*J. Chronomics, human time estimation, and aging*. Clinical Interventions in Aging 2008; 3 (4): 749-760. http://www.dovepress.com/articles.php?article_id=2608

[5] Halberg F, Cornélissen G, Spector NH, Sonkowsky RP, Otsuka K, Baciu I, Hriscu M, Schwartzkopff O, Bakken EE. Stress/strain/life revisited. Quantification by blood pressure chronomics: benetensive, transtensive or maletensivechrono-vasculo-neuro-immuno-modulation. *Biomed and Pharmacother* 2003; 57 (Suppl 1): 136s-163s.

[6] Maschke C, Harder J, Cornélissen G, Hecht K, Otsuka K, Halberg F. Chronoecoepidemiology of "strain": infradianchronomics of urinary cortisol and catecholamines during nightly exposure to noise. *Biomed and Pharmacother* 2003; 57 (Suppl 1): 126s-135s.

[7] Halberg F. Organisms as circadian systems; temporal analysis of their physiologic and pathologic reponses, including injury and death. In: Symposium on Medical Aspects of Stress in the Military Climate, Walter Reed Army Institute of Research (Col. William D. Tigertt, Medical Corps, Director and Commandant), *Walter Reed Army Medical Center* (Maj. Gen. A.L. Tynes, Medical Corps, Commanding), Washington DC, 22-24 April 1964. Washington DC: U.S. Government Printing Office 1965-778-714; 1965. p. 1-36.

[8] Halberg F, Cornélissen F, Halberg Francine, Kessler T, Otsuka K. Measuring mental strain by duration of blood pressure overswing (CHAT): case report. *World Heart J.* 2010; 2 (2): 141-167.

[9] Cornélissen G, De Meester F, Wilczynska-Kwiatek A, Halberg F. Changes with age in the relative prominence of the week and the day in blood pressure. International Congress "*Natural Cataclysms and Global Problems of the Modern Civilization*", Istanbul, Turkey, September 19-21, 2011, in preparation.

[10] Maeder MT, Kaye DM. Heart failure with normal left ventricular ejection fraction. *J. Am. Coll. Cardiol* 2009; 53: 905-918.

[11] Kumagai Y, Shiga T, Sunaga K, Cornélissen G, Ebihara A, Halberg F. Usefulness of circadian amplitude of blood pressure in predicting hypertensive cardiac involvement. *Chronobiologia* 1992; 19: 43-58.

In: Cardiovascular Health and Chronomics
Editor: Ram B. Singh
ISBN: 978-1-62948-976-6

Chapter 17

EARLY GLIMPSES OF LUNISOLAR AND RECIPROCAL BIOSPHERIC CYCLES WITH OVERLAPPING UNCERTAINTIES

Franz Halberg, Germaine Cornélissen and Othild Schwartzkopff
Halberg Chronobiology Center, University of Minnesota, Minneapolis, MN, US
Support: U.S. National Institutes of Health , (GM-13981) (FH),
University of Minnesota Supercomputing Institute (GC, FH)

1. PREAMBLE

According to the late Reinhart Koselleck's "Futures Past" [1], history falls into the categories of "The Space of Experience" and "The Horizon of Expectation". We here add "and into the category of an inferential statistical estimation of the uncertainties involved in any claim of experience and/or expectation whenever possible". Koselleck concludes that lasting structures are concealed by conventions and must be discovered and investigated if historical experience is to be transformed into historical science. The view adapted herein is that for history to be scientific, first it has to examine claims, whenever the data permit and when time series are involved, meta-analyses are an opportunity to re-examine the past. This can be done by testing the hypotheses postulated and estimating the parameters involved in the light of all the information available at a given time in history and re-examining further data accumulated in the interim. The task of validating or invalidating claims is particularly pertinent in the case of cycles, in the light of accumulating added evidence from lengthening time series, notably those in space weather in an era of satellites and computers. From this practical viewpoint, we find that, as yet, a consideration of any history of effects of space weather on human affairs is included in neither the space of experience nor the horizon of expectation of most contemporary historians, the late William L. Langer [2] being an exception. Second in interpreting history completely requires scholars to start learning the simple fact (with wide-reaching consequences) that, as Yohsuke Kamide puts it [3], we live in the sun's atmosphere and there is much that we cannot see in it, and (we add) it importantly involves, as a yet unknown factor, many human affairs, including whether or not we die suddenly by an electrical accident of the heart [4] (which brings the topic immediately within the scope of this journal), by our own hand [5-7] or by the hand of others [8]. Here lies a challenge for those concerned with the heart and circulation, reflecting in time series the cycles of our cosmos, in which we evolved.

2. Deep History

We can image in time the effects of our cosmos in living things that may have taken place much earlier than in Lucy's time some millions of years ago [9] and have made us what we are, with all our ills as individuals and as societies.

As yet, we may be in an early stage of development, if we compare the assumed span of intelligent life with the discussed very much longer existence of the earth itself. Solar-biospheric associations in this context are hardly surprising from the viewpoint of deep or big history, when we find some of the cycles of the lunar tide in the forms of life believed to be the oldest (archaea), Figure 1 [10], perhaps because life originated in the sea.

After a long battle and a committee's decree that about 24-hour (circadian) systems are limited to cells with a nucleus [11], a near 24-hour period in bacteria, Figure 2 [12-14] is amply documented by many laboratories in forms of life without a nucleus (in prokaryotes as in eukaryotes). Figure 3A also shows a built-in about 7-day period in a eukaryotic cell that has been on earth for an estimated 500 million years. Figure 3B shows a hint of a transyear and Figures 3C and 3D indicate decadal periods in bacterial sectoring. Thus, one can try to trace by cycles the origin of life broadly and the adversities it encountered, planetary and extraplanetary, as prehuman and eventually human history. We must indeed uncover and map previously concealed cycles, including those acquired during the relatively short appearance of humans with a language, so that we can know and possibly predict, as Herschel put it [15; cf. 16], when the sun and the broader universe(s) are ill-disposed, before we can then try to do something about countermeasures that may be as important as heating and air conditioning against extremes of temperature. The relative importance of untoward effects of thermic, photic and non-photic cycles may differ in dealing with various conditions, such as myocardial infarctions versus sudden cardiac death, versus terrorism, as can be gauged perhaps by the relative prominence of yearly versus paraannual cyclic components.

3. Glocality in Time as in Space

Today, the time structure of decades-, centuries- and, in the case of the aurora, nearly millennium-long observations in the form of serial data can be imaged by an armamentarium of glocal methods that apply to geo- and space- and to corresponding biospheric-science history. As the portmanteau word "glocal" implies, these procedures are global, such as a spectrum of all (auroral or other) data, and local, insofar as each time series is also analyzed in sections of interest or preferably in a set of systematic sections of different length satisfying both of the competing aspects of optimal resolution in time and in frequency, by a systematic shortening and lengthening of the sections investigated, Figure 4. The slogan "think globally, act locally" gains temporo-spatial dimensions from imaging in time, revealing mostly unseen components in a heretofore unknown broad transdisciplinary spectrum, consisting of periods shared by the biosphere with its cosmos, for a scrutiny of mechanisms of biospheric-solar associations, the topic of follow-up historical examinations by meta-analyses of the available pertinent time series.

3.1. Minimal Sampling Requirements

By superposed epochs and other procedures, we found during three years of intense solar activity an increased incidence of myocardial infarctions in Moscow after the change from positive to negative of the north-south vector of the interplanetary magnetic field (Bz), Figure 5 [17]. This result was confirmed by others in the same extensive data representing diagnoses made in cases gathered in response to a call for an ambulance [17; cf. 18], and extended by us to another location in space and in a broader span of time in Minnesota, USA. The nonstationary aspects of the phenomenon on hand became apparent [20].

In the light of a 29-year-long series from the USA, an unwarranted east-west controversy was resolved, as documented elsewhere [19].

The controversy was based on an inference from abundant data which covered less than a decadal cycle [21, 22], an always-hazardous and often unnecessary limitation of those who decide to disregard or eliminate cycles, as will be elaborated under Technical Comments. Long epidemiological and physiological time series led, beyond the finding of time-varying cross-spectral coherence [23], to a subtractive and/or an additive approach, like the "remove and replace", in dealing with a gland in endocrinology, except that the manipulation is done by the cosmos if and when the latter pleases, and it cannot necessarily be repeated any time, at the investigator's will [4]. Thus, resonances, past and/or present, in the biosphere with solar spatio-temporal components, become apparent, helped by satellites, computers and inferential statistics. When the environment cooperates, the lessons learned may be interesting and, we trust, useful, at least as history in its own right, its raison d'etre, and, we anticipate, beyond that aim as a tangible expectation and perhaps prediction of the future, as a horizon of expectation by hindsight.

4. THE NOT-SO-PERFECT SUN

Two years before Schwabe's 1838 paper [24], Joseph Johann von Littrow [25] scholarly reviewed early information on sunspots and located a slightly longer series than that of Schwabe. It was by Horrebow, whose work was published by Thiele [26], as noted by Yu.I. Vitinskii [27], as we were kindly informed by Dr. Ilya Usoskin.

Coexisting Built-in Luni-Solar Resonance in an Archaeon

From Tidal Cycle (top), over Lengthening Period (middle) to Free-Running Circadian Rhythmic (bottom) Transcription of Previously LD12:12 Synchronized Halophilic Archaeaon *Halobacterium salnarum NRC-1* Released into Continuous Darkness *

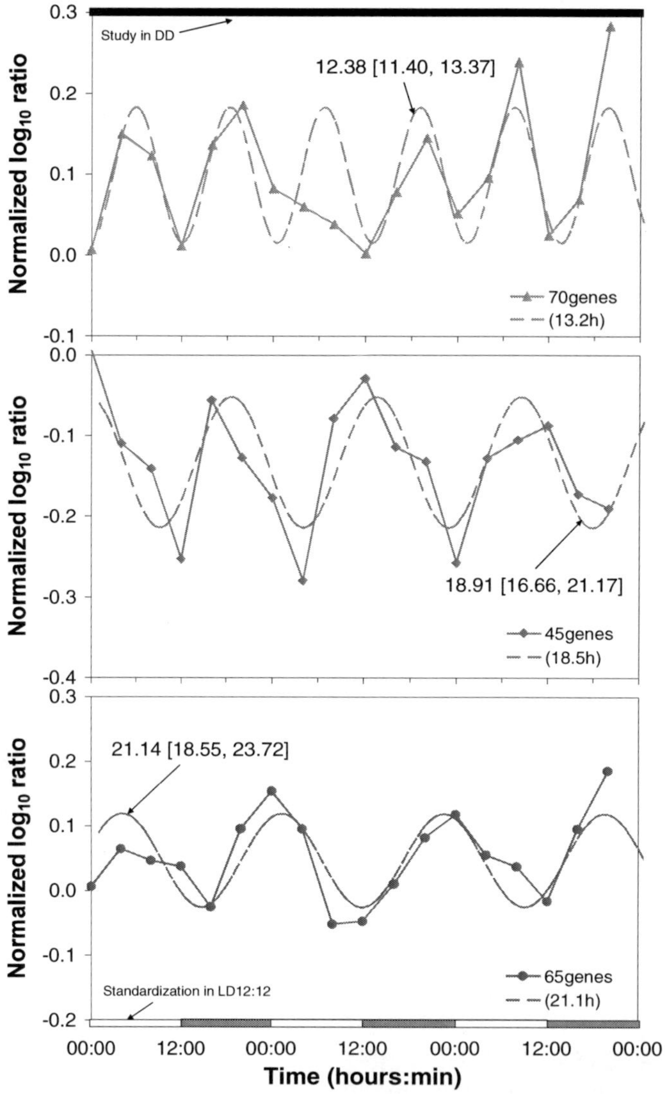

* Periodicity Persists for at least 72 hours in Continuous Darkness in Three Classes of mRNA Profiles for 180 of the 290 Genes Detected as Cyclers. Chronometaanalysis of data taken off Figure 3A from PLoS one 2009; 4(5): e5485 (K Whitehead et al. Diurnally entrained anticipatory behavior in Archae). Key lists period reported by authors, assessed by Lomb-Scargle periodogram. Arrow pointing to fitted curve reports period assessed nonlinearly by the extended cosinor together with its 95% confidence interval in [ ].

Figure 1. Earliest life may well be characterized by a lunar (tidal, top) component, the average period lengthening (middle) toward a solar (circadian, bottom) cycle. Original data of K. Whitehead et al. in PLoS one 2009; 4 (5); e5485; meta-analysis in Halberg et al. (10).

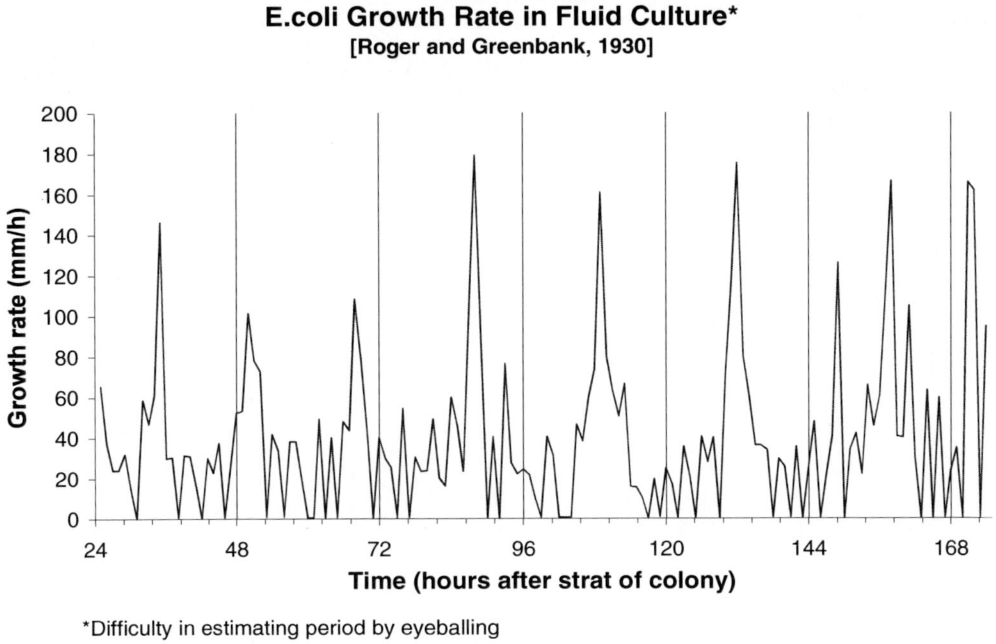

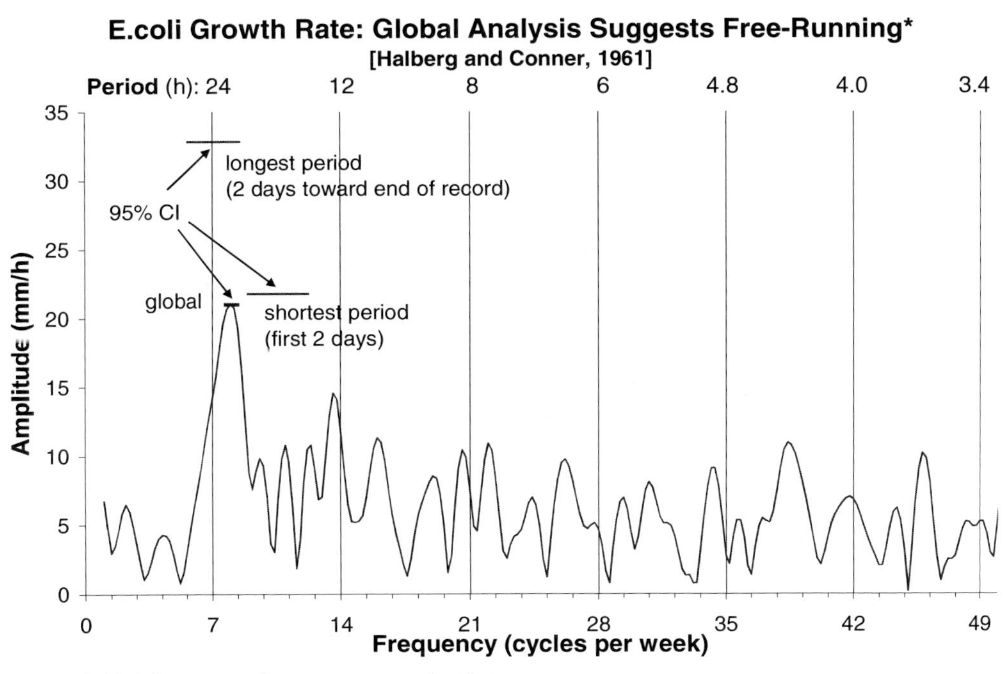

(A)

Figure 2. (Continued)

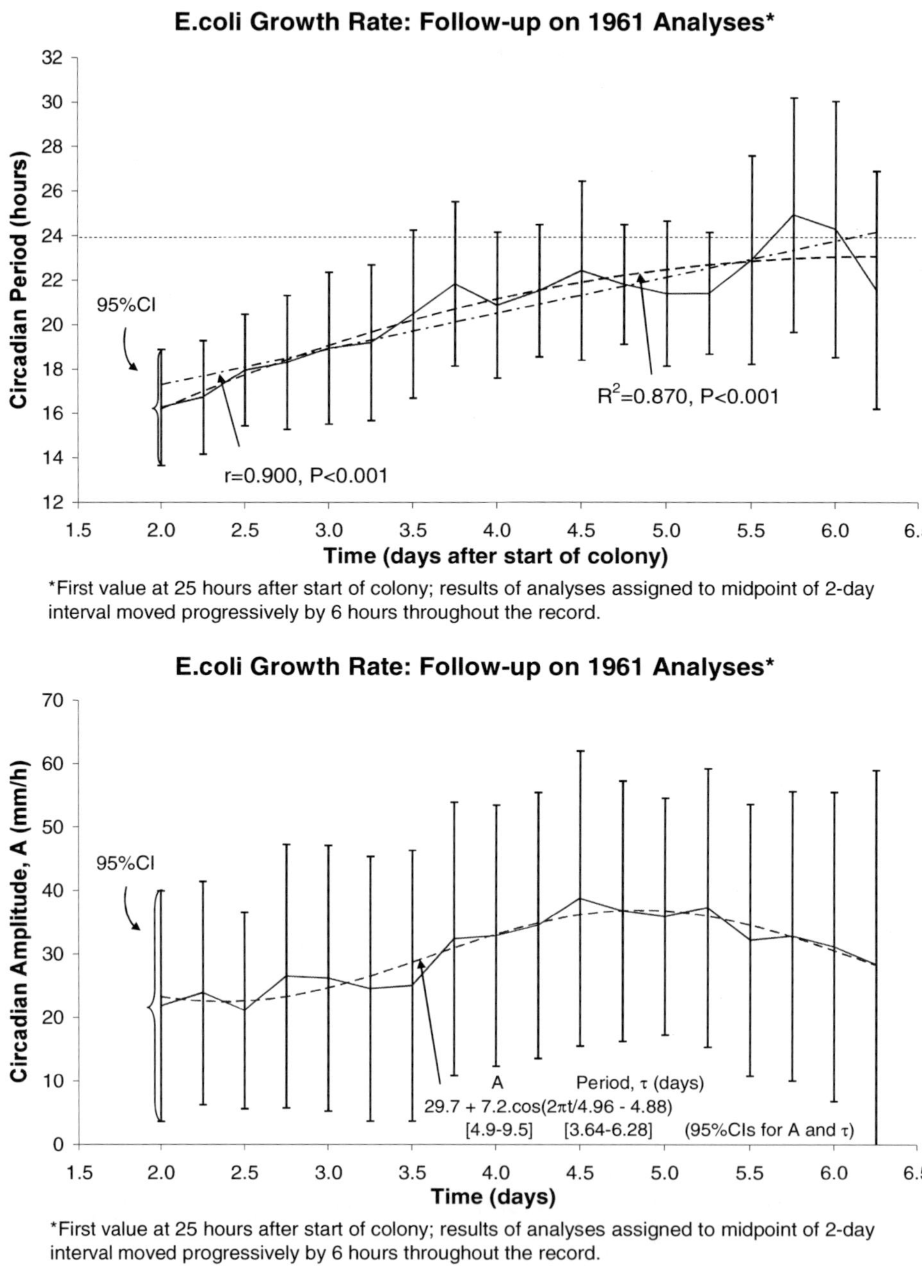

(B)

Figure 2. A) Top: Time plot of the 1930 data of Lore Alford Rogers and G.R. Greenbank, a noted bacteriologist described by a Cosmos Club (Washington, DC) Vignette of December 1967 as "the bright star in the [US Department of Agriculture] scientific horizon before World War II". In 1961, the data on growth rate (actually colony advance) of E. coli assessed at hourly intervals for slightly over 6 days were taken off the graph published by Rogers and Greenbank in 1930. Some regularity in the time series is apparent from the spikes in colony growth occurring at intervals shorter than 24 hours, as seen by reference to the vertical dashed lines drawn at daily intervals, yet eyeballing can hardly serve for an estimate of a period and certainly forestalls any evaluation of uncertainties. Rogers and Greenbank wrote: "Figure 1 [on top], which is representative of a number of experiments, shows that there is a considerable degree of periodicity to the alternation of rapid and slow growth." In considering this graph, it should be kept in mind that the curve represents not extent of growth, but rate of advance of growth for definite time periods. The number of bacteria in the different periods could not be determined. Bottom: Results from least squares spectrum of the entire series. A spectral peak corresponds to a period slightly shorter than 24 hours. Analyses in (12, 14). B) Top: Periods and corresponding 95% CIs characterizing the data shown in Figure 2A are here estimated for intervals of 2 days progressively moved throughout the time series by 6 hours. An interval of 2 days was selected to have at least one replication to secure a more reliable estimate of the period while also allowing the investigation of any change in period length with time. Separate non-overlapping 1-day intervals analyzed similarly yield similar estimates. The period is seen to lengthen as a function of time. As a first approximation, the dashed dotted line is in keeping with a linear increase. The period estimate is only of about 18 hours at the beginning of the series, but approaches 24 hours by the end of the first week of monitoring. With the qualification that consecutive estimates are not independent, the lengthening of the period has an ordering $P < 0.001$ ($r = 0.90$). The fit of a second-order polynomial (dashed line) suggests that the period levels off. In this model, the quadratic term significantly reduces the residual variance (ordering $P < 0.02$) beyond the linear term. The results could be interpreted as an originally free-running circadian component that may be pulled by (if not become synchronized with) the 24-hour day. Bottom: Corresponding changes In the circadian amplitude.

LARGER *ABOUT-WEEKLY* THAN *ABOUT-DAILY* CYCLE in ELECTRICAL POTENTIAL of a UNICELLULAR ALGA *ACETABULARIA ACETABULUM* EVOLVED 500 MILLION YEARS AGO *

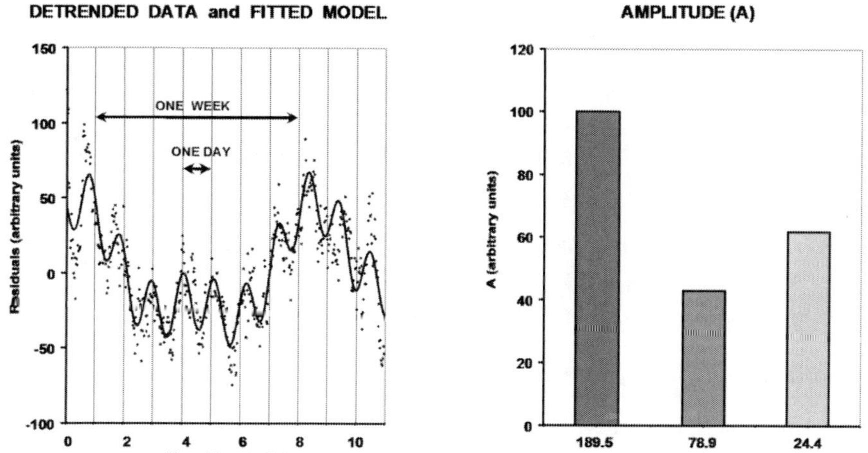

*** Nonlinear spectral analysis on signal-averaged data from 20 single cells, released (zero time) into continuous light (LL), after prior standardization in light and darkness alternating at 12-hour intervals (LD12:12) for up to one week. Total number of observations: 38,578; experimental span: 376 days. Note a more prominent amplitude (A) for a component with a period near a week, than the As of the about daily and about half-weekly components (all free-running). The circaseptan A is eqated to 100 and the other As are expressed as percentage of the circaseptan. Metaanalyzed data of Dr. Sigrid Berger, Dr. Lübbo von Lindern and the late Dr. Hans-Georg Schweiger.**

(A)

Figure 3. (Continued)

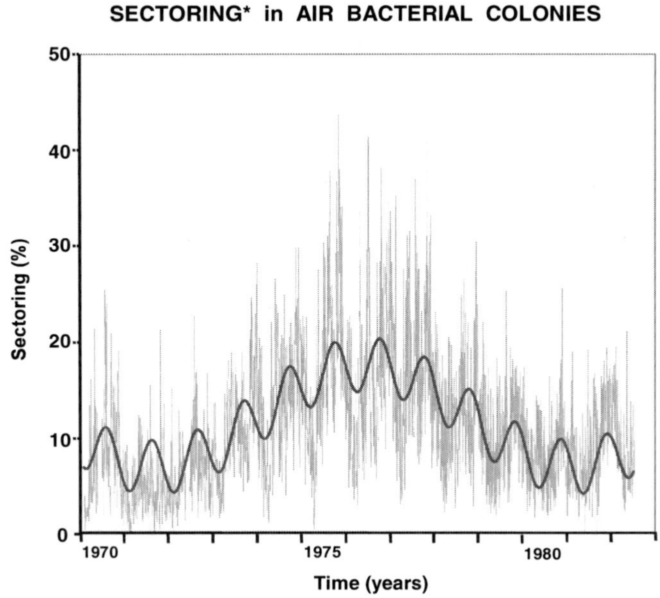

* Percent of mutations and/or other genetic changes, daily data.
Parameters of the fitted curve: MESOR = 12.3 %,

Period (y)	Amplitude (%)	Acrophase (degrees)
9.445	5.37	-245.7
1.026	2.84	-223.4

Phase reference from Jan 1, 1970.

(B)

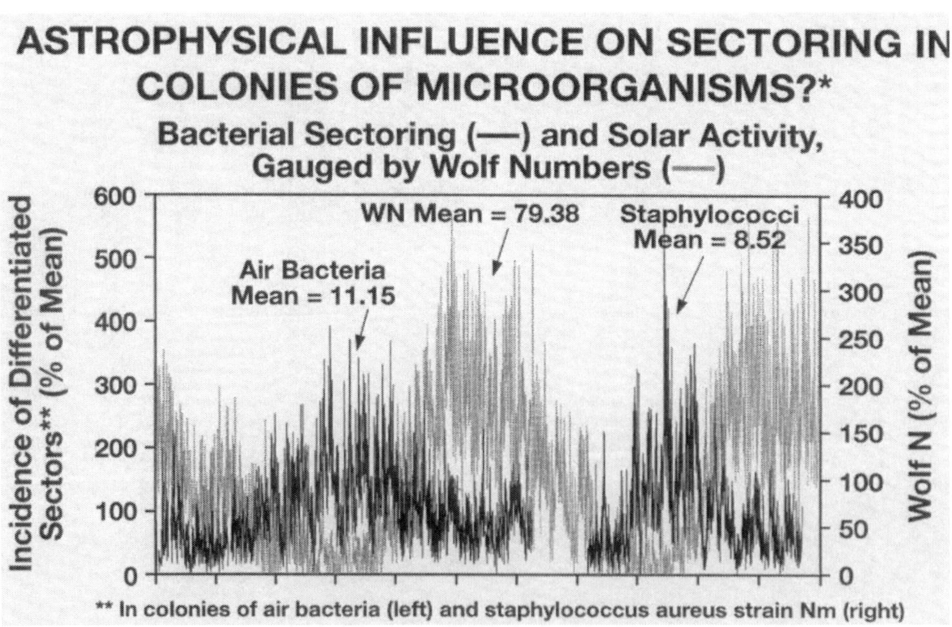

(C)

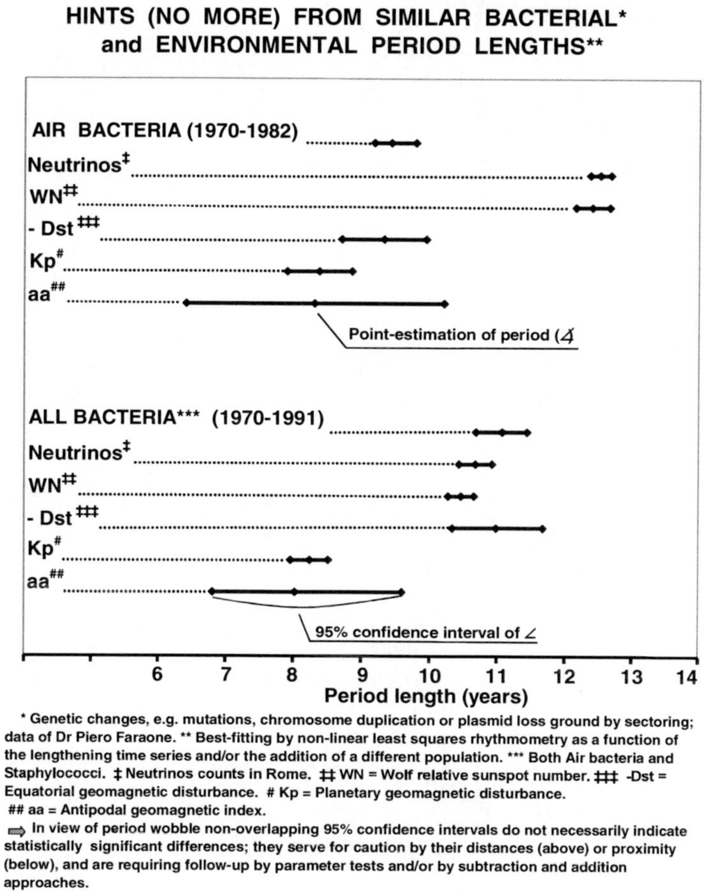

(D)

Figure 3. A) Left: The naked eye sees that the about 7-day (circaseptan) component is more prominent than the circadian rhythm in the electrical potential of *Acetabularia acetabulum* after release into continuous light (data and fitted model). Right: Amplitude ratios indicate the relative prominence of the week over the half-week and day. A spectral component with 95% CIs not overlapping a precise week, a hint of endogenicity, is present early in phylogeny, in a unicellular eukaryotic alga's oxygen production as well as in its electrical potential, as it was also in *E. coli*. The circaseptan-to-circadian amplitude ratio is above unity in the one eukaryotic unicell studied over 14 years from this viewpoint *(Acetabularia)*. Edited data from the same source as those here analyzed are in the public domain, thanks to Dr. Lübbo von Lindern. B) A spectral analysis of data on sectoring in air bacterial colonies quantifies several components that the naked eye can try to discern, including a candidate near-transyear, which, however, turns out to be a broad peak, covering the precise 1-year trial period useful as a hint; it is no evidence of a mechanism. Data from Dr. Piero Faraone (14). C) An about 10-year cycle in air bacteria and in all bacteria is readily apparent to the naked eye and is compared with Wolf numbers. Cross-spectral coherence appears between sectoring and geomagnetics (not shown). Data on bacteria from Dr. Piero Faraone. D) Chart of reciprocal candidate periods: time series analysis shows different results in the length of the cycles resolved when air bacteria are analyzed separately or combined with data from staphylococci. Similar periods characterize air bacteria and the equatorial index Dst. When all microbial data are pooled, more of the variables show similar periods. But the similarity of periods, useful as a hint, is no evidence of mechanisms without added information by addition and/or subtraction.

According to Littrow [25], Christoph Scheiner, a Jesuit priest in Swabia, collected, "with continued diligence for 19 years, observations of sunspots and properly interpreted their movements". (With 19 years of systematic observation, some cyclicity must have been apparent.) Scheiner claimed to have seen the spots in March 1611, but was prevented from communicating his discovery by his superior (the Provinzial) Busaeus, who believed that it was impossible for the sun, the model of highest purity, to be tainted with spots, since this finding contradicted all of Aristotle's philosophy. Galileo claimed to have shown the spots with his telescope to several friends in the garden of Cardinal Bandini in April 1611, and that he saw them first at the beginning of that year. The first work on sunspots was published by Johann Fabricius, also called Frisius [28], who also observed their movement on the sun and thence concluded that the sun rotated. In papers found posthumously from Thomas Harriot, his observations of sunspots started on December 8, 1610, but remained unknown for a long time in continental Europe.

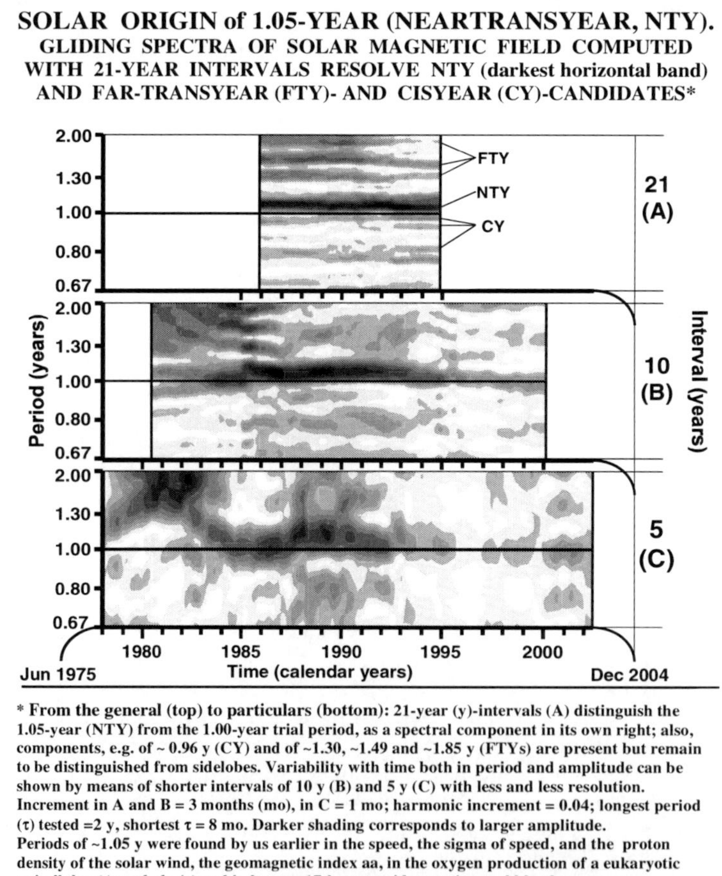

Figure 4. The resolution in frequency increases (and that in time decreases) with the lengthening of the interval used for analysis (bottom to top) in time series of terrestrial telescope-recorded magnetism, with focus on a near-transyear (A) and of human diastolic blood pressure, with focus on an about 5-month periodicity (B), as separate illustrations of glocality in time.

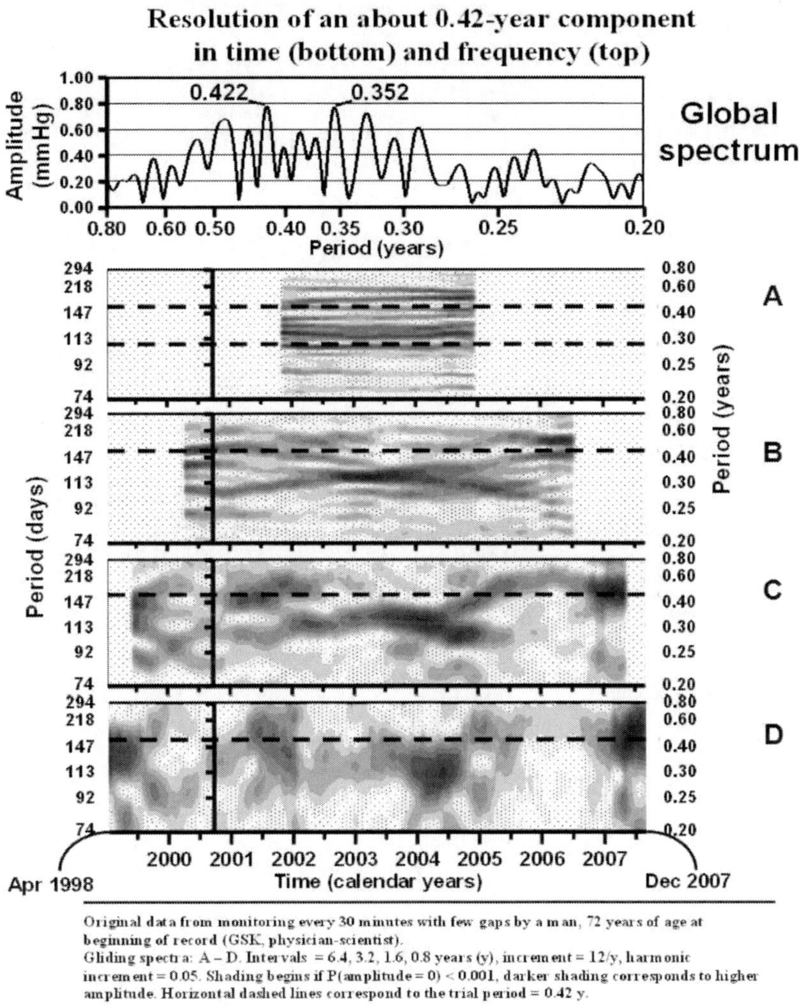

© Halberg.

Figure 5. There is a direct association between events taking place in the IMF and the daily incidence of myocardial infarction. Southward turns of Bz have been associated with aurorae and magnetic storms. A biologically relevant southward turn of Bz is defined herein as a change between a daily average of Bz of $\geq$1 nanoTesla (nT) to one of $\leq$-1.5 nT. Once such an event is identified, the daily incidence of myocardial infarctions is recorded for these two days as well as for the two days preceding the day when Bz $\geq$1.0 nT (days -1 and -2) and the two days following the day when Bz $\leq$-1.5 nT (days +1 and +2). The daily incidence of myocardial infarctions for these 6 days is then averaged over all events identified during the 3-year span (1979-1981). After a southward turn of Bz, there is a statistically significant increase in myocardial infarctions (P=0.027). A statistically significant difference is found by a paired t-test comparing incidence of myocardial infarctions on day +1 vs. that on day -1.

The Arab historian Albufaradge [29] wrote reportedly that the sunlight was damped for 14 days and that from October 626 to June 627 half of the sun appeared to be black, perhaps an indication of sunspots. Similar observations according to Littrow [25] are found in Pliny, Virgil, Plutarch and others: "in fact, one sees them usually coming in great numbers and apparently periodically, while at other times the sun is free of them". Against this background, we focus herein on a few pioneers who, willingly or in De Mairan's case unwillingly, started a transdisciplinary unified science with much of the follow-up solar-biospheric interactions, alluded to elsewhere [8, 30-33], to be the topic of separate reviews of the various components of a novel transdisciplinary spectrum that constitutes parts of a global map of past and present interactions of space weather with biospheric elements.

5. Endpoints from Historical Imaging in Time

In scholarly books, Rudolf Wolf [34] and Wilfried Schröder [35] both indicate the difficulty of writing history for the intelligent public in a way acceptable to the specialist in the field. The task is even more complicated when the past is scrutinized in inferential statistical terms in long time series analyzed with the ever-present dangers of artifacts (shared by imaging in time with that in space, both effects of changes in period or phase having consequences reminiscent of the aberrations found in the development of light microscopy). The principles and endpoints of time structure microscopy are easy to visualize, as is the method of least-squares itself (see Technical Comment). The MESOR, M, short for midline-estimating statistic of rhythm, is usually more precise and more accurate than the arithmetic mean. The period, τ, of any cycle is obtained and, for each τ, a double amplitude, 2A, and an acrophase, ϕ, measures of the extent and timing of change, and the (A,ϕ) pairs of harmonics indicate the waveform. Multiple τs can be detected by some of the available methods, as is also apparent from the simulations in Table 1.

5.1. Considerations of Uncertainties

Changes measured as time structures, or chronomes, fall into the categories of deterministic and other chaos, trends and cycles. Cycles can characterize endpoints of deterministic chaos [36] which can generate them [37]; cycles can be assessed with trends that can be parts of cycles longer than a given time series. Shared cycles in the biosphere and in its physical environment can be used as a key to the history of associations between solar disturbances and biological systems [38]. A photic reciprocal cycle pair is seen as the environmental alternations of light with darkness and as corresponding organismic changes from wakefulness to sleep. Other paired environmental-biospheric cycles are unseen (nonphotic), related, i.a., to possibly interacting gravity and magnetism, which are also mimicked in biological systems, *quod est demonstrandum.* Built-in mechanisms, documented at the molecular level for counterparts of the photic day [39-42], are suggested for biologic non-photics as well [23]

Some of these cycles -- dubbed aeolian, after Aeolus, the capricious ruler of the winds in Greek mythology -- around us, and their counterparts in us, reciprocal only in their similar length, are nonstationary; they wax and wane in amplitude to the point of disappearing and reappearing; they drift in phase or period, bifurcate or split into even more parts, and reunite.

5.2. Illustration of Imaging in Time

Using the principle of least-squares, cycles can be assessed by curve-fitting with a computer, which also yields dynamic endpoints, with methods of differing resolving power, notably when the components to be resolved differ little in period and much in amplitude (cf. Table 1).

Imaging in time is done glocally, i.e., globally, in a window of the longest available series as a whole at a given time as shown in Figure 6A, right, and locally, in sections of the time series such as in the gliding spectral window in Figure 6A, left.

In (extraterrestrial) space, measurements from satellites extend glocality to the interplanetary magnetic field. There is a need for glocality in time as well as in space; both, enabled by the computer, that have allowed a remove-and-replace approach by chronomics, e.g., Figure 6B.

Physical cycles were, however, observed in the era before computers, satellites and inferential statistics (albeit all three approaches as used in Figures 6A and 6B are invaluable).

The history of the Saros cycle in solar eclipses of a little over 18 years is a case in point: it is said to have been discovered by Babylonian astronomers in the last several centuries BC (Chinese records of solar eclipses date back over 4,000 years, and an eclipse recorded in stone in Ireland is dated November 30, 3340 BC [43, 44]). Human observation and intuition achieved the concept of a heliocentric system proposed by Aristarchus of Samos (310 BC-c. 230 BC) as well as by Copernicus (1473-1543) without computers.

6. ARCHILOCHUS

The scholarship of the late Jürgen Aschoff suggested that the history of biospheric cycle research began ~650 years before Christ, when the Greek lyric poet and mercenary Archilochus of Paros, "a very dynamic person as you will expect from someone who is interested in rhythms" [45, 46], wrote "Recognize which rhythm governs man", Figure 7. Aschoff did not comment on another fragment by Archilochus, written when he witnessed a solar eclipse (on April 6, 647 BC) [47]. The latter fragment reveals that Archilochus was sufficiently interested to note in verse that the eclipse "made night from mid-day, hiding the light of the shining Sun". This latter fragment authenticates the contribution in the fragment on rhythm as dating from the 7th century BC, and identifies Archilochus as a lyric link, not only to the study of biological rhythms, but also to that of the sun and moon, and to their visible, as well as their subtler but no less important effects. In this perspective, different fragments from the same poet link rhythms, the broader chronobiology, and thus the biosphere, to astronomy, to lunisolar and other space physics, and to the transdisciplinary mapping of time structures, or chronomes.

Table 1. Comparison of Methods Applied to Noisy Test Series with Two Components with Close Periods -- Only Two Approaches (7 And 8) Resolve Both Components†

$Y_i = 100 + 10\cos(2\pi t/24 - \pi) + 2\cos(2\pi t/24.8 - \pi) + 5R$, where i = 1, …, 336 ($\Delta t$=1 hour; T=14 days) and R is uniformly distributed with zero mean and range =1 (±0.5)

Method		Period (h)#	MESOR	Amplitude	Acrophase (degrees)	Detection		N of components detected
N	Kind					Test	P	
1#	"Cosinor"-SC[1]	24.00	100.18	10.47	-98	F(2,333) = 2951.51	<0.001	1
2#	Cosine_fit	24.00	100.18	11.53	-98	GoodFit¶ = 0.001	<0.001	1
3*	Fourier	23.93					<0.001	1
4*	Lomb-Scargle	24.00				Robustness = 96%	<0.001	1
5*	Enright	24.00				Robustness = 96%	<0.001	1
6	ARIMA.MLE in SPLUS§	24.00	100.15 (99.8, 100.5)	10.45 (10.11, 10.80)	-96.5 (-94.5, -101)		<0.001	>1[2]
7a	Linear step 1 (L)[3]	24.00	100.18 (99.98, 100.37)	10.47 (10.20, 10.74)	-98 (-97,-100)	PR=95%	<0.001	
7b	Nonlinear step 2 (NL)	23.97 (23.83, 24.10)	100.13 (99.83, 100.43)	9.75 (7.67, 11.84)	-91 (-78,-105)		<0.05@	
		24.63 (24.04, 25.21)		2.49 (0.34, 4.63)	-98 (-47,-149)		<0.05@	2
8	Simulated Annealing (GOSA software)[4]	23.96 (23.89, 24.03)	100.4 (100.1, 100.7)	9.6 (8.4,10.8)	-94 (-83,-105)		<0.05	
		24.6 (24.3, 24.9)		2.5 (1.3,3.7)	-94 (-60,-128)		<0.05	2
9	Wavelets	24.00	Time evolution noted but not quantified by point-and-interval estimates.					

Acrophase expressed in (negative) degrees, with 360°≡period length, and 0° set to 00:00 at start of series.

Period is anticipated trial period. * Only one peak detected. GoodFit: Goodness of Fit. @ From non-overlap of zero-amplitude by 95% Confidence Interval (CI).

†Items 1-8 from Refinetti R, Cornélissen G, Halberg F. Procedures for numerical analysis of circadian rhythms. Biological Rhythm Research 2007; 38 (4): 275-325. http://dx.doi.org/10.1080/09291010600903692; item 9 courtesy of Prof. S.R. Prabhakaran Nayar, Department of Physics, Kerala University, Kariavattom, Trivandrum, India.

1SC: South Carolina: cosinor restricted to 24-hour component, by contrast to Minnesota's cosinor, which includes least squares fit of multiple components consisting of cosine functions with anticipated components not necessarily equal to 24 hours, originally called 'single cosinor' to indicate analyses of individual time series (and NOT of a single component fit), soon thereafter extended to include the chronobiologic serial section and the combined linear-nonlinear rhythmometry, among others; all procedures involving the least squares fit of COSINe curves often displayed vectORially along polar coordinates and checked against chronograms and data stacked for the period(s) found, i.e., as plexograms, for each cosinor-isolated component (periods are estimated with their respective confidence intervals).

2 Structure of residuals found to be correlated and modeled with an ARMA (1,1) model, with AR coefficient = 0.78 (0.49, 0.89) and MA coefficient = 0.54 (0.23, 0.75). Results reported at fixed anticipated period of 24 hours.

3Least squares spectrum detects peak at anticipated period of 24 hours and small sidelobes with periods of 26.7 and 21.8 hours resolved nonlinearly as presence of second component with period slightly longer than 24 hours.

6, 7 and 8: 95% CIs in parentheses; 7a and 7b are part of the more extensive, as need be, MN-cosinor.

§Chatfield C. The analysis of time series: an introduction. London: Chapman and Hall; 1984.

4Czaplicki J, Cornélissen G, Halberg F. GOSA, a simulated annealing-based program for global optimization of nonlinear problems, also reveals transyears.

J Applied Biomedicine 2006; 4: 87-93. http://www.zsf.jcu.cz/vyzkum/jab/4_2/czaplicki.pdf.

GLIDING (left) AND GLOBAL (right) SPECTRAL WINDOWS OF SYSTOLIC BLOOD PRESSURE OF AN ELDERLY* SHOW A CONSISTENT (left) CIRCADIENNIAN (2-year) COMPONENT AS MAJOR PEAK (right) AND SEVERAL OTHER PARTLY TRANSIENT AEOLIAN TRANS- AND CIS-YEARS AND -HALFYEARS

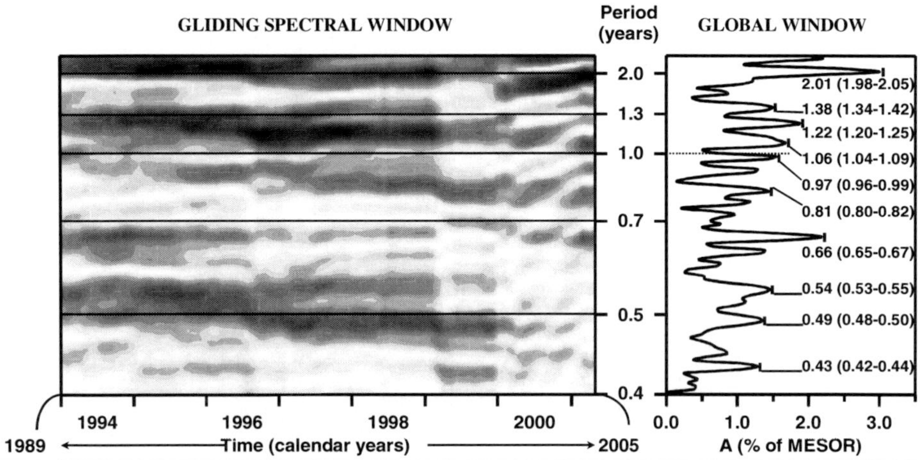

*** FH, man, 72 years (y) of age at start of automatic half-hourly around the clock measurements for ~16 y (N= 2418 daily averages, totally ~ 55000). Gliding spectrum computed with interval = 8 y, increment = 1 month, trial periods from 2.5 to 0.4 y, with harmonic increment = 0.05. Darker shading corresponds to larger amplitude (A). Vertical bars close to the spectral peaks in global spectrum = 95% confidence intervals of periods; they are shown also in digits.**

(A)

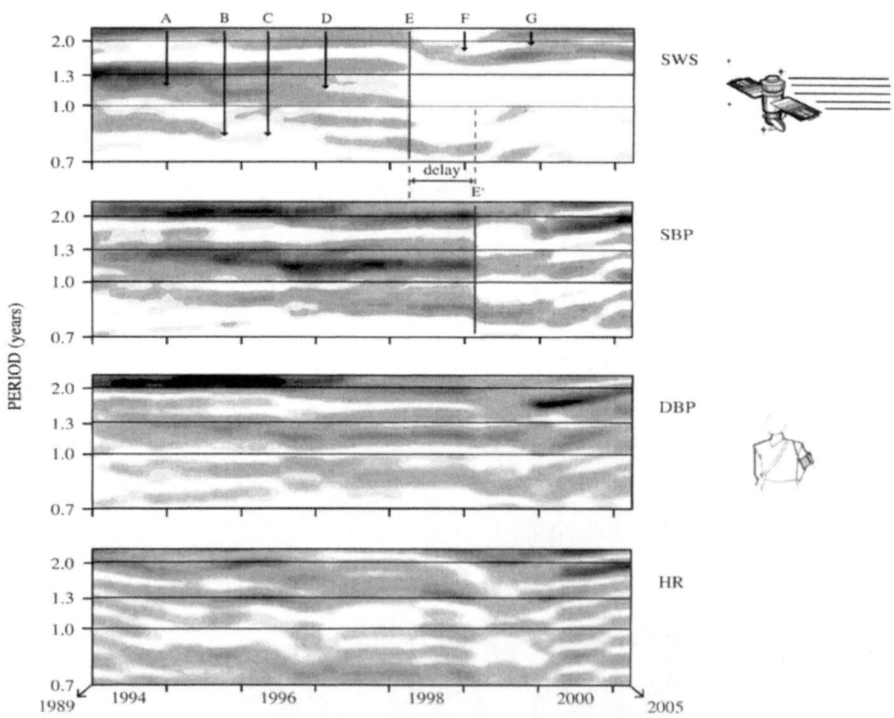

(B)

Figure 6. A) Gliding (left) and global (right) spectral windows of systolic blood pressure show a consistent (left) circadiennian (2-year) component as major peak (right) and several other partly transient aeolian trans- and cis-years and cis–half-years. Global spectral window provides a numerical indication of the candidate periods found, with tentative ordering 95% confidence intervals given in parentheses. This perspective reveals the variability in amplitude not only around the year but also around the half-year and beyond. Since only anticipated components can be taken at face value, and since there were gaps in the record, some of the periods are possibly spurious. B) Time courses of the frequency structures of the speed of the solar wind (SWS) (top) and of an elderly man's (FH) systolic and diastolic blood pressure and heart rate, SBP, DBP and HR (rows 2-4, respectively), examined by gliding spectral windows. Human SBP selectively resonates with SWS (top 2 sections). No obvious resonance, only minor coincident changes are seen in DBP or HR (bottom 2 sections). Aeolian rhythms* in gliding spectra of SWS and SBP change in frequency (smoothly [A] or abruptly [B,C,D], bifurcating [D,F] and rejoining [G], they also change in amplitude (B) (up to disappearing [C,E] and reappearing). During a nearly 16-year span, there are no consistent components with a period averaging precisely 1 year in the three physiologic variables. While *post hoc ergo propter hoc* reasoning can never be ruled out, an abrupt change on top in SWS is followed in the second row in SBP by the disappearance of some components, suggesting that as a first demonstration, some of FH's cis- and trans-year components were driven by the SW [since they disappeared with a lag of about a transyear following the disappearance (subtraction) of the same components from the SWS spectrum]. The persistence of other spectral features in turn suggests partial endogenicity, i.e., an evolutionary acquisition of solar transyear oscillations that may reflect solar dynamics for the past billions of years. *BP and HR data are from a man 70 years of age at start of around-the-clock monitoring, mostly at 30-min intervals, with interruptions for nearly 16 years (N=2,418 daily averages, total about 55,000). Gliding spectra computed with interval = 8 years, resolution low in time but high in frequency, increment = 1 month, trial periods from 2.5 to 0.4 year(s), with harmonic increment = 0.05. Darker shading corresponds to larger amplitude. When several of these broad bands disappear in the SWS, at E, parts of the bands in SBP also disappear, with a lag (delay) at E', while other parts persist. These components are presumably built into organisms over billions of years; as persistence without corresponding components in SWS shows, but can be driven in part by the solar wind, as their disappearance after loss of corresponding components in SWS suggests. "Aeolian", derived from Aeolus, Greek god of winds, who packed the winds up, then let them loose and had them change. © Halberg.

7. ANDROSTHENES

During the fourth century BC, the opening by day and closing by night of the leaves of the tamarind tree (Tamarindus indicus) was described as the "nyktitropic movement" (now nyctinasty). Hugo Bretzl of Göttingen in 1903 [48] refers to Androsthenes of Thasos, who in 324 BC commanded a ship with 30 oars in the campaigns of Alexander the Great, explored Bahrain and its flora and the Arabian coast, and circumnavigated India. His book, now apparently lost, was cited by Theophrastus [49]. According to Bretzl, "Androsthenes was the lucky one who ... discovered for the first time the important fact that plants are capable of movement, a characteristic previously attributed only to the animal world. Like all scientific observations which were ordered by Alexander the Great himself, the description of the daily periodic [italics Bretzl's] movements of the leaflets in their four stages is written so clearly and so succinctly that until the time of our new physiological works, it remains the best (description) of the sleep of plants, even if they are not noted and forgotten." The daily changes from light to darkness rather than those in environmental temperature were imputed by Androsthenes as the environmental factor to which the plant may respond by the "nyktitropic" movements of its leaves. Light-sensitive molecules, photoreceptors, indeed are identified today as intermediates between the aspect of the environment and the plant's or animal's circadian system [50]. But they may not be the entire story: light is an important "switch" [51-54], but not the only pertinent cyclic factor.

8. The Biological Week

Aschoff's gallery of ancestors [45] in the study of rhythms includes Hippocrates, Aristotle and, in particular, Galen. He does not, however, mention Abn Ali Husain ibn Abdallah Ibn Sina, or Avicenna (980-1032), who found, as had Hippocrates and Galen before him, that the week was an important unit of biological time, usually elapsing between the start and resolution of disease: before sulfonamides and antibiotics became available in cases of pneumonia, it took 7 days to learn whether the patient survived or died. That the circaseptan (about 1-week) interval is built into us [56-62] was recognized in antiquity [57]. The week is not merely a reactive adjustment to a single stimulus that carries no 7-day information, such as the removal of a kidney or of part of the liver [56, 61], a transplant [60], or even a stay in a spa (57). The latter proposition is supported by a scholarly book by the late Gunther Hildebrandt [57; cf. 62].

In looking for a counterpart for the built-in biological week, we found the circaseptan in geomagnetics [17], as confirmed by physicists [63, 64].

9. William Gilberd: Preparing for Vernadsky's Unified Science

A physician's treatise on geomagnetism continues tearing down occupational, if not disciplinary barriers. Geomagnetism took a relatively early formal lead in 1600 due to William Gilberd (so spelled on his tombstone; also spelled as "Gilbert" or "Gylberde": 1544-1603) [65]. Gilberd studied medicine at Cambridge University and set up practice in London; he became president of the College of Physicians in 1599, while he also experimentally documented geomagnetism in his still-famous treatise "De Magnete". Gilberd also coined the word "electricity", from the Greek for "amber", and apparently first distinguished electric from magnetic phenomena.

For each of their two classic volumes on "Geomagnetism", Sydney Chapman and Julius Bartels [66] use as their motto Gilberd's "Magnus magnes ipse est globus terrestris". Chapman and Bartels, the physicists, describe the contribution of Gilberd as "... the fruit of long years of thought and experiment [by] an amateur in physical science [sic], for his profession was medical; in this also, he rose to eminence, being appointed in 1601 chief physician in personal attendance on Queen Elizabeth [I; and thereafter of King James I]. His book has been described as the first modern scientific treatise." Thereafter, Chapman and Bartels provide an index of chapters, which reveals that chapter 15 of book 1 deals with "the medicinal virtue of iron". Thus by 1600, the crossing and removal of disciplinary barriers, started much earlier by Androsthenes for Alexander the Great, from military science to plant physiology, had led to the crossing from medicine to physics by a pioneering treatise of physics as the contribution of a physician.

Gilberd's book is thus a monument for transdisciplinary endeavors. In this perspective, the history of associations between solar disturbances and biological systems is here presented.

Editor Schlegel noted the need to examine Russian contributions in particular; again, it is a pleasure to emphasize Vasily Vasilievich Dokuchaev's and Vladimir Ivanovich Vernadsky's attempt to eliminate disciplinary barriers [66], a goal also pursued by investigators in the West [67, 68] as steps toward a unified science. For Dokuchaev and his mentee Vernadsky, a great discovery in one's field, such as that of Charles Darwin, was not enough; it was necessary to be transdisciplinary. Accordingly, in Russian culture, Vernadsky rates higher than Darwin or Lavoisier, since the latter embraced only two sciences, biology and chemistry.

Figure 7. Archilochus, a mercenary poet, interested in both a solar eclipse and human rhythms, was a forefather of solar-terrestrial-biospheric associations and of individuals ignoring occupational barriers.

10. Santorio: Physiological Monitoring Started in the 17th Century

Gilberd's book preceded by only a few years the other major milestone in transdisciplinary science. In 1603, William Harvey noted: "The movement of blood occurs constantly in a circular manner and is the result of the beating of the heart" [70]. Before Harvey, Ibn al-Nafis (c. 1210-1280) had postulated the pulmonary circulation [71]. Contemporary to Gilberd was Santorio Santorio (1561-1636) [72-78], who kept a 30-year record of his bodily functions, which allowed him to discover "insensible perspiration" and many details about this phenomenon. While aphorisms abound, the search for any time series data from Santorio was as yet unsuccessful [72-78].

10.1. Timing Treatment: Benefit from a Reciprocal Cycle in a Marker of Tumor Activity

Also in the seventeenth century, Thomas Sydenham (1624-1689), credited with the discovery of the first effective drug for a specific disorder -- "Peruvian bark", whose active ingredient was quinine -- advocated the administration of opiates with a specific timing for the treatment of pain, in the evening [79]. With a few notable exceptions, as in the case of the electrocardiogram, heart rate, blood pressure, temperature and activity, but not yet as a routine, except perhaps for blood sugar in patients with severe diabetes, Santorio's self-monitoring or Sydenham's and Virey's timing [80] still await entry into biomedicine [81].

In most aspects of treatment, Sydenham's recommendation of what books one should read to qualify as a physician remains timely. In one interpretation, Sydenham's advice was that whether one reads Cervantes or Hippocrates, one would be equally unqualified for (medical) practice, and equally unsuccessful in it. Cervantes would still be timely in the third millennium AD if, in any one of the modern textbooks of pharmacology, physiology or internal medicine, one sought for generally accepted ways of utilizing the normal range of physiological function for health care. It is the more important to note that concern about cycles in tumor markers, in the range of physiological cyclic (circadian) variability, doubled the two-year disease-free survival rate of patients with perioral cancers [82]. The change from a single sample spotcheck to a continuous monitoring endeavor on ourselves (of cycles) for health care (read chronobiology) linked to monitoring around us (read chronomics) remains our challenge toward a unified science that can serve physics as well as medicine.

11. Jean-Jacques D'Ortous De Mairan: A Missed Opportunity in the Eighteenth Century

The physicist Jean-Jacques D'Ortous de Mairan (1678-1771), a towering figure in eighteenth-century science, made his own important transdisciplinary contribution [83]. Born into the minor nobility, orphaned at sixteen and lacking a complete formal education, he won the Bordeaux Academy of Science's prize in three consecutive years at a relatively early age, before being asked not to apply again but to become a member. He studied physics and mathematics, among other fields, in Paris from 1698-1702, and became a professional astronomer who wrote about a variety of topics in physics and became perpetual secretary of the French Academy of Science.

De Mairan reported incidentally that he moved a "sensitive" heliotrope plant (perhaps Mimosa pudica) to a place where sunlight could not reach it. He found that the plant still opened its leaves during the day and folded them during the night [83]. His finding can be interpreted as demonstrating the persistence of what we now call circadian rhythms [84]. It seems noteworthy that the circadian rhythm continued in the very absence of the daily change in light and darkness (and, we add), interpreted as the cause of nyctinasty by Androsthenes. Credit is due to de Mairan for this finding as such. He deserves even greater credit for adding that the persisting rhythm was not only endogenous, but also of interest to medicine. De Mairan concluded, however, that it was not his job to pursue the problem. This line of thought, in a different contemporary context, is rightly criticized by Juan Roederer [68], who as a physicist cites Georg Christoph Lichtenburg's (1742-1799) aphorism, equally apropos of other sciences: "He who understands nothing but chemistry does not fully understand chemistry either".

Erwin Bünning [85] had the scholarship to find de Mairan's physiological contribution [83] and to interpret the observation on the plant from a "clock" viewpoint. He may have been unaware or unimpressed by de Mairan's interest in the aurora, and suggested, as have many others, that the sensitive plant continued nyctinasty "... in the absence of environmental time cues ...". Apparently not stated, or perhaps not known by scholars in the field of clocks, is the fact that de Mairan, as a physicist, wrote specifically on the topic (including the history) of the aurora borealis, the northern lights that coincide with magnetic storms and occur primarily around the magnetic poles, prominently visible with the naked eye by night, the aurorae themselves thus undergoing a "circadian" appearance.

Without scrutiny or even knowledge of de Mairan's book on the aurora borealis, it would have seemed tempting to speculate that he may have suspected geomagnetics as possibly providing a 24-hourly periodic signal for the sleep movements of the sensitive plant kept in continuous darkness, an as-yet unsolved problem. In looking into de Mairan's history, the senior author tracked down a micro-card copy of his book [86], reading it page by page under a magnifying viewer, until he found the buried statement that de Mairan ruled out magnetism as the phenomenon underlying the aurora borealis because he believed that "magnetic material is constant". It was disappointing to read that he had indeed considered the possibility that geomagnetics contributed to aurorae (if not to the sleep of plants) but ruled it out, although de Mairan does deserve further credit for identifying the role of "particles from the sun" being involved in the aurora (centuries before a solar wind was identified).

De Mairan refers to Descartes, to whom it occurred to consider a solar atmosphere consisting of very rarefied matter, which he called "air", comparable to the air that surrounds our earth; he supposed it to extend from the sun, to the sphere of Mercury and even beyond, and ascribed to it a similar origin. What appears to favor this idea is that for the last 5 or 6 years (i.e., the maximum half of the 11-year cycle, adds Schove [87]), the aurorae, as often happens, have become very frequent, and, according to our (de Mairan's) hypothesis, should be associated with great extensions of this air (italics in original), which have likewise been numerous. Schove points out further that de Mairan emphasized the frequency of aurorae and sunspots in the early seventeenth century, until the 1620s, and their rarity in the period from 1650 to 1670.

Having postulated that particles from the sun may reach the polar regions, de Mairan could also have postulated and checked whether the timing of the rhythm in leaf movement persisting in continuous darkness may have been synchronized by 24-hourly changes in geomagnetic intensity or disturbance. Had he measured the latter, a task in his very field of physics and near to the major goal assigned to him by his academy, to clarify the mechanism of the aurora, he would have discovered as a minimum that magnetic material is not constant [86]. In other words, de Mairan could have changed his attitude that he must not follow-up on something outside his field, and could have concurrently pursued both the sleep of plants and the physical problem of the aurora borealis. In so doing, he might have unquestionably improved at least the interpretation of his own topic. Whether or not there is possibly also a subtle association between geomagnetics and the sleep-movement of plants requires further study. Rather than assuming constancy, de Mairan would have discovered geomagnetic periodicity in any event, i.e., the changes along the 24-hour scale in terrestrial magnetism, as did eventually François Arago (1786-1853) in 1827, according to a letter by Humboldt to Gauss (88). In the interim, the spectrum of auroral incidence has been mapped and reveals the components encountered in sunspots, Figure 8 [35, 89; cf. 90, 91]. The test of the speculation whether geomagnetic changes relate in any way to the sleep of plants remains a challenge for physicists and plant physiologists alike.

12. Christian Pedersen Horrebow: Sunspots

Shortly after 1600, the telescope was invented; Harriot, Fabricius, Scheiner, Galileo and eventually Newton began observing sunspots. But the statistically validated discovery of the about-decadal cycle in sunspots began with Christian Pedersen Horrebow (1718-1776), a Danish astronomer. He was the son of Peder (Nielsen) Horrebow (1679-1764), whom he succeeded as head of the astronomical observatory associated with the University of Copenhagen. Christian Horrebow deserves credit for a thorough sunspot data collection between 1761 and the time of his death (16 years), continued by his successor only until 1777 [26]. Some of Horrebow's data were lost "during the Napoleonic wars' shelling of Copenhagen", according to Yu.I. Vitinskii in Solar Activity Forecasting [27], who cites Wolfgang Gleissberg (92) in this context. The rejection of the "no-cycle", i.e., "zero-amplitude" assumption, in sunspots with data covering, with gaps, not more than a single cycle was nonetheless documented by a chrono-meta-analysis as a statistically and, in retrospect, transdisciplinarily significant phenomenon [89].

Carl Friedrich Gauss (1777-1855), whose least-squares procedures were used for curve-fitting, was born the year after Christian Horrebow died (in 1776); Joseph Fourier at that time was 8 years of age. Horrebow provided a first series of sunspot numbers, with gaps, at the time probably examined only by eyeballing, without curve-fitting. Horrebow's counting of sunspots was continued for only a little while after his death, perhaps a year, since Clar. Baggio, Horrebow's immediate successor, had other interests, according to Thorwald Nicolai Thiele (1838-1910), then a student of astronomy (later an actuarial mathematician and statistician in addition to being an astronomer, and eventually Horrebow's successor as director of the astronomical observatory of the University of Copenhagen, from 1875 to 1907) [26].

Horrebow, cited by Thiele, had written that "... astronomers to date have taken too little care to make frequent observations of sun spots, undoubtedly because they think nothing can be accomplished thereby which would be of great importance to astronomy or physics. ... It is, however, to be hoped, by making diligent observations in this matter, that a definite cycle of heavenly bodies will be discovered; then at last it will be time to investigate as to how [heavenly] bodies, which are governed and illuminated by the sun, are brought into existence by the sun spots." (... astronomis adhuc parum curae fuit crebras macularum observationes facere, sine dubio quod nihil inde effici posse iis videbatur, quod astronomiae aut physicae magno opere interesset. Sperandum tamen est, diligenter observando etiam hac in re, ut in motibus ceterorum corporum coelestium periodum certam inventum iri; tum demum tempus erit inquirere, quonam modo corpora, quae a sole reguntur et illustrantur, maculis solis efficiantur."

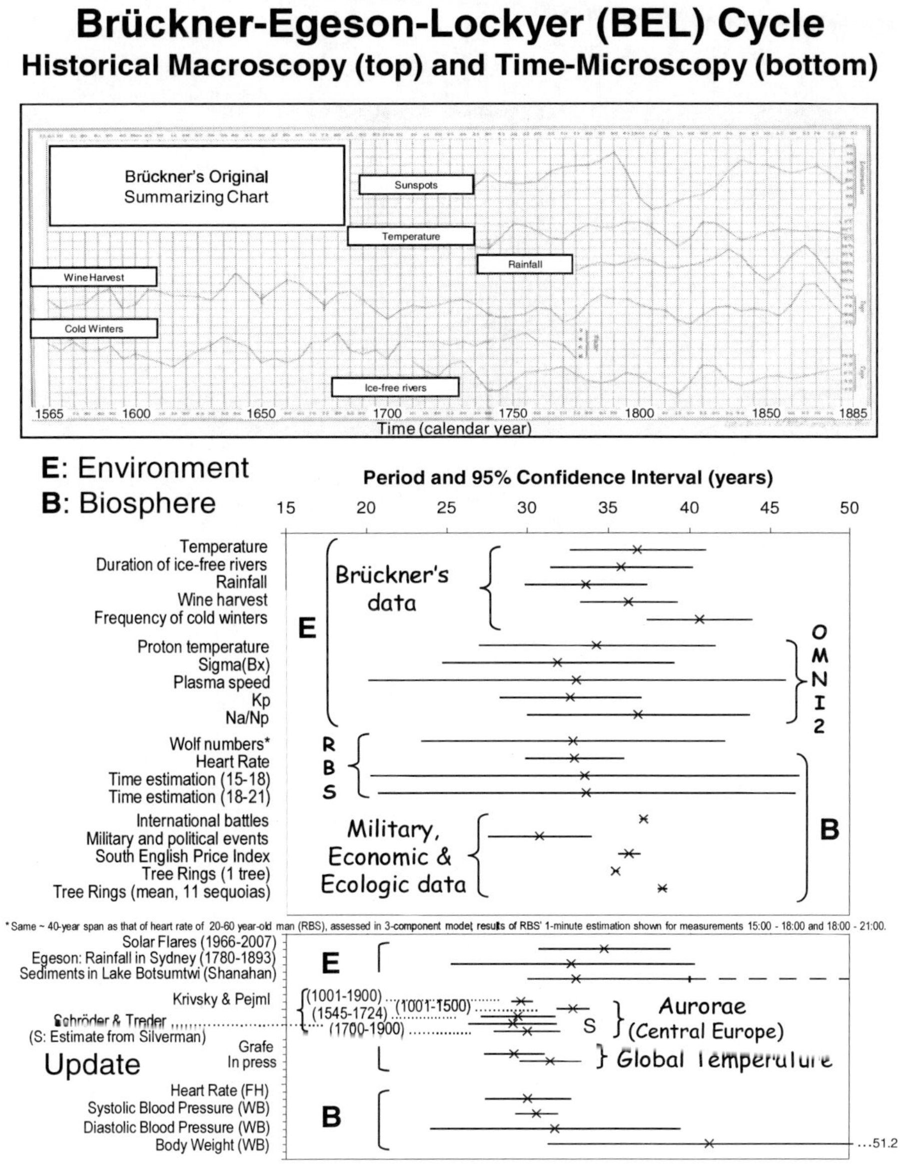

© Halberg.

Figure 8. The components described for sunspots are individually apparent in the spectrum of the aurorae (bottom); we focus in particular on the para-tridecadal cycle of Brückner-Egeson-Lockyer (BEL) found in the aurorae first by Silverman and Shapiro (90).

A time plot (chronogram) on top in Figure 9A of the data tabulated by Thiele shows about 1.5 sunspot (solar activity) cycles (with gaps), guesstimated from Horrebow's 204 monthly values between 1761 and 1776 and the addition of Baggio's data from 1777 [26]. A least squares spectrum (Figure A middle) has peaks around 9.5, 4.2 and 2.0 years.

Horrebow's Data from 18th Century (top) Allow Detection (middle) and Model Fitting (bottom) of Circadecadal Sunspot Cycle

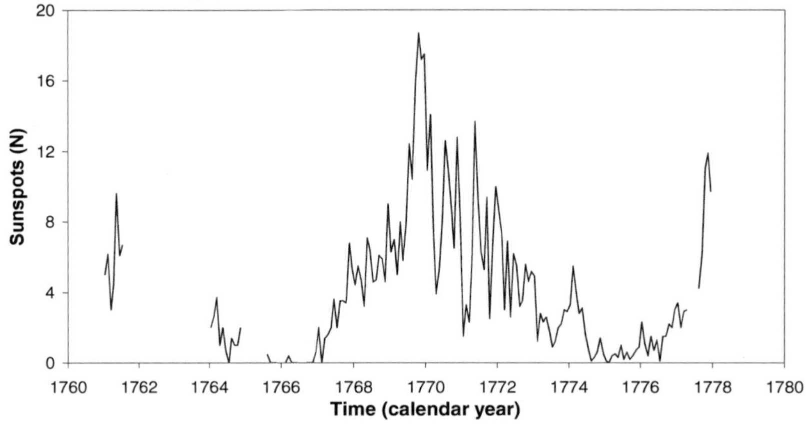

Astronomische Nachrichten No 1193. De Macularum Solis antiquioribus quibusdam observationibus Hafniae institutis, scripsit Th. N. Thiele, astr. Stud. John G. Wolbach Library, Harvard-Smithsonian Center for Astrophysics - Provided by the NASA Astrophysics Data System

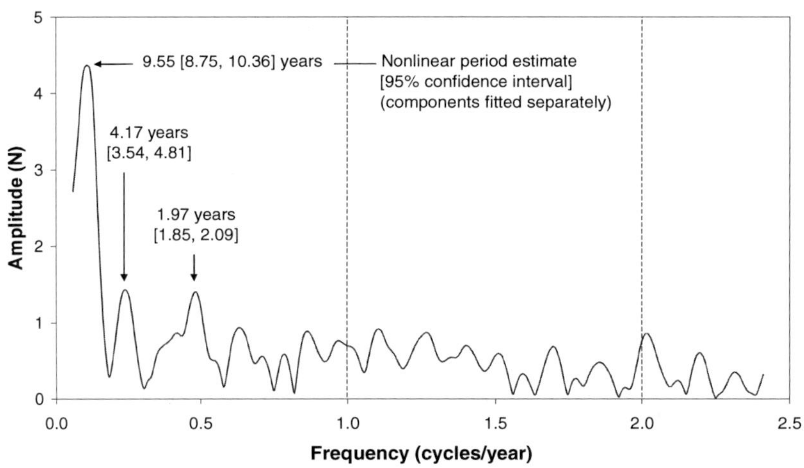

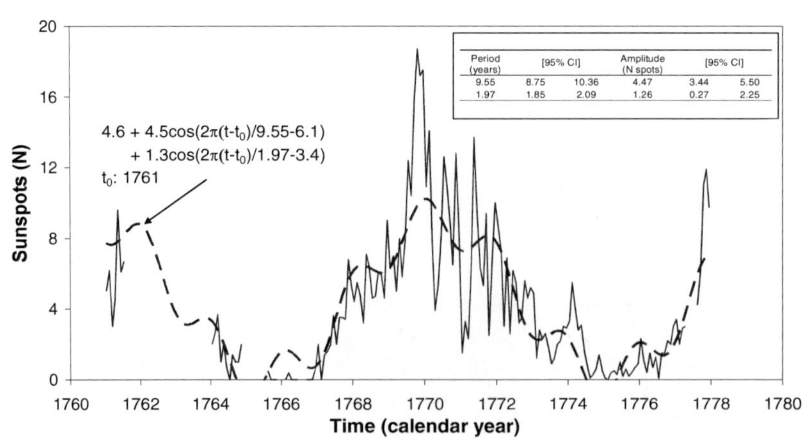

(A)

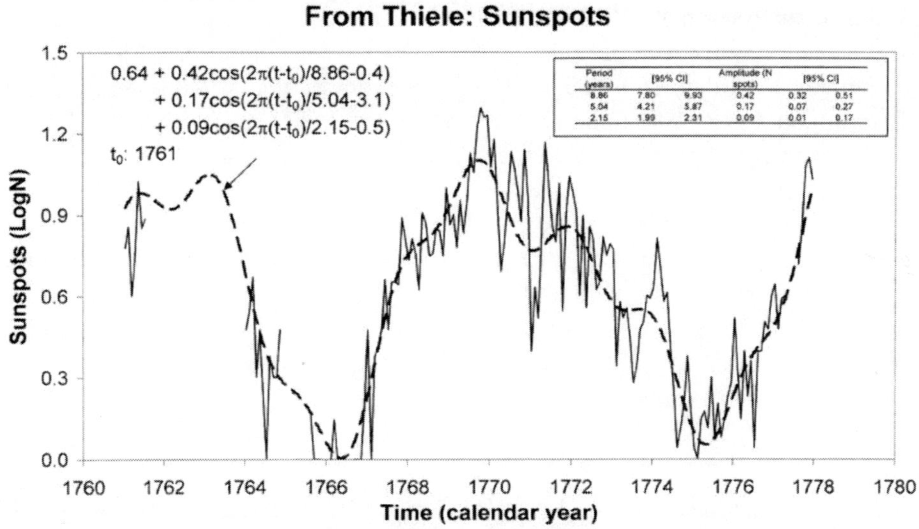

Astronomische Nachrichten No 1193. De Macularum Solis antiquioribus quibusdam observationibus Hafniae institutis, scripsit Th. N. Thiele, astr. Stud. John G. Wolbach Library, Harvard-Smithsonian Center for Astrophysics - Provided by the NASA Astrophysics Data System

(B)

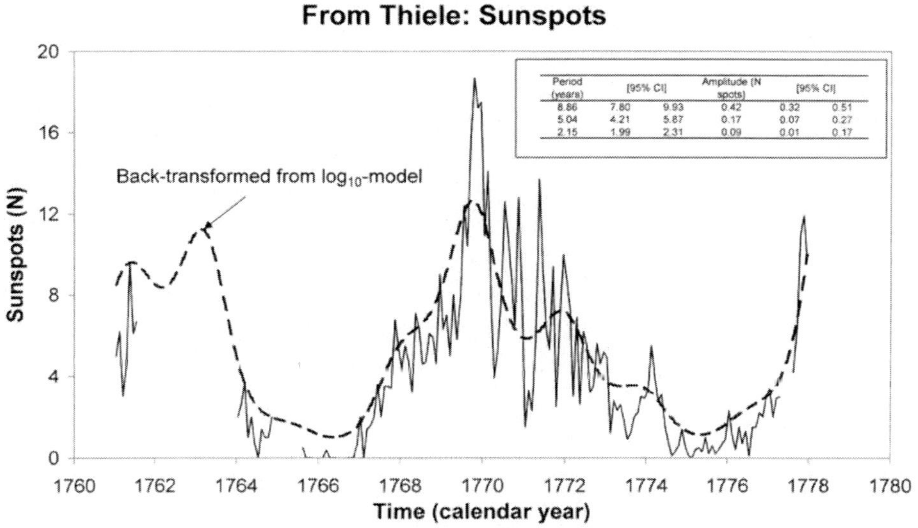

Astronomische Nachrichten No 1193. De Macularum Solis antiquioribus quibusdam observationibus Hafniae institutis, scripsit Th. N. Thiele, astr. Stud. John G. Wolbach Library, Harvard-Smithsonian Center for Astrophysics - Provided by the NASA Astrophysics Data System

(C)

© Halberg.

Figure 9. A) Yearly sunspot data (with interruptions) from Horrebow (published by Thiele) (top) are characterized in a linear spectrum by three peaks corresponding to periods of about 9.55, 4.17, and 1.97 years validated nonlinearly when considered separately (middle). A composite two-component model is fitted to the data (bottom). B) After log-transformation, a three-component model, fitted to the sunspot data from Horrebow, is validated nonlinearly. C) The three-component model fitted to the log-transformed sunspot data from Horrebow is back-transformed to be displayed with the original data. Note slight overfit at the beginning of the record when data are missing.

Nonlinear analyses that allow the period to vary as a parameter, on the original data resolve 9.4- and 2.1-year components but not the 4.2-year component. The two-component model is fitted to the data and displayed as a function of time at the bottom of Figure 9A. Periods and their CIs (95% confidence intervals) are 9.554 [8.75, 10.36] and 1.971 [1.85, 2.09] years, with corresponding amplitudes of 4.47 [3.44, 5.50] and 1.26 [0.27, 2.25] years, Figure 9A (bottom).

After log-transformation ($y=\log_{10}(x+1)$), all three components can be resolved nonlinearly with periods and 95% CIs of 8.863 [7.80, 9.93], 5.037 [4.21, 5.87] and 2.153 [1.99, 2.31] years and corresponding amplitudes of 0.42 [0.32, 0.51], 0.17 [0.07, 0.27], and 0.09 [0.01, 0.17], when fitted concomitantly. Figure 9B illustrates the three-component model fitted to the log-transformed data; Figure 9C shows the same model after back-transformation together with the original data.

13. Samuel Heinrich Schwabe: Intellectual Climate Regarding Periodicity and Barrier Crossing From Pharmacy and Botany to Astronomy

The trained pharmacist and amateur botanist/astronomer Samuel Heinrich Schwabe crossed disciplinary barriers in the nineteenth century [24, 93] with a report of explicit decadal periodicity, repeating the feat of another amateur botanist, Androsthenes. While Schwabe, in counting sunspots in 1838, did not dare to extrapolate to the importance of the solar activity cycle on earth, and indeed, while conclusions must not be drawn from the finding of variations with similar periods, it is nonetheless most interesting that in the very year 1838, when Schwabe published the first 12 years of his observations [24; cf. 94], without even daring suggest a cycle in the face of an obvious up-and-down trend, Clarke described an about 10- to 11-year trade cycle [94; cf. 95]. Revolutions and the climate as well as the economy, notably crops, were already earlier related to solar activity: in 1801, William Herschel had described the weather as subject to a "solar indisposition" [15; cf. 94-97]. Manuel J. Johnson, the president of the Royal Astronomical Society, said in his address on presenting the Society's medal to Samuel Heinrich Schwabe in 1857 [96; cf. 97]:

> M. Schwabe is a gentleman resident in ... Dessau, the capital of the Duchy of Anhalt Dessau, who, having no professional duties, has devoted much time and attention to scientific studies, and has attained considerable reputation in other departments of knowledge than that in which he is known to this Society. Among other accomplishments, I am informed that M. Schwabe is an excellent botanist, and has composed, though, I believe, not published, a flora of the neighbourhood of Dessau.
>
> It was in 1826 that he entered upon those researches which are now to engage our attention. I am not aware of the motive that induced him -- whether any particular views had suggested themselves to his own mind--or whether it was a general desire of investigating, more thoroughly than his predecessors had done, the laws of a remarkable phenomenon, which it had long been the fashion to neglect. [Author's note: He did a pointed description of the view that periodic phenomena were not profitable to study, perhaps conveying the thought of all those who seek solutions, mathematical or biomedical, without accounting for time.] He could hardly have anticipated the kind of result at which he has arrived; at the same time we

cannot imagine a course of proceeding better calculated for its detection, even if his mind had been prepared for it, than that which he has pursued from the very commencement of his career. Assuredly if he entertained such an idea, it was not borrowed from the authorities of the last century [author's note: including Galileo and Newton], to whom the solar spots were objects of more attention than they have been of late years.

"Nulla constanti temporum lege apparent aut evanescunt," says Keill in 1739. -- Introduct. ad. Physic. Astronom. p. 253.

"Il est manifest par ce que nous venons de rapporter qu'il n'y a point de règle certaine de leur formation, ni de leur nombre et de leur figure," says Cassini II. in 1740. -- Elem. d'Astron. vol. i, p. 82.

"Il semble qu'elles ne suivent aucune loi dans leur apparitions," says Le Monnier in 1746. -- Instit. Astron. p. 83.

"Solar spots observe no regularity in their shape, magnitude, number, or in the time of their appearance or continuance," says Long in 1764. -- Astron. vol. ii., p. 472.

"Les apparitions des tâches du soleil n'ont rien de regulier," says Lalande in 1771. -- Astron., vol. iii, § 3131, 2nd edit.

And Delambre's opinion may be inferred from a well-known passage in the third volume of his Astronomy (p. 20), published in 1814, where treating of the solar spots he says, "Il est vrai qu'elles sont plus curieuses que vraiment utiles."

I cite these passages to show, whatever might have been Schwabe's own view of the matter, that a periodicity of the spots was not a current idea at the time he entered upon the subject, and therefore that he has not merely developed a law for which men's minds were prepared, but that he has been, to all intents and purposes, the discoverer of it.

The first suggestion I have met with (and that a very vague one) of anything of the kind occurs in the article "Sonnenflecken", in Gehler's Physikalisches Wörterbuch, written by the elder Littrow, and published in 1836, long after Schwabe had been at work.

It was in 1826, as I have said, that Schwabe began his observations, but I am not aware of any published results before those in No. 350 of the Astronomische Nachrichten, which appeared in April 1838. Here he gives a summary of twelve years' labour, and though at this time he must have begun to foresee the issue, for the indications of periodicity are plain, he makes no remark, nor does the memoir appear to have attracted the attention of astronomers.

From this time commence his annual contributions to the Astronomische Nachrichten; but it was not until the end of 1843, when he had passed through two periods of maximum and minimum, that he very modestly remarks [in 1844] that his observations heretofore had given indications of periodicity which that year's result tended to confirm. Still the subject attracted little attention.

As far as I can discover, the only person who had taken it up was Julius Schmidt, the present indefatigable Director of the Observatory at Olmütz, then residing near Hamburg. But the philosopher of Dessau was not disconcerted; he went on accumulating fresh proofs of his great discovery, which, when announced in 1851, by Alexander Von Humboldt in the third volume of his immortal Cosmos, came upon the world with all the freshness of novelty, though in reality the secret had been revealed eight years before. [Or even earlier by Horrebow, and perhaps with his 19 years of observation Scheiner encountered a cycle even earlier]...

The exact period Schwabe does not pretend to have determined. That it is liable to perturbation is evident. During twenty-seven years of the series the results were extremely regular; during the last three years they have shown symptoms of disturbance.

The epoch of minimum, which, consistently with earlier indications, should have happened in 1853, did not occur until 1856.

14. François Jean Dominique Arago (1786-1863): Magnetic Variation Binding 'Earth with Other Worlds'

After making a number of other comments on Schwabe's observations of sunspots, Johnson continued [96]:

> There is, however, one more topic too closely connected with our subject, and too remarkable to pass unnoticed.
>
> You are aware, gentlemen, it has been long known that the intensity of magnetic declination is subject to a daily variation; and this variation was also known to be in some way connected with the sun, attaining its maximum western limit when that body is at its upper and lower culmination, and its maximum eastern limit about 6 o'clock in the morning and evening. These are not the exact times, but they are sufficiently near for our purpose. About the year 1850, Professor Lamont announced that this variation was again liable to another variation, which observed a period, from maximum to minimum, and from minimum to maximum, of about 10 years. He was led to this result by a discussion of his own observations at Munich. It was shortly after fully confirmed by our colleague, General Sabine, by the observations made at the Magnetic Observatories of Toronto and Hobarton. But in the course of the latter discussions General Sabine detected another important circumstance which had escaped Lamont, viz., that the periods of maximum variation of the declination-needle corresponded exactly with those of the maximum frequency of the solar spots, and vice versa, the minimum of the one with the minimum of the other. Hardly had General Sabine's paper been read before the Royal Society on March 18, 1852, when intelligence was received that two Swiss physicists, Professor Gautier of Geneva, and Professor Wolf of Berne, had arrived at the same conclusion, independently of each other, from a perusal of Lamont's results.
>
> All the observations I have mentioned, were made during nearly the same time -- between the years 1840 and 1851 -- comprehending only a single period of change. Obviously, therefore, in this state of the inquiry it was of the highest importance to obtain a trustworthy series made at another time and under different circumstances. Though hardly to be expected, it happened, most fortunately, that such a series was at hand. The diurnal variation was one of those subjects which, many years ago, had particularly interested the vigorous mind of the lamented Arago; and among his papers were found the records of a most laborious course of experiments, conducted with all the care which no one knew better than he how to bestow on a delicate investigation. These experiments, extending from 1820 to 1831, have been rigorously discussed by M. Thoman, and exhibit the prevalence of the same law that obtains in the later series. Nor does the evidence of connexion rest here. In a recent paper sent to the Royal Society, General Sabine has shown that at Toronto (the only series hitherto fully discussed) all the magnetic elements subjected to observation undergo similar variations. In the face of these coincidences, to doubt the relationship between the two phenomena seems to me to be almost as unreasonable as to doubt the influence of the moon on the tides of the ocean.
>
> M. Schwabe's service has been the discovery of an important astronomical fact, falling strictly within the cognizance of this Society. His merits are in no way affected, though, no doubt, the value of his discovery is enhanced, by this remarkable episode. No longer is its scope confined to the disclosure of a physical peculiarity in the constitution of the sun. It promises to be the means of revealing the prevalence of a principle, throughout the solar system, co-extensive with gravitation, and of establishing another link in the chain of analogies binding Earth with other worlds, the ultimate effect of which on the progress of physical research no one can foresee.

15. ANALYSES OF SCHWABE'S DATA

Two series (1826-1837 and 1838-1843) of yearly sunspot numbers, Figure 10A, were combined into one and analyzed by linear-nonlinear least squares rhythmometry [91].

We wished to empirically find the minimal number of yearly data on whose basis the sunspot cycle could have been documented by rejecting the no-cycle (zero-amplitude) assumption, had Schwabe used the method of least-squares that had been available since 1801, when Gauss had developed it to locate Ceres, the asteroid that was lost after Giuseppe Piazzi (1746-1826) initially observed it. A trial period of 11 years was used. The results in Figure 10B need to be qualified by the fact that using a trial period of 11 years assumes the existence of prior information, which was not the case at the time of the original documentation of a cycle in solar activity, unless Schwabe knew about Horrebow's data, which, judging from Johnson's address, he did not. Moreover, had he known about Horrebow's observations, he could have used Gauss' least-squares, e.g., the equivalent of a cosinor spectrum, and would have located a peak around 9.4 years.

An about 9-year cycle corresponds to the distance between two peaks in Figure 10A and can be documented by the cosinor rejection of the zero-amplitude assumption with as few as 9 values (9 years) [91].

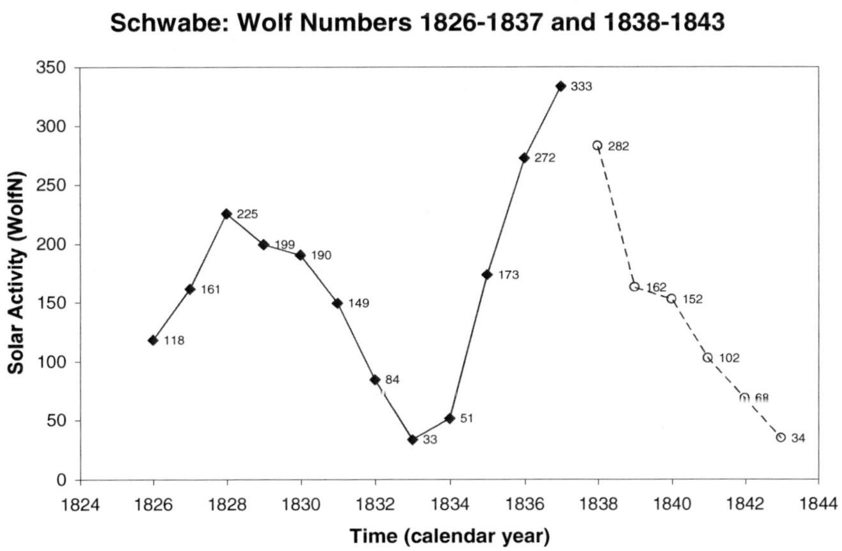

(A)

Figure 10. (Continued)

Using Gauss' Least Squares, Schwabe Could Have Validated a Periodicity in Sunspots with His First 9 Years of Data *

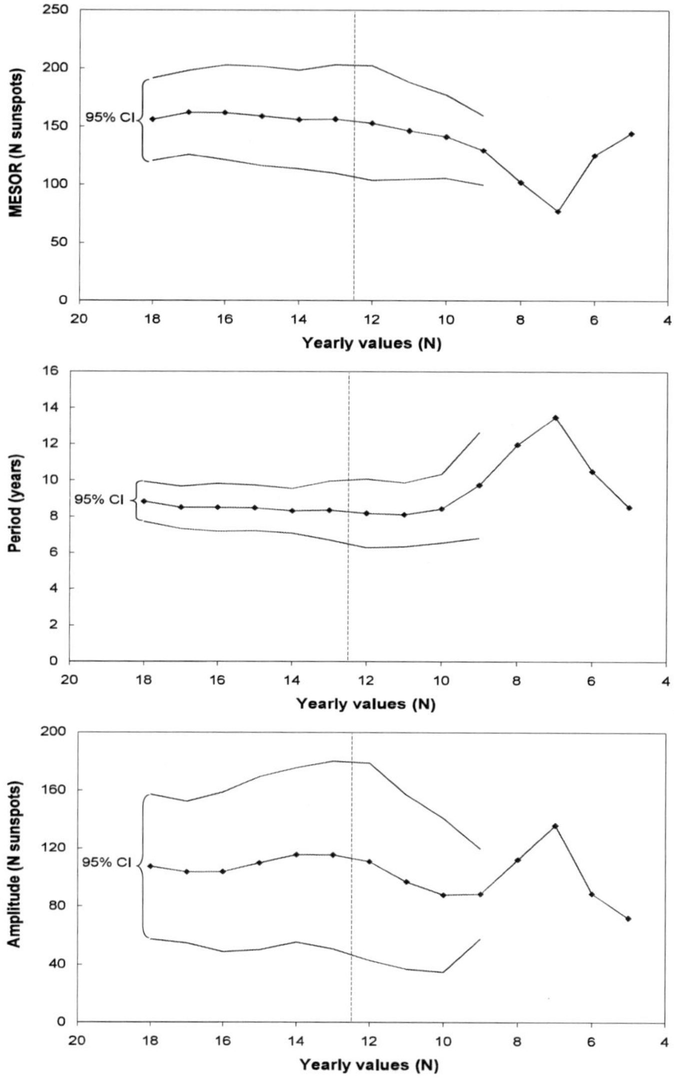

* Time series of yearly sunspot numbers (1826-1837 and 1838-1843) were combined for the nonlinear fit of a single cosine curve with a trial period of 11 years (that could not then but can now be anticipated) to all 18 values and after the progressive removal of the latest value. Nonlinear estimates of the MESOR (top), period (middle) and amplitude (bottom) with their respective 95% confidence intervals (95%CI) (when the zero-amplitude assumption could be rejected with $P<0.05$) shown as a function of the decreasing number of data indicate that relatively stable estimates are obtained with 9 or more years of data.

(B)

Figure 10. A) Chronogram of Schwabe's sunspots data. B) Results from nonlinear analyses of Schwabe's data, analyzed over the entire 18-year span and on progressively shorter series obtained by removing each time the last yearly value, in an attempt to determine what is the shortest series suitable for an assessment of the solar activity cycle at the time. Even with only the first 9 yearly data, it is possible to derive point and 95% confidence intervals for all parameters. Results shown for the MESOR (top), period length (middle), and amplitude (bottom).

For Figure 10B, a set of time series was prepared using all of Schwabe's 18 yearly data and progressively removing the latest available in each prior series (the early data were always kept). The shortest series contains the first 5 values. An 11-year cosine curve was fitted nonlinearly to each time series, covering the entire 18-year span or only the first 17, 16, 15, or 5 years, to obtain estimates of the period as well as of the MESOR, amplitude and acrophase. Results are shown in Figure 10B as a function of the decreasing observation span, for the MESOR (top), the estimated period length (middle), and the corresponding amplitude, each with a measure of its CI. The result would be the same if Horrebow had taken a trial period of 9 years, the distance between the two maxima of Figure 10A. Note that only with an 8-year series, there is a lack of a 95% confidence interval for the estimated parameters in Figure 10B [91].

Thorvald Nicolai Thiele (1838-1910) -- of Thiele differential equation fame ("the fundament of modern life insurance mathematics"; 26) -- concluded: "But until a subsequent age adds more light to this matter, it seems necessary to be satisfied with a regular period of about ten years" ("Sed antequam postera aetas huic rei plus lucis attulerit, in periodo regulari decem fere annorum acquiescendum esse videtur") [26]. It seems pertinent to note that both Horrebow's and Schwabe's (if not Scheiner's) data allowed by 1838 the demonstration of a circadecadal cycle with a measure of its uncertainty on the basis of data covering only 9 years. In Thiele's and Schwabe's time, the circadecadal cycle in sunspot numbers was shorter than its average length in all available data (in 1826, the solar activity cycle was actually the shortest between 1799 and 1947).

16. Technical Comment

Figures 11A-E visualize the approach by least-squares (A), its endpoints, albeit without uncertainties (B) and sources of artifacts (C-E), but as shown in the foregoing, much work could be done without such time-microscopy. It is desirable in historical and other time series analyses to cover, whenever possible, more than several cycles of a given rhythmic phenomenon investigated. The demonstration of a statistically significant cycle in itself requires added indications of scientific clinical or other significance [98] and, in itself, must not encourage premature extrapolation from insufficient data. But with indispensable caveats, data covering a full cycle are not invariably required for the demonstration of cycles that characterize major features of corresponding solar and human variability. With the linear-nonlinear extended cosinor, the zero-amplitude assumption could be rejected for a circatridecadal BEL in the heart rate of an elderly man (FH) based on only 22 years of measurements, i.e., with only ~three-fourths (~73.3%) of the cycle's total length covered by the data [99].

On other occasions, with emphasis on necessary caution, extrapolation as-one-goes becomes desirable before the data cover a full cycle, as in the case of a clinically healthy man, RBS, when the output of computer analyses yielded an ~50-year cycle (that was not part of the input, but has precedents in human affairs [30]), in ~6 self-ratings per day of mood recorded during 43.5 years, starting when he was 20.5 years of age. Waiting for a maximum of three covered cycles, stipulated by some as a minimum, would mean waiting until RBS reaches 170 years of age. This may be highly desirable since he is a motivated, careful scientist, but may not be assuredly possible (cf. 40, 100).

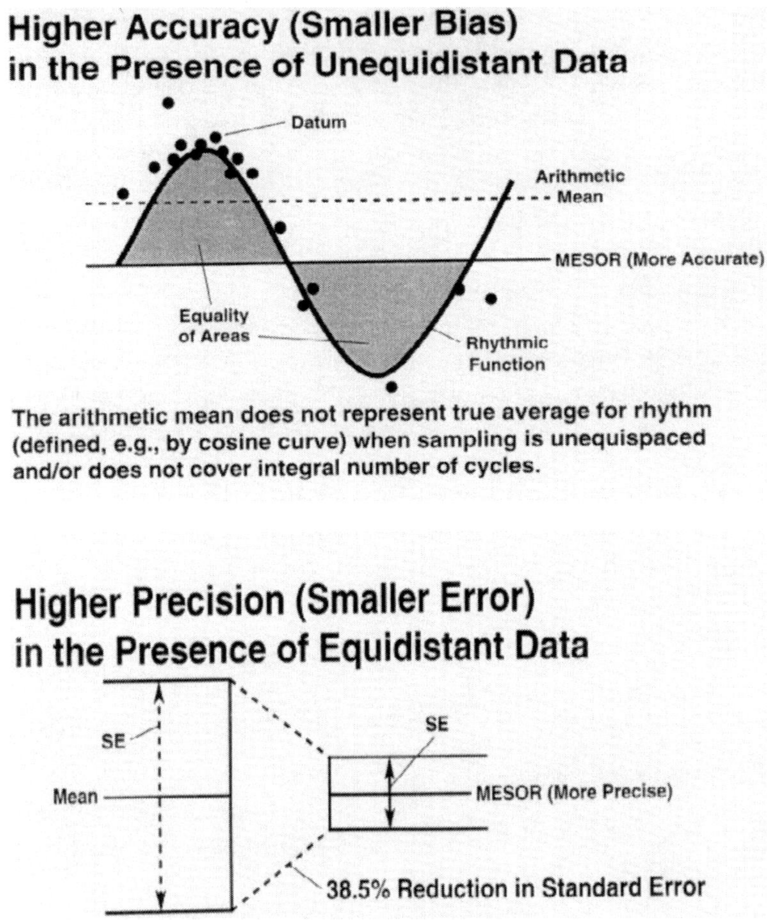

(A)

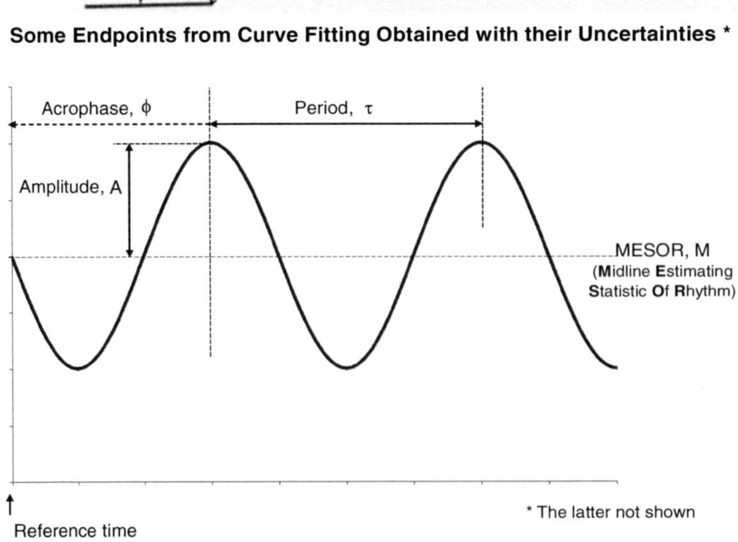

(B)

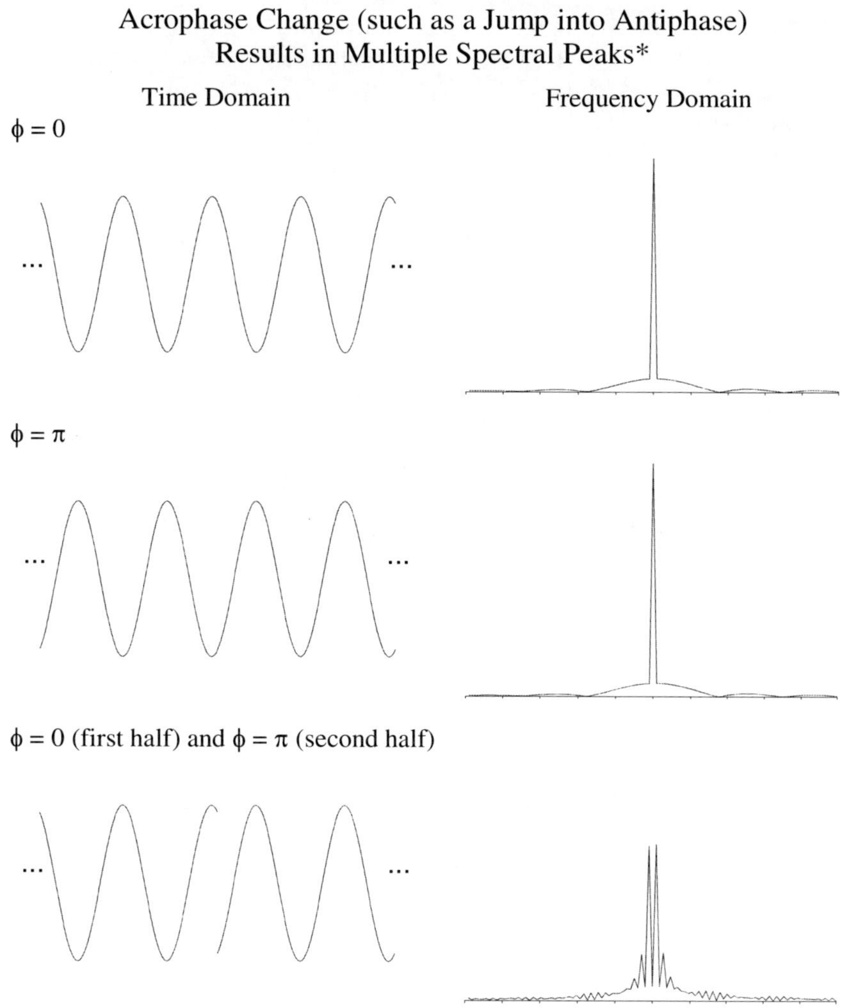

* Observed, for instance, after return from a transmeridian flight across 7 or more time zones in the circaseptan component of blood pressure, revealed by chronobiological serial section.
Demonstration by simulation of two peaks in the spectrum (bottom right) resulting from addition of two sinusoidal components with the same period but in antiphase (top and middle left) during different spans (bottom left). Note the ambiguity of the two spectral components indicating only a difference in phase of a single component. Very small sidelobes result from slight departure of record length from integer multiple of the cycle length.

(C)

Figure 11. A) Gauss' least-squares method in the inferentially extended cosinor provides a usually more accurate and more precise average than the arithmetic mean. It lends itself well to hypothesis testing (rhythm detection) and to parameter estimation, with results compared to those of other approaches in Table 1. B) Diagnostically useful endpoints from the cosinor. The MESOR is a rhythm-adjusted mean (see Figure 11A). The double amplitude is a measure of the extent of predictable change within a cycle. The acrophase is a measure of the timing of overall high values recurring in each cycle. The period is the duration of one cycle. C) The presence of a phase jump (e.g., by 180°) of a cosine curve with a fixed period is spuriously associated with two major spectral peaks.

Frequency Change Results in Multiple Spectral Peaks with Reduced Amplitude*

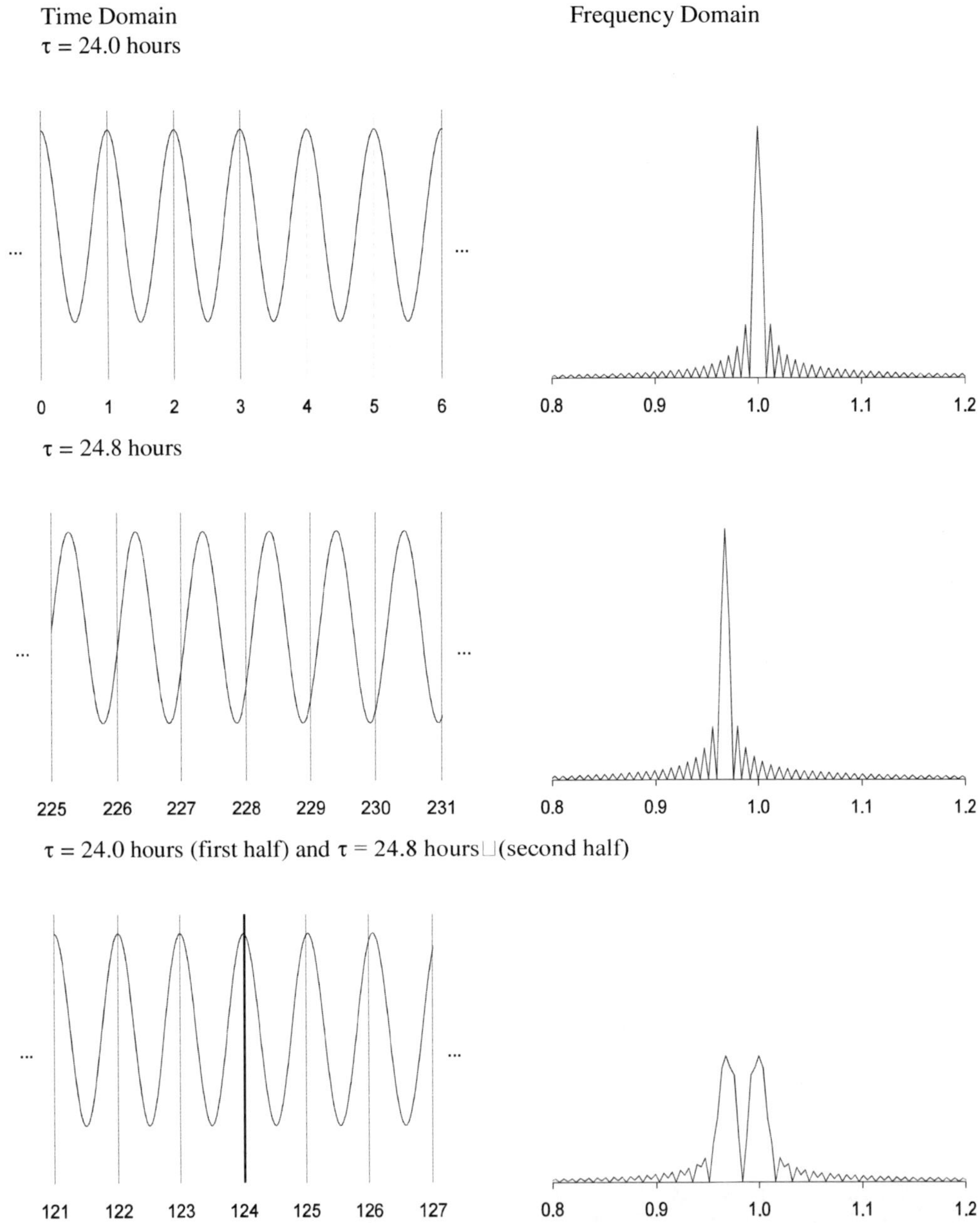

* Observed, for instance, in relation to a desynchronization of a circadian rhythm for about 2 months followed by resynchronization with the 24-hour routine, revealed by chronobiological serial section.

Demonstration by simulation of two peaks in the spectrum (bottom right) resulting from juxtaposition (bottom left) of two sinusoidal components with two slightly different periods during first (top left) and second (middle left) half of the series. Note that amplitude of double peak is half the actual amplitude of each component, albeit present only during half the total span. Very small sidelobes result from use of fractionated harmonic increment.

Figure 11D. The occurrence of a change in period length of a single cosine component (e.g., from 24.0 to 24.8 hours in mid-course of a 124-day record) is spuriously associated with two major spectral peaks with amplitudes smaller than that of actual signal.

Two models were fitted to RBS' heart rate. As judged by the naked eye or by the % of variance, accounted for by the two different fitted models or preferably both ways, there is a better fit with a model that was not anticipated and a poorer fit with an anticipated period. Nonetheless, the anticipated period was preferred (Figure 12). The anticipation, in this case, was based on a previous analysis of the same accumulating time series of RBS when it was some years shorter; it then yielded a paratridecadal, very close (32.9-year) period with a confidence arc from 29.25 to 35.96 years (100) for RBS's heart rate and a 32.82 [23.40, 42.16]-year period for Wolf's sunspot numbers.

The result on Wolf numbers during the same span investigated in RBS, although it is of low amplitude and not demonstrable in other spans, strengthens the BEL in heart rate by the agreement of the reciprocal periods in the biosphere and its cosmos.

There is the added fact that a paratridecadal period was also found thus far in one variable of each of three other men, a total of four men with decades-long data being available to us. These earlier findings of a BEL cycle [101, 102] were reasons for the choice of the predicted, four times replicated, albeit still sparsely documented model in this case, and any others. Mapped and hence anticipated periods were given precedence historically whenever one had to choose between alternative models, even when the mapped cycle was the best fit neither to the unaided eye nor by the criterion of the extent of residual variance reduction or both.

We do not generalize but suggest (for historians, who deal with accumulating time series, such as geophysical ones, as for the clinic) the merit of, if not the need for, watchful continuously analyzed monitoring until at least one full cycle is documented, when possible. The deviation of the data from a fitted model, a result of both the cycle's waveform [103] and of the noise in the time series are likely to be critical factors in determining the minimal length for tentatively stipulating the possible presence of a periodicity. In view of the aeolian nature of solar activity, in any event the minimal length for rejecting as for accepting an association, if not an effect, should be much longer than a (decadal or other) cycle length. Figure 13 shows that in the absence of noise, current tools such as the extended cosinor can reconstruct a cycle when the data cover just one-eighth of its length.

Single Component Signals (top) Have Single Spectral Peaks at Corresponding Frequencies; Average of Two Signals Shows Both Spectral Peaks; Single Component with Period Change in Mid-course Can Show Broad Single Peak *

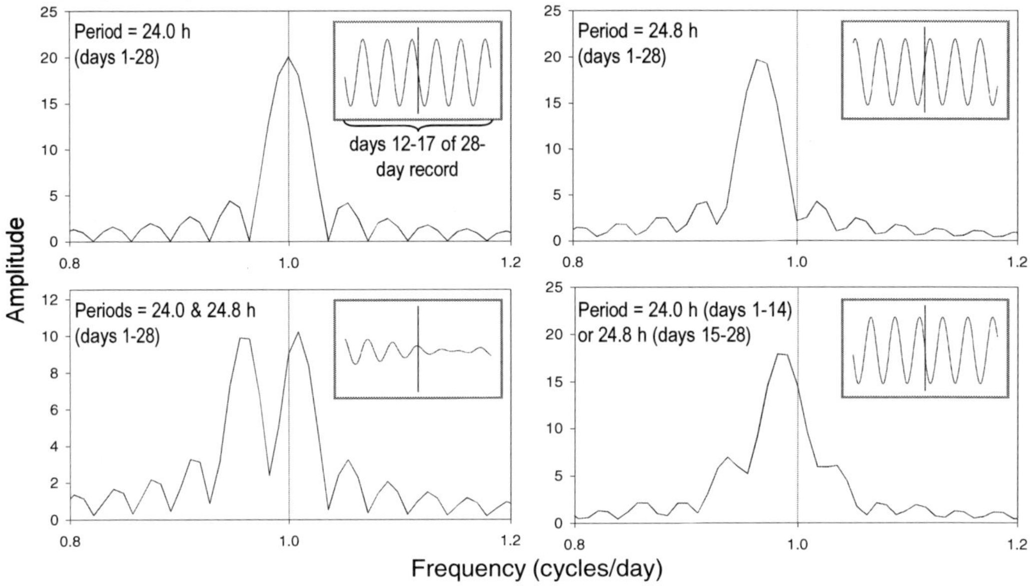

* Depending on overall record length, closeness of the two periods, and on when the change in period length occurs, the single broad spectral peak can become two separate peaks.

Figure 11E. A change in period length (e.g., from 24.0 to 24.8 hours in the mid-course of a 28-day record) of a single cosine component is spuriously associated with a single broad spectral peak with a reduced amplitude and sidelobes. Whether a single broad peak or two spectral peaks are observed depends on the length of the record, the extent of the difference in period and the time when the switch in period occurs in the record.

Table 2. The problem of heliobiology in the "West": negative results on millions of people in 2 North American papers*

Daily deaths	**Estimated totals 1964-71 (8 y)**
All	14,493,120
Coronary heart disease	4,383,000
Stroke	1,577,880

From Feinleib et al. (21) 1964-1966 (3 years) daily; 1964-1971 (8 years) monthly

	Average N of deaths per day		
	Coronary heart disease		Stroke
Total (1962-1966)	2,750,880		1,002,246

From Lipa et al. (22) 1962-1966 (5 years) = 1826 days

*Totals estimated based on infradians in the original papers on highest and lowest numbers/day that led to an estimated value/day multiplied by the number of days (21) or based on weekly means (22).

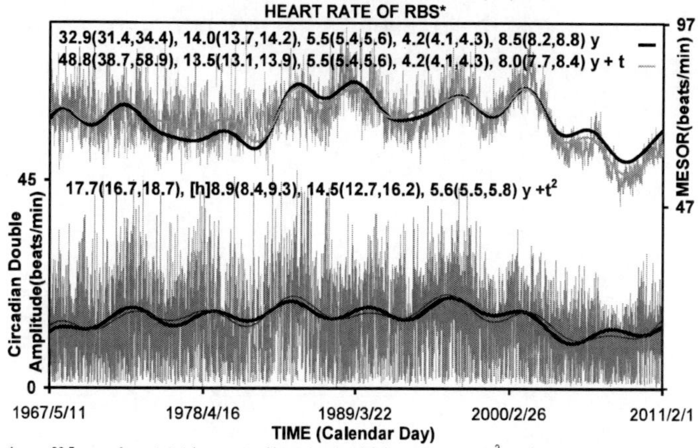

Figure 12. Two models fit the MESOR (mean) of RBS's heart rate. An about 50-year cycle has been reported for epidemiological data on stroke incidence (19) and a tridecadal period for RBS's heart rate when the series was a few years shorter (100, 101). Until further data change the choice, the tridecadal BEL is preferred since it was also found earlier in RBS and in the circulation of three other men studied longitudinally for decades. As yet, there is no other previously reported near-50-year physiological cycles assessed longitudinally and hence no precedent in other individuals as there was for a para-tridecadal cycle. © Halberg.

Table 3. Relative incidence of myocardial infarction (MI) and sudden cardiac death (SCD) before and after a southward turn of the vertical component (Bz) of the interplanetary magnetic field: diagnoses made between January 1, 1979, and December 31, 1981, in Moscow hospitals

			day			
	-2	-1	Bz≥1nT	Bz≥1.5 nT*	+1	+2
*MI (N=85,819**)*						
Mean ± SE	100.37 ± 3.41	98.08 ± 2.55	98.85 ± 2.59	101.69 ± 3.12	107.61 ± 3.28	94.13 ± 2.56
t	0.1085	0.7329	0.4440	0.5417	2.3180	2.2930
P	0.9143	0.4572	0.6601	0.5919	0.0272	0.0288
SCD (N=71,525)						
Mean ± SE	102.17 ± 2.08	98.93 ± 2.24	100.03 ± 1.93	101.23 ± 1.93	103.04 ± 2.52	94.60± 2.02
t	1.0452	0.4756	0.0157	0.6350	1.2036	2.6741
P	0.3040	0.6377	0.9876	0.5301	0.2379	0.0119

Incidence espressed as a percentage of the mean over 6 days bracketing an event; t = Student t-testing H0: mean = 100%. *Event. **Total number of cases extracted from the data base over the 3-year span. A statistically significant increase in the incidence of MI on day +1 and a decrease in SCD (ICD-9) on day +2 is qualified by a similar overall time course in both MI and SCD. Reference to cycles provides added criteria for a comparison insofar as the absence of a yearly component and the presence of a component also found in the speed of the solar wind suggest an overwhelming nonphotic effect as seen for some geographic regions in Figure 16. From Cornélissen G, Wendt HW, Guillaume F, Bingham C, Halberg F, Breus TK, Rapoport S, Komarov F. Disturbances of the interplanetary magnetic field and human pathology. *Chronobiologia* 1994; 21: 151-154.

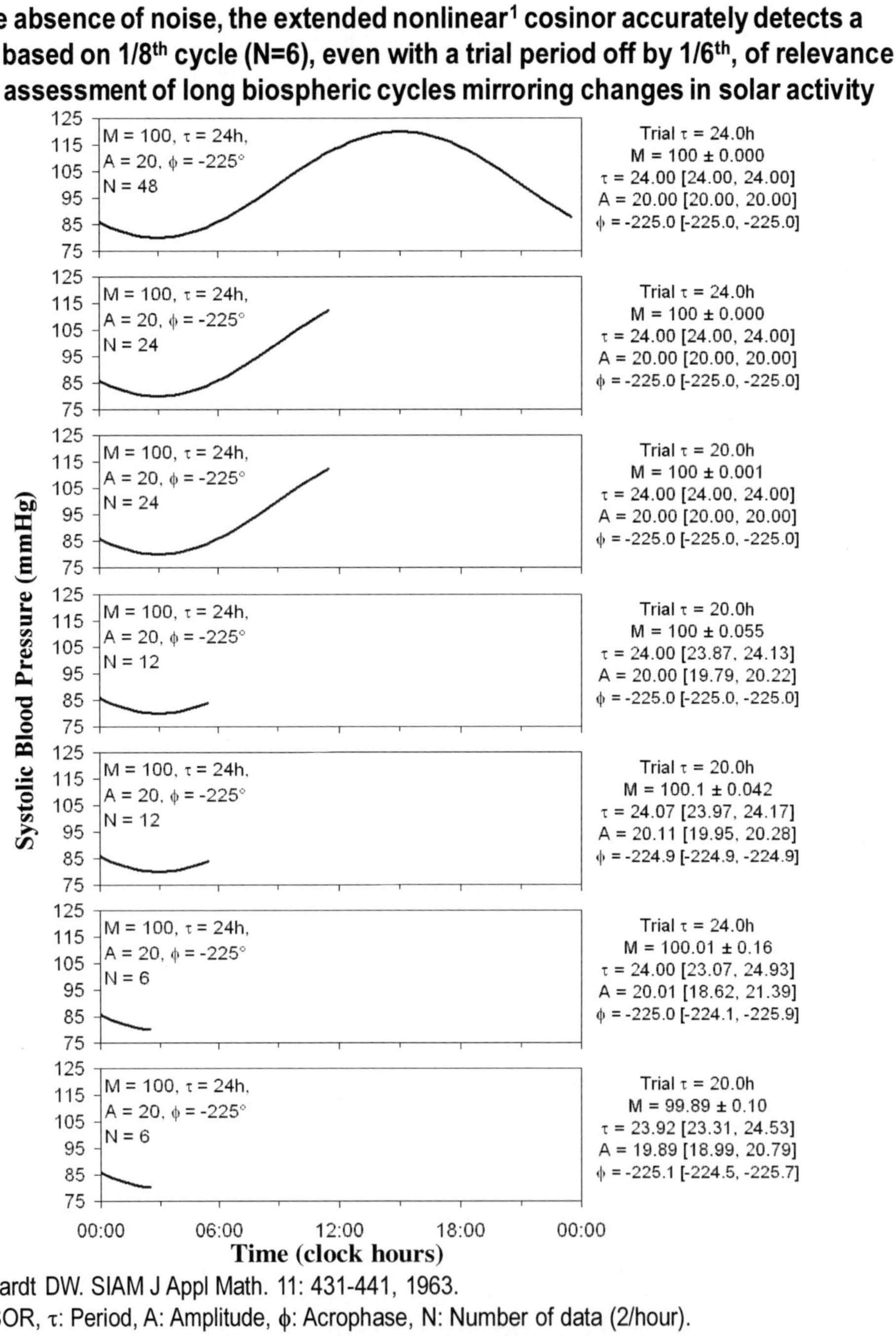

Figure 13. In the absence of noise, the extended nonlinear cosinor resolves a 24-hour cycle in 6 equidistant values covering only 3 hours, one-eighth of the cycle's length, even when a trial period of 20 instead of 24 hours is used.

GEOGRAPHIC LOCATION and DIFFERENCES in OUTCOMES in PARTLY CONTEMPORANEOUS DATA* LIMITED TO FRACTIONS OF A SOLAR CYCLE;
CONTROVERSY RESOLVED BY MINNESOTAN DATA COVERING SEVERAL SOLAR CYCLES**

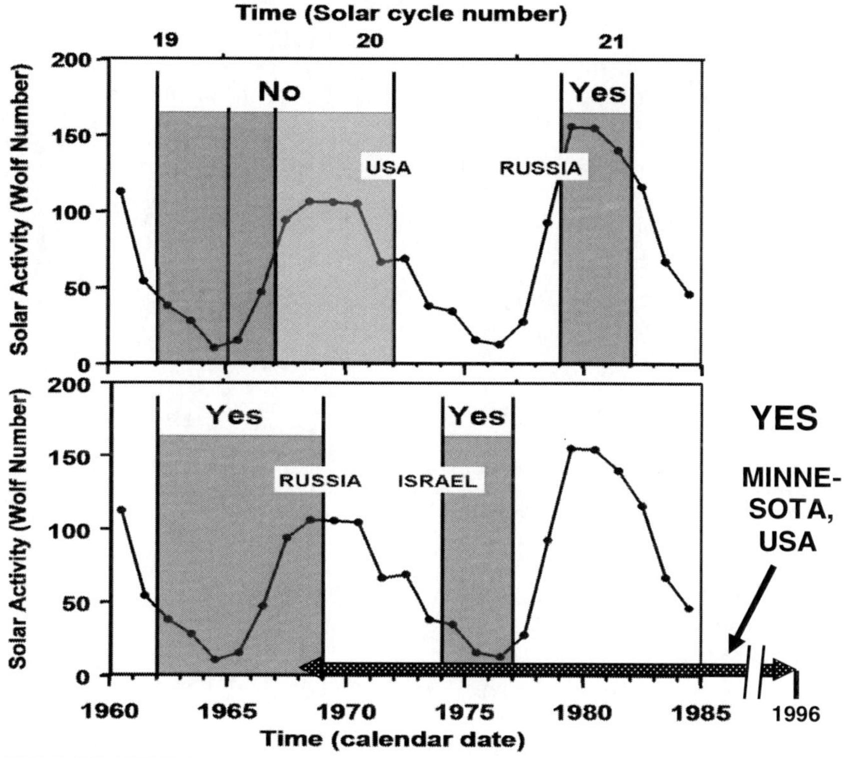

* Association of magnetic storms and mortality from myocardial infarction (MI) not detected in the USA (1962-1966 daily data and only monthly to 1971) but detected in Russia (including morbidity) in 1979-1981 (top) and earlier. Bottom: association also detected in 1962-1968 in Russia and in 1974 – 1976 in Israel, at minima of Schwabe (~10.5 y) and Hale (~21-y) cycles, unless data in the late sixties in Russia contribute the outcome at the preceding minimum. ** Minnesotan data over several solar cycles (1968 – 1996) document excess of 220 MI/year during solar maxima vs. solar minima (black horizontal arrow along abscissa.

Figure 14. Resolution of a controversy when time- and site-specific changes are anticipated and documented. The analogy of blood pressure measurement over a single circadian cycle comes to mind: such a spotcheck may lead to a lifetime of unneeded treatment (when, in several repeated sessions, the patient was excited and the spotchecks showed a MESOR-elevation for another 24-hour vascular variability anomaly that disappeared in subsequent days of continued surveillance). Likewise, a hypertensive patient, during several 24-hour measurement series, may happen to have occasional normotensive mean values and hence will not be treated until a stroke occurs. Furthermore, an individual with blood pressure overswinging limited to the normal range may not be recognized as being at a very high risk of stroke by limitation to a single cycle. Studies that cover a single solar cycle may not be conclusive and may miss a relation that can be documented in the same solar cycle elsewhere, e.g., over several solar cycles in Minnesota, USA. Taking evidence over a single solar cycle is equivalent to taking the pulse for one second. We are here dealing with geographic as well as temporal differences since in the USA as well, an excess of death can be demonstrated in a sufficiently long (29-year) series. Time-varying associations revealed further by changing phase-synchronizations and coherences (23) account for the negative site- and time-specific results.

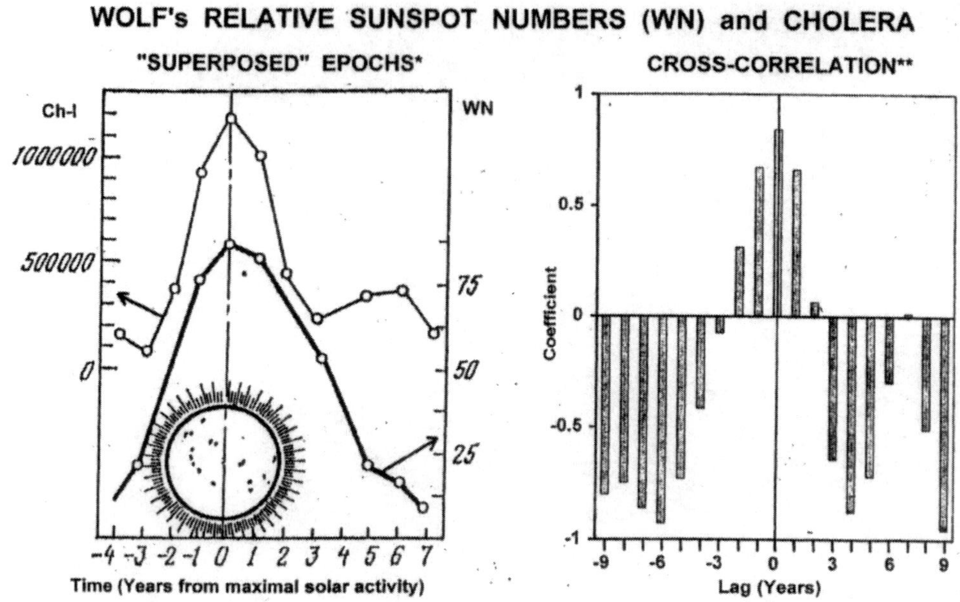

* Data taken off published graph from A.L. Chizhevsky: Epidemische Katastrophen und periodische Taetigkeit der Sonne. Moscow, 1930. ** Input: WN; output: Ch-I in Russia; data from 1823 to 1923 folded over an about 11-year solar activity cycle. Note in-phaseness in both summaries.

Figure 15. Cholera in Moscow characterized by a decadal cycle revealed by superposed epochs summarized by Alexander Leonidovich Chizhevsky, according to Sigel (108). Data from Chizhevsky on cases of cholera recorded in Moscow between 1823 and 1923, originally folded over an about 11-year solar activity cycle, showing a peak incidence coinciding with maximal solar activity (left). Data taken off the published graph, analyzed by cross-correlation, corroborate that the largest correlation corresponds to lag of zero (right). Should the original 100-year series have been available for analysis, a measure of uncertainty could have complemented the average curves (left).

Extrapolating from part of a solar cycle has, however, already led to controversy. Inferences from very extensive data in the USA regarding the lack of an effect of geomagnetics on morbidity are an important case in point; no effect of magnetic storms upon the cardiovascular system was detected based on data covering less than a circadecadal solar cycle, Figure 14 [8].

Classic statistical methods were used on very large samples, Table 2, one totaling an estimated 1 million and another over 14 million cases [21, 22]. The controversy, however, could be readily resolved by longer series from the USA [18, 38] and elsewhere by the use of chronomics, combining glocal inferential statistical time series analyses, Table 3. It may be realized that basing a study on less than a single solar cycle is equivalent to taking the pulse for less than a single cardiac cycle (less than 1 second)!

The components of the spectrum of the aurora in Figure 8 were documented with nearly a millennium of data, while Charles Egeson (aka J.J. George) looked at 117 years of data to derive his paratridecadal cycle [104; cf. 101]; Eduard Brückner again had a much longer series for the same paratridecadal spectral component [105], particularly pertinent to discussions of global warming [106, 107].

Alexander Leonidovich Chizhevsky used data covering 100 years to summarize an about-decadal cycle by superposed epochs in the incidence of cholera in Moscow, Figure 15 [108-115; cf. 31]. The histories of these cyclic components (to follow) in a broad spectrum are the histories of their discoverers, in keeping with Goethe's dictum that "the history of science, the science itself, is the story of the individual, is the individual" [116].

17. A Sphere of Expectations for Follow-Up Glimpses

A search in Medline reveals (as of March 7, 2011) 61,557 papers referring to the best investigated period among paired environmental-biospheric cycles, not here explored, the circadian system [40-42]. These publications focus on the signature of the daily rotational change of the earth around its axis for reasons other than the mere fact that most responses change along the 24-hour scale, to an extent that they can be beneficial or toxic, as in the case of anti-cancer drugs [82]. Along the scale of human aging, a study of time estimation at the daily maximum vs. that of the minimum yields opposite effects [117]. Accordingly, biospheric-solar associations are best investigated by focus upon all aspects characterizing the data and the cycles are assessed, rather than ignored or eliminated.

Admittedly, history and everything else can be done and is done without considering cycles. Bill Bryson's well-written *A Short History of Nearly Everything* (118) does not refer to cycles, and wittily states: "We really are at the beginning of it all [i.e., life on earth]. The trick, of course, is to make sure we never find the end."

A recent Russian study, also ignoring cycles, reports [119] that "The influence of weather factors (atmospheric pressure and temperature) and the geomagnetic activity on the development of severe cardiological pathologies has been studied using the daily data from two Moscow clinics, accumulated over a period of 12 and 7 years. It was shown that the most biotropic factors are variations of atmospheric temperature.

The relative contribution of the geomagnetic activity to the development of diseases is only 20%; however, its action is combined with the effect of ordinary weather because both these factors affect the vascular tonus of humans" [119].

Possibly different responses of myocardial infarction and sudden cardiac death to events in the interplanetary magnetic field, Table 4, may be pertinent and may also apply to geomagnetics.

Moreover, in some series from several locations at certain times, Figure 16 [4], there is no yearly component in the incidence pattern of sudden cardiac death comparable to the prominent yearly peaks in the spectrum of myocardial infarctions [120, 121]. Spectra are particularly informative in the case of the incidence pattern of violence.

There is no yearly component in the incidence of terrorism during 39 years; one finds instead only a longer than yearly (far-transyear) component that corresponds in length to one in the solar wind's speed (SWS) and in geomagnetism (aa) and becomes statistically significant with a lag after that in the SWS and in aa.

Focus on circadian cycles as marker rhythms already served to better treat some of the cancers of individuals in that the 2-year disease-free survival rate was doubled by timing [82]. Focus on infradians may serve to elucidate solar-biospheric associations mimicked in the incidence patterns of sudden cardiac death, Figure 16 [107], suicide [5-7], and terrorism, Figures 17 and 18.

To develop countermeasures, we need to learn more about the mechanisms involved. For this purpose, we follow the advice of Sir Norman Lockyer, the co-discoverer of helium and founder of the journal Nature [122]:

> Surely in Meteorology, as in Astronomy, the thing to hunt down is a cycle, and if that is not to be found in the temperate zone, then go to the frigid zones or to the torrid zone to look for it, and if found, then above all things, and in whatever manner, lay hold of, study it, record it and see what it means.

Much was thereafter discovered and reported elsewhere [31, 123-126]. Rohde and Muller [127] complemented Milankovitch cycles in thousands of years [128, 129] associated with the Earth's movement and its climate by myriadennians, possibly related to volcanism [127, 130].

As seen in Figure 19 (top), confirming the earlier results of Rohde and Muller [127], two major peaks corresponding to periods of about 62 and 140 million years (Myr) are present in our least squares spectral analysis. A third component with a period of about 37.6 Myr is also detected and validated by nonlinear least squares when fitted concomitantly with the other two components (130), as shown by the model fitted to the original detrended data in Figure 19 (bottom).

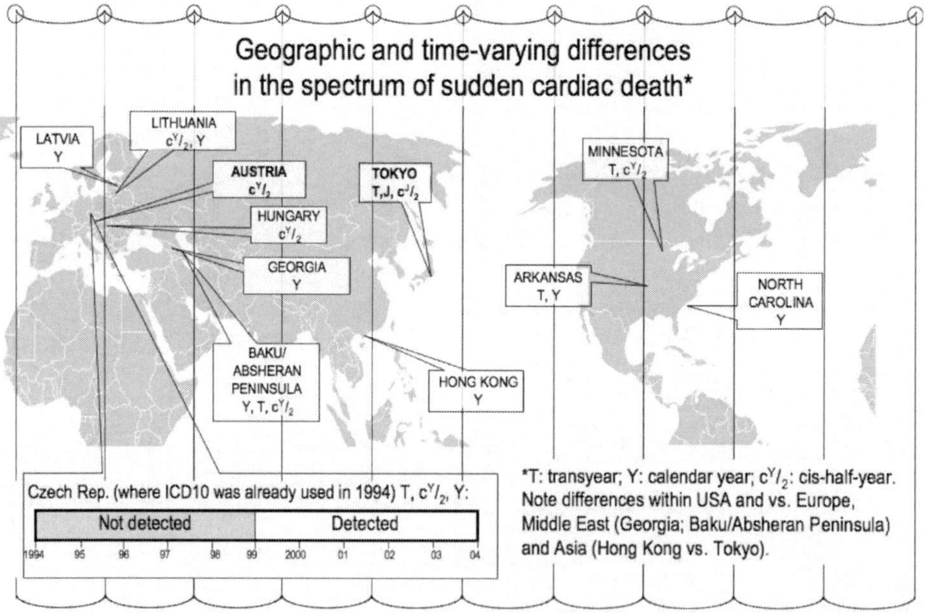

Figure 16. Albeit on as-yet limited data, geographic and temporal differences are shown for the incidence of sudden cardiac death in the form of rhythmic patterns with different frequencies characterizing this condition at different sites. Note site-specificity in the frequency domain displayed by patterns of sudden cardiac death worldwide, and at bottom left as a function of time in the Czech Republic.

Table 4. Geomagnetic/geographic differences among cycles with periods in the range of 0.8-2.0 years, characterizing the incidence of sudden cardiac death (SCD)[1]* and myocardial infarction (MI)

Sudden Cardiac Death (SCD)[1]							
Site	**Span**	**T, Δt, N**	**SCD (N)**	**Period (y) (95% CI)**	**Amplitude (95% CI)**	**A (% MESOR)**	**P-value[2]**
				----- Transyear (TY) or Candidate Transyear (cTY) Detected -----			
Minnesota	1999-2003	5y, 1d, 1826	343	1.392 (TY) (1.173, 1.611)	0.042 (0.00, 0.09)	22.0	0.014
Arkansas	1999-2003	5y, 1d, 1826	273	1.095(0.939, 1.251)	0.032 (0.00, 0.07)	21.1	0.040
				1,686 (cTY) (1.293, 2.071)	0.031 (0.00, 0.07)	20.7	0.044
Czech Rep.	1999-2003	5y, 1d, 1826	1006	0.974(0.856, 1.091)	0.078 (0.00, 0.16)	14.2	0.007
				1.759 (cTY) (1.408, 2.110)	0.077 (0.00, 0.15)	13.9	0.01
	1994-2003	10y, 1d, 3652	1792	1.726 (TY) (1.605, 1.848)	0.074 (0.02, 0.13)	15.1	<0.001
				1(0.944, 1.056)	0.052 (0.00, 0.10)	10.6	0.01
				----- Candidate Transyear Not Detected -----			
North Carolina	1999-2003	5y, 1d, 1826	752	0.929(0.834, 1.023)	0.069 (0.00, 0.14)	16.9	0.007
Tbilisi, Georgia	Nov 99-2003	4.1y, 1d, 1505	130	0.988(0.862, 1.114)	0.035 (0.00, 0.07)	40.7	0.007
Hong Kong	2001-2003	3y, 1m, 36	52	0.843(0.651, 1.036)	(NS)	44.9	0.077
Myocardial Infarction (MI)							
Site	**Span**	**T, Δt, N**	**MI (N)**	**Period (y) (95% CI)**	**Amplitude (95% CI)**	**A (% MESOR)**	**P-value[2]**
				----- Coexisting Year (Circannual) and Transyear (TY) -----			
Czech Rep.	1999-2003	5y, 1d, 1826	52598	1.014(0.989, 1.038)	2.85 (2.22, 3.48)	9.88	<0.001
				1.354 (TY)(1.252, 1.456)	1.35 (0.69, 2.02)	4.68	<0.001
	1994-2003	10y, 1d, 3652	115520	0.998 (0.988, 1.009)	3.03 (2.47, 3.60)	9.58	<0.001
				1.453 (TY)(1.417, 1.489)	1.91 (1.34, 2.49)	6.04	<0.001
				1.15 (TY) (1.116, 1.184)	1.23 (0.64, 1.82)	3.88	<0.001

*With focus on transyears, with periods longer than 1.0 year (underlined; double underline for near-transyear).

[1]International Classification of Diseases (ICD10), Code I46.1, excluding MI and sudden death of unknown or unspecified cause (except before 1999). T: length of data series (y=years); Δt: sampling interval (d=day, m=month); N: number of data (including 0s). Period and 95% confidence interval (CI) estimated by nonlinear least squares. In longer (10-y) series, a near-transyear (cycle with a period between 1.0 and 1.2 y) is detected for MIs in addition to a far-transyear. Brevity of series and lack of ordering statistical significance qualify results from Hong Kong. Note that transyears are found in 3 of 6 locations ($P<0.05$ by linear least-squares) with a relative amplitude >12 (% of MESOR).

[2]From linear least-squares analyses, not corrected for multiple testing. Amplitude expressed in N/day.

Daily Incidence of International Terrorist Acts (top) and Cosinor Periodogram (bottom) *

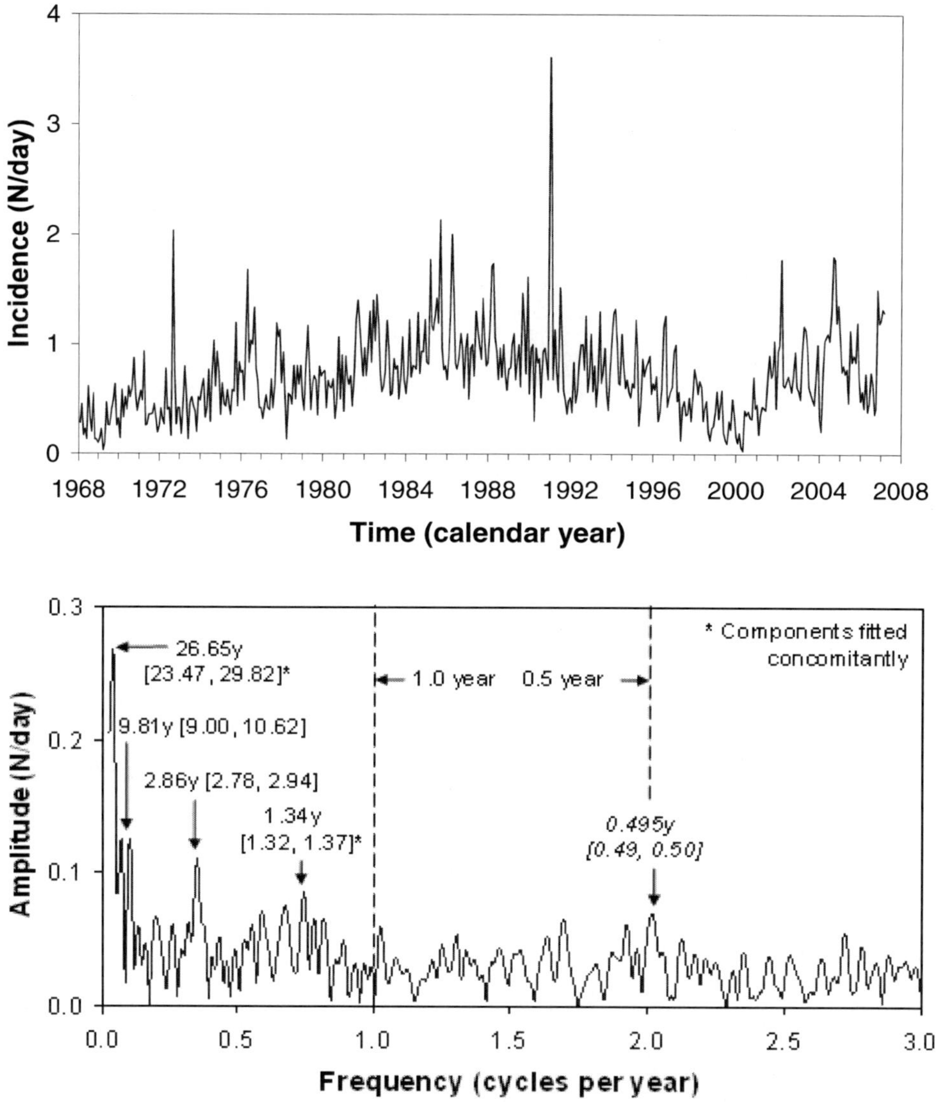

* From February 1968 to March 2007, daily incidence computerd from monthly totals (adjusted for differences in the bumber of days per month). Data from MIPT Terrorism Knowledge Base (http:www.tkb.org/).

Figure 17. In the absence of a calendar year component, a transyear with a period of about 1.3 years is detected in data from the Memorial Institute for the Prevention of Terrorism's (MIPT) Terrorism Knowledge Base (TKB, which ceased operations on 31 March 2008). Top: Plot of the number of cases of terrorism as a function of time. Bottom: Results from least squares spectrum.

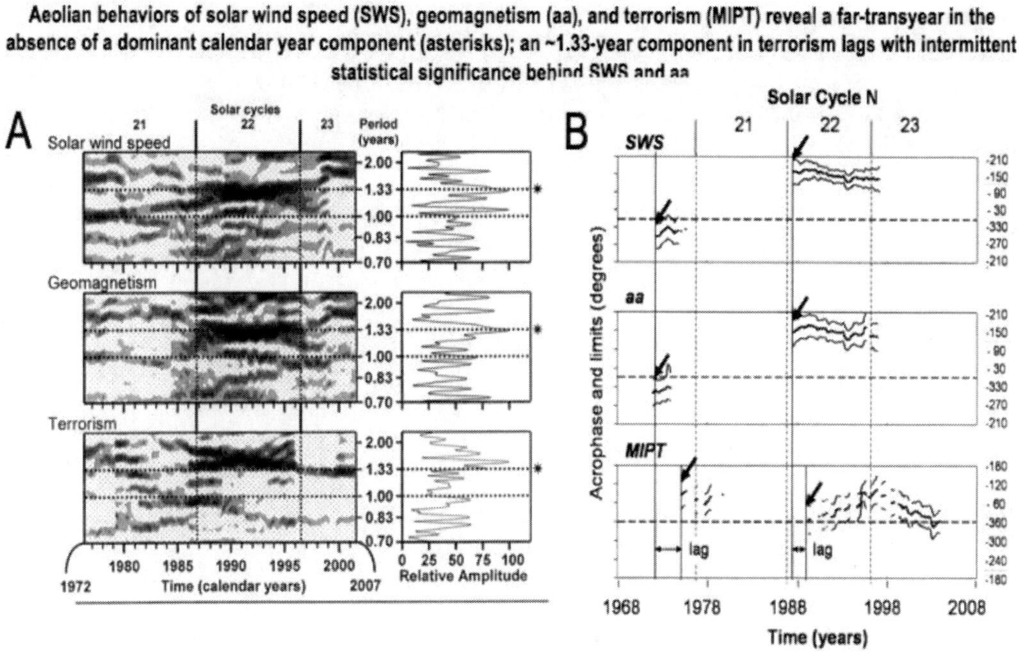

Figure 18. Temporal relations of interplanetary (SWS) and planetary (aa) magnetism and the incidence of worldwide terrorism. A (left): Gliding and global spectral windows indicate the presence of an about 1.3-year spectral component common to all three variables. B (right): Time course of acrophases at the average transyear period illustrates that after each of two spans of the appearance of a statistically significant (far-transyear) component in SWS and aa, the same spectral component (of about 1.3 years) appears, with a lag, in the incidence of terrorism (14,579 cases recorded during 39 years).

CONCLUSION

Earliest pre-human life mirrors cycles of its environment. A solar eclipse and human rhythmicity, both in fragments from the soldier-poet Archilochus, may be solar-biospheric associations. Alexander the Great's ship captain Androsthenes connected the sleep movement of plants with the alternation of light and darkness; Gilberd, the physician to Queen Elizabeth I and King James I of England, was a scholar of magnetism; and the scientist de Mairan, a scholar studying the aurora, wondered why the leaf movement of plants continued in the absence of light. The aurora and, in particular, sunspots, the latter thanks to Horrebow and Schwabe, teach us about a category of cycles that may come about as a result of many interacting, some unseen factors including magnetism and gravity, summarized as weather in extraterrestrial space. This historical note leads to the fact that space weather disturbs not only communications systems around us, but also has dramatic effects in the biosphere, including human health and other affairs.

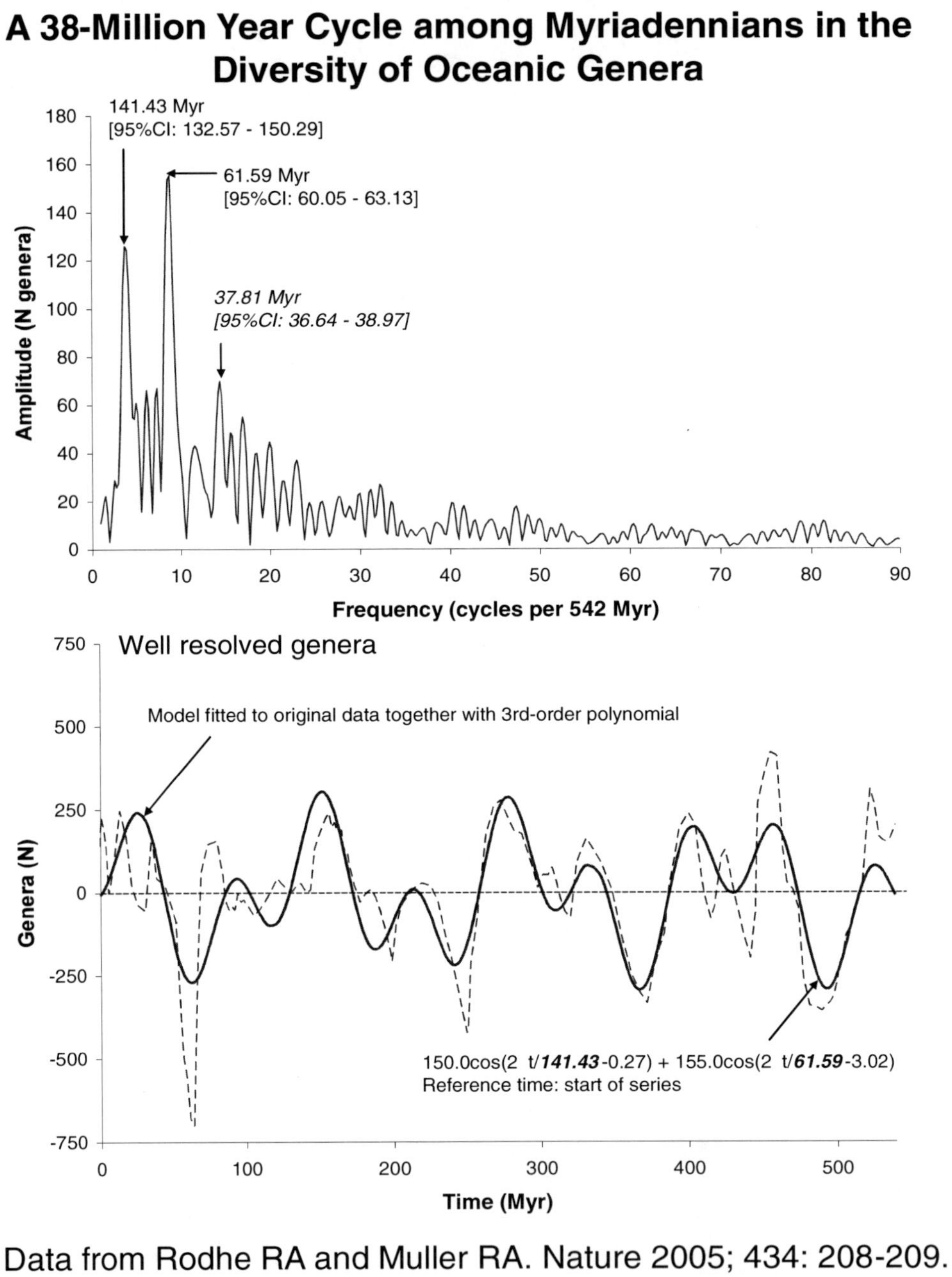

Figure 19. Cycles with periods of about 62 and 140 million years (Myr) were reported by Rohde and Muller (127) in fossil diversity during the Phanerozoic. A least squares spectrum of the data (top) identifies three peaks, including the two originally reported components. All three components, including the newly detected about 37.6-Myr one, are validated by nonlinear least squares, also yielding a measure of uncertainty for the period estimates, shown in [ ]. The composite model fitted to the detrended data (bottom) provides an acceptable approximation of changes with time in fossil diversity. © Halberg.

REFERENCES

[1] Koselleck R. Vergangene Zukunft: Zur Semantik geschichtlicher Zeiten. Frankfurt am Main: Suhrkamp Verlag; 1979. 389 pp. English translation by Tribe K. *Futures Past: On the Semantics of Historical Time.* Cambridge, MA: MIT Press; 1985. 330 pp.

[2] Langer WL. The next assignment. *Am. Hist. Rev.* 1958; 63: 283-304.

[3] Kamide Y. We reside in the sun's atmosphere. *Biomed and Pharmacother* 2005; 59 (Suppl 1): S1-S4.

[4] Halberg F, Cornélissen G, Katinas G, Tvildiani L, Gigolashvili M, Janashia K, Toba T, Revilla M, Regal P, Sothern RB, Wendt HW, Wang ZR, Zeman M, Jozsa R, Singh RB, Mitsutake G, Chibisov SM, Lee J, Holley D, Holte JE, Sonkowsky RP, Schwartzkopff O, Delmore P, Otsuka K, Bakken EE, Czaplicki J, International BIOCOS Group. Chronobiology's progress: season's appreciations 2004-2005. Time-, frequency-, phase-, variable-, individual-, age- and site-specific chronomics. *J. App. Biomed.* 2006; 4: 1-38. http://www.zsf.jcu.cz/vyzkum/jab/4_1/halberg.pdf

[5] Halberg F, Cornélissen G, Panksepp J, Otsuka K, Johnson D. Chronomics of autism and suicide. *Biomed and Pharmacother* 2005; 59 (Suppl 1): S100-S108.

[6] Cornélissen G, Halberg F. Chronomics of suicides and the solar wind. *Br. J. Psychiatry* 2006; 189: 567-568.

[7] Halberg F, Cornélissen G, Berk M, Dodd S, Henry M, Wetterberg L, Nolley E, Beaty L, Siegelova J, Fiser B, Wolff C, BIOCOS project. Solar signatures in Australian suicide incidence: gender differences in prominence of photic vs. nonphotic spectral components. In: Halberg F, Kenner T, Fiser B, Siegelova J. (Eds.) *Proceedings, Noninvasive Methods in Cardiology,* Brno, Czech Republic, October 4-7, 2008. p. 44-62. http://web.fnusa.cz/files/kfdr2008/sbornik_2008.pdf

[8] Halberg F, Cornélissen G, Sothern RB, Katinas GS, Schwartzkopff O, Otsuka K. Cycles tipping the scale between death and survival (= "life"). *Progress of Theoretical Physics* 2008; Suppl. 173: 153-181.

[9] Edgar B. Lucy up close: a paleo-celebrity's contributions to evolutionary science. Archaeology 2007; 60 (6). www.archaeology.org/0711/etc/lucy.html

[10] Halberg F, Cornélissen G, Sothern RB, Barnwell F, Cegielski N, Ilyia E, Siegelova J. The moon's and the genes' tides and double tides pulling the biosphere. In: Halberg F, Kenner T, Fiser B, Siegelova J. (Eds.) *Noninvasive Methods in Cardiology*, September 16-17, 2010, Brno, Czech Republic. Brno: Faculty of Medicine, Masaryk University. p. 23-45.

[11] Hastings JW, Schweiger HG. (Eds.) Life Sciences Research Report 1: The Molecular Basis of Circadian Rhythms. *Report of the Dahlem Workshop on the Molecular Basis of Circadian Rhythms,* Berlin, November 3-7, 1975. Berlin: Dahlem Konferenzen; 1976. 462 pp.

[12] Halberg F, Conner RL. Circadian organization and microbiology: Variance spectra and a periodogram on behavior of *Escherichia coli* growing in fluid culture. *Proc. Minn. Acad. Sci.* 1961; 29, 227-239.

[13] Sturtevant R. Circadian variability in *Klebsiella* demonstrated by cosinor analysis. *Int. J. Chronobiol.* 1973; 1: 141-146.

[14] Halberg F, Cornélissen G, Faraone P, Poeggeler B, Hardeland R, Katinas G, Schwartzkopff O, Otsuka K, Bakken EE. Prokaryotic and eukaryotic unicellular chronomics. *Biomed and Pharmacother* 2005; 59 (Suppl 1): S192-S202.

[15] Herschel W. XIII. Observations tending to investigate the Nature of the Sun, in order to find the Causes or Symptoms of its variable Emission of Light and Heat; with Remarks on the Use that may possibly be drawn from Solar Observations. *Royal Soc London Philos Trans* 1801; 91: 265-318. Excerpts from pp. 306, 310 and 315 reprinted in Schove DJ (Ed.), 1983.

[16] Clark S. The Sun Kings: The Unexpected *Tragedy of Richard Carrington and the Tale of How Modern Astronomy Began*. Princeton, NJ: Princeton University Press; 2007. 224 pp.

[17] Halberg F, Breus TK, Cornélissen G, Bingham C, Hillman DC, Rigatuso J, Delmore P, Bakken E, International Womb-to-Tomb Chronome Initiative Group: Chronobiology in space. Keynote, *37th Ann. Mtg. Japan Soc. for Aerospace and Environmental Medicine*, Nagoya, Japan, November 8-9, 1991. University of Minnesota/Medtronic Chronobiology Seminar Series, #1, December 1991, 21 pp. of text, 70 figures.

[18] Villoresi G, Breus TK, Iucci N, Dorman LI, Rapoport SI. The influence of geophysical and social effects on the incidences of clinically important pathologies (Moscow 1979-1981). *Physica Medica* 1994; 10: 79-91.

[19] Cornélissen G, Halberg F, Breus T, Syutkina EV, Baevsky R, Weydahl A, Watanabe Y, Otsuka K, Siegelova J, Fiser B, Bakken EE. Non-photic solar associations of heart rate variability and myocardial infarction. *J. Atmos Solar-Terr. Phys.* 2002; 64: 707-720.

[20] Halberg F, Cornélissen G, Regal P, Otsuka K, Wang ZR, Katinas GS, Siegelova J, Homolka P, Prikryl P, Chibisov SM, Holley DC, Wendt RW, Bingham C, Palm SL, Sonkowsky RP, Sothern RB, Pales E, Mikulecky M, Tarquini R, Perfetto F, Salti R, Maggioni C, Jozsa R, Konradov AA, Kharlitskaya EV, Revilla M, Wan CM, Herold M, Syutkina EV, Masalov AV, Faraone P, Singh RB, Singh RK, Kumar A, Singh R, Sundaram S, Sarabandi T, Pantaleoni GC, Watanabe Y, Kumagai Y, Gubin D, Uezono K, Olah A, Borer K, Kanabrocki EA, Bathina S, Haus E, Hillman D, Schwartzkopff O, Bakken EE, Zeman M. Chronoastrobiology: proposal, nine conferences, heliogeomagnetics, transyears, near-weeks, near-decades, phylogenetic and ontogenetic memories. *Biomed and Pharmacother* 2004; 58 (Suppl 1): S150-S187.

[21] Feinleib M, Rogot E, Sturrock PA. Solar activity and mortality in the United States. *Int. J. Epidemiol.* 1975; 4: 227-229.

[22] Lipa BJ, Sturrock PA, Rogot E. Search for correlation between geomagnetic disturbances and mortality. *Nature* 1976; 259: 302-304.

[23] Halberg F, Cornélissen G, Bingham C, Witte H, Ribary U, Hesse W, Petsche H, Engebretson M, Geissler H-G, Weiss S, Klimesch W, Rappelsberger P, Katinas G, Schwartzkopff O. Chronomics: Imaging in time by phase synchronization reveals wide spectral-biospheric resonances beyond short rhythms. ("Wenn man über kurze Rhythmen hinausgeht") In memoriam – lost future: Dr.-Ing. habil. Dr. rer. nat. Barbara Schack: 1952-2003. *Neuroendocrinol. Lett.* 2003; 24: 355-380.

[24] Schwabe H. Über die Flecken der Sonne. *Astronomische Nachrichten* 1838; 15: 244-248 (no. 350).

[25] Littrow JJv. Sonnenflecken. In: Gehler JST. *Physikalisches Wörterbuch.* Bd. 9: abt. 2. Leipzig: E.B. Schweikert; 1836 (series 1825-1845). p. 850-865.

[26] Thiele ThN. De Macularum Solis antiquioribus quibusdam observationibus Hafniae institutis. *Astronomische Nachrichten* 1859; 50: 259-261.

[27] Vitinskii YuI (Lerman Z, trans; Hardin R, Ed.) Solar-activity forecasting. Jerusalem: Israel *Program for Scientific Translations*; 1965. 129 pp.

[28] Fabricius J. De maculis in sole observatis: et apparente earum cum sole conversione narratio cui adjecta est de modo eductionis specierum visibilium dubitatio. Witebergae: Typis Laurentij Seuberlichij, impensis Iohan. Borneri senioris and Eliæ Rehifeldij. bibliop. Lips., Anno 1611. With a dedication of 13 June of that year.

[29] Albufaradge. *Historia Dynast*., p. 94. Cited by Littrow (2).

[30] Halberg F, Cornélissen G, Otsuka K, Watanabe Y, Katinas GS, Burioka N, Delyukov A, Gorgo Y, Zhao ZY, Weydahl A, Sothern RB, Siegelova J, Fiser B, Dusek J, Syutkina EV, Perfetto F, Tarquini R, Singh RB, Rhees B, Lofstrom D, Lofstrom P, Johnson PWC, Schwartzkopff O, International BIOCOS Study Group. Cross-spectrally coherent ~10.5- and 21-year biological and physical cycles, magnetic storms and myocardial infarctions. *Neuroendocrinol. Lett.* 2000; 21: 233-258.

[31] Halberg F, Cornélissen G, Otsuka K, Katinas G, Schwartzkopff O. Essays on chronomics spawned by transdisciplinary chronobiology: Witness in time: Earl Elmer Bakken. *Neuroendocrinol. Lett*. 2001; 22: 359-384.

[32] Halberg F, Cornélissen G, Otsuka K, Schwartzkopff O, Halberg J, Bakken EE. Chronomics. Biomed and Pharmacother 2001; 55 (Suppl 1): 153s-190s.

[33] Halberg F, Cornélissen G, Schack B, Wendt HW, Minne H, Sothern RB, Watanabe Y, Katinas G, Otsuka K, Bakken EE. Blood pressure self-surveillance for health also reflects 1.3-year Richardson solar wind variation: spin-off from chronomics. *Biomed and Pharmacother* 2003; 57 (Suppl 1): 58s-76s.

[34] Wolf R. Geschichte der Astronomie. Münich: Druck und Verlag von M. Oldenbourg; 1877.

[35] Schröder W. Das Phänomen des Polarlichts: Geschichtsschreibung, Forschungsergebnisse und Probleme. Darmstadt: Wissenschaftliche Buchgesellschaft; 1984. 156 pp.

[36] Otsuka K, Cornélissen G, Halberg F. Circadian rhythmic fractal scaling of heart rate variability in health and coronary artery disease. *Clinical Cardiology* 1997; 20: 631-638.

[37] Goldbeter A. Biochemical oscillations and cellular rhythms. Cambridge: Cambridge University Press, 1996, 605 pp. Foreword to the French Edition by I. Prigogine, pp. xix-xxii (see p. xxi).

[38] Halberg F, Bakken EE, Katinas GS, Cornélissen G, Zaslavskaya RM, Blank MA, Syutkina EV, Breus TK, Watanabe Y, Masalov A, Chibisov SM. Chronoastrobiology: Vernadsky's future science? Benefits from spectra of circadians and promise of a new transdisciplinary spectrum of near-matching cycles in and around us. Opening keynote, Proceedings, III International Conference, Civilization diseases in the spirit of V.I. Vernadsky, People's Friendship University of Russia, Moscow, Oct. 10-12, 2005, p. 4-22.

[39] Sothern RB, Yamamoto T, Cornélissen G, Takumi T, Halberg F. Central and peripheral circadian clock genes, their statistical analysis for rhythms, and relationship to health and disease. *Scripta medica* (Brno) 2009; 82: 133-163.

[40] Koukkari WL, Sothern RB. Introducing Biological Rhythms: A primer on the temporal organization of life, with implications for health, society, reproduction and the natural environment. New York: Springer; 2006. 655 pp.

[41] Refinetti R. *Circadian Physiology*. 2nd ed. Boca Raton, Florida: CRC Press; 2006. 700 pp.

[42] Halberg Franz, Cornélissen G, Katinas G, Syutkina EV, Sothern RB, Zaslavskaya R, Halberg Francine, Watanabe Y, Schwartzkopff O, Otsuka K, Tarquini R, Perfetto P, Siegelova J. Transdisciplinary unifying implications of circadian findings in the 1950s. *J. Circadian Rhythms* 2003; 1: 2. 61 pp. www.JCircadianRhythms.com/content/pdf/1740-3391-2-3.pdf

[43] Total solar eclipse, 29 March 2006. Solar eclipses in history and mythology: Historical observations of solar eclipses. Bibliotheca Alexandrina (access 3 May 2007) http://www.bibalex.org/eclipse2006/HistoricalObservationsofSolarEclipses.htm

[44] Griffin P. Loughcrew Cairn L Megalithic Monument, Oldcastle, Co. Meath, Ireland: solar eclipse 3340 BC Nov 30th: confirmation of world's oldest solar eclipse recorded in stone. Copyright © Paul Griffin 2002. http://www.astronomy.ca/3340eclipse/ (access 2 May 2007)

[45] Aschoff J. Speech after dinner. 1974 Capri Symposium on timing and toxicity. In: Aschoff J, Ceresa F, Halberg F. (Eds.) *Chronobiological Aspects of Endocrinology*. Stuttgart: F.K. Schattauer Verlag, 1974/Chronobiologia 1974; 1 (Suppl. 1): 483-495.

[46] Archilochus. Fragmente, 67 a D. See also Pedas T, Sigler M. Eclipses in History: From Fear to Fascination. http://www.nauticom.net/www/planet/files/EclipseHistory-FearToFascination.html

[47] Archilochus. Encyclopaedia Britannica. http://www.britannica.com/EBchecked/topic/32798/Archilochus; see also http://www.britannica.com/EBchecked /topic/178098/eclipse/11220/Greek?anchor=ref21512

[48] Bretzl H. *Botanische Forschungen des Alexanderzuges*. Leipzig: B.G. Teubner; 1903. p. 120-132.

[49] Theophrastus (Hort A. trans.). Enquiry into plants, and minor works on odours and weather signs. Cambridge, Mass.: Harvard University Press / London: W. Heinemann; 1980.

[50] Ensminger PA. Life under the Sun. New Haven, Connecticut: Yale University Press, 2001. A SAD Tale, pp. 56–68; *Blue Moons and Red Tides,* pp. 149–160.

[51] Halberg F. Chronobiology. *Annu. Rev. Physiol.* 1969; 31: 675-725.

[52] Halberg F, Barnum CP, Silber RH, Bittner JJ. 24-hour rhythms at several levels of integration in mice on different lighting regimens. Proc Soc exp Biol (NY) 1958; 97, 897-900; cf. Halberg F, Halberg E, Barnum CP, Bittner JJ. Physiologic 24-hour periodicity in human beings and mice, the lighting regimen and daily routine. In: Withrow RB. (Ed.) *Photoperiodism and Related Phenomena in Plants and Animals.* Ed. Publ. No. 55. Washington DC: AAAS; 1959. p. 803-878.

[53] Pennisi E. Multiple clocks keep time in fruit fly tissues. *Science* 1997; 278: 1560-1561.

[54] Plautz JD, Kaneko M, Hall JC, Kay SA. Independent photoreceptive circadian clocks throughout Drosophila. *Science* 1997; 278: 1632–1635.

[55] Halberg F, Engeli M, Hamburger C, Hillman D. Spectral resolution of low-frequency, small-amplitude rhythms in excreted 17-ketosteroid; probable androgen induced circaseptan desychronization. *Acta endocrinol* (Kbh) 1965; 50 (Suppl 103): 5-54.

[56] Hübner K. Kompensatorische Hypertrophie, Wachstum und Regeneration der Rattenniere. Ergebnisse der allgemeinen Pathologie und pathologischen *Anatomie* 1967; 100: 1-80.

[57] Hildebrandt G, Bandt-Reges I. Chronobiologie in der Naturheilkunde: Grundlagen der Circaseptanperiodik. Heidelberg: Haug; 1992. 102 pp.

[58] Reimann H. *Periodic diseases*. Philadelphia: F.A. Davis; 1963. 189 pp.

[59] Richter CP. *Biological Clocks in Medicine and Psychiatry*. Springfield, Illinois: Charles C. Thomas; 1965. 109 pp.

[60] DeVecchi A, Halberg F, Sothern RB, Cantaluppi A, Ponticelli C. Circaseptan rhythmic aspects of rejection in treated patients with kidney transplant. In: Walker CA, Winget CM, Soliman KFA. (Eds.) *Chronopharmacology and Chronotherapeutics. Tallahassee,* Florida: Florida A and M University Foundation; 1981. p. 339-353.

[61] Halberg F. *The week in phylogeny and ontogeny: opportunities for oncology*. In vivo 1995; 9: 269-278.

[62] Halberg F. Historical encounters between geophysics and biomedicine leading to the Cornélissen-series and chronoastrobiology. In: Schröder W. (Ed.) *Long- and Short-Term Variability in Sun's History and Global Change*. Bremen: Science Edition; 2000. p. 271-301.

[63] Roederer JG. Are magnetic storms hazardous to your health? *Eos, Transactions, American Geophysical Union* 1995; 76: 441, 444-445.

[64] Vladimirskii BM, Narmanskii VYa, Temuriantz NA. Global rhythmics of the solar system in the terrestrial habitat. *Biophysics* 1995; 40: 731-736.

[65] Gilberd (also spelled Gilbert) W. De magnete, magneticisque corporibus, et de magno magnete tellure; physiologia noua, plurimis et argumentis, et experimentis demonstrata [On magnetism, magnetic bodies, and the great magnet Earth]. London; Short: 1600. 204 pp.

[66] Chapman S, Bartels *J. Geomagnetism*. 3rd ed. Oxford: Clarendon Press; 1962. 1049 pp.

[67] Bailes KE. Science and Russian Culture in an Age of Revolutions: V.I. Vernadsky and his Scientific School, 1863-1945. Bloomington/Indianapolis: Indiana University Press; 1990. 238 pp. Dokuchaev, p. 182-183.

[68] Roederer JG. Tearing down disciplinary barriers. *Eos, Transactions, American Geophysical Union* 1985; 66: 681, 684-685.

[69] Gregori GP. Galaxy - Sun - Earth relations: The origin of the magnetic field and of the endogenous energy of the Earth, with implications for volcanism, geodynamics and climate control, and related items of concern for stars, planets, satellites, and other planetary objects. A discussion in a prologue and two parts. Beiträge zur Geschichte der Geophysik und Kosmischen Physik, Band 3, Heft 3, 2002, 471 pp. ISNN: 1615 - 2824 – NE: Gregori, Giovanni P. © Science Edition, Arbeitskreis Geschichte Geophysik / W. Schröder, Germany 2002.

[70] Sutcliffe J, Duin N. A *History of Medicine*. New York: Barnes and Noble; 1992. 256 pp.

[71] Sarton G. Introduction to the History of Science. Vol II: From Rabbi Ben Ezra to Roger Bacon. Part II. Baltimore: Williams and Wilkins (for the Carnegie Institution of Washington); 1931. p. 1099-1101.

[72] Santorio S. Methodi vitandorum errorum omnium qui in arte medica contingunt libri quindecim, quorum principia sunt ab auctoritate medicorum and philosophorum

principum desumpta, eaque omnia experimenis, and rationibus analyticis comprobata. Sanctorio Sanctorio ... auctore ... Cum tripli indice vno librorum, altero capitum omnium, tertio rerum notabilium. Venetiis: apud Franciscum Barilettum, 1603.

[73] Santorio S. Commentaria in artem medicinalem Galeni. Ad illustrissimum, and excellentissimum Andream Maurocenum ... Venetiis: apud Marcum Antonium Brogiollum, 1630.

[74] Santorio S. Methodi vitandorum errorum omnium qui in arte medica contingunt libri quindecim, quorum principia sunt ab auctoritate medicorum and philosophorum principum desumpta ... Nunc primum accessit ejusdem authoris De inventione remediorum libre. Cum tripli indice ... Genevae: apud Petrum Aubertum, Reipublicae and academiae typographum, 1631.

[75] Santorio S. De *Statica Medicina*. Hagae-Comitis, ex typographia A. Vlaco, 1657.

[76] Santorio S. Medicina statica: being the aphorisms of Sanctorius, translated into English, with large explanations. To which is added, Dr. Keil's Medicina statica Britannica, with comparative remarks and explanations. As also Medico-physical essays ... by John Quincy, M.D. London: printed for J. Osborn and T. Longman, in Paternoster-row: and J. Newton, in Little Britain, 1728.

[77] Ettari LS, Procopio M. *Santorio Santorio: la vita e le opere.* Rome: Istituto nazionale della nutrizione; 1968. 165 pp.

[78] Fidanza F. The "De Medicina Statica" by Santorio Santorio. *Nutrition* 1989; 5 (4): 226-228.

[79] Sydenham T (Swan J, trans). The Entire Works of Dr. Thomas Sydenham, Newly made English from the Originals: Wherein the History of acute and chronic Diseases, and the safest and most effectual Methods of treating them, are faithfully, clearly, and accurately delivered. London: Edward Cave, St. John's Gate, 1742. *The Life of Dr. Sydenham*, pp. v-xi.

[80] Virey JJ. Ephémérides de la vie humaine ou recherches sur la révolution journalière et la périodicité de ses phénomènes dans la santé et les maladies. Thesis, University of Paris, 1814, 39 pp.

[81] Halberg F. From aniatrotoxicosis and aniatrosepsis toward chronotherapy: Introductory remarks to the 1974 Capri Symposium on timing and toxicity: the necessity for relating treatment to bodily rhythms. In: Aschoff J, Ceresa F, Halberg F. (Eds.) *Chronobiological Aspects of Endocrinology.* Stuttgart: F.K. Schattauer Verlag; 1974. p. 1-34.

[82] Halberg F, Cornélissen G, Wang ZR, Wan C, Ulmer W, Katinas G, Singh Ranjana, Singh RK, Singh Rajesh, Gupta BD, Singh RB, Kumar A, Kanabrocki E, Sothern RB, Rao G, Bhatt MLBD, Srivastava M, Rai G, Singh S, Pati AK, Nath P, Halberg Francine, Halberg J, Schwartzkopff O, Bakken E, Shastri VK. Chronomics: circadian and circaseptan timing of radiotherapy, drugs, calories, perhaps nutriceuticals and beyond. *J. Exp. Therapeutics Oncol.* 2003; 3: 223-260.

[83] de Mairan JJ d'O. Observation botanique. Histoire de l'Académie Royale des Sciences 1729. pp. 35-36.

[84] Halberg F. Physiologic 24-hour periodicity; general and procedural considerations with reference to the adrenal cycle. Z Vitamin-, Hormon-u Fermentforsch 1959; 10: 225-296.

[85] Bünning E. Die physiologische Uhr. Berlin: *Springer,* 1958, 105 pp.

[86] de Mairan JJ d'O. Traité physique et historique de l'aurore boréale. Suite des Mém. *Acad. Roy. Sci.* 570 pp. Paris, 1731; 2nd ed., 1754.

[87] Schove DJ. (Ed). Sunspot Cycles. Benchmark Papers in Geology/68. Stroudsburg, PA: Hutchinson Ross; 1983. 397 pp.

[88] Biermann K-R. (Ed). Briefwechsel zwischen Alexander von Humboldt und Carl Friedrich Gauss: Zum 200. Geburtstag von C.F. Gauss im Auftrage des Gauss-Komitees bei der Akademie der Wissenschaften der DDR. Berlin: Akademie-Verlag; 1977. 202 pp. + plates.

[89] Cornélissen G, Halberg F, Revilla M, Siegelova J, Schwartzkopff O, Czaplicki J, Sanchez de la Peña S, Schröder W, BIOCOS project. Auroral aeolian periodicity during 1545-1724, including the Maunder minimum. In: Halberg F, Kenner T, Fiser B, Siegelova J. (Eds.) *Proceedings, Noninvasive Methods in Cardiology,* Brno, Czech Republic, October 4-7, 2008. p. 124-127. http://web.fnusa.cz/files/kfdr2008/sbornik_2008.pdf

[90] Silverman SM, Shapiro R. Power spectral analysis of auroral occurrence frequency. *J. Geophys. Res.* 1983; 88: 6310-6316.

[91] Cornélissen G, Halberg F, Sonkowsky R, Siegelova J, Homolka P, Dusek J, Fiser B. Meta-analysis of Horrebow's and Schwabe's scholarship with a view of sampling requirements. In: Proceedings, Noninvasive Methods in Cardiology, Brno, Czech Republic, July 7-10, 2009. p. 141-158. http://web.fnusa.cz/files/kfdr2009/sbornik_2009.pdf

[92] Gleissberg W. Die Häufigkeit der Sonnenflecken. Berlin: Akademie-Verlag; 1952. 91 pp.

[93] Schwabe H. Sonnen-Beobachtungen im Jahre 1843. *Astronomische Nachrichten* 1844; 21: 254-256 (no. 495).

[94] Clarke H. On the political economy and capital of joint stock banks. Railway Mag 1838; 27: 288–293. Cf. also Clarke H, Railway Mag. 1847 (cited in Editor's Comments on Papers 29 through 36, in Schove DJ (Ed.). *Sunspot Cycles*. Benchmark Papers in Geology/68. Stroudsburg, PA: Hutchinson Ross, 1983: 226–238).

[95] Jevons WS. The periodicity of commercial crises, and its physical explanation. Brit Assoc Adv Sci Rep 1878, pp. 666–667. Reprinted in Schove DJ (Ed.). *Sunspot Cycles* Benchmark Papers in Geology/68. Stroudsburg, PA: Hutchinson Ross, 1983: 241.

[96] Johnson MJ. [Address delivered by the President, M.J. Johnson, Esq., on presenting the Medal of the Society to M. Schwabe.] *Monthly Notices,* Royal Astronomical Society 1857; 17: 126-132.

[97] Schwabe SH. [Extract of a letter from M. Schwabe to Mr. Carrington. (Translated freely.)] *Monthly Notices*, Royal Astronomical Society 1859; 20: 15.

[98] Cornélissen G, Sothern RB, Wendt HW, Tarquini B, Antuñano M, Siegelova J, Fiser B, Dusek J, Prikryl P, Halberg F. Statistical significance without biologic signification is not enough: illustrative example. *Chronobiologi*a 1994; 21: 315-320.

[99] Halberg F, Sothern RB, Watanabe Y, Hillman D, Best WR, Schwartzkopff O, Cornélissen G. Signatures of space weather in the aging human blood circulation? *World Heart Journal*, in press.

[100] Sothern RB, Katinas GS, Fiser B, Siegelova J, Cornélissen G, Halberg F. A transtridecadal cycle in human heart rate: Selective infradian, notably multidecadal solar-physiologic BEL congruences. In: Halberg F, Kenner T, Fiser B, Siegelova J.

(Eds.) *Proceedings, Noninvasive Methods in Cardiology,* Brno, Czech Republic, October 4-7, 2008. p. 204-213. http://web.fnusa.cz/files/kfdr2008/sbornik_2008.pdf

[101] Halberg F, Cornélissen G, Bernhardt K-H, Sampson M, Schwartzkopff O, Sonntag D. Egeson's (George's) transtridecadal weather cycling and sunspots. *Hist Geo Space Sci* 2010; 1: 49-61.

[102] Halberg F, Cornélissen G, Sothern RB, Czaplicki J, Schwartzkopff O. Thirty-five-year climatic cycle in heliogeophysics, psychophysiology, military politics, and economics. *Izvestiya, Atmospheric and Oceanic Physics* 2010; 46 (7): 844-864. (Back-translation from Geophysical Processes and Biosphere 2009; 8 [2]: 13-42.)

[103] Halberg F, Cornélissen G, Sothern RB, Wallach LA, Halberg E, Ahlgren A, Kuzel M, Radke A, Barbosa J, Goetz F, Buckley J, Mandel J, Schuman L, Haus E, Lakatua D, Sackett L, Berg H, Wendt HW, Kawasaki T, Ueno M, Uezono K, Matsuoka M, Omae T, Tarquini B, Cagnoni M, Garcia Sainz M, Perez Vega E, Wilson D, Griffiths K, Donati L, Tatti P, Vasta M, Locatelli I, Camagna A, Lauro R, Tritsch G, Wetterberg L. International geographic studies of oncological interest on chronobiological variables. In: Kaiser H. (Ed.) *Neoplasms—Comparative Pathology of Growth in Animals, Plants and Man.* Baltimore: Williams and Wilkins; 1981. p. 553-596.

[104] Egeson C. Egeson's weather system of sun-spot causality: being original researches in solar and terrestrial meteorology. Sydney: Turner and Henderson; 1889. 63 pp.

[105] Brückner E. Klimaschwankungen seit 1700 nebst Beobachtungen über die Klimaschwankungen der Diluvialzeit. Wien und Olmütz: E. Hölzel; 1890. 324 pp. (Penck A, Hrsg. Geographische Abhandlungen, Band IV.)

[106] Akasofu S-i. On the recovery from the Little Ice *Age. Natural Sci* 2010; 2: 1211-1224.

[107] Halberg F, Cornélissen G, Grambsch P, McCraty R, Beaty L, Siegelova J, Homolka P, Hillman DC, Finley J, Thomas F, Kino T, Revilla M, Schwartzkopff O. Personalized chronobiologic cybercare; other chronomics' progress by transdisciplinary cycles' congruences: Season's Appreciations 2009. *J. Appl. Biomed.* 2011; 9: 1-34. DOI 10.2478/v10136-009-0022-8.

[108] Sigel F (Dreier W, Lerche D, Übers.; Göring H, Wissenschaftl. Red. der deutschsprachigen). Schuld ist die Sonne. Thun/Frankfurt am Main: Harri Deutsch; 1979. 215 pp.

[109] Chizhevsky AL. *Astronomy, psychology and history.* Moscow: M.A. Institute; 1921. 78 pp.

[110] Chizhevsky AL. Kosmische Einflüsse, die die Entstehung von Massenpsychosen begünstigen. Deutsch-Russische Medizinische Zeitschrift 1928; 4: 86-116.

[111] Chizhevsky AL. Über die Veränderung der Nervenerregbarkeit unter dem Einfluß der Perturbationen in der äußeren chemisch-physikalischen Umwelt. Versuch zum Studium der Kollektiv-Psychoneurologie. Deutsch-Russische Medizinische Zeitschrift 1928; 4: 405-430 and 479-500.

[112] Chizhevsky AL. Action de l'ionisation de l'atmosphère et de l'ionisation artificielle de l'air sur les organismes sains et les organismes malades. In: Piéry M. (Ed.) Traité de Climatologie: Biologique et médicale. Tome premier. Paris: Masson et Cie; 1934. p. 662-673.

[113] Chizhevsky AL. Les épidémies et les perturbations électromagnétiques du milieu extérieur. Paris: Éditions Hippocrate; 1938. 239 pp.

[114] Chizhevsky AL (Fedynsky VV, Ed.) The Earth in the Universe. Translated from Russian [and edited by IRST staff]. NASA TT F-345. Jerusalem: Israel Program for Scientific Translations [available from US Dept of Commerce, Clearinghouse for Federal Scientific and Technical Information, Springfield, Virginia]; 1968. 280 pp.

[115] Chizhevsky [Tchijevsky] AL (de Smitt VP, trans and condensed). Physical factors of the historical process. *Cycles* 1971; 22: 11-27. http://www.cyclesresearchinstitute.org/chizhevsky/chizhevsky1.pdf

[116] Goethe JWv. Die Schriften zur Naturwissenschaft, Leopoldina-Ausgabe, I. Abteilung: Texte, Vol. 1-11, Weimar, hier Vol. 2, Seite 164, 1947-1970.

[117] Halberg F, Sothern RB, Cornélissen G, Czaplicki J. Chronomics, human time estimation, and aging. Clinical Interventions in Aging 2008; 3 (4): 749-760. http://www.dovepress.com/articles.php?article_id=2608

[118] Bryson B. A *Short History of Nearly Everything*. New York: Broadway Books; 2003. p. 478.

[119] Ozheredov VA, Breus TK, Gurfinkel YuI, Revich DA, Mitrofanova TA. Influence of some weather factors and geomagnetic activity on the development of severe cardiological pathologies. *Biophysics* 2010; 55; 110-119. (Original Russian text Biofizika 2010; 55: 133-144.)

[120] Cornélissen G, Tamura K, Tarquini B, Germanò G, Fersini C, Rostagno C, Zaslavskaya RM, Carandente O, Carandente F, Halberg F. Differences in some circadian patterns of cardiac arrhythmia, myocardial infarction and other adverse vascular events. *Chronobiologia* 1994; 21: 79-88.

[121] Cornélissen G, Wendt HW, Guillaume F, Bingham C, Halberg F, Breus TK, Rapoport S, Komarov F. Disturbances of the interplanetary magnetic field and human pathology. *Chronobiologia* 1994; 21: 151-154.

[122] Lockyer N. Contributions to Solar Physics. London: Macmillan; 1874. 676 pp. "*The thing to hunt down is a cycle*", p. 424-425.

[123] Stetson HT. Sunspots in action. New York: Ronald Press Co.; 1947. 252 p.

[124] Düll B, Düll T. Erd- und Sonnenphysikalische Vorgänge in ihrer Bedeutung für Krankheits- und Todauslosung. *Nosokomeion* 1938; 9: 103-120.

[125] Dubrov AP (Sinclair FL, trans; Brown FA Jr, translation ed). *The geomagnetic field and life: geomagnetobiology*. New York: Plenum Press; 1978. 318 pp.

[126] Breus TK, Chibisov SM, Baevsky RM, Schebzuhov KV. The chronostructure of heart biorhythms under effects of external environment factors. Moscow: Russian University of People's Friendship, Polygraph Service; 2002. 231 pp.

[127] Rohde RA, Muller RA. Cycles in fossil diversity. *Nature* 2005 (March 10) 434; 147: 208-209.

[128] Hays JD, Imbrie J, Shackleton NJ. Variations in the Earth's orbit: pacemaker of the ice ages. *Science* 1976; 194: 1121-1132.

[129] Shopov Y, Stoykova D, Cornélissen G, Halberg F. Further cyclicities in speleothems waiting for transdisciplinary counterparts. *Abstract, III International Conference, Civilization diseases in the spirit of V.I. Vernadsky*, People's Friendship University of Russia, Moscow, Oct. 10-12, 2005, p. 120-123.

[130] Cornélissen G, Bakken EE, Sonkowsky RP, Halberg F. A 38-million-year cycle among myriadennians in the diversity of oceanic genera. *Abstract, III International Conference, Civilization diseases in the spirit of V.I. Vernadsky,* People's Friendship University of Russia, Moscow, Oct. 10-12, 2005, p. 47-49.

In: Cardiovascular Health and Chronomics
Editor: Ram B. Singh

ISBN: 978-1-62948-976-6

Chapter 18

In Memoriam, Salvador Sanchez Alfonso De La Pena, MD, PhD: A Comrade in the Chronousphere (November 14, 1948 - November 2, 2011)

R. B. Singh[1*], Germaine Cornélissen[2], Robert B. Sothern[3], Jonathan Levi Rito Medina[4], Douglas W. Wilson[5], Miguel Revilla[6], Fabien De Meester[7†], Sergey Chibisov[8], Radzhesh Agarval[8], Othild Schwartzkopff[9] and Franz Halberg[10]

[1]MD, Halberg Hospital and Research Institute, Civil Lines, Moradabad, India
[2]Halberg Chronobiology Center, University of Minnesota, Minneapolis, MN, US
[3]PhD, Research Asst Professor, The Rhythmometry Lab, University of Minnesota, St. Paul, MN, US
[4]3rd year medical student, Escuela Nacional de Medecina y Homeopatía, Instituto Politécnico Nacional, Mexico City, Mexico
[5]PhD, School of Medicine and Health, University of Durham, UK
[6]PhD, University of Valladolid, Spain
[7]The Tsim Tsoum Institute, Krokow, Poland
[8]Russian Peoples, Friendship University, Moscow, Russia
[9]MD, FAAP, Halberg Chronobiology Center, University of Minnesota, Minneapolis, MN, US
[10]M.D., Dr. h.c. (mult.): Halberg Chronobiology Center, University of Minnesota, Minneapolis, MN, US

* Correspondence: RB Singh, MD, Halberg Hospital and Research Institute, Civil Lines, Moradabad, India. rbs@tsimtsoum.net.
† Email: fdm@tsimtsoum.net.

SALVADOR A. SANCHEZ DE LA PEÑA

Graduated from the National Autonomous University of Mexico (UNAM), September 18, 1973

Specialization in biomedical investigation from the Instituto Mexicano del Seguro Social (IMSS), 1976-1979, CMN (National Medical Center) Oncology Hospital

Master in Science, December 17, 1985 (Laboratory Medicine and Pathology, University of Minnesota)

Doctorate in Science from the Instituto Politécnico Nacional, March 31, 1986 (CINVESTAV-IPN)

Member of System of National Investigators, Category I, 1994-1997

Director of Chronobiology and Chronogeriatrics Laboratory and Professor, Escuela Nacional de Medicina y Homeopatia (ENMyH)

Director of Medical Chronobiology Laboratory, 1998-2011

Editor, Revista Internacional Bilingüe: GERONTO-GERIATRICS

COMMENTS BY HIS AMERICAN MENTOR AND SOME OF HIS CLOSE ASSOCIATES

It seems that there are no better words to remember Salvador than those written by the Lord of Time - Dr Franz Halberg: For members of the BIOCOS worldwide, including Minnesota friends, Salvador remains an integral part of what we do. After Mexico City, Minnesota was his second home, whether he enthusiastically described what he did to Earl E. Bakken, the developer of the implantable cardiac pacemaker or analyzed data from an Italian woman isolated from society for over 100 days, to find what Earl had described as "free-running". His real home, of course, was and remains worldwide and beyond, the sphere of the human mind, what Edouard Le Roy, Pierre Teilhard de Chardin and above all Vladimir Ivanovich Vernadsky called the noösphere, the sphere of human creativity, complementing Eduard Suess's (the Viennese geophysist) biosphere. Salvador helped take us a step beyond the noösphere by examining both the basic rules and their applications. These were steps toward introducing time (chronos) into the noösphere. By seeking new transdisciplinary cycles to quantitatively assess the otherwise qualitative noosphere, we can measure, apparently for the first time in statistical inferential terms, the domain of the human mind by its mood and its creativity, manifested in the occurrence at about 500-year intervals of prominent historians, poets, and physicians. In classical Greek, "noos" and "nous" both describe the human mind; the Attic Greek "nous" serves to portmanteau it into "chronos", time, and "sphairos", sphere, into a chronousphere. On the basic side, Sal's enthusiasm and his stamina supporting his inquiry into the unknown laid the basis of a time-qualified feedsideward (rather than time-unqualified feedback) in the pineal-hypothalamic-pituitary-adrenal-cellular network first and in the cosmos next. His last contributions were data for examining the role of the cosmos in modulating human blood pressure.

Whatever anybody does in time, by selecting a partial system from the universe for investigation, he/she has to consider contributions by the complementary system of the cosmos and may best account for the latter's cycles. Sal showed that the same dose of the same molecule can enhance or delay the development of a malignancy. He contributed many research articles on chronooncology (e.g., 2-9). His publications assure him a lasting place in science and health care through which he will continue to contribute second to none (1-182). May the premetabolic syndrome, on the applied side be his legacy's highlight (179-181).

Dr. Wilson: Salvador was a steadfast friend who believed in the warmth of friendship. Sal gave his all to those that knew him. He was a shining light that guided us with his happy enthusiasm, visionary chronobiological thinking and his exemplary work ethos and energy.

In the last 12 months, he co-authored an important classic article that has been accepted for publication (1) and he made contributions to presentations in Moscow and earlier in Taipei. He lives in and beyond the noosphere. We will always think of him as a great friend and his parting is a great loss. I knew him for the last 30 years, planned great adventures with him in Mexico on premetabolic syndrome in children and adults and somehow these should be kept alive.

Dr. Sothern: it was truly with great sadness and shock that I learned of Salvador's passing earlier this November. Even though many of us had been in touch quite often over the past several years about mutual research interests, most of us had absolutely no idea that he had a recurrence of non-Hodgkin's lymphoma and was undergoing treatment with cis-platinum that irreparably damaged his kidneys. In an October 15 email from him Sal wrote that he was collecting patient data. He mentioned being in the hospital, but didn't say why, so I assumed something routine (like when he broke his arm in Sept 2010 or wrote in May 2010 that he had some health problems in April 2010 but was better). It was wrong to assume there was nothing to worry about with his recent hospitalization. I did not know about his life-threatening illness.

Sal and I were friends since his days in the late 1970's and into the 1980's in Minnesota in the Chronobiology Laboratory of Franz Halberg and we continued to stay in touch as he followed Bill Hrushesky to Albany, New York and later returned to Mexico City in the 1990s to continue his medical and research career. He was always such an interesting character, always optimistic with grand ideas, a gracious attitude in the face of great pressures, and eager to engage in as much research as possible, including editing his "modest" (his word!) "International Journal of Geronoto-Geriatrics", serving as a reviewer for several international journals, and collaborating with so many others around the world on research projects.

Along these lines, in July of this year we published a paper together, which may be his last as a first author, but the first (we think) to combine the areas of homeopathy and chronotherapy (chrono-homeopathy), two areas that he was passionate about that each continue to generate a certain amount of skepticism (182). We were very pleased to find a rhythmic aspect of aconite's effect of mouse body temperature and were hoping that this would be of interest to others in the field of homeopathy (i.e., timing is everything, but is still widely overlooked in most fields).

As recently as October of this year, Sal mentioned that all the research that he was tirelessly planning to do and/or had already started, including monitoring heart patients who received homeopathy treatment with aconite, and measuring clock genes in these and cancer patients. Of course he had many, many other research interests, including chronomics and the premetabolic syndrome, for which he was honored with lifetime achievement awards for excellence in medicine and health applications by the International Organization for Medical Teaching and Research for all of Latin America, Central America and the Caribbean that he received in Bogota, Colombia in November 2010. He was very happy to receive this recognition and was eager to push forward with his pursuits. Sadly, this was not to be.

With a heavy heart, but a smile from having known such a nice person, my sympathy goes out to all who knew and cared about Sal, including his children Sal Jr. and Amaranda, and Elizabeth, their mother. "Dost thou love life? Then do not squander time, for that's the stuff life is made of." Benjamin Franklin (1706-1790). Cand. Med. Rito was his student and collaborator for the last 5 years and was very useful in the setting of the Research Laboratory on Chronomics, a critical step beyond Chronobiology since the cosmos, part of chronomics and the chronousphere, is the ever-present control for whatever is being done in time.

Germaine (Professor Cornelissen) and her husband, Francis (Dr Guillaume) to his son: Please accept our heartfelt condolences. We lost a great friend with the passing of your father. Salvador came to the lab soon after I joined Franz's lab, so our friendship dates a long time. We had a very close cooperation not only when we were on the same team but also after he left.

We learned the sad news yesterday but even today it is hard to realize that this was the last good-bye. Your father came to the lab shortly after I did and we shared many fond memories of hard work at the lab and great discussions, excitement at findings that came almost everyday thanks to the hard work of Sal. We traveled to meetings together, eager to present our results and meet new friends who came from the four corners of the world. His mother came to visit him in Minnesota as did my parents, a small family away from home. I can still remember how proud your father was when you were born. And he remained proud of you over the years as you grew up and became the handsome man we saw in the pictures of your wedding. Whereas nothing can replace your father, his legacy will live on.

There were several ongoing projects we had the privilege to cooperate on. Please be sure that we shall do our utmost to complete the tasks so that the efforts he put into these endeavors to the very end, despite his failing health, will come to fruition. The last time we saw your father was on the occasion of Franz's 90th birthday. He was in good spirits then.

Attached are two pictures taken on that occasion, one with Franz at the lab (the orange shirt he is wearing was a special gift from Professor Keiko Uezono on the occasion of his 90th birthday), and one with me (to his right), Miguel Revilla from Spain (to my right), RB Singh from India (to his left) and Doug Wilson whom you may have met at your wedding (far right in the picture).We loved your father. He will be greatly missed. With deepest sympathy, Germaine and Francis.

Dr. Singh met Salvador in 2004 during the 3rd International Congress in Cardiovascular diseases, Taipei, Taiwan, 26-28 Nov 2004 and got the opportunity to be his friend (144). Dr Singh had another opportunity to enjoy his friendship during the 90th birthday of Dr Halberg, when Professor Revilla from Spain, Dr DeMeester from Belgium and Drs Sergey Chibisov and Radzhesh Agarval from Russia had also come for the celebration, hosted by Franz's daughters Francine and Julia, his wife Othild, and Germaine.

We all in BIOCOS send our deepest condolences to his family herewith - the words do not give justice to Salvador, but the heart is deeply touched and the mind is sorrowful but knows of his great achievements. He has merely moved to another place, where all of us will be moving eventually to greet him.

REFERENCES

Ayala D., Hermida R.C., Sanchez de la Peña S., Halberg F. Circadian cardiovascular profile during two trimesters of human pregnancy reveals steadily increasing heart rate MESOR. *Chronobiologia* 16: 110-111, 1989.

Ayala D.E., Hermida R.C., Fernández J.R., Sanchez S., Halberg F. Variaciones de las caracteristicas circadianas de la presion sanguinea entre el tercer y el cuarto mes de embarazo. Abstract, II Reunion Nacional de Cronobiologia, Santiago de Compostela, Spain, December 6-7, 1989, p. 15.

Ayala D.E., Hermida R.C., Sanchez S., Halberg F. Analysis of blood pressure assessed by ambulatory noninvasive monitoring in a healthy pregnancy. In: Proc. 40th ISMM Int. Conf. Mini- and Microcomputes and their Applications, Lugano, Switzerland, June 19-21, 1990, pp. 32-35.

Bingham C., Halberg F., Wu J., Cornélissen G., Sanchez S., Halberg E., Harvey J., Lakatua D., Bazin H. "Treatment effects" before "treatment"—need for repeated checks of bias specified a priori for variable and timing and, when indicated, rerandomization. *Chronobiologia* 14: 152-153, 1987.

Brown G.M., Grota L.J., Sanchez de la Peña S., Halberg F., Halberg E. Circadian melatonin rhythm: part of stimulatory feed-sideward in pineal modulation of adrenal response to ACTH 1-17? Abstracts of papers, *Minn. Acad. Sci.* 51st Annual Spring Meeting, University of Minnesota-Duluth, April 29-30, 1983, p. 12.

Brown G.M., Sanchez de la Peña S., Marques N., Grota L.J, Ungar F., Halberg F. More on the circadian melatonin rhythm in pineals from domesticated B6D2F1 mice. *Chronobiologia* 13: 335-338, 1986.

Cecchettin M., Sanchez de la Peña S., Halberg F., Cornélissen G., Ungar F., Brown G., Mantero F., Vecsei P., Simmons D., Walker R.F., Scheving L.E. Reproducibility of pineal feed-sideward upon the ACTH 1-17 effect on adrenocortical function. In: *Neuroimmunomodulation,* Proc. 1st Int. Workshop on NIM, Bethesda, MD, Nov. 27-30, 1984, IWGN, Bethesda, 1985, pp. 139-141.

Cecchettin M., Sanchez de la Peña S., Halberg F., Cornélissen G., Ungar F., Brown G., Mantero F., Vecsei P., Simmons D., Walker R.F., Scheving L.E. Reproducibility of pineal feed-sideward upon the ACTH 1-17 effect on adrenocortical function. *Chronobiologia* 13: 71-73, 1986.

Conneeley O., Sanchez de la Peña S., Halberg F., Ungar F., Anaya C., Yunis E. Seasonal reproducibility on staggered LD12:12 lighting regimen of corticosterone chronobioassay for ACTH 1-17. *Chronobiologia* 10: 116-117, 1983.

Cornélissen G, Delcourt A, Toussaint G, Otsuka K, Watanabe Y, Siegelova J, Fiser B, Dusek J, Homolka P, Singh RB, Kumar A, Singh RK, Sanchez S, Gonzalez C, Holley D, Sundaram B, Zhao Z, Tomlinson B, Fok B, Zeman M, Dulkova K, Halberg F. Opportunity of detecting pre-hypertension: worldwide data on blood pressure overswinging. *Biomedicine and Pharmacotherapy* 2005; 59 (Suppl 1): S152-S157.

Cornélissen G, Delcourt A, Toussaint G, Otsuka K, Watanabe Y, Siegelova J, Fiser B, Dusek J, Homolka P, Singh RB, Kumar A, Singh RK, Sanchez S, Gonzalez C, Holley D, Sundaram B, Zhao Z, Tomlinson B, Fok B, Zeman M, Dulkova K, Halberg F. Opportunity of detecting pre-hypertension: worldwide data on blood pressure overswinging. (Reprint of Biomedicine and Pharmacotherapy 2005; 59 [Suppl 1]: S152-S157. *Proceedings, International Conference on the Frontiers of Biomedical Science: Chronobiology,* Chengdu, China, September 24-26, 2006, p. 123-129.

Cornélissen G, Halberg F, Katinas G, Schwartzkopff O, Holley D, Borer K, Homolka P, Siegelova J, Fiser B, Otsuka K, Yano A, Delcourt A, Toussaint G, Sanchez de la Pena S,

Gonzalez C, Zhao Z, Aslanian N, Singh RB, Kumar A, Tarquini R, Perfetto F, and collaborators in the BIOCOS project. Stroke and other vascular disease prevention by chronomics for prehabilitation. Abstract #10, MEFA, Brno, Czech Republic, 04-07 Nov 2003. p. 14-15.

Cornélissen G, Halberg F, Katinas G, Schwartzkopff O, Holley D, Borer K, Homolka P, Siegelova J, Fiser B, Dusek J, Otsuka K, Yano A, Delcourt A, Toussaint G, Sanchez de la Peña S, Gonzalez C, Zhao Z, Aslanian N, Singh RB, Kumar A, Tarquini R, Perfetto F. Stroke and other vascular disease prevention by chronomics. *Scripta medica* (Brno) 2006; 79 (3): 141-146.

Cornélissen G, Halberg F, Kaufman SC, Sanchez de la Peña S, Otsuka K, Siegelova J, Fiser B, BIOCOS project. Should 7-day/24-hour chronobiologically interpreted blood pressure monitoring replace single measurements before cataract surgery ? In: Halberg F, Kenner T, Fiser B, Siegelova J, eds. Proceedings, Noninvasive Methods in Cardiology, Brno, Czech Republic, October 4-7, 2008. p. 90-105. http://web.fnusa.cz/files/kfdr2008/sbornik_2008.pdf.

Cornélissen G, Halberg F, Perfetto F, Tarquini R, Salti R, Maggioni C, Sanchez de la Pena S, Schwartzkopff O, Bakken EE. The chronome, time structure of melatonin. Abstractband, 11. Gothaer Tage der Kinderheilkunde, HELIOS Kreiskrankenhaus, Gotha/Ohrdruf GmbH, Germany, 28. November 2003. p. 1-5.

Cornélissen G, Halberg F, Revilla M, Siegelova J, Schwartzkopff O, Czaplicki J, Sanchez de la Peña S, Schröder W, BIOCOS project. Auroral aeolian periodicity during 1545-1724, including the Maunder minimum. In: Halberg F, Kenner T, Fiser B, Siegelova J, eds. *Proceedings, Noninvasive Methods in Cardiology,* Brno, Czech Republic, October 4-7, 2008. p. 124-127. http://web.fnusa.cz/files/kfdr2008/sbornik_2008.pdf.

Cornélissen G, Halberg F, Schwartzkopff O, Sanchez de la Peña S, Chibisov SM, Radysh IV, Blagonravov M, Katinas GS. Economics and biology: the BEL. In: *Proceedings, 14th International Symposium, Ecologo-Physiological Problems of Adaptation,* 9-10 April 2009. Moscow: People's Friendship University of Russia; 2009. p. 486-489.

Cornélissen G, Halberg F, Wendt HW, Nelson RD, Schwartzkopff O, Wang ZR, Otsuka K, Oinuma S, Revilla M, Sanchez de la Peña S, Beaty L, Sonkowsky R, Katinas GS, Blagonravov ML, Chibisov SM. Transyears, about 17-month cycles in opinion polls about US president. *Proceedings, 1st International Workshop, Physiology of adaptation and quality of life: problems of traditional medicine and innovation, People's Friendship University of Russia,* Moscow, Russia, May 14-16, 2008. p. 387-390.

Cornélissen G, Halberg J, Halberg F, Sanchez de la Pena S, Nelson W, Schwartzkopff O, Stoynev A, Haus E. Schedule shifts, cancer and longevity: good, bad or indifferent? *J. Experimental Therapeutics Oncol.* 2008; 7 (4): 263-274.

Cornélissen G, Schwartzkopff O, Katinas G, Halberg F, Bakken EE, Holley D, Sundaram S, Sanchez S, Gonzalez C, Delcourt A, Toussaint G, Siegelova J, Fiser B, Dusek J, Homolka P, Zeman M, Dulkova K, Otsuka K, Watanabe Y, Tomlinson B, Fok B, Zhao ZY, Singh RB, Singh RK, Kumar A, Chibisov SM. Diagnosing blood pressure overswinging chronomically worldwide. *Abstract, III International Conference, Civilization diseases in the spirit of V.I. Vernadsky, People's Friendship University of Russia*, Moscow, Oct. 10-12, 2005, p. 34-35.

Cornélissen G., Halberg F., Sanchez de la Peña S., Wu J., Carandente F. The need for both macroscopy and microscopy in dealing with spectral structure. Chronobiologia 15: 323-327, 1988. (Comment on Powell E.W. The master clock illusion. *Chronobiologia* 15: 321-322, 1988).

Cornélissen G., Sanchez de la Peña S., Halberg F. From drug synergism and antagonism to endocrine chronoimmunomodulators. J. Interdiscipl. *Cycle Res.* 16: 245-246, 1985.

Fernández J.R., Hermida R.C., Ayala D.E., Sanchez de la Peña S., Halberg F. Nonparametric testing of melatonin characteristics as potential markers of breast cancer risk. *Chronobiologia* 16: 133, 1989.

Fernandez-Cerdeno A., Valverde C., Sanchez de la Peña S., Halberg F. Circadian variation in urinary thyroxine (T4) excretion by patients with emotional disorder. *Chronobiologia* 8: 185, 1981.

Galvagno A., Montalbini M., Sanchez de la Peña S., Scheving L.E., Degioanni J., Kutyna F., Halberg F. About 7-day (circaseptan) cardiovascular in the social isolation of a cave. Abstract, 1st International Design for Extreme Environments Assembly, Houston, TX, Nov. 12-15, 1991, HF 2.3.2.

Garcia L., Lodeiro C., Castro H., Gomez Bermudez M.C., Barreiro M., Hermida R.C., Sanchez de la Peña S., Hillman D.C., Wu J., Halberg E., Cornélissen G., Halberg F. Serum triglyceride rhythm alteration detected chronobiologically but not conventionally. Abstract, 2nd World Conference on Clinical Chronobiology, Monte Carlo, April 10-13, 1990, ALM s.r.l., Florence, 1990, p. 71, AND *Chronobiologia* 17: 167-168, 1990.

Gonzalez V. RC, Sanchez de la Peña S, Siegelova J, Svacinova H, Cornélissen G, Zhou S, Holte J, Katinas GS, Halberg F. Putative blood pressure overswinging, circadian hyper-amplitude-tension (CHAT), in type 2 diabetes. *Scripta medica* (Brno) 2002; 75: 121-124.

Gonzalez V. RC, Sanchez de la Peña S, Siegelova J, Svacinova H, Zhou S, Holte J, Katinas GS, Cornélissen G, Halberg F. Putative circadian hyper-amplitude-tension (CHAT) with MESOR-hypotension in type 2 diabetes in Mexico City. *MEFA 8th International Fair of Medical Technology and Pharmacy,* Brno, Czech Republic, 7-10 Nov 2000, abstract 15.

Halberg E., Brockway B., Osgood D., Nelson W., Halberg J., Sanchez de la Pena S., Halberg F. Microcomputer-related modules for experimental standardization and temperature-activity-rhythmometry in pharmacology, toxicology and biology. In: Annual Review of Chronopharmacology, *Proc. lst Int. Montreux Conf. of Biological Rhythms and Medications, Montreux, Switzerland,* March 26-30, 1984, A. Reinberg, M. Smolensky, G. Labrecque, eds., Pergamon Press, Oxford, 1984, pp. 201-204.

Halberg E., Halberg F., Cornélissen G., Sanchez S., Herold M., Günther R. Amplification by alsactide of the adrenocortical cycle in rheumatoid arthritics—feedsideward in disease. *Chronobiologia* 14: 181, 1987.

Halberg E., Halberg F., Sanchez de la Peña S., Keane W., Kasiske B., Jassim A. Circadian rhythms in urine volume of SHR-SP and LOU rats; increase in circadian mesor and amplitude with >90% reduction in kidney mass. Abstracts of papers, Minn. Acad. Sci. 51st Annual Spring Meeting, University of Minnesota-Duluth, April 29-30, 1983, p. 12.

Halberg E., Sanchez de la Peña S., Halberg F. Host and tumor temperature, motor activity and cyclosporine (Cs) immunopharmacology. Biological Rhythms and Medications, *Proc. 1st Montreux Conf. Chronopharmacol.,* Montreux, Switzerland, 1984, #303.

Halberg E., Sothern R.B., Sanchez S., Halberg F. Transplantable plasmacytoma growth in LOU rats with 24-hour synchronized or free-running metabolism gauged by i.p. temperature telemetry. *Chronobiologia* 8: 185, 1981.

Halberg F, Cornélissen F, Schwartzkopff O, Blagonravov ML, Chibisov SM, Otsuka K, Siegelova J, Beaty L, Nolley E, Sanchez de la Peña S, Zaslavskaya R, Radysh IV. Vascular variability disorders (VVDs) and syndromes (VVSs): MESOR-hypertension, CHAT and other. *Proceedings, 1st International Workshop, Physiology of adaptation and quality of life: problems of traditional medicine and innovation, People's Friendship University of Russia,* Moscow, Russia, May 14-16, 2008. p. 401-403.

Halberg F, Cornélissen G, Bingham C, Hillman D, Katinas G, Sampson M, Revilla M, Prikryl P Sr, Prikryl P Jr, Sanchez de la Pena S, Gonzalez C, Amory-Mazaudier C, Bouvet J,

Barnwell F, Maggioni C, Sothern RB, Wang ZR, Schwartzkopff O, Bakken E. Season's Appreciations 2001. *Neuroendocrinol. Lett.* 2002; 23: 170-187.

Halberg F, Cornélissen G, Czaplicki J, Schwartzkopff O, Sanchez de la Peña S, Finley J, Thomas F, Kino T, Chrousos G. Coexisting wrestling lunisolar periods in a selenosensitive circulation rather than circadian free-running? Leibniz-Online: Zeitschrift der Leibniz-Sozietät e.V. 2011; 9. 17 pp. http://www.leibniz-sozietaet.de/journal/archive/09_10/Halberg_et_al.pdf.

Halberg F, Cornélissen G, Hillman D, Ilyia E, Cegielski N, el-Khoury M, McCraty R, Strestik J, Finley J, Thomas F, Kino T, Sanchez de la Pena S, Chrousos GP, de Meester F, Singh RB, Mikulecky M. Half-yearly recurrent adynamic loss of 24-hour synchronization. *Int. J. Geronto-Geriatrics* 2010; 13: 35-62. [Dated 2010 but contains a reference dated 2011].

Halberg F, Cornélissen G, Katinas GS, Hillman D, Berger S, Woolum JC, Sanchez de la Peña S, Siegelova J. The cosmos in a eukaryotic unicell. In: Halberg F, Kenner T, Fiser B, Siegelova J, eds. Proceedings, Noninvasive Methods in Cardiology, Brno, Czech Republic, October 4-7, 2008. p. 32-43. http://web.fnusa.cz/files/kfdr2008/sbornik_2008.pdf.

Halberg F, Cornélissen G, Katinas GS, Siegelova J, Watanabe Y, Sanchez de la Peña S. Introduction: Internet- and self-help-based, scientific, preventive and curative chronobiologic health care: much more for less. Geronto-Geriatrics: *Int. J. Gerontology-ChronomeGeriatrics* 2008; 11 (14): 110-118.

Halberg F, Cornélissen G, Kaufman SC, Sanchez de la Peña S, Otsuka K, Siegelova J, Fiser B, BIOCOS group. Seven-day/24-hour chronobiologically interpreted blood pressure monitoring and single measurements before cataract surgery. *Scripta medica* (Brno) 2008; 81 (3): 141-158.

Halberg F, Cornélissen G, Matalova A, Sonkowsky R, Chibisov SM, Ulmer W, Wilson D, Revilla M, Katinas GS, Homolka P, Dusek J, Fiser B, Sanchez de la Pena S, Schwartzkopff O, Beaty LA. Omnis cyclus e cyclo: Mendel's chronoastrobiological legacy for transdisciplinary science in personalized health care. In: Halberg F, Kenner T, Fiser B, Siegelova J, eds. *Proceedings, Noninvasive Methods in Cardiology*, Brno, Czech Republic, July 7-10, 2009. (Dedicated to the 90th Anniversary of Prof. Franz Halberg.) p. 77-111. http://web.fnusa.cz/files/kfdr2009/sbornik_2009.pdf.

Halberg F, Cornélissen G, Otsuka K, Sanchez de la Peña S, Schwartzkopff O, Watanabe Y, Pati AK, Wall DG, Delmore P, Borer K, Beaty LA, Nolley ES, Adams C, Siegelova J, Homolka P, Dusek J, Fiser B, Prikryl P. Why and how to implement 7-day/24 hour blood pressure monitoring? Geronto-Geriatrics: *Int. J. Gerontology-ChronomeGeriatrics* 2005; 8 (1): 1-31. [Dated 2005 but published in June 2008.].

Halberg F, Cornelissen G, Otsuka K, Siegelova J, Fiser B, Dusek J, Homolka P, Sanchez de la Pena S, Singh RB, BIOCOS project. Extended consensus on need and means to detect vascular variability disorders (VVDs) and vascular variability syndromes (VVSs). Geronto-Geriatrics: *Int. J. Gerontology-ChronomeGeriatrics* 2008; 11 (14): 119-146.

Halberg F, Cornelissen G, Otsuka K, Siegelova J, Fiser B, Dusek J, Homolka P, Sanchez de la Pena S, Singh RB, BIOCOS project. Extended consensus on means and need to detect vascular variability disorders (VVDs) and vascular variability syndromes (VVSs). Leibniz-Online Nr. 5, 2009 (http://www2.hu-berlin.de/leibniz-so zietaet/journal/archiv_5_09.html). 35 pp, AND *World Heart J.*, in press.

Halberg F, Cornélissen G, Otsuka K, Siegelova J, Fiser B, Dusek J, Homolka P, Sanchez de la Peña S, Singh RB, BIOCOS project. Extended consensus on need and means to detect vascular variability disorders (VVDs) and vascular variability syndromes (VVSs). [Reprinted from Geronto-Geriatrics: *Int. J. Gerontology-ChronomeGeriatrics* 2008; 11 (14): 119-146.] Conference Guide, Third International Conference on Advanced Cardiac

Sciences "King of Organs 2010", September 27-30, 2010, Al Ahsa, Saudi Arabia. p. 145-172.

Halberg F, Cornélissen G, Otsuka K, Siegelova J, Homolka P, Fiser B, Dusek J, Sanchez de la Pena S, Schwartzkopff O, Singh RB. Home ambulatory blood pressure and heart rate monitoring in practice and for science: transyears and magnetoperiodism vs. photoperiodism. Abstract S8-02, 3rd Int Congress on Cardiovascular Disease, Taipei, Taiwan, 26-28 Nov 2004. *Int. J. Cardiol.* 2004; 97 (Suppl 2): S12.

Halberg F, Cornélissen G, Otsuka K, Watanabe Y, Singh RB, Revilla M, Sanchez de la Pena S, Gonzalez C, Siegelova J, Homolka P, Dusek J, Zeman M, Singh RK, Johnson D, Fiser B. Home C-ABPM for preventive and curative health care and transdisciplinary science. *World Heart J.* 2008; 1 (3): 233-261.

Halberg F, Cornélissen G, Sanchez de la Peña S, Katinas GS, Schwartzkopff O. Beyond Mikulecky's circasemimillennials back to selenosensitivity. *Int. J. Geronto-Geriatrics* 2010; 13 (1): 1-8.

Halberg F, Cornélissen G, Sanchez de la Peña S, Schwartzkopff O, Wall DG, Kysylyczyn J, Sarkozy S, Delmore P, Borer K, Otsuka K, Siegelova J, Homolka P, Dusek J, Fiser B. [Why and how to implement 7-day/24-hour blood pressure monitoring.] Addendum 1, *Proceedings, 4th International Symposium of Chronobiology and Chronomedicine*, Yantai, China, Oct. 7-14, 2000, 37 pp. [Translated into Chinese.].

Halberg F, Cornélissen G, Sanchez de la Peña S. Conclusion. Geronto-Geriatrics: *Int. J. Gerontology-ChronomeGeriatrics* 2008; 11 (14): 184-185.

Halberg F, Cornélissen G, Wall D, Otsuka K, Halberg J, Katinas G, Watanabe Y, Halhuber M, Müller-Bohn T, Delmore P, Siegelova J, Homolka P, Fiser B, Dusek J, Sanchez de la Peña S, Maggioni C, Delyukov A, Gorgo Y, Gubin D, Carandente F, Schaffer E, Rhodus N, Borer K, Sonkowsky RP, Schwartzkopff O. Engineering and governmental challenge: 7-day/24-hour chronobiologic blood pressure and heart rate screening: Part I. *Biomedical Instrumentation and Technology* 2002; 36: 89-122.

Halberg F, Cornélissen G, Wall D, Otsuka K, Halberg J, Katinas G, Watanabe Y, Halhuber M, Müller-Bohn T, Delmore P, Siegelova J, Homolka P, Fiser B, Dusek J, Sanchez de la Peña S, Maggioni C, Delyukov A, Gorgo Y, Gubin D, Carandente F, Schaffer E, Rhodus N, Borer K, Sonkowsky RP, Schwartzkopff O. Engineering and governmental challenge: 7-day/24-hour chronobiologic blood pressure and heart rate screening: Part II. *Biomedical Instrumentation and Technology* 2002; 36: 183-197.

Halberg F, Cornélissen G, with Beaty LA, Otsuka K, Watanabe Y, Sothern RB, Katinas GS, Czaplicki J, Sanchez de la Pena S, Ulmer W, Revilla M, Zeman M, Schwartzkopff O, Singh RB, the Phoenix Study Group, and the BIOCOS Project Working Group. Chronomics' progress in 2006-2008: Part I. Congruence of biospheric and heliogeophysical cycles. *Geophysical Processes and Biosphere* 2009; 8 (2): 43-74. [In Russian with English summary].

Halberg F, Katinas GS, Cornélissen G, Sanchez de la Peña S. Continued self-surveillance for treating a frequent vascular variability syndrome. Geronto-Geriatrics: *Int. J. Gerontology-ChronomeGeriatrics* 2008; 11 (14): 147-154.

Halberg F., Bakken E., Cornélissen G., Halberg J., Halberg E., Wu J., Sanchez de la Peña S., Delmore P., Tarquini B. Chronobiologic blood pressure assessment with a cardiovascular summary, the sphygmochron. In: *Blood Pressure Measurements,* Meyer-Sabellek W., Anlauf M., Gotzen R., Steinfeld L., eds., Steinkopff Verlag, Darmstadt, FRG, 1990, pp. 297-326.

Halberg F., Cornélissen G., Halberg J., Bakken E., Delmore P., Wu J., Sanchez de la Peña S., Halberg E. The sphygmochron for blood pressure and heart rate assessment: a chronobiologic approach. In: *Medical Monitoring in the Home and Work Environment,* Miles L.E., Broughton R.J. eds., Raven Press, New York, 1990, pp. 85-98.

Halberg F., Cornélissen G., Lemire Rodger G., Sanchez de la Peña S. Cronobiología y el mejoramiento de la salud, un reto a enfermería. *Desarollo Científico de Enfermeria* 5: 7-20, 1997.

Halberg F., Cornélissen G., Tarquini B., Grafe A., Syutkina E.V., Otsuka K., Watanabe Y., Siegelova J., Sanchez de la Pena S., Carandente F., Schwartzkopff O. Pineals, cancers and feedsidewards in the Biosphere and the Cosmoi (BIOCOS). Abstract, Pineal Gland and Cancer, Blaubeuren, Germany, October 2-5, 1997, unpaginated (in alphabetical order of first author), AND *Cancer Biotherapy and Radiopharmaceuticals* 12: 421-422, 1997.

Halberg F., Delmore P., Bakken E., Cornélissen G., Shinoda M., Tarquini B., Scarpelli P.T., Sanchez de la Peña S., Wu J.Y., Halberg J., Halberg E., Germanó G., Cagnoni M. Chronobiologic blood pressure and heart rate assessment in clinical practice: today and tomorrow. Handout accompanying invited main lecture in Monte Carlo, Monaco, March 17, 1988, at 1st *World Congress on Chronobiology.*

Halberg F., Galvagno A., Montalbini M., Follini S., Hillman D., Sanchez S., Holte J., Cornélissen G. *Implications for schedules in extraterrestrial (e.g., lunar) habitats of psychophysiologic rhythmicity in human social isolation.* In: IX College Park Colloquium: A Lunar-Based Chemical Analysis Laboratory (LBCAL), October 29-31, 1989, pp. 38-39.

Halberg F., Guillaume F., Sanchez de la Peña S., Cavallini M., Cornélissen G. Cephalo-adrenal interactions in the broader context of pragmatic and theoretical rhythm models. *Chronobiologia* 13: 137-154, 1986.

Halberg F., Halberg E., Halberg J., Cornélissen G., Sanchez de la Peña S., Wu J., Cagnoni M., Scarpelli P., Tarquini B., Mainardi G., Panero C., Cugini P., Sorice V, Cariddi A., Delmore P., Bakken E. Chronobiology, the indispensable control in bioscience: chronobioengineering for social medicine. In: Social Diseases and Chronobiology: Proc. III Int. Symp. Social Diseases and Chronobiology, Florence, Nov. 29, 1986, Tarquini B., ed., Società Editrice Esculapio, Bologna, 1987, pp. 7-35.

Halberg F., Halberg J., Cornélissen G., Sanchez de la Peña S., Wu J., Halberg F., Halberg E. A glance at scope, tools, mechanisms, aims and promise of chronobiology. *Brain Dysfunction* 1: 227-244, 1988.

Halberg F., Marques N., Cornélissen G., Bingham C., Sanchez de la Peña S., Halberg J., Marques M., Wu J., Halberg E. Circaseptan biologic time structure reviewed in the light of contributions by Laurence K. Cutkomp and Ladislav Dérer. *Acta entomol. bohemoslov.* 87: 1-29, 1990.

Halberg F., Otsuka K., Delmore P., Bakken E., Cornélissen G., Shinoda M., Tarquini B., Scarpelli P.T., Sanchez de la Peña S., Wu J., Halberg J., Halberg E., Germano G., Cagnoni M. [Clinical aspects of Holter blood pressure monitoring and Holter electrocardiography: chronobiological analyses and clinical significance of blood pressure and heart rate in general clinical practice.] *Erekutoronikusu no Rinsho* [Clinical Electronics] 60: 21-28, 1990. (In Japanese).

Halberg F., Sanchez de la Peña S. Intermittent systolic with or without diastolic circadian amplitude-hypertension in the 7th decade of life. *Chronobiologia* 16: 139, 1989.

Halberg F., Sanchez de la Peña S., Cornélissen G. Circadian rhythms and the central nervous system. *In: Circadian Rhythms in the Central Nervous System, Proc. IX Conf. IUPHAR, Sat. Symp.*, Bath, England, August 4-5, 1984, P.H. Redfern, I. Campbell, J.A. Xavier, K.F. Martin eds., Macmillan, 1985, pp. 57-79.

Halberg F., Sanchez de la Peña S., Cornélissen G., Gubin D., Otsuka K., Halberg Francine, Ikonomov O., Stoynev A., Madjirova N., Schwartzkopff O. Time structures, chronomes, broaden the base of gerontology and geriatrics/Estructuras temporales, cronomas, la base amplia de la gerontologia y geriatria. Geronto-Geriatrics 1: 25-46, 1998. [In English and Spanish].

Halberg F., Sanchez de la Peña S., Fernandes G. Immunochronopharmacology. In: *Advances in Immunopharmacology,* J. Hadden, L. Chedid, P. Dukor, F. Spreafico, D. Willoughby, eds., Pergamon Press, Oxford, 1983, pp. 463-478.

Halberg F., Sanchez de la Peña S., Wetterberg L., Halberg J., Halberg Francine, Wrba H., Dutter A., Hermida Dominguez R.C. Chronobiology as a tool for research, notably on melatonin and tumor development. In: *Biorhythms and Stress in the Physiopathology of Reproduction*, Pancheri P., Zichella L., eds., Hemisphere, New York, 1988, pp. 131-175.

Halberg F., Sanchez S., Brown H., Haus E., Lakatua D., Melby J., Wilson T., Sothern R., Berg H., Scheving L.E. Pretreatment with time-dependently active short-chain adrenocorticotropin (ACTH 1-17; HOE 433) for convenient optimization of murine adriamycin chronotolerance. *Proc. AACR/ASCO* 21: 307 (abstract 233), 1980.

Halberg F., Sanchez S., Nelson W. Rhythm scrambling and tumorigenesis in CD2F1 mice. *III Int. Symposium, Social Diseases and Chronobiology,* Florence, Nov. 29, 1986, B. Tarquini, R. Vergassola eds., pp. 59-61.

Halberg F., Wu J., Cornélissen G., Sanchez de la Peña S., Hayes D.K. Feedsideward mechanisms coordinating hormone concentration: delta-rhythms. *Beltsville Symposia in Agricultural Research XII: Biomechanisms Regulating Growth and Development: Keys to Progress*, Beltsville, Maryland, May 3-7, 1987, abstract #20.

Halberg F., Wu J., Sanchez de la Peña S., Halberg J., Cornélissen G., Marques N., Marques M.D., Harvey J., Halberg E. Rhythm-scrambling (dissimilation) on simulated shift-schedules may be innocuous or beneficial. Abstract, 8th Int. Symp. on Night- and Shift-Work, Kraków, June 1987, p. 9.

Halberg Francine, Cornélissen G, Halberg F, Ulmer W, Sanchez de la Peña S, Siegelova J, Schwartzkopff O, BIOCOS project. Reasons for a protocol for radiation treatment aimed at exploiting weekly rhythms. In: Halberg F, Kenner T, Fiser B, Siegelova J, eds. *Proceedings, Noninvasive Methods in Cardiology*, Brno, Czech Republic, October 4-7, 2008. p. 63-73. http://web.fnusa.cz/files/kfdr2008/sbornik_2008.pdf.

Halberg Francine, Halberg E., Halberg J., Cornélissen G., Sanchez S., Halberg F. Timing by clock-hour and season: prelude to marker rhythmometry for cancer chronoradiotherapy. *III Int. Symposium, Social Diseases and Chronobiology,* Florence, Nov. 29, 1986, B. Tarquini, R. Vergassola eds., pp. 11-13.

Halberg J., Halberg E., Cornélissen G., Wu J.Y., Sanchez de la Peña S., Hillman D., Zhou S.L., Otto S., Halberg F. Cardiovascular rhythms, their adjustment to schedule change and shift work. *Proc. 2nd Ann. IEEE Symp. on Computer-Based Medical Systems,* Minneapolis, June 26-27, 1989, Computer Society Press, Washington DC, 1989, pp. 260-266.'

Halberg J., Halberg E., Sanchez S., Wu J., Anderson S.M., Hillman D., Cornélissen G., Halberg F. Blood pressure (BP) and heart rate (HR) chronobiology in shift working police officers vs. non-shift workers. Abstract, 2nd World Conference on Clinical Chronobiology, Monte Carlo, April 10-13, 1990, ALM s.r.l., Florence, 1990, p. 39, AND *Chronobiologia* 17: 170, 1990.

Halberg J., Hillman D., Halberg F., Sanchez de la Peña S., Wu J., Halberg F. Systolic, mean arterial and diastolic blood pressure and heart rate infradians after schedule shifts. *Chronobiologia* 15: 267, 1988.

Halberg J.U., Halberg E., Cornélissen G., Hillman D., Sanchez de la Peña S., Wu J., Otto S., Halberg F. Chronobiologically deviant blood pressure in shift working police on metropolitan street duty. *Progress in Clinical and Biological Research* 341B: 281-290, 1990.

Haus E., Halberg F., Sothern R.B., Lakatua D., Scheving L.E., Sanchez de la Peña S., Sanchez E., Melby J., Wilson T., Brown H., Berg H., Levi F., Culley D., Halberg E., Hrushesky W., Pauly J. Time-varying effects in mice and rats of several synthetic ACTH preparations. *Chronobiologia* 7: 211-226, 1980.

Hermida Dominguez R.C., Ayala Garcia D.E., Sanchez de la Peña S., Halberg F., Chavarria Crespo F. Comprobación no-paramétrica de las características de la melatonina como marcadores potenciales del riesgo de desarrollo de cáncer de mama. In: *Cronocancerología,* Chavarría F. (ed.), Fundación Científica de la AECC, Madrid, 1989, pp. 259-271.

Hermida R.C., Fernández J.R., Ayala D.E., Sanchez de la Peña S., Halberg F. Nonparametric testing of melatonin characteristics as potential markers of breast cancer risk. *Progress in Clinical and Biological Research* 341A: 21-30, 1990.

Hermida R.C., Halberg F., Sanchez de la Peña S. Bootstrapping strengthens relations of plasma dehydroepiandrosterone sulfate to personality—illustrative rationale for prophylactic chronotherapy. *Ann. Rev. Chronopharmacol.,* Vol. 3, Reinberg A., Smolensky M., Labrecque G., eds., Pergamon Press, Oxford, 1986, pp. 277-280.

Hrushesky W., Kubicek W., Levi F., Olson M., Beck R., Bliss D., Sanchez S., Halberg F., Kennedy B.J. Impedance cardiography (IC) detects adriamycin-induced cardiac dysfunction. *Clin. Res.* 28: 416A, 1980.

Hrushesky W., Levi F., Halberg F., Haus E., Scheving L.E., Sanchez S., Medini E., Brown H., Kennedy B.J. Clinical chrono-oncology. In: *Chronobiology: Principles and Applications to Shifts in Schedules,* L.E. Scheving, F. Halberg eds., Sijthoff and Noordhoff, Alphen aan den Rijn, The Netherlands, 1980, pp. 513-534.

Hrushesky W., Sanchez S., Levi F., Halberg F., Kennedy B.J. The RNA content of circulating mononuclear cells (MC) demonstrates marked circadian rhythmicity (CR). *Clin. Res.* 28: 349A, 1980.

Hrushesky W.J., Sanchez S., Levi F., Brown H., Halberg F., Haus E., Sothern R., Kennedy B.J. Total RNA content of cancer patient's mononuclear cells demonstrates marked circadian rhythmicity. *Blood* 54 (#5221a): 596, 1979.

Jo D.H., Sanchez de la Peña S., Halberg F., Ungar F., Baulieu E.E., Robel P. Circadian-infradian rhythmic variation of brain neurosteroids in the female rat. *Progress in Clinical and Biological Research* 341B. 125-134, 1990.

Jo D.H., Sanchez de la Peña S., Synguelakis M., Halberg F., Baulieu E.E., Robel P. Chronobiologic dynamics of neurosteroids in rat brain. *Chronobiologia* 16: 147-148, 1989.

Katinas GS, Cornélissen G, Halberg F, Sanchez de la Peña S, Czaplicki J, Siegelova J, BIOCOS project. Pain and the cardiovascular system revisited in a long-term monitoring. In: Halberg F, Kenner T, Fiser B, Siegelova J, eds. Proceedings, Noninvasive Methods in Cardiology, Brno, Czech Republic, October 4-7, 2008. p. 147-148. http://web.fnusa.cz/files/kfdr2008/sbornik_2008.pdf.

Katinas GS, Cornélissen G, Sanchez de la Peña S, Watanabe Y, Halberg J, Halberg F. Cerebro-cardiovascular risk assessment for chronobiological analyses. World Forum "*Natural Cataclysms and Global Problems of the Modern Civilization*", 19-21 September, 2011, Istanbul, Turkey. London: SWB; 2011. p. 123-124.

Katinas GS, Halberg F, Cornélissen G, Sanchez de la Peña S, Czaplicki J, Siegelova J, BIOCOS project. C-ABPM reveals solar cis-halfyear and transyear signatures in human diastolic blood pressure. In: Halberg F, Kenner T, Fiser B, Siegelova J, eds. Proceedings,

Noninvasive Methods in Cardiology, Brno, Czech Republic, October 4-7, 2008. p. 144-146. http://web.fnusa.cz/files/kfdr2008/sbornik_2008.pdf.

Katinas GS, McCraty R, Cornélissen G, Sanchez de la Peña S, Watanabe Y, Halberg F. The wrong trade-off, when a blood pressure overswing (CHAT, circadian hyper-amplitude-tension) replaces MESOR-hypertension, unknown to caregiver and receiver, almost certainly in many millions of people. In: *Proceedings, 14th International Symposium, Ecologo-Physiological Problems of Adaptation,* 9-10 April 2009. Moscow: People's Friendship University of Russia; 2009. p. 483-486.

Kennedy B.W., Denny R., Halberg F., Sanchez de la Peña S., Cornélissen G. Cosinor-quantified chronopathology in leaf movement of soybean (Glycine max) infected with tobacco streak virus. *J. Minn. Acad. Sci.* 53: 18, 1988.

Kennedy B.W., Denny R., Halberg F., Sanchez de la Peña S., Fetzer J., Cornélissen G. Harbingers of plant pathology: potential models for applications in agriculture and medicine. 3rd UofM Research Poster Session, Minneapolis, May 25, 1988, #063, p. 23, AND *Chronobiologia* 15: 272, 1988.

Kreze A., Spirová E., Sanchez de la Peña S., Cugini P., Mikulecky M., Halberg E., Halberg F. Altered circadian plasma cortisol and aldosterone group rhythms in Cushing's syndrome vs. obesity and health. *Progress in Clinical and Biological Research* 227B: 203-217, 1987.

Kreze A., Spirová E., Sanchez de la Peña S., Halberg F. Circadian pattern of circulating human cortisol in health and Cushing's syndrome. Chronobiologia 12: 253-254, 1985, AND *Biochim. Clin.* 9: 1058, 1985.

Lakatua D., Sanchez de la Peña S., Halberg F., Marques N., Wu J., Carandente F. Complementary circadian and circaseptan intermodulation in ACTH 1-17 effect upon adrenocortial function. *Chronobiologia* 14: 58, 1987.

Langevin T., Hrushesky W., Sanchez S., Halberg F. Melatonin (M) modulates survival of CD2F1 mice with L1210 leukemia. *Chronobiologia* 10: 173-174, 1983.

Lester B., Sanchez de la Peña S., Wu J., Halberg F. Chronoimmunomodulation by cefodizime (CDZ) of the susceptibility of B6D2F1 mice to a melanoma B16 (MB16). *Chronobiologia* 16: 154, 1989.

Levi F., Halberg F., Haus E., Sanchez de la Peña S., Sothern R.B., Halberg E., Hrushesky W., Brown H., Scheving L.E., Kennedy B.J. Synthetic adrenocorticotropin for optimizing murine circadian chronotolerance for adriamycin. *Chronobiologia* 7: 227-244, 1980.

Levi F., Hrushesky W., Holtzman J., Halberg F., Sanchez S., Kennedy B.J. ACTH 1-17 raises cardiac-reduced glutathione and protects mice from lethal cardiotoxicity of doxorubicin. *Int. J. Chronobiol.* 7: 277, 1981.

Levi F., Sanchez de la Peña S., Hrushesky W., Bixby L., Haus E., Lakatua D., Halberg F., Kennedy B.J. Adriamycin (ADR) cardiotoxicity: early assessment by creatinine phosphokinase (CPK) isoenzymes. *Proc. AACR/ASCO* 22: 259, 1981.

Little J., Sanchez de la Peña S., Cornélissen G., Abramowitz P., Halberg F. Slopes guide antihypertensive chronotherapy during trends of washout on placebo or on various drug schedules. *Chronobiologia* 16: 156, 1989.

Little J., Sanchez de la Peña S., Cornélissen G., Abramowitz P., Tuna N., Halberg F. Longitudinal chronobiologic blood pressure monitoring for assessing the need and timing of antihypertensive treatment. *Progress in Clinical and Biological Research* 341B: 601-611, 1990.

Marques N., Cornélissen G., Halberg F., Bingham C., Sanchez de la Peña S., Marques M.D., Hurwitz A., Hamblen M., Hall J. Meta-chronanalysis reveals circadian-to-ultradian shift in Drosophila mutant and altered circadian structure after gene transfer. *Chronobiologia* 14: 203, 1987.

Marques N., Sanchez de la Peña S., Brown G.M., Grota L.J., Yunis E., Scheving L.E., Halberg F. Circadian and circaseptan rhythm in murine pineal melatonin content assessed by chronobiologic pilot studies. Laboratory Animals, ICLAS-CEMIB-FESBE Meeting, Aguas de Lindóia, Nov. 18-21, 1986, #P-3.2, p. 86.

Marques N., Sanchez de la Peña S., Mushiya T., Ungar F., Halberg F. Infradian modulation of liver nucleic acid and lipid content of adult female Lewis/S rats. *Braz. J. Med. Biol. Res.* 22: 1479-1483, 1989.

Marques N., Sanchez de la Peña S., Mushiya T., Ungar F., Halberg F. Modulação infradiana de ácidos nucléicos e lipídeos em fígado de ratos Lewis/S. II *Simposio Brasiliero de Cronobiologia,* February 1-3, 1989, São Sebastião, SP, Brazil, p. 16.

Marques N., Sanchez de la Peña S., Mushiya T., Yasmineh W.G., Cornélissen G., Halberg F. Circadian-infradian variation of heart creatinine phosphokinase activity in male Holtzman rats. *Abstract, III Simposio Brasileiro de Cronobiologia/I Simposio Latino Americano de Cronobiologia,* Itatiaia National Park, RJ, Brazil, February 18-21, 1991, p. 39.

Marques N., Sanchez de la Peña S., Mushiya T., Yasmineh W.G., Cornélissen G., Halberg F. Ultradian-infradian variation of cardiac creatine phosphokinase (CPK) activity in male Holtzman rats. *Chronobiologia* 21: 241-250, 1994.

Marques N., Sanchez de la Peña S., Tsai T.H., Scheving L.E., Halberg F. Intermodulação circadiana-infradiana de mitoses do epitélio corneal de ratos Holtzman adultos. II Simposio Brasiliero de Cronobiologia, February 1-3, 1989, São Sebastião, SP, Brazil, pp. 17-18.

Marques N., Sanchez de la Peña S., Ungar F., Halberg F. Circadian stage-dependent effect of ACTH upon protein synthesis in rat adrenal cells. *Chronobiologia* 14: 59, 1987.

Marques N., Sanchez de la Peña S., Ungar F., Halberg F. Circadian stage-dependent effect of ACTH and melatonin on protein synthesis by rat adrenal cells. *Braz. J. Med. Biol. Res.* 21: 759-762, 1988.

Marques N., Sanchez de la Peña S., Ungar F., Halberg F. Os efeitos do ACTH e da melatonina sobre a sintese proteica das celulas adrenais do rato sao dependentes do estagio circadiano. Programa e resumos, III Reunião Anual da Fedaração das Sociedades de Biologia Experimental, Caxambu, June 29-July 3, 1988, p. 443.

Marques N., Sanchez de la Peña S., Ungar F., Halberg F. Resposia ritmica circadiana das proteinas da glandula adrenal de ratos a estimulacão in vitro por ACTH e melatonina. *Ciência e Cultura* 40: 996, 1988.

Marques N., Tsai T.H., Van Pilsum J., Sanchez de la Peña S., Ungar F., Towle H., Scheving L.E., Halberg F. Infradian modulation of adult female Lewis rat corneal mitoses and kidney transaminidase. *Chronobiologia* 12: 258, 1985.

McDonald K., Sanchez de la Peña S., Cavallini M., Olivari M.T., Cohn J.N., Halberg F. Chronobiologic blood pressure and heart rate assessment of patients with heart transplants. *Progress in Clinical and Biological Research* 341B: 471-480, 1990.

McDonald K., Sanchez de la Peña S., Olivari M.T., Cavallini M., Halberg F. Sampling requirements for chronobiologic ambulatory cardiovascular assessment of patients after heart transplantation (HT). *Chronobiologia* 16: 158, 1989.

Merchesan de Ionescu M.M., Sanchez de la Peña S., Dumitriu M., Alonso Andraca G., Cornélissen G., Halberg F. Time structure (chronome) of blood pressure (BP) and heart rate (HR) in elderly Mexicans. Proc. XXXIII Int. *Cong. International Union of Physiological Sciences*, St. Petersburg, Russia, June 30-July 5, 1997, abstract P055.23.

Montalbini M., Galvagno A., Follini S., Sanchez de la Peña S., Hillman D.C., Wu J., Degioanni J., Kawabata Y., Kutyna F., Bungo M., Halberg F. Cardiovascular about-24-hour (circadian) and about-7-day (circaseptan) rhythms in human social isolation.

Abstract, Health of Inner Cities and Urban Areas, International Conference, Cardiff, Wales, September 4-7, 1989, The Institute of Statisticians/SIRMCE, 1989, p. 101.

Mushiya T., Sanchez (de la Peña) S., Sothern R.B., Halberg F. Rapid human cardiovascular transmeridian and transequatorial adjustment with transient hyperbaric impact (I). *Chronobiologia* 12: 262, 1985.

Rivera Medina R., Sanchez de la Peña S., Islas Andradé S., Cornélissen G., Ayala D.E., Hermida Dominguez R.C., Arechiga H., Halberg F. Circasemiseptan (about half-weekly) chrones in the diastolic (D) blood pressure (BP) chronome and preeclampsia. Proc. XXXIII Int. *Cong. International Union of Physiological Sciences*, St. Petersburg, Russia, June 30-July 5, 1997, abstract P055.22.

Robyn C., Sanchez de la Peña S., Halberg E., Halberg F. Meta(chron)analyses of prolactin in human cord blood. *Chronobiologia* 14: 226-227, 1987.

Sanchez de la Peña S, Cornélissen G, Gonzalez C, Halberg F. Action prompted by the premetabolic syndrome in Latin Americans in North America. Abstract, III International Conference, Civilization diseases in the spirit of V.I. Vernadsky, People's Friendship University of Russia, Moscow, Oct. 10-12, 2005, p. 421-423.

Sanchez de la Peña S, Cornélissen G, Gonzalez Villalpando C, Halberg F. Blood pressure and heart rate chronomics based on monitoring for about 4.5 days. Proceedings, International Conference on the Frontiers of Biomedical Science: Chronobiology, Chengdu, China, September 24-26, 2006, p. 204-207.

Sanchez de la Pena S, Gonzalez C, Cornélissen G, Halberg F. Blood pressure (BP), heart rate (HR) and non-insulin-dependent diabetes mellitus (NIDDM) chronobiology. Abstract S8-06, 3rd Int Congress on Cardiovascular Disease, Taipei, Taiwan, 26-28 Nov 2004. *Int. J. Cardiol.* 2004; 97 (Suppl 2): S14.

Sanchez de la Peña S, Gonzalez C, Cornélissen G, Halberg F. Chronomics: time structure (chronome) of blood pressure and heart rate in health and diabetes. *Abstract 26, 2nd International Symposium: Workshop on Chronoastrobiology and Chronotherapy,* Tokyo Kasei University, Tokyo, Japan, November 2001, unpaginated (2 pp).

Sanchez de la Peña S, Halberg F, Mendoza-Lujambio I, Lopez-Fiesco A, Cornélissen G, Rito-Medina JL. Chronomic aspects of pre-metabolic syndrome (CPMS) on Mexico City. Poster presentation, Semana de la Ciencia y la Innovacion 2010, Mexico City, November 24, 2010.

Sanchez de la Peña S, Halberg F, Rito-Medina JL, Mendoza-Lujambio I, Cornélissen G, Lopez-Fiesco A. Detection of chrono premetabolic syndrome in different Mexican populations. P061, *Proceedings, 3rd World Congress of Chronobiology,* Puebla, Mexico, 5-9 May 2011, p. 177-178.

Sanchez de la Peña S, Rito-Medina J, Halberg F, Peña A, Mendoza-Lujambio I, Lopez-Fiesco A, Flores-Guevara E, Cornélissen G. Mapping chrono pre-metabolic syndrome in Mexico City. P1-19, *Proceedings, XI Latin American Symposium on Chronobiology*, Cholula, May 4-5, 2011, p. 87.

Sánchez de la Peña S, Sothern RB, López FS, Lujambio IM, Waizel-Bucay J, Sánchez CO, Monroy CP, Betancourt ET. Circadian aspects of hyperthermia in mice induced by Aconitum napellus. *Pharmacog. Mag.* 2011; 7(27): 235-242.

Sanchez de la Peña S, Wilson D, Halberg F, Cornélissen G, Guzman Felix CE. Chrono-pre-metabolic syndrome (CPMS) in Mexican populations. 32nd Annual Conference, International Society for Clinical Biostatistics, Ottawa, Canada, 21-25 August 2011.

Sanchez de la Peña S., Ayala D., Hermida Dominguez R.C., Cornélissen G., Halberg F. Individualized assessment of blood pressure and heart rate late in human pregnancy. *Progress in Clinical and Biological Research* 341A: 615-624, 1990.

Sanchez de la Peña S., Brown G.M., Ungar F., Marques N., Scheving L.E., Grota L.J., Halberg F. Chronobiologic lead study cost-effectively assesses circadian-circaseptan intermodulation in murine pineal melatonin content. *Chronobiologia* 13: 329-333, 1986.

Sanchez de la Peña S., Ganten D., Halberg E., Sanchez E., Shankaraiah K., Halberg F. Circadian rhythm of angiotensin I activity in pineal gland of rats. *Chronobiologia* 9: 454-455, 1982.

Sanchez de la Peña S., Halberg E., Halberg F., Cornélissen G., Sothern R.B., Yunis E. Circadian stage-dependence of L1210 leukemia growth and antileukemic effects of cyclosporine. *Proc. AACR*, Toronto, May 9-12, 1984, #937, p. 237.

Sanchez de la Peña S., Halberg E., Halberg F., Cornélissen G., Sothern R.B., Yunis E. Circadian modulation of the anti-tumor effect of cyclosporine. In: *Neuroimmunomodulation,* Proc. 1st Int. Workshop on NIM, Bethesda, MD, Nov. 27-30, 1984, IWGN, Bethesda, 1985, pp. 252-257.

Sanchez de la Peña S., Halberg E., Halberg F., Sanchez E., Cornélissen G. Chronomodulatory effect of cyclosporine upon survival of DBA mice with L1210 leukemia. *Chronobiologia* 13: 129-135, 1986.

Sanchez de la Peña S., Halberg F., Cornélissen G., Halberg E., Kawasaki T., Scheving L., Woodward D., Schuman L., Lucas E., Hambley B., Kennedy B., Cardoso S., Ginn L.J., Spector N.H., Farr L. Circadian cardiovascular monitoring during a conference at high altitude and thereafter. *Abstract, 1st International Conference of Chronobiology and Chronomedicine*, Chengdu, People's Republic of China, October 2-7, 1988, pp. 24-26.

Sanchez de la Peña S., Halberg F., Cornélissen G., Wu J., Wood P., Hrushesky W., Eaton J. Circaseptan aspects of mouse malaria: agreement of original and new protochronanalyses with metachronanalyses. *Chronobiologia* 14: 231, 1987.

Sanchez de la Peña S., Halberg F., Galvagno A., Montalbini M., Follini S., Wu J., Degioanni J., Kutyna F., Hillman D.C., Kawabata Y., Cornélissen G. Circadian and circaseptan (about-7-day) free-running physiologic rhythms of a woman in social isolation. *Proc. 2nd Ann. IEEE Symp. on Computer-Based Medical Systems, Minneapolis*, June 26-27, 1989, Computer Society Press, Washington DC, pp. 273-278.

Sanchez de la Peña S., Halberg F., Halberg E., Ungar F., Cornélissen G., Sanchez E., Brown G., Scheving L.E., Yunis E.G., Vecsei P. Pineal modulation of ACTH 1-17 effect upon murine corticosterone production. *Brain Res. Bull.* 11: 117-125, 1983.

Sanchez de la Peña S., Halberg F., Haus E. Circadian rhythms in unstimulated, ACTH 1-17-stimulated and puromycin-inhibited adrenocortical RNA content. *Chronobiologia* 8: 183, 1981.

Sanchez de la Peña S., Halberg F., Haus E., Lakatua D., Halberg E., Ungar F. Circadian modulation by melatonin of murine ACTH secretion and corticosterone production in vitro. *Abstract 2122, 7th International Congress of Endocrinology,* July 1-7, 1984, Quebec City, Canada. Excerpta Medica, Amsterdam/Oxford/Princeton, 1984, p. 1321.

Sanchez de la Peña S., Halberg F., Lakatua D., Ungar F. Rhythmic pineal-pituitary-adrenal intermodulations ex vivo. *Progress in Clinical and Biological Research* 227A: 421-449, 1987.

Sanchez de la Peña S., Halberg F., Schweiger H.-G., Eaton J., Sheppard J.R. Circadian and circaseptan rhythmic aspects of murine mortality from Plasmodium berghei inoculation. *Chronobiologia* 10: 154-155, 1983.

Sanchez de la Peña S., Halberg F., Schweiger H-G., Eaton J., Sheppard J. Circadian temperature rhythm and circadian-circaseptan (about 7-day) aspects of murine death from malaria. Proc. Soc. exp. *Biol. Med.* 175: 196-204, 1984.

Sanchez de la Peña S., Halberg F., Ungar F. Pineal chronomodulation—the feed-sideward. *Clin. Chem. Newsletter* 2: 129-130, 1982.

Sanchez de la Peña S., Halberg F., Ungar F., Haus E., Lakatua D., Scheving L.E., Sanchez E., Vecsei P. Circadian pineal modulation of pituitary effect on murine corticosterone in vitro. *Brain Res. Bull.* 10: 559-565, 1983.

Sanchez de la Peña S., Halberg F., Ungar F., Lakatua D. Ex vivo hierarchy of circadian-infradian rhythmic pineal-pituitary-adrenal intermodulations in rodents. In: *Biorhythms and Stress in the Physiopathology of Reproduction*, Pancheri P., Zichella L., eds., Hemisphere, New York, 1988, pp. 177-214.

Sanchez de la Peña S., Halberg F., Ungar F., Sanchez E. Circadian response of adrenals incubated with pituitaries (Pt), hypothalami (Ht), pineal homogenate (PH) or ACTH 1-17, Synchrodyn® (Sy). *Abstracts of papers, Minn. Acad. Sci., 50th Annual Spring Meeting*, Gustavus Adolphus College, St. Peter, Minnesota, April 23-24, 1982, p. 6, AND *Chronobiologia* 9: 348, 1982.

Sanchez de la Peña S., Halberg F., Ungar F., Sanchez E., Anaya C., Yunis E., Halberg E. Further infradian variation in circadian rhythmic murine corticosterone response to ACTH 1-17. *Abstracts of papers, Minn. Acad. Sci. 51st Annual Spring Meeting*, University of Minnesota-Duluth, April 29-30, 1983, p. 11.

Sanchez de la Peña S., Halberg F., Wetterberg L., Wu J., Dammacco F., Carandente F. Melatonin and cefodizime may broaden defense by blocking murine circadian hypothalamic and pituitary melatonin rhythms. 3rd UofM Research Poster Session, Minneapolis, May 25, 1988, #109, p. 31, AND *Chronobiologia* 15: 273, 1988.

Sanchez de la Peña S., Lakatua D., Wu J., Marques N., Halberg F. Circadian rhythmic corticosterone (B) production by aged stroke-prone spontaneously hypertensive (SHR-SP) rats. *Chronobiologia* 14: 59-60, 1987.

Sanchez de la Peña S., Marques N., Halberg F. Infradian modulation of corticosterone production by hamster adrenals in vitro. *Proc. 2nd Int. Conf. Medico-Social Aspects of Chronobiology*, Florence, Oct. 2, 1984, Halberg F., Reale L., Tarquini B., eds., Istituto Italiano di Medicina Sociale, Rome, 1986, pp. 771-775.

Sanchez de la Peña S., Marques N., Halberg F., Marques M., Ryba R., Weydahl A. Chronobiologically assessed penbutolol and saline effects upon blood pressure of SHR-SP rats. *Chronobiologia* 11: 408, 1984.

Sanchez de la Peña S., Rivera Medina R., Pons O., Islas Andradé S., Revilla Monsalve C., Mendez J.D., Ikonomov O., Cornélissen G., Halberg F. The chronome initiative's blood pressure (BP) and heart rate (HR) study of pregnant Mexican women. Proc. XXXIII *Int. Cong. International Union of Physiological Sciences*, St. Petersburg, Russia, June 30-July 5, 1997, abstract P055.21.

Sanchez de la Peña S., Sanchez E., Halberg F., Haus E., Hrushesky W. Circadian rhythmic response of murine adrenal RNA to in vitro stimulation by ACTH 1-17 inhibited by puromycin. *Int. J. Chronobiol.* 7: 306, 1981.

Sanchez de la Peña S., Sanchez E., Halberg F., Lakatua D., Ungar F., Halberg E., Berg H., Yunis E. Circadian modulation of prolactin release from mouse adenohypophysis by melatonin. *Chronobiologia* 12: 271, 1985.

Sanchez de la Peña S., Ungar F., Halberg F., Sanchez E., Yunis E., Vecsei P. Circadian and infradian variation in murine adrenal corticosterone production: a test system for Synchrodyn. In: Toward Chronopharmacology, Proc. 8th IUPHAR Cong. and Sat. Symposia, Nagasaki, July 27-28, 1981, R. Takahashi, F. Halberg and C. Walker, eds., Pergamon Press, Oxford/New York, 1982, pp. 203-209.

Sanchez de la Peña S., Van Pilsum J.F., Halberg F., Marques N., Ungar F. Infradian rhythms characterize transaminidase activity in rat kidney. *Chronobiologia* 12: 183-184, 1985.

Sanchez de la Peña S., Wood P., Halberg E., Cornélissen G., Eaton J., Halberg F. Circadian and circaseptan aspects of malarial infection and cyclosporine treatment in mice. In:

Annual Review of Chronopharmacology, Proc. lst Int. Montreux Conf. of Biological Rhythms and Medications, Montreux, Switzerland, March 26-30, 1984, A. Reinberg, M. Smolensky, G. Labrecque, eds., Pergamon Press, Oxford, 1984, pp. 411-414.

Sanchez de la Peña S., Wood P., Halberg E., Cornélissen G., Eaton J., Halberg F. Circadian and circaseptan aspects of malarial infection and cyclosporine (Cs) treatment in mice. Biological Rhythms and Medications, Proc. 1st Montreux Conf. Chronopharmacol., Montreux, Switzerland, 1984, #812.

Sanchez de la Peña S., Wu J., Halberg F., Cornélissen G., Wetterberg L., Halberg E., Lakatua D., Marques N., Harvey J., Bazin H., Tran B., Carandente F. Immunomodulation by an antibiotic in malaria and aging: internal endocrine schedule-shift and chronopharmacologic feedsidewards. (Chronoimmuno modulation by an Antimicrobial Agent, 1989, F. Carandente, F. Dammacco, F. Halberg, eds.) *Chronobiologia* 15: 129-182, 1988.

Schwartzkopff O, Katinas G, Singh RB, Sanchez de la Pena S, Zhao ZY, Uezono K, Prikryl P, Siegelova J, Fiser B, Dusek J, Otsuka K, Cornélissen G, Halberg F. Self-surveillance and self-experimentation (SSSE) for preventing vascular diseases needed to follow established successes of anesthesia and vaccine development. Abstract #13, MEFA, Brno, Czech Republic, 04-07 Nov 2003. p. 18.

Sorice V., Maggioni C., Halberg F., Cornélissen G., Tarquini B., Mainardi G., Panero C., Cariddi A., Scarpelli P., Sanchez S., Wu J., Romano S., Haus E., Cagnoni M. Infradian postnatal structure of mean arterial blood pressure (MAP) and human cardiovascular disease risk. *Chronobiologia* 14: 242, 1987.

Sundaram B, Hanumansetty R, Cornélissen G, Otsuka K, Katinas G, Siegelova J, Homolka P, Sanchez de la Peña S, Borer K, Schaffer E, Holley DC, Halberg F. Blood pressure and pulse dynamics quantify everyday life's emotions -- if excessive by circadian overswinging, CHAT. *Am. J. Hypertens* 2004; 17 (5 Part 2): 57a-58a.

Tsai T.H., Scheving L.E., Marques N., Sanchez de la Peña S., Halberg F. Circadian-infradian intermodulation of corneal epithelial mitoses in adult female rats. *Progress in Clinical and Biological Research* 227A: 193-198, 1987.

Tsai T.H., Scheving L.E., Sanchez de la Peña S., Bazin H., Halberg F. Circaseptan-circadian rhythmic changes in corneal mitoses of Louvain rats. *Progress in Clinical and Biological Research* 341A: 91-99, 1990.

Tsai T.H., Scheving L.E., Sanchez de la Peña S., Marques N., Bazin H., Halberg F. Strain differences in infradian modulation of the circadian cellular system gauged by murine corneal mitoses. *Chronobiologia* 16: 191-192, 1989.

Tsai T.H., Scheving L.E., Sanchez de la Peña S., Marques N., Halberg F. Circaseptan (about 7-day) modulation of circadian rhythm in corneal mitoses of Holtzman rats. *Anat. Rec.* 225: 181-188, 1989.

Ungar F., Sanchez de la Peña S., Halberg F., Sanchez E., Halberg E., Yunis E. Circadian and infradian pineal chronomodulation of ACTH 1-17 effect upon corticosterone production in vitro. *Chronobiologia* 12: 281, 1985.

Valverde C., Legarreta G., Aceves C., Sanchez de la Peña S., Halberg F. Circadian rhythmic urinary excretion of thyroxine (T4) in healthy subjects. *Chronobiologia* 8: 184, 1981.

Wetterberg L., Sanchez de la Peña S., Halberg F. Circadian rhythm of melatonin release from pineal, hypothalamus and pituitary in hypertensive rats. *Progress in Clinical and Biological Research* 341B:145-153, 1990.

Wilson, D.W.1, Hungin, A.P.S.1, Griffiths, K.2, Singh, R.B.3,4, De Meester, F.3, Sanchez de la Pena, S.5,Fedacko,J.6 and A.Wilczyńska3. Perspectives of human development:time, space, and chrononutrition. *World Heart J.* 2011,3: (In press).

Wrba H., Dutter A., Sanchez de la Peña S., Cornélissen G., Halberg F. Secular or circadian effects of placebo and melatonin on murine breast cancer?. *Progress in Clinical and Biological Research* 341A: 31-40, 1990.

Wrba H., Dutter A., Sanchez de la Peña S., Wu J., Carandente F., Halberg F. Secular or circannual immunomodulation of placebo (Pl) and melatonin (Mt) effects on murine breast cancer? *Chronobiologia* 16: 197, 1989.

Wu J., Cornélissen G., Halberg E., Sanchez de la Peña S., Halberg F. Infradian (estral) and circadian timing of motor activity (MA) and core temperature (CT) of female LOU rats. *Chronobiologia* 14: 257, 1987.

Wu J., Sanchez de la Peña S., Halberg F., Cornélissen G., Wetterberg L., Halberg E., Lakatua D., Bingham C., Harvey J., Bazin H., Zheng T., Leung B., Tran B. Chronosynergistic effects of lighting schedule-shift and cefodizime on plasmacytoma growth and host survival time. (Chronoimmunomodulation by an Antimicrobial Agent, 1989, F. Carandente, F. Dammacco, F. Halberg, eds.) *Chronobiologia* 15: 105-128, 1988.

Wu J., Sanchez de la Peña S., Pham C., Halberg F. Design of circadian approach to blood pressure variation in progeny of calcium-treated pregnant rats. In: Social Diseases and Chronobiology: Proc. III Int. Symp. Social Diseases and Chronobiology, Florence, Nov. 29, 1986, Tarquini B., ed., *Società Editrice Esculapio*, Bologna, 1987, pp. 471-475.

In: Cardiovascular Health and Chronomics
Editor: Ram B. Singh
ISBN: 978-1-62948-976-6

Chapter 19

BOHUMIL FISER (*22.10.1943-†21.03.2011): CHRONOBIOLOGIST, EMERITUS HEAD OF THE PHYSIOLOGY DEPARTMENT AT MASARYK UNIVERSITY (BRNO, CZECH REPUBLIC), CZECH MINISTER OF HEALTH, AND EXECUTIVE BOARD MEMBER OF THE WORLD HEALTH ORGANIZATION: HIS LEGACIES FOR PUBLIC AND PERSONALIZED HEALTH CARE

Franz Halberg[1], Germaine Cornélissen[1], Thomas Kenner[2], Jiri Dusek[3], Brigitte Kenner[2], Othild Schwartzkopff[1] and Jarmila Siegelova[4]

[1]Halberg Chronobiology Center, University of Minnesota, Minneapolis, MN, US
[2]Department of Physiology, Medical University, Graz, Austria
[3]Health Medical Center, South Moravia Region, Brno, Czech Republic
[4]Department of Functional Diagnostics and Rehabilitation, St. Anna Teaching Hospital, Brno, Czech Republic

INTRODUCTION

In 2003, one of us (JS) dedicated a special volume of papers dealing with non-invasive cardiology [1] to Bohumil Fiser, MD, our dear friend, Figure 1, and, his relatively young age notwithstanding, our mentor and supporter. In that volume, GC and JS laid the basis for a chrononeonatology. Another of us (OS) emphasized the need for long-term, preferably lifetime monitoring of blood pressure and heart rate as a concern for everybody and hence for government and ethics committees. Bohumil Fiser was on many decision-making bodies, always there whenever and wherever he was needed, whether the task on hand was small or of nationwide or worldwide scope.

Figure 1. Bohumil Fiser (1943-2011).

In the honorable tradition of his fellow-countryman Jan Evangelista Purkinje, he served as a subject and as a principal investigator.

He was there in Brno with Jan Penaz as a young scientist-subject when the first continuous beat-to-beat blood pressure recorder on the finger for automatic use was developed. He was there (BF in Tables 1-3) with some of us (JS, JD and GC) as a self-surveilling colleague and fellow investigator in Minnesota, when the transition from an ambulatory monitor (ABPM-630) of Colin Medical Instruments (founded by Masayuki Shinoda, starting in a chicken house, yet requiring an advanced degree for its use) to the less cumbersome performance of an instrument from the A and D Company (TM-2421) was examined. He was there when the Roseville, Minnesota project was implemented in Brno, Czech Republic, the home town of Johann Gregor Mendel, the founder of genetics. There, Bohumil helped Jarmila and Jiri develop the sister science to diversity in space, that of diversity in time.

He was there in Prague as Czech minister of health providing data [2] for the documentation of an about-50-year cycle in stroke incidence, Figure 2 [3, 4]. And he was there, again in Minnesota, lecturing at the Supercomputing Institute and advising on tests of sensitivity and specificity of a chronobiologic approach to stroke prevention, as illustrated in Figure 3 [5]. Not long ago, back in Brno, he documented the problem of exclusive reliance upon a classification by dipping [6, 7] when a chronobiologic approach to blood pressure was available [8], pointing to the merits of the latter and the shortcomings of the former.

Bohumil was there in Minnesota, insisting on washing the dishes as a house guest, just as he was glad to advocate or support new projects and do analyses in a unified science, bringing Brno into the international project on The BIOsphere and the COSmos, BIOCOS.

Table 1. Agreement in circadian double amplitude (2A) and acrophase (ϕ) of blood pressure (BP) and heart rate (HR) between auscultatory (Aus) and oscillometric (Osc) measurements taken with the TM-2421 ambulatory monitor from A and D*

			Aus				Osc				P-values from comparison of			
ID	Variable**	N	P	M	2A	ϕ	P	M	2A	ϕ	M	2A	ϕ	(A. ϕ)
JS	SBP	95	<0.001	126.2	28.5	-267°	<0.001	133.1↑	28.2	-257°	0.001	0.957	0.338	0.628
	DBP		<0.001	83.6	22.6	-260°	<0.001	90.1↑	25.2	-250°	<0.001	0.586	0.344	0.535
	HR		0.057	80.2	6.3	-327°	0.019	81.1	8.3	-309°	0.527	0.625	0.569	0.750
JD	SBP	172	0.110	134.8	10.3	-250°	0.009	135.7	12.7	-241°	0.654	0.700	0.759	0.884
	DBP		0.544	72.1	4.2	-70°	0.358	77.9↑	4.0	-222°	<0.001	0.971	—	0.240
	HR		<0.001	77.9	12.2	-238°	<0.001	79.2	17.5	-236°	0.332	0.178	0.874	0.394
BF	SBP	287	<0.001	120.7	30.0	-263°	<0.001	125.7↑	25.4	-253°	<0.001	0.152	0.107	0.109
	DBP		<0.001	79.1	28.1	-254°	<0.001	82.5↑	27.5	-251°	<0.001	0.689	0.549	0.775
	HR		<0.001	65.8	9.5	-228°	<0.001	66.0	10.3	-227°	0.802	0.765	0.951	0.955

*Notwithstanding a statistically significant technical difference in the rhythm-adjusted mean (MESOR, M) of BP, but not of HR in data of three physician-scientists measuring automatically at 30- to 60-minute intervals for 4 days.

**SBP: systolic BP; DBP: diastolic BP.

N: number of concomitant valid measurements (excluding all sets of measurements including a "zero" for any one of the variables); P: P-value from zero-amplitude (no-rhythm) test; MESOR: Midline-Estimating Statistic Of Rhythm; 2A: measure of extent of predictable change within a cycle; ϕ: a measure of the timing of overall high values recurring in each cycle; M and 2A expressed in mmHg (BP) and beats/min (HR); ϕ expressed in (negative) degrees, with 360° ≡ 24 hours; 0°=00:00. Analyses by Germaine Cornélissen.

Note: In these three subjects, SBP and DBP measured oscillometrically are higher than the corresponding auscultatory values. With the ABPM-630 from Colin Medical, SBP measured oscillometrically was found to be on the average higher than the corresponding auscultatory value, but DBP measured oscillometrically was on the average statistically significantly lower than the corresponding auscultatory value. The differences between the two monitors may stem from a difference in algorithm used to derive the oscillometric readings of SBP and DBP from the pressure wave envelope and mean arterial pressure. It would be desirable to retrieve mean arterial pressure (measured oscillometrically) from the A and D instrument for a direct comparison with the ABPM-630 of Colin Medical Instruments.

Table 2. Circadian rhythm characteristics of blood pressure (BP) and heart rate (HR) assessed oscillometrically by two different ambulatory monitors

		A and D***					Colin***					P-values from comparison of			
ID*	Variable**	N	P	M	2A	ϕ	N	P	M	2A	ϕ	M	2A	ϕ	(A, ϕ)
JS	SBP	95	<0.001	133.1	28.2	-257°	166	<0.001	133.0	29.5	-249°	0.943	0.796	0.455	0.736
	DBP		<0.001	90.1	25.2	-250°		<0.001	85.0↓	21.5	-253°	<0.001	0.331	0.699	0.461
	HR		0.019	81.1	8.3	-309°		0.033	80.6	4.6	-297°	0.604	0.286	0.696	0.475
JD	SBP	172	0.009	135.7	12.7	-241°	149	<0.001	139.0	23.6↑	-237°	0.074	0.041	0.514	0.102
	DBP		0.358	77.9	4.0	-222°		<0.001	79.8	12.6↑	-245°	0.145	0.027	0.575	0.068
	HR		<0.001	79.2	17.5	-236°		<0.001	80.5	16.5	-247°	0.282	0.761	0.311	0.561
BF	SBP	287	<0.001	125.7	25.4	-253°	154	<0.001	120.2↓	23.3	-252°	<0.001	0.484	0.915	0.752
	DBP		<0.001	82.5	27.2	-251°		<0.001	76.2↓	20.4↓	-245°	<0.001	0.003	0.326	0.012
	HR		<0.001	66.0	10.3	-227°		0.007	66.2	8.3	-221°	0.887	0.535	0.805	0.801

*Three physician-scientists.

**SBP: systolic BP; DBP: diastolic BP.

N: number of measurements; whereas the monitoring spans and sampling rates are the same for both instruments, the numbers of measurements differ since retrials and measurement errors differed between the two monitors which measured only approximately at the same time (± 1 minute) and on different arms; P: P-value from zero-amplitude (no-rhythm) test; MESOR: Midline-Estimating Statistic Of Rhythm; 2A: measure of extent of predictable change within a cycle; ϕ: a measure of the timing of overall high values recurring in each cycle; M and 2A expressed in mmHg (BP) and beats/min (HR); ϕ expressed in (negative) degrees, with 360° ≡ 24 hours; 0°=00:00. Analyses by Germaine Cornélissen.

Note: For subjects JS and BF, the lower MESOR of SBP observed with the ABPM-630 monitor from Colin Medical as compared to the TM-2421 of A and D is in keeping with the earlier documentation that oscillometric values of DBP are on the average lower than auscultatory values of DBP with the Colin instrument whereas the opposite result is found with the A and D device for the three subjects examined herein (see Table 1).

Table 3. Auscultatory (Aus) and oscillometric (Osc) measurements obtained with the TM-2421 ambulatory monitor of A and D on four subjects*

	r	P	Intercept (a_0) 00:00 Jan 13-20, 1995	Slope (a_1)
GC (F, 45 y): N=564; Δt=15 min				
SBP	0.751	<0.001	9.97 ± 3.74*	0.842 ± 0.031*
DBP	0.698	<0.001	14.86 ± 2.22*	0.770 ± 0.033*
HR	0.929	<0.001	7.11 ± 1.17*	0.893 ± 0.015*
JS (F, 53 y); N=95; Δt=30 min				
SBP	0.654	<0.001	30.52 ± 11.29*	0.714 ± 0.086*
DBP	0.643	<0.001	24.11 ± 7.21*	0.654 ± 0.081*
HR	0.818	<0.001	19.40 ± 4.46*	0.751 ± 0.055*
JD (M, 48 y): N=172; Δt=30 min				
SBP	0.498	<0.001	54.99 ± 10.78*	0.588 ± 0.079*
DBP	0.393	<0.001	30.56 ± 7.53*	0.531 ± 0.095*
HR	0.613	<0.001	21.63 ± 5.67*	0.710 ± 0.070*
BF (M, 51 y): N=287; Δt=30 min				
SBP	0.627	<0.001	13.13 ± 7.93	0.854 ± 0.063+
DBP	0.828	<0.001	4.51 ± 3.02	0.903 ± 0.036*
HR	0.952	<0.001	3.68 ± 1.21*	0.940 ± 0.018*

*Data collected January 13-20, 1995; r: Pearson product-moment correlation coefficient; P: P-value from test of H_0: r=0; a_0 and a_1 are regression coefficients (with their standard error) from Aus = a_0 + a_1 Osc, where Aus and Osc refer to auscultatory and oscillometric readings (of SBP, DBP and HR), respectively.

+P<0.05; *P<0.01 in test of H_0: $a_0 = 0$ and H_0: $a_1 = 1$. Analyses by Germaine Cornélissen.

Note: The correlation between oscillometric and auscultatory readings is invariably highly statistically significant (P<0.001), even when assessed on an individualized basis, so that the correlation does not depend upon extreme inter-individual differences. Nonetheless, the two kinds of measurements are not necessarily equivalent; the slopes can drastically differ from unity and the intercepts depart statistically significantly from zero. A possible lack of interchangeability prompts the consistent use of one or the other kind of measurement, perhaps the one in which measurement failures are least common (from our limited experience with the A and D instrument so far, there seem to be fewer measurement errors for the oscillometric than for the auscultatory readings; if movement is limited during measurement, the oscillometric may be preferred, notably if acoustic noise cannot be controlled). On the positive side, there may also be much to learn from the difference between (concomitant) oscillometric and auscultatory readings. For this purpose, a model (or artificial limb) where, among other factors, compliance, muscle tone and fat can all be varied could contain additional information regarding the vascular state of a given person.

A list of joint publications from Brno and Minneapolis consists of over 350 titles relating mainly to the heart and the circulation, his field of specialization. His curriculum vitae has been described in the proceedings of the symposium "Chronobiologic analyses in pathophysiology of cardiovascular systems", dedicated to him on his 60th birthday, edited by all of us [1]. His untimely death notwithstanding, he will be with us in spirit in our efforts to change a medicine based on occasional professional spotchecks with sometimes long intervals between successive office visits into one relying on less costly since computer-implemented self-help-based, but more efficient automated around-the-clock self-surveillance.

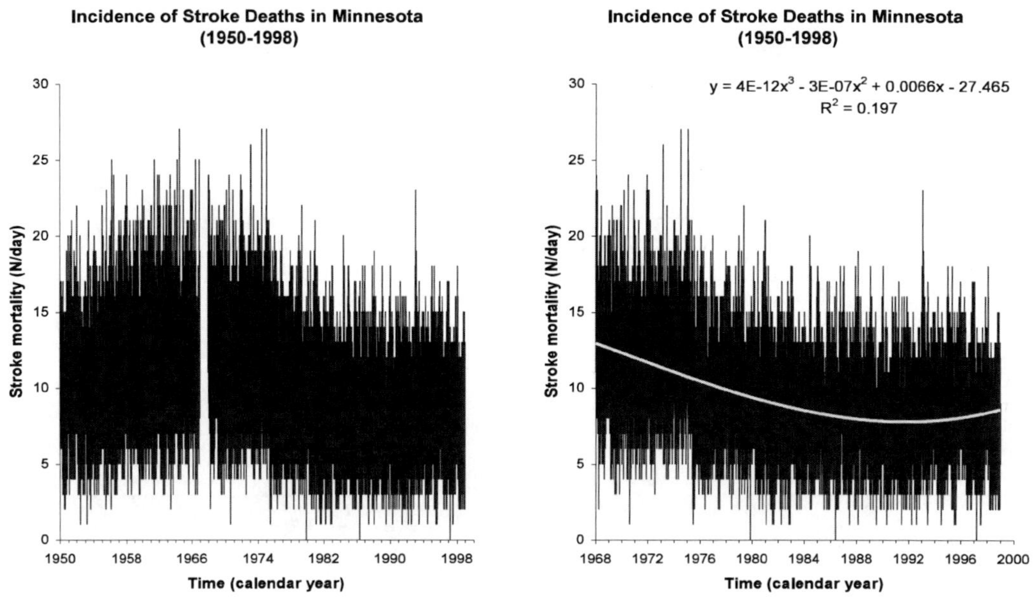

Figure 2. With the qualification that the classification of stroke mortality was changed over the years, an about 50-year pattern (left) may account for the recent increase in the incidence of stroke deaths observed by the fit of a third-order polynomial to data since 1968, using the same disease classification (right). © Halberg.

Thereby, health care can change from one based on sheer intuition and hard-to-define experience listing a number of diagnoses to an objective assessment of many novel rhythm and other variability alterations, usually already occurring within the limits of the normal range, before overt disease sets in.

These novel diagnoses, Figure 4 [9, 10], made in the light of inferential statistical estimations with a measure of uncertainty, serve as a basis for pre-habilitation, to save on rehabilitation.

What is generally accepted as indispensable for research – uncertainties in the form of probabilities and confidence intervals – can enter personalized health care where the likelihood of a condition and the effects of treatment are the more important. This requirement is rendered practicable by time series replacing spotchecks. The latter's single values do not allow inferential statistical tests relating to the given individual, who is the only concern of a truly temporally personalized medicine.

BIOCOS has lost a friend. It will be hard to find a better one. But Bohumil will live on in the unified science that he helped create, transcending disciplinary barriers. His legacy includes a document guided by him with another cardiac physiologist and physiology department head, TK, emeritus President of the Medical University, Graz, Austria, constituting a consensus [9]. If his advocacy of a chronobiologic approach is implemented, novel diagnoses, Figure 4 [cf. 9 and 10], may be made on millions of people who currently are not recognized as being at a very high risk of cardiovascular disease. Bohumil cared to detect Vascular Variability Disorders (VVDs) and Vascular Variability Syndromes (VVSs), Figure 4, and to use them as gauges of loads and as harbingers of risk for the prevention of both the individual's and society's ills [11].

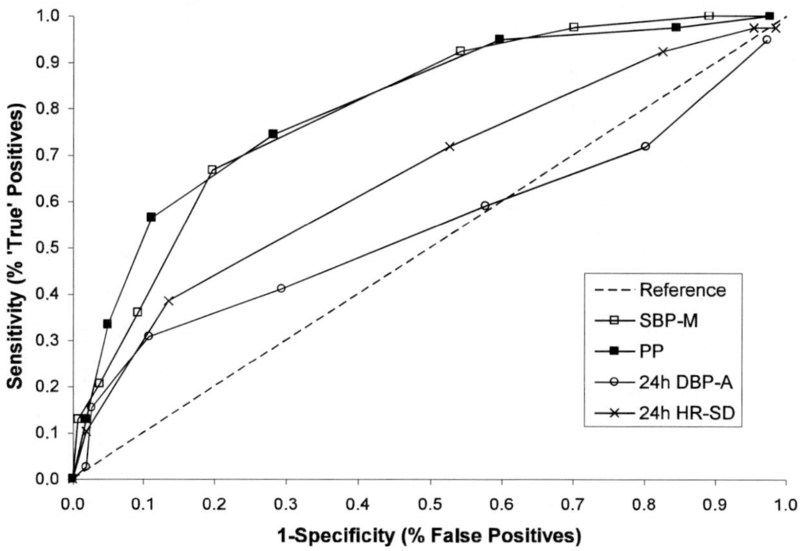

Figure 3. The percentage of morbid outcomes (such as a stroke) among subjects testing positive (e.g., for CHAT or for MESOR-hypertension) is known as the sensitivity of the test (e.g., circadian amplitude or MESOR of blood pressure). CHAT has a sensitivity of 28.2%, as compared to 92.3% for the case of MESOR-hypertension. The specificity of a given test is the percentage of subjects with no adverse outcome among those testing negative. CHAT has a specificity of 94.6%, as compared to 45.7% for the case of MESOR-hypertension. The efficiency coefficient is the percentage of correct diagnoses in the light of outcomes. For CHAT it is 85.9%, as compared to 51.9% for MESOR-hypertension. These values strongly suggest the merit of assessing the blood pressure variability, notably the circadian amplitude of blood pressure, since the diagnosis of CHAT is not meant to be used as a substitute, but rather as a complement to the diagnosis of MESOR-hypertension. The same comment applies to the assessment of heart rate variability, gauged by the standard deviation (sensitivity: 25.6%; specificity: 95.7%; coefficient of efficiency: 86.5%), and to that of an elevated pulse pressure (sensitivity: 56.4%; specificity: 94.6%; coefficient of efficiency: 84.5%) [5]. CHAT, Circadian Hyper-Amplitude-Tension, a condition defined as a circadian amplitude of blood pressure above the upper 95% prediction limit of clinically healthy peers matched by gender and age. © Halberg.

The sequential analysis of data from ongoing physiological monitoring coupled with the daily surveillance of natality, morbidity and mortality should become the mandate of every health department. Sequential testing, e.g., by yet-to-be-improved cumulative sum procedures (CUSUMs) [12] can now be applied to both individuals and populations. The currently as-yet missing analyses of the cycles of our cosmos in psychophysiology will eventually replace the vexing (if cycle lengths are not specified) secular variations, so named to actually disregard their existence by accepting single samples as imaginary baselines.

With support of self-surveillance by the individual in order to detect and treat VVDs and VVSs in a truly personalized health care, a health department, offering a cost-free analysis, or preferably an international multilingual website, could use the same data as an added source of potential harbingers for finding out when the cosmos is ill-disposed, as Herschel put it in 1801 [13]. An international meeting on Geochange in Istanbul, Turkey, September 19-21, 2011, has on its agenda the development of a consensus on the specifications for an international multilingual educative and automatically analyzing website. This meeting is being organized by Acad. Elchin Khalilov of Baku, Azerbaijan (admin@geo-change.org, www.geo-change.org).

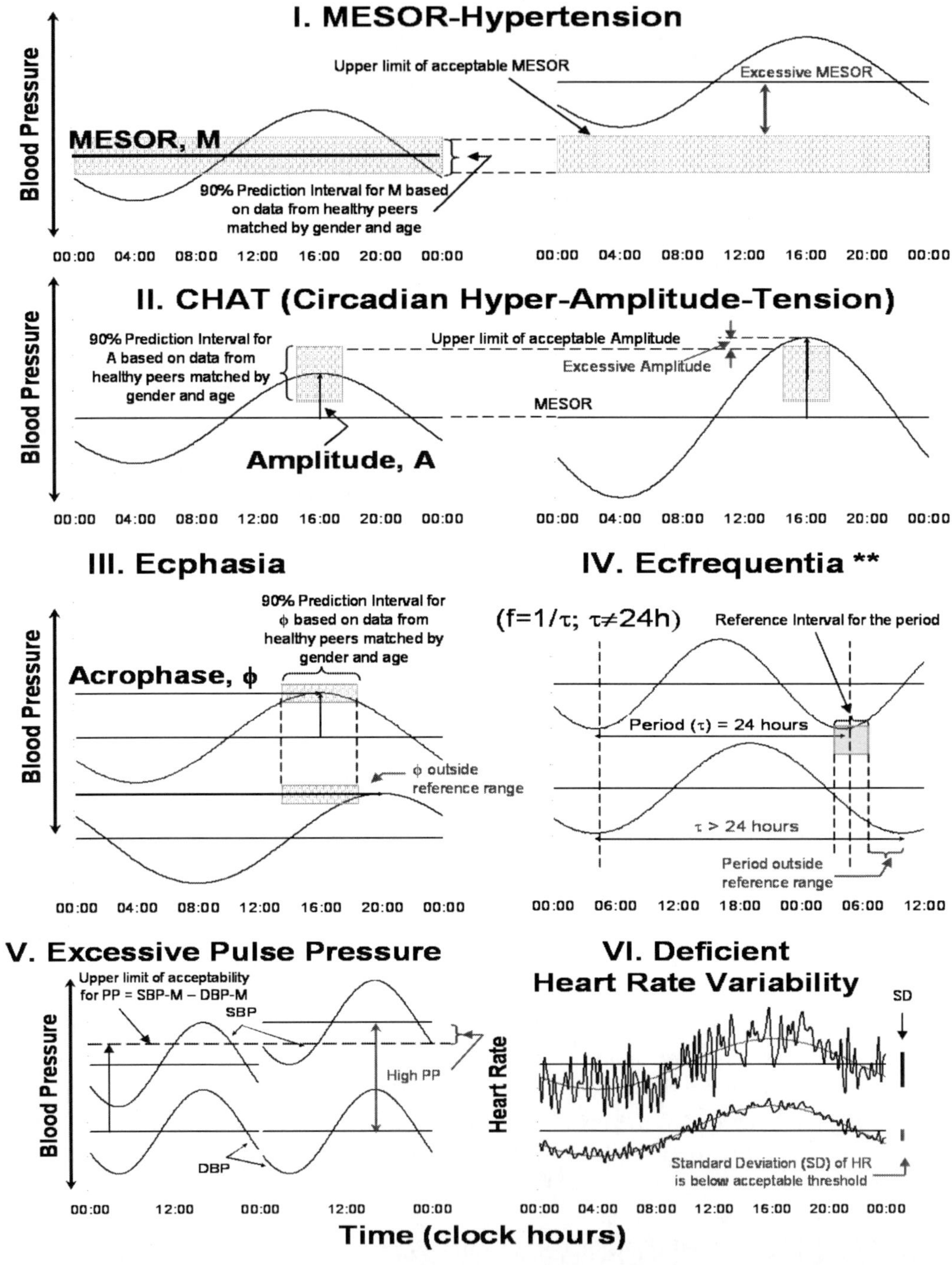

Figure 4. Vascular Variability Anomalies (VVAs) picked up by chronobiologically interpreted 7-day around-the-clock records of blood pressure and heart rate monitoring become Vascular Variability Disorders (VVDs) when they are replicated in successive 24-hour/7-day records. If several VVDs coexist, the risk of an ischemic stroke within 6 years increases from about 5% to near 100%. To the five VVDs in the consensus, we can add a sixth, a circadian desynchronization of the endocrines and the circulation more recently documented as ecfrequentia in association with adynamia and depression recurring mostly twice-yearly in an extensively studied 62-year-old woman [10]. © Halberg.

An international multilingual, partly automatically analyzing website for computer-aided psychophysiological self-surveillance, can also screen weekly sequentially accumulating other data from health departments, i.a, on the incidence of natality, morbidity and mortality from various causes and from other government sources on crime and terrorism. Cycles in all these statistics concerning us and the environment, usually ignored, are tools for understanding, prediction, alerts and countermeasures. Mapping the time structures of health and aligning them with one of weather in space and on earth, including terrorist acts and a host of environmental variables, leads to putative triggers affecting the mind of populations, analyzable as alterations of cycles in individuals. Automatized weekly sequential (e.g., by CUSUM) testing of changes in psychophysiology and epidemiological statistics, sadly unrealized mandates of public health systems, should elevate the latter into more than mere data deposits. They could provide for automatic analyses of weekly around-the-clock data for personal and societal health, the latter aim including international security, alerting and preferably avoiding September-11-like events. To start with, blood pressure and heart rate monitoring and analysis are warranted.

Marked seasonal environmental light and temperature changes notwithstanding, analyses of databases on worldwide terrorism reveal, instead of any annual component, transyears (cycles with periods of about 1.3 years) (and show further prominent cross-spectral coherences between solar activity, gauged by Wolf's sunspot numbers and the incidence of terrorism during the past 42 years). Different, albeit overlapping sets of psychophysiologic cycles, with some periods shared only with solar activity, others shared only with geomagnetic disturbance, some shared with both, are found. All of them are novel in the biosphere, albeit some, like para-tridecadals, found long ago, are now mostly forgotten. They are validated in inferential statistical terms by several approaches in different biospheric data sets. A transyear characterizing the incidence of international terrorism is prominent when this component is more strongly expressed in non-photic environmental variables. It is also detected in individual records of physiological variables such as blood pressure and heart rate that serve for personal self-surveillance aiming at the prevention of adverse cerebrovascular events. These same self-surveilling data, analyzed automatically cost-effectively by a website, contribute immediately useful marker variables for as-one-goes associations with forecasts and nowcasts of space and terrestrial weather in the broadest sense, including magnetism, and for the recognition of potential threats to national and international security.

Bohumil Fiser's legacy is a system of better since continuous computer-aided self-surveillance with cost-free analyses for personalized health and international security. If it can be developed, Bohumil will actually help us, as he wanted during his lifetime [9], to prevent or avoid, respectively, the personal cataclysm of a stroke, societal crime and terrorism, or the natural physical cataclysms, such as earthquakes, some affecting large populations.

REFERENCES

[1] Halberg, F., Kenner, T., Siegelova, J. (Eds.) *Proceedings, Symposium, Chronobiological Analysis in Pathophysiology of Cardiovascular System.* Brno: Masaryk University; 2003. [Dedicated to the 60th Anniversary of Prof. Bohumil Fiser].

[2] Fiser, B., Cornélissen, G., Siegelova, J., Dusek, J., Homolka, P., Mazankova, V., Halberg, F. Increase in stroke deaths after 1997 in the Czech Republic. *MEFA 8th International Fair of Medical Technology and Pharmacy,* Brno, Czech Republic, 7-10 Nov. 2000, abstract 4.

[3] Halberg, F., Cornélissen, G., Otsuka, K., Watanabe, Y., Katinas, G. S., Burioka, N., Delyukov, A., Gorgo, Y., Zhao, Z. Y., Weydahl, A., Sothern, R. B., Siegelova, J., Fiser, B., Dusek, J., Syutkina, E. V., Perfetto, F., Tarquini, R., Singh, R. B., Rhees, B., Lofstrom, D., Lofstrom, P., Johnson, P. W. C., Schwartzkopff, O., International BIOCOS Study Group. Cross-spectrally coherent ~10.5- and 21-year biological and physical cycles, magnetic storms and myocardial infarctions. *Neuroendocrinol. Lett.* 2000; 21: 233-258.

[4] Cornélissen, G., Halberg, F., Breus, T., Syutkina, E. V., Baevsky, R., Weydahl, A., Watanabe, Y., Otsuka, K., Siegelova, J., Fiser, B., Bakken, E. E. Non-photic solar associations of heart rate variability and myocardial infarction. *J. Atmos. Solar-Terr. Phys.* 2002; 64: 707-720.

[5] Cornélissen, G., Halberg, F., Bakken, E. E., Singh, R. B., Otsuka, K., Tomlinson, B., Delcourt, A., Toussaint, G., Bathina, S., Schwartzkopff, O., Wang, Z. R., Tarquini, R., Perfetto, F., Pantaleoni, G. C., Jozsa, R., Delmore, P. A., Nolley, E. 100 or 30 years after Janeway or Bartter, Healthwatch helps avoid "flying blind". *Biomed. and Pharmacother.* 2004; 58 (Suppl. 1): S69-S86.

[6] Fiser, B., Havelkova, A., Siegelova, J., Dusek, J., Pohanka, M., Cornélissen, G., Halberg, F. Night-to-day blood pressure ratio during seven-day ambulatory blood pressure monitoring. In: Halberg, F., Kenner, T., Fiser, B., Siegelova, J. (Eds.) *Noninvasive Methods in Cardiology*, September 16-17, 2010, Brno, Czech Republic. Brno: Faculty of Medicine, Masaryk University. p. 128-132.

[7] Siegelova, J., Havelkova, A., Fiser, B., Dusek, J., Pohanka, M., Dunklerova, L., Cornélissen, G., Halberg, F. Day and night blood pressure variability during seven-day ambulatory blood pressure monitoring. In: Halberg, F., Kenner, T., Fiser, B., Siegelova, J. (Eds.) *Noninvasive Methods in Cardiology,* September 16-17, 2010, Brno, Czech Republic. Brno: Faculty of Medicine, Masaryk University. p. 133-138.

[8] Cornélissen, G., Halberg, F., Otsuka, K., Singh, R. B., Chen, C. H. Chronobiology predicts actual and proxy outcomes when dipping fails. *Hypertension* 2007; 49: 237-239. doi:10.1161/01.HYP.0000250392.51418.64.

[9] Halberg, F., Cornélissen, G., Otsuka, K., Siegelova, J., Fiser, B., Dusek, J., Homolka, P., Sanchez de la Pena, S., Singh, R. B., BIOCOS project. Extended consensus on means and need to detect vascular variability disorders (VVDs) and vascular variability syndromes (VVSs). *World Heart J.,* in press (cf. Leibniz-Online Nr. 5, 2009, http://www2.hu-berlin.de/leibniz-sozietaet/journal/archiv _5_09.html. 35 pp).

[10] Halberg, F., Cornélissen, G., Sothern, R. B., Barnwell, F., Cegielski, N., Ilyia, E., Siegelova, J. The moon's and the genes' tides and double tides pulling the biosphere. In: Halberg, F., Kenner, T., Fiser, B., Siegelova, J. (Eds.) *Noninvasive Methods in Cardiology,* September 16-17, 2010, Brno, Czech Republic. Brno: Faculty of Medicine, Masaryk University. p. 23-45.

[11] Halberg, F., Cornélissen, G., Grambsch, P., McCraty, R., Beaty, L., Siegelova, J., Homolka, P., Hillman, D. C., Finley, J., Thomas, F., Kino, T., Revilla, M., Schwartzkopff, O. Personalized chronobiologic cybercare; other chronomics' progress

by transdisciplinary cycles' congruences: Season's Appreciations 2009. *J. Appl. Biomed.* 2011; 9: 1-34. DOI 10.2478/v10136-009-0022-8.

[12] Cornélissen, G., Halberg, F., Hawkins, D., Otsuka, K., Henke, W. Individual assessment of antihypertensive response by self-starting cumulative sums. *J. Medical Engineering and Technology* 1997; 21: 111-120.

[13] Herschel, W. Observations tending to investigate the nature of the sun, in order to find the causes or symptoms of its variable emission of light and heat; with remarks on the use that may possibly be drawn from solar observations. *Phil. Trans. Roy. Soc. London* 1801; 91: 265-318.

INDEX

B

C

D

E

H

I

N

O

P

Q

R

S

T

U

V

W

Y

Z